Raimo Alén

Chemistry for Biomass Utilization

Also of Interest

BioProducts.
Green Materials for an Emerging Circular and Sustainable Economy
Vijayendran (Ed.), 2023
ISBN 978-3-11-079121-1, e-ISBN (PDF) 978-3-11-079122-8

Gasification.
Sustainable Decarbonization
Silva, Cardoso, Chavando, 2023
ISBN 978-3-11-075820-7, e-ISBN (PDF) 978-3-11-075821-4

Industrial Green Chemistry
Kaliaguine, Dubois (Eds.), 2020
ISBN 978-3-11-064684-9, e-ISBN (PDF) 978-3-11-064685-6

Carbon Dioxide Utilisation.
Volume 1: Fundamentals
North, Styring (Eds.), 2019
ISBN 978-3-11-056309-2, e-ISBN 978-3-11-056319-1

Carbon Dioxide Utilisation.
Volume 2: Transformation
North, Styring (Eds.), 2019
ISBN 978-3-11-066503-1, e-ISBN (PDF) 978-3-11-066514-7

Raimo Alén

Chemistry for Biomass Utilization

—

DE GRUYTER

Author
Prof. Dr. Raimo Alén
University of Jyväskylä
Department of Chemistry
Applied Chemistry
FI-40014 Jyväskylä
Finland
E-Mail: raimo.j.alen@jyu.fi

ISBN 978-3-11-060834-2
e-ISBN (PDF) 978-3-11-060836-6
e-ISBN (EPUB) 978-3-11-060848-9

Library of Congress Control Number: 2023941098

Bibliographic information published by the Deutsche Nationalbibliothek
The Deutsche Nationalbibliothek lists this publication in the Deutsche Nationalbibliografie;
detailed bibliographic data are available on the Internet at http://dnb.dnb.de.

© 2024 Walter de Gruyter GmbH, Berlin/Boston
Cover image: Austin Park/Unsplash
Typesetting: Integra Software Services Pvt. Ltd.
Printing and binding: CPI books GmbH, Leck

www.degruyter.com

Preface

Much interest has been directed to the versatile possibilities of using lignocellulosic biomass resources (*i.e.*, "renewable raw materials") for the full-scale production of various chemicals and other bioproducts together with solid, liquid, and gaseous fuels. This kind of approach has traditionally been an integral part of organic chemistry, including wood chemistry and carbohydrate chemistry, as well as of pulping chemistry and chemical technology. The recent practice has been adopted of expressing these biorefinery concepts in terms of "green chemistry" or/and "green technology" comprising the essential building blocks of a new potential technology platform with a significant range of technological breakthroughs. However, it is also the well-recognized fact that the effective utilization of wood and other biomass feedstocks already has a long history, involving a tremendous interest in creating versatile technologies, and that presently the products based on renewable raw materials are seen as one of the most promising future markets with a high potential for innovation. Despite these positive prospects, the future strategic decisions are still rather complicated in view of the general challenges in society and the world, for example, including the looming climate change and several other factors, such as food production, water resources, and resource depletion. Hence, it is important that the biorefining processes are not only technically feasible but also economically sustainable.

The basic intention of this book was to present in a reader-friendly way a comprehensive approach concerning all the important areas related to relevant chemistry of common lignocellulosic biomass-utilizing processes. The book illustrates the general principles of the main methods described, rather than going into experimental and theoretical details that are necessarily not essential for understanding each method. One basic aim was also to include plenty of the carefully selected literature references to be useful, especially if the reader wants to go further into any topic in greater detail. In many cases, the short historical perspective ranging from fundamental studies to practical applications is also given. Of course, because of the expanse of the subject, the idea to write a book, fully covering this diversified topic, is rather ambitious. However, if needed there are a great number of specialized books available that deal with the detailed data on each topic or address specific topics within this expanded field. This book is intended to meet a simple need for condensed textbook that can be primarily used by a wide circle of readers in many traditional academic disciplines.

The book was designed in the belief to be simple and understandable; its organization comprises seven chapters. It was believed that the straightforward approach selected (*i.e.*, the conversion of lignocellulosic biomass feedstocks directly to useful products or obtained as by-products of other processing, such as chemical pulping) will help the reader to see the essential differences between different conversion techniques and to understand their actual potential. The purpose of the two first chapters (1 and 2) is to give some background and illustrative data on biorefining and the chemical composition of wood and some non-wood materials. Chapter 3 deals with

https://doi.org/10.1515/9783110608366-202

the various chemical and biochemical ways to convert lignocellulosic feedstocks into a wide range of useful products; it mainly describes an array that organic chemicals can be made from the simple sugars first obtained from feedstock by different hydrolysis methods. Chapter 4 is devoted to the reactions of chemical pulping, whereas Chapter 5 mainly discusses the by-products from the integrating forest biorefining. Chapter 6 gives examples of cellulose derivatives as well as of cellulose solvents and nanotechnological aspects on cellulose. Finally, Chapter 7 outlines several potential ways to produce biofuels with different properties from lignocellulosic biomass feedstocks.

Helsinki, December 2022
Raimo Alén

Contents

1 Introduction to biorefining

1.1 General approach

The present production of energy, chemicals, and a huge number of various materials is significantly based on fossil-derived resources and, for example, petroleum-based distillates, are used in a wide range of industrial applications [1, 2]. Additionally, it is obvious that the existence and further development of the prevailing civilization requires expanded consumption of these commodities [3]. However, the increasing global warming and other environmental concerns associated with the excessive consumption of fossil-based carbon reserves have led to urgent needs to utilize renewable feedstock carbon sources as feedstock materials and hence, for example, the chemical industry is gradually shifting to a more efficient utilization of new carbon dioxide (CO_2)-neutral lignocellulosic biomass raw materials [1, 4–7]. It can be further concluded that the importance of wood (as well as non-wood) resources is also increasing globally, making up a strong value chain "from forest to products" and represents one of the promising ways to overcome the dependence of society on fossil carbon resources, especially in the production of fuels [8–10].

The biorefinery concept can be simply defined as a process for fractionating and/ or converting lignocellulosic biomass materials in an ecosystem-friendly way through advanced technologies into useful solid, liquid, and gaseous bioproducts [8, 11–26]. Hence, the main principle is, according to the idea of circular economy, to maximize the value of lignocellulosic biomass and, on the other hand, to minimize the production of waste; that is, the bio-based economy primarily targets to replace fossil resources with renewable biological resources. Furthermore, this type of production is analogous to that of petrorefineries that utilize fossil resources for manufacturing a variety of fractionated and refined products. However, in comparison to petroleum refineries, biorefineries utilize, in most cases, a wider range of feedstocks and process technologies. In general, the concept of bioeconomy links together the sectors responsible for sustainable primary production (*i.e.*, agriculture, forestry, fisheries, and aquaculture) and the sectors producing final consumables.

A circular bioeconomy is a straightforward route to a genuinely sustainable economy in which, material, information, and value circulate together [27]. Thus, the term "bioeconomy" means that we are dealing with renewable resources (*i.e.*, biomass in all forms) and states that we are focusing on operations that generate economic actions, resulting in the growth of the economy and wellbeing of the society. When optimizing resources, the integrating supply and value chains generally aim at wasteless and emissionless circulation and the cascading use of materials and products. Hence, the use and reuse of both non-renewables and renewables can be also seen as elements of the new materials economy that should also emphasize recycling critical raw materials over the utilization of virgin alternatives. Additionally, the evolution of

https://doi.org/10.1515/9783110608366-001

bioeconomy is enabled by the fast development of digital and information technology solutions. Improved processes, materials, and product design rely on the use of big data and artificial intelligence to improve resource efficiency.

The carbon capture and utilization is the process by which carbon is captured from a source and is either utilized on site or transported elsewhere to be used [28]. Utilization can be the direct use of CO_2 separated, for example, from flue gases, as such, or as a source of carbon for the synthesis of chemical and biochemical products. The process does not necessarily decrease atmospheric CO_2, but delays CO_2 release. Some CO_2-based chemicals are already produced by chemical routes, for example, urea, certain organic carbonates (*e.g.*, dimethyl, ethylene, and propylene carbonates), polycarbonates, polycarbonate polyols, and polyurethanes as well as inorganic carbonates (*e.g.*, with cations of calcium ($Ca^{2\oplus}$) and magnesium ($Mg^{2\oplus}$)). On the other hand, carbon capture and storage are processes by which carbon is captured from a source and stored on site or often off site, for example, at a depleted gas or oil field, or other geological formation.

Three main drivers can be identified for the carbon reuse economy [28]: *i)* the need to reduce CO_2 emissions into the atmosphere, *ii)* expanding regional resource bases and securing energy demand for carbon-dependent industries, and *iii)* the potential for developing new businesses, based on the sustainable supply and the use of carbon. Thus, a circular economy has many shared goals with a low-carbon economy, which battles climate change by reducing greenhouse gas (GHG) emissions with non-carbon energy sources (including solar, wind, water (H_2O), and geothermal energy) and aims to improve resource efficiency through energy intelligence and carbon capture and utilization [27].

It should be also pointed out that the atmosphere recognizes no national boundaries, and the products should be made where this production can be carried out at the lowest possible level of emissions [29]. In principle, this general approach helps mitigate climate change by replacing products that are manufactured with high levels of fossil emissions and responds also to the problems caused by population growth. In the carbon reuse economy, the aim is that fossil carbon would remain in the ground, whereas carbon in biomaterials would circulate, without accumulating in the atmosphere [28]. It is also evident that the carbon reuse economy is strongly linked to energy and therefore, energy policies [27]. This fact agrees well with the general principle and approach of a circular economy, which considers the sufficiency of materials and energy.

Forest industry companies have already significantly reduced their emissions and are already replacing many products that cause harmful emissions [29, 30]. The demand for wood-based products used for construction, fiber-based packaging materials, special products for improving our quality of life, and several product innovations is expected to increase and the demand for printing paper to decline. However, chemical pulp manufacturing is also often the process behind new innovative products that enables the harnessing of the different qualities of wood for a variety of uses. For example, certain chemical pulp fibers as such are utilized textile fibers, where pulp fibers can replace cotton and synthetic fiber materials (*e.g.*, polyester) made from fossil resources and have great environmental impacts. The manufacturing process of ecological textile fibers involves only moderate H_2O consumption as the involved liquids can be recycled. Furthermore, the textiles can be recycled in the same process again or together with other textile waste after use. Additionally, the chemical components in lignocellulosic biomass materials can be used as components in the production of bottles, packaging, washing agents and detergents, and cosmetics or in industrial applications, such as glues and resins, automotive and electronic batteries, and energy storage.

Forests, as well as a variety of agricultural wastes, are receiving increasing attention as a versatile resource and many potential new uses are being developed for these raw materials [31]. However, it still is reasonable to assume that, for example, the main functions of forests will remain relatively unchanged and that they will continue to have a decisive role in the foreseeable future as a source of construction materials and chemical fiber. Hence, to maintain forest resources is of great importance and, on the other hand, the CO_2 sink of forests (*i.e.*, carbon sequestration) [2] is one of the critical factors. This key role of forests can be even improved by a better management of seedlings and increase in young stands as well as by the improvement of the nutrient management. In this way, for example, forests act as both carbon sinks and as an important source of carbon [28]. The most important carbon sinks have been fossil fuel deposits buried deep inside the Earth and thus, they have been separated from the carbon cycle in the atmosphere [2]. However, this situation ended when humans started to utilize these fossil resources, mainly *via* returning their carbon as CO_2 into the atmosphere.

1.2 Versatile conversion methods of lignocellulosic biomass materials

There are several ways to classify wood and non-wood biomass feedstocks, but usually they are divided according to their origin as shown in Table 1.1. Presently, it seems that, for example, in the first development phase of diversifying the production of forest-based industry, the more effective and versatile utilization of various harvesting residues would be needed, but it is also necessary to find novel by-products, especially in kraft pulping (see Section 4.2 "Kraft pulping" and Chapter 5 "Integrated forest biorefining").

Before utilization, the lignocellulosic biomass materials are typically upgraded by some form of pretreatment, including physical (or mechanical), physicochemical, chemical, and biochemical processing to enhance their general reactivity and digestibility [2, 32–37]. In practice, in most cases, this means drying, crushing, or grinding to an appropriate particle size distribution (*i.e.*, by physical treatment of feedstock) and/ or densifying to more compact and regular shapes as briquettes and pellets. Examples

Table 1.1: Classification of wood and non-wood biomass feedstocks [2].

Feedstock type	Examples
Wood-based materials	Softwoods, hardwoods, sawdust, shavings, bark, and harvesting residues
Agricultural residues	Sugarcane, bagasse, sorghum, corn stalks, cotton stalks, rice straw, and cereal straws
Natural-growing plants	Bamboo, esparto, sabia, elephant grass, and reeds, such as reed canary grass
Non-wood crops[a]	
Bast (stem) fibers	Jute, ramie, hemp, kenaf, and flax tow
Leaf fibers	Abaca and sisal
Seed hair fibers	Cotton linter

[a]Grown primarily for their fiber content.

of chemical pretreatment methods include acidic and alkaline treatments and of biochemical pretreatment methods, various enzymatic or bacterial treatments.

The production of lignocellulosic biomass-derived liquid fuels is clearly the primary driver for development in green chemistry or green engineering, although the current and rapid technological progress in the biorefining area that aims at more multiple processes will gradually result in many novel bioproducts [21]. Some basic principles should be generally considered when designing novel processes and products that are based on renewable resources [38]. These clean technological principles focus on evaluating and improving current methods as well as developing new ones that will have a limited impact on the environment, and are more sustainable and economical in the long run. The most important issues of green chemistry mainly include *i*) the use of renewable feedstock materials (together with safer chemicals and auxiliaries as well as selective catalysts) rather than depleting ones, *ii*) straightforward and well-designed syntheses with minimum energy requirements and environmental impacts, *iii*) development of on-line (real-time) analyses for production, *iv*) the manufacturing of degradative products in the environment, and *v*) the prevention of possible chemical accidents in the production process. The additional principles concerning green technology primarily comprise, besides the use of renewable material and energy inputs, the maximization of process efficiency with respect to feedstock and requirements, process conditions, and energy consumption as well as the minimization of waste formation.

The use of catalysts is an obvious way to reduce energy requirements of chemical reactions [21, 39]. A catalyst is usually defined as a material that changes the rate of chemical reaction without itself being consumed in the process. Catalytic reactions have a lower rate-limiting free energy of activation than the corresponding uncatalyzed reactions, leading to a lower overall energy required and hence, a higher reaction rate at the same temperature. Additionally, due to the use of catalysts, reaction selectivity is generally increased. The catalyst can be either homogeneous (*i.e.*, it exists in the same phase as a substrate) or heterogeneous, and different biocatalysts are often considered a separate group. In addition to these "positive catalysts", there are "negative catalysts" (*i.e.*, inhibitors) whose harmful impact may be important, for example, in fermentation processes. The application of catalytic chemistry rather than conventional stoichiometric chemistry is one of the key research areas in green chemistry when producing renewable feedstocks-based chemicals [38].

The main chemical components of lignocellulosic biomass materials are carbohydrates (cellulose and hemicelluloses) and lignin, together with minor amounts of extractives and inorganic material [40–45]. The conversion of these materials into a great variety of products can be performed *via* different routes, depending on the feedstock characteristics (Table 1.2). It is also possible to recover extractives directly from the feedstock by different techniques (see Section 2.4 "Nonstructural constituents") [46]. However, the most common way of utilizing chemically virgin lignocellulosic biomass materials is acid hydrolysis of carbohydrates (*i.e.*, saccharification) to convert them *via* selective reactions into H_2O-soluble low-molar-mass carbohydrates (*i.e.*, mainly monosaccharides and disaccharides), which then can be used for making a multitude of value-added products (see Chapter 3 "Chemical and biochemical conversion") [21, 33]. Additionally, the production of solid fuels from biomass is becoming more important [21, 47]. Although many raw lignocellulosic biomass feedstocks are suitable to be burned as such to provide heat, various thermochemical methods can be also used *via* unselective reactions for producing gases, condensable liquids (tars), and solid char for different purposes (see Chapter 7 "Thermochemical conversion"). The relative proportions of these products depend on the chosen method and the specific reaction conditions applied.

Table 1.2: The main conversion routes of lignocellulosic biomass materials for a variety of products (energy, chemicals, and other bioproducts) [21].

Conversion method	Process types
Chemical and biochemical methods	Extraction, hydrolysis, and anaerobic digestion
Thermochemical methods	Direct combustion, pyrolysis, gasification, hydrothermal carbonization, liquefaction, torrefaction, and steam explosion
Other methods	Densification[a] and other mechanical processing

[a]Refers to the manufacture of pellets and briquettes.

Analogous to the general biorefinery concept, wood material is fractionated in the kraft process into pulp (mainly cellulose), extractives (turpentine and tall oil), and spent cooking liquor ("black liquor") (see Chapter 4 "Chemical pulping-based methods") [21, 48–54]. Black liquor is the most important by-product consisting, in addition to inorganic substances (contain mainly the residual cooking chemicals and sodium bound to organics), degraded lignin, primarily carbohydrates-derived (mostly from hemicelluloses) aliphatic carboxylic acids, and residual extractives (see Section 5.3 "Black liquor"). It is conventionally burnt in a recovery furnace, after evaporation, to recover energy and inorganic cooking chemicals [55]. The full-scale fiber line for producing bleached kraft pulp comprises, after kraft pulping, the subsequent oxygen-alkali delignification stage (see Section 4.5.1 "Oxygen-alkali delignification") and finally, the bleaching stage for removing the residual lignin and other "impurities" from the kraft pulp (see Section 4.5.2 "Delignifying bleaching") [56, 57]. The integrated biorefinery concepts in the pulp and paper industry are of increasing importance, creating new technologies (see Chapter 5 "Integrated forest biorefining") [21, 31, 58–66].

Kraft pulping is the most important delignification process (corresponding to about 95% of chemical pulp production) and still is principally the same method as that in its early days about 160 years ago [21, 54]. There are also other chemical delignification methods such as sulfite, organosolv, soda and soda-anthraquinone (AQ) pulping (see Chapter 4 "Chemical pulping-based methods"). The liberating of fibers from the wood matrix can be also accomplished mechanically without removing lignin (*i.e.*, by means of refining) or by combining chemical and mechanical treatments [67]. The term "high-yield pulp" is

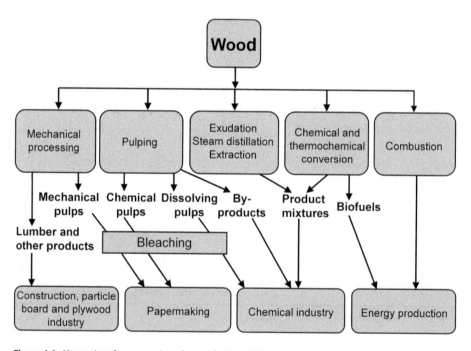

Figure 1.1: Alternatives for conversion of wood feedstocks into a wide range of products [21].

often used for different types of lignin-rich pulps (mainly from neutral sulfite pulping) that need mechanical defibration.

Wood can be traditionally processed (*i.e.*, partly fractionated) in several ways, using mechanical, chemical (including biochemical), and thermochemical methods (Figure 1.1). The average proportions of the total wood utilization are as follows [21]: fuel 50–55%, construction 25–30%, fiber 10–15%, and others 5%. For example, wood obtained from forest trees is a major source of cellulosic fiber for pulp and paper manufacture, corresponding to about 90% of the total, the rest being based on non-wood resources.

References

[1] Maher, K.D., and Bressler, D.C. 2007. Pyrolysis of triglyceric materials for the production of renewable fuels and chemicals. *Bioresource Technology* 98:2351–2368.

[2] Alén, R. 2018. *Carbohydrate Chemistry – Fundamentals and Applications*. World Scientific, Singapore. Pp. 472–496.

[3] Ioelovich, M. 2015. Recent findings and energetic potential of plant biomass as a renewable source of biofuels – a review. *BioResources* 10(1):1879–1914.

[4] Gallezot, P. 2012. Conversion of biomass to selected chemical products. *Chemical Society Reviews* 41:1538–1558.

[5] Kucherov, F.A., Romashov, L.G., Galkin, K.I., and Ananikov, V.P. 2018. Chemical transformations of biomass-derived C6-furanic platform chemicals for sustainable energy research, material science, and synthetic building blocks. *ACS Sustainable Chemistry & Engineering* 6:8064–8092.

[6] Wang, Y., Zhao, D., Rodriguez-Padrón, D., and Len, C. 2019. Recent advances in catalytic hydrogenation of furfural. *Catalysts* 9(10):796–828.

[7] An, Z., and Li, J. 2022. Recent advances in the catalytic transfer hydrogenation of furfural to furfuryl alcohol over heterogeneous catalysts. *Green Chemistry* 24:1780–1808.

[8] Goldstein, I.S. (Ed.). 1981. *Organic Chemicals from Biomass*. CRC Press, Boca Raton, FL, USA. 310 p.

[9] Argyropoulos, D.S. (Ed.). 2006. *Materials, Chemicals, and Energy from Forest Biomass*. ACS Symposium Series 954. American Chemical Society, Washington, DC, USA. 591 p.

[10] Hill, C.A.S. 2006. *Wood Modification – Chemical, Thermal and Other Processes*. John Wiley & Sons, Chichester, England. 239 p.

[11] Herrick, F.W., and Hergert, H.L. 1977. Utilization of chemicals from wood: retrospect and prospect. In: Loewus, F.A. and Runecles, V.C. (Eds.). *The Structure, Biosynthesis, and Degradation of Wood, Recent Advances in Phytochemistry, Vol. 11*. Plenium Press, New York, NY, USA. Pp. 443–515.

[12] Wolf, O. (Ed.). 2005. Techno-economic feasibility of large-scale production of bio-based polymers in Europe. *Technical Report EUR 22103 EN*. European Commission, Joint Research Centre, Institute for Prospective Technological Studies, Spain. 27 p.

[13] Koukoulas, A.A. 2007. Cellulosic biorefineries – charting a new course for wood use. *Pulp & Paper Canada* 108(6):17–19.

[14] Rosillo-Calle, F., de Groot, P., Hemstock, S.L., and Woods, J. (Eds.). 2007. *The Biomass Assessment Handbook – Bioenergy for a Sustainable Environment*. Earthscan, Sterling, VA, USA. 269 p.

[15] Clark, J.H., and Deswarte, E.I. (Eds.). 2008. *Introduction to Chemicals from Biomass*. John Wiley & Sons, New York, NY, USA. 184 p.

[16] Höfer, R. (Ed.). 2009. *Sustainable Solutions for Modern Economies*. RSC Publishing, The Royal Society of Chemistry, Cambridge, United Kingdom. 497 p.

[17]   Singh nee' Nigam, P., and Pandey, A. (Eds.). 2009. *Biotechnology for Agro-Industrial Residues Utilisation*. Springer, Heidelberg, Germany. 630 p.

[18]   Cherubini, F. 2010. The biorefinery concept: using biomass instead of oil for producing energy and chemicals. *Energy Conversion Management* 51(7):1412–1421.

[19]   Demirbas, A. 2010. *Biorefineries – For Biomass Upgrading Facilities*. Springer, Heidelberg, Germany. 240 p.

[20]   Kamm, B., Gruber, P.R., and Kamm, M. (Eds.). 2010. *Biorefineries – Industrial Processes and Products*. Wiley-VCH, Weinheim, Germany. 903 p.

[21]   Alén, R. 2011. Principles of biorefining. In: Alén, R. (Ed.). *Biorefining of Forest Resources*. Paper Engineers' Association, Helsinki, Finland. Pp. 55–104.

[22]   Liu, S., Abrahamson, L.P., and Scott, G.M. 2012. Biorefinery: ensuring biomass as a sustainable renewable source of chemicals, materials and energy. *Biomass and Bioenergy* 39(4):1–4.

[23]   Liu, S., Lu, H., Hu, R., Shupe, A., Lin, L., and Liang, B. 2012. A sustainable woody biomass biorefinery. *Biotechnology Advances* 30(4):785–810.

[24]   Pandey, A., Höfer, R., Taherzadeh, M., Nampoothiri, K.M., and Larroche, C. (Eds.). 2015. *Industrial Biorefineries & White Biotechnology*. Elsevier, Amsterdam, The Netherlands. 710 p.

[25]   Kruus, K., and Hakala, T. (Eds.). 2017. *The Making of Bioeconomy Transformation*. VTT Technical Research Centre of Finland Ltd., Espoo, Finland. 82 p.

[26]   Rastegari, A.A., Yadav, A.N., and Gupta, A. (Eds.). 2019. *Prospects of Renewable Bioprocessing in Future Energy Systems*. Springer, Heidelberg, Germany. 261 p.

[27]   Lantto, R., Järnefelt, V., and Tähtinen, M. (Eds.). 2018. *Going Beyond a Circular Economy – A Vision of a Sustainable Economy in which Material, Value and Information are Integrated and Circulate Together*. VTT Technical Research Centre of Finland Ltd., Espoo, Finland. 60 p.

[28]   Lehtonen, J., and Järnefelt, V. (Eds.). 2019. *The Carbon Reuse Economy – Transforming CO_2 from a Pollutant into a Resource*. VTT Technical Research Centre of Finland Ltd., Espoo, Finland. 50 p.

[29]   Anon. 2022. *Green and Vibrant Economy – the Climate Roadmap for the Forest Industry 2035*. Finnish Forest Industries, Helsinki, Finland. 19 p.

[30]   Anon. 2022. *Success Grows on Trees – Wood-based Innovations for a Sustainable Future*. Finnish Forest Industries, Helsinki, Finland. 15 p.

[31]   Alén, R. 2015. Pulp mills and wood-based biorefineries. In: Pandey, A., Höfer, R., Taherzadeh, M., Nampoothiri, K.M., and Larroche, C. (Eds.). *Industrial Biorefineries & White Biotechnology*. Elsevier, Amsterdam, The Netherlands. Pp. 91–126.

[32]   Naik, S.N., Goud, V.V., Rout, P.K., and Dalai, A.K. 2010. Production of first and second generation biofuels: a comprehensive review. *Renewable and Sustainable, Energy Reviews* 14(2):578–597.

[33]   Viikari, L., and Alén, R. 2011. In: Alén, R. (Ed.). *Biorefining of forest resources*. Paper Engineers' Association, Helsinki, Finland. Pp. 225–261.

[34]   Pandey, A., Bhaskar, T., Stöcker, M., and Sukumaran, R. (Eds.). 2015. *Recent Advances in Thermochemical Conversion of Biomass*. Elsevier, Amsterdam, The Netherlands. 504 p.

[35]   Guerriero, G., Hausman, J.-F., Strauss, J., Ertan, H., and Siddiqui, K.S. 2016. Lignocellulosic biomass: biosynthesis, degradation, and industrial utilization. *Engineering in Life Sciences* 16:1–16.

[36]   Mendoza, C.L.M. 2021. *Assessment of Agro-Forest and Industrial Residues Potential as an Alternative Energy Source*. Doctoral Thesis. Lappeenranta-Lahti University of Technology LUT, Laboratory of Sustainable Energy Systems, Lappeenranta, Finland. 91 p.

[37]   Salami, A. 2021. *Biorefining of Lignocellulosic Biomass and Chemical Characterization of Slow Pyrolysis Distillates*. Doctoral Thesis. University of Eastern Finland, Kuopio, Finland. 143 p.

[38]   Kerton, F.M. 2008. Green chemical technologies. In: Clark, J.H., and Deswarte, F.E.I. (Eds.). *Introduction to Chemicals from Biomass*. John Wiley & Sons, Chichester, United Kingdom. Pp. 47–76.

[39]   Lancaster, M. 2002. *Green Chemistry: An Introduction Text*. Royal Society of Chemistry, Cambridge, England. 310 p.

[40] Fengel, D., and Wegener, G. 1989. *Wood – Chemistry, Ultrastructure, Reactions*. Walter de Gruyter, Berlin, Germany. 613 p.

[41] Lewin, M., and Goldstein, I.S. (Eds.). 1991. *Wood Structure and Composition*. Marcel Dekker, New York, NY, USA. 488 p.

[42] Sjöström, E. 1993. *Wood Chemistry – Fundamentals and Applications*. 2nd edition. Academic Press, San Diego, CA, USA. 293 p.

[43] Alén, R. 2000. Structure and chemical composition of wood. In: Stenius, P. (Ed.). *Forest Products Chemistry*. Fapet, Helsinki, Finland. Pp. 11–57.

[44] Hon, D.N.-S., and Shiraishi, N. (Eds.). 2001. *Wood and Cellulosic Chemistry*. 2nd edition. Marcel Dekker, New York, NY, USA. 914 p.

[45] Alén, R. 2011. Structure and chemical composition of biomass feedstocks. In: Alén, R. (Ed.). *Biorefining of Forest Resources*. Paper Engineers' Association, Helsinki, Finland. Pp. 17–54.

[46] Holmbom, B. 2011. Extraction and utilization of non-structural wood and bark components. In: Alén, R. (Ed.). *Biorefining of Forest Resources*. Paper Engineers' Association, Helsinki, Finland. Pp. 178–224.

[47] Konttinen, J., Reinikainen, M., Oasmaa, A., and Solantausta, Y. 2011. Thermochemical conversion of forest biomass. In: Alén, R. (Ed.). *Biorefining of Forest Resources*. Paper Engineers' Association, Helsinki, Finland. Pp. 262–304.

[48] Rydholm, S. 1965. *Pulping Processes*. Interscience Publishers, New York, NY, USA. 1269 p.

[49] Casey, J.P. (Ed.). 1980. *Pulp and Paper – Chemistry and Chemical Technology, Volume 1*. 3rd edition. John Wiley & Sons, New York, NY, USA. 820 p.

[50] Biermann, C.J. 1993. *Essentials of Pulping and Papermaking*. Academic Press, San Diego, CA, USA. 472 p.

[51] Biermann, C.J. 1996. *Handbook of Pulping and Papermaking*. 2nd edition. Academic Press, San Diego, CA, USA. 754 p.

[52] Alén, R. 2000. Basic chemistry of wood delignification. In: Stenius, P. (Ed.). *Forest Products Chemistry*. Fapet, Helsinki, Finland. Pp. 58–104.

[53] Bajpai, B. 2010. *Environmentally Friendly Production of Pulp and Paper*. John Wiley & Sons, New York, NY, USA. 365 p.

[54] Alén, R. 2018. Manufacturing cellulose fibres for making paper: A historical perspective. In: Särkkä, T., Gutiérrez-Poch, M., and Kuhlberg, M. (Eds.). *Technological Transformation in the Global Pulp and Paper Industry 1800–2018*. Springer Nature, Switzerland. Pp. 13–34.

[55] Tran, H. (Ed.). 2020. *Kraft Recovery Boilers*. 3rd edition. TAPPI Press, Atlanta, GA, USA. 375 p.

[56] Patrick, K. (Ed.). 1991. *Bleaching Technology for Chemical and Mechanical Pulps*. Miller Freeman, San Francisco, CA, USA. 169 p.

[57] Dence; C.W., and Reeve, D.W. (Eds.). 1996. *Pulp Bleaching – Principles and Practice*. TAPPI Press, Atlanta, GA, USA. 868 p.

[58] Ragauskas, A.J., Nagy, M., and Kim, D.H. 2006. From wood to fuels – integrating biofuels and pulp production. *Industrial Biotechnology* 2(1):55–65.

[59] Mendes, C.V.T., Carvalho, M.G.V.S., Baptista, C.M.S.G., Rocha, J.M.S., Soares, B.I.G., and Sousa, G.D.A. 2009. Valorisation of hardwood hemicelluloses in the kraft pulping process by using an integrated biorefinery concept. *Food and Bioproducts Processing* 87(3):197–207.

[60] Huang, H.-J., Ramaswamy, S., Al-Dajani, W.W., and Tschirner, U. 2010. Process modeling and analysis of pulp mill-based integrated biorefinery with hemicellulose pre-extraction for ethanol production: a comparative study. *Bioresource Technology* 101(2):624–631.

[61] Bajpai, P. 2013. *Biorefinery in the Pulp and Paper Industry*. Elsevier, Amsterdam, The Netherlands. 103 p.

[62] Huang, F., and Ragauskas, A. 2013. Integration of hemicellulose pre-extraction in the bleach-grade pulp production process. *Tappi Journal* 12(10):55–61.

[63] Moshkelani, M., Marinova, M., Perrier, M., and Paris, J. 2013. The forest biorefinery and its implementation in the pulp and paper industry: energy overview. *Applied Thermal Engineering* 50(2):1427–1436.

[64] Sanglard, M., Chirat, C., Jarman, B., and Lachenal, D. 2013. Biorefinery in a pulp mill: simultaneous production of cellulosic fibers from *Eucalyptus* globulus by soda-anthraquinone cooking and surface-active agents. *Holzforschung* 67(5):481–488.

[65] Martin-Sampedro, R., Eugenio, M.E., Moreno, J.A., Revilla, E., and Villar, J.C. 2014. Integration of a kraft pulping mill into a forest biorefinery: pre-extraction of hemicellulose by steam explosion versus steam treatment. *Bioresource Technology* 153(2):236–244.

[66] Lehto, J. 2015. *Advanced Biorefinery Concepts Integrated to Chemical Pulping*. Doctoral Thesis. University of Jyväskylä, Laboratory of Applied Chemistry, Jyväskylä, Finland. 173 p.

[67] Kappel, J. 1999. *Mechanical Pulps: From Wood to Bleached Pulp*. TAPPI Press, Atlanta, GA, USA. 396 p.

2 Chemical composition of lignocellulosic biomass materials

2.1 Content of the main chemical constituents

The major chemical constituents of all wood species are so-called "structural substances" (*i.e.*, cellulose, hemicelluloses, and lignin) and other polymeric constituents present in lesser and varying quantities are pectins, starch, and proteins [1]. Besides these macromolecular components, various nonstructural and mostly low-molar-mass compounds (*i.e.*, extractives, water (H_2O)-soluble organics, and inorganics) can be found in small quantities in both hardwoods and softwoods. The gross chemical composition of the stemwood (*i.e.*, based on average results of several chemical wood analyses) differs to some extent from that of the other macroscopic parts of the tree. Additionally, it is known that there are some variations in the chemical composition within the same stem, especially in the radial direction as well as differences between the normal wood (*i.e.*, xylem including pith, heartwood, and sapwood) and reaction wood (*i.e.*, tension wood in hardwoods and compression wood in softwoods). The following discussion is mainly restricted to stemwood, which is normally used, for example, for pulping. Bark is also a potential feedstock for many purposes and its chemical composition differs somewhat from that of the wood material and mostly depends on the wood species.

Both hardwoods and softwoods are widely distributed on the Earth, ranging from tropical to Arctic regions [1]. The total number of known wood species is about 58,500 and the number of softwood species is relatively low (less than 1,000) when compared to that of hardwood species. However, due to the more extensive exploitation of tropical forests, only a minor part of these wood species is currently utilized commercially, although this principal feedstock basis may diversify in the future. For example, in North America, about 1,200 species exist naturally and of these some 100 are of commercial importance, whereas in Europe the approximate numbers are, respectively, 100 and 20. It is generally obvious that wood possesses unique chemical and structural characteristics that render this material desirable for a broad variety of end uses. Hence, it can be also concluded that to optimize the selection of a particular tree species to various end uses, basic knowledge of its chemical composition and structure is of great importance.

The moisture content of the living tree (*i.e.*, the average values are in the range of 40–50% of the total wood mass) varies seasonally and even diurnally depending on the weather [1, 2]. In woods, the cellulose content is about the same (*i.e.*, 40–45% of the wood dry solids), but hardwoods usually contain more hemicelluloses and less lignin than softwoods (Figure 2.1). The typical content of hemicelluloses in hardwoods (mainly glucuronoxylan or simply xylan) and in softwoods (mainly galactoglucomannan or simply glucomannan) is, respectively, 30–35% and 25–30% of the wood dry solids. On the other hand, the lignin content of hardwoods varies typically between 20% and 25% of the

https://doi.org/10.1515/9783110608366-002

wood dry solids, whereas that of softwoods is normally in the range of 25–30% of the wood dry solids.

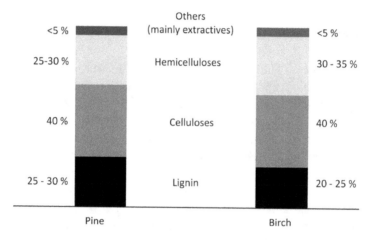

Figure 2.1: Average chemical composition of Scots pine (*Pinus sylvestris*) and silver birch (*Betula pendula*). Figures are given as percentages of wood dry solids.

The other compounds, mainly extractives, in woods from temperate zones usually make up about 5% of the wood dry solids; tropical species often exceed this value [1, 2]. In woods from temperate zones, elements other than carbon, oxygen, and hydrogen account for between 0.1% and 0.5% of the dry solids and the generally measured ash content with metal oxides may vary in the range of 0.3–1.5% of the dry solids. In contrast, in tropical and subtropical regions the ash content can make up even to 5% of the wood dry solids. Hence, in woods from temperate zones the macromolecular structural substances building up the cell walls practically account for about 95% of the wood dry solids (Figure 2.2) and for tropical woods, the corresponding average value may decrease to 90%.

The structure of bark is essentially different from that of wood and differences between species are much greater in bark than in wood [4, 5]. In boreal countries, in the traditional forest industry, the annual consumption of wood-derived lignocellulosic biomass together with the generation of by-products, including bark from debarking and logging residues, is substantial. For example, bark contains a massive amount of various extractives, which might be of commercial importance. However, today, bark is mainly used almost only for the high-volume and straightforward production of energy when combusting in a bark-burning furnace often with bio sludge or for some non-energy purposes, such as landscaping. Additionally, bark could be a potential feedstock material for vacuum pyrolysis, for example, for producing phenols [6, 7] and activated carbon [8].

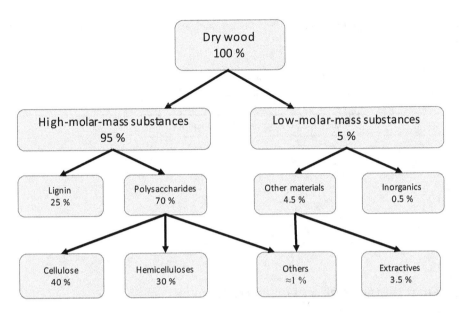

Figure 2.2: General classification and content of chemical wood components (% of the wood dry solids) [3].

In general, bark has, compared to wood, higher amounts of extractives (5–20% of the dry solids) and inorganics (2–5% of the dry solids), but also so-called "other organics" (5–20%, in wood about 1% of the dry solids) [1]. The chemical composition of the fraction of other organics depends greatly on the wood species, but it may typically contain suberin (2–8% of the dry solids), polyphenols (2–7% of the dry solids) as well as starch and proteins (2–5% of the dry solids).

The chemical compositions of wood and non-wood lignocellulosic biomass materials are principally rather close to each other, although data reported on the chemical properties of non-wood fibers may vary greatly [1, 9–16]. Normally, non-wood feedstocks contain higher amounts of extractives and proteins and especially, inorganics (Table 2.1). For example, the content of silicon dioxide or silica (SiO_2) in rice straw can be even about 14% of the dry solids. The high concentration of inorganics typically accounts for the most harmful properties of non-wood lignocellulosic biomass materials in alkaline pulp production; SiO_2 as well as other sparingly soluble materials promote scale formation in the evaporators of the recovery cycle of pulping chemicals. Additionally, although the mutual ratios of cellulose, hemicelluloses, and lignin are quite similar in wood and non-wood materials, there are still some characteristic chemical differences, for example, between wood lignin and non-wood lignin. Hence, also in the case of non-wood lignocellulosic biomass materials, a deep knowledge of the initial composition of these feedstocks is one of the most important factors when aiming at their effective utilization. In practice, the specific selection as well as the accurate quality control of the feedstock materials used for different purposes is of importance.

Table 2.1: Comparison between the typical chemical composition of non-wood and wood feedstocks used for pulping (% of the dry solids) [1].

Component	Non-wood feedstock	Wood feedstock
Carbohydrates	50–80	65–80
Cellulose	30–45	40–45
Hemicelluloses	20–35	25–35
Lignin	10–25	20–30
Extractives	5–15	2–5
Proteins	5–10	<0.5
Inorganics	0.5–10	0.1–1

2.2 Fiber dimensions and ultrastructure

Table 2.2 presents examples of the fiber dimensions of typical wood and non-wood fibers. It is obvious that non-wood fibers are typically shorter than softwood fibers, but similar in length to hardwood fibers. Another general feature of non-wood fibers is that they contain a high proportion of non-fiber fine cells that are significantly different from wood fine cells [17]. The average content of fibers in softwoods and hardwoods is, respectively, 90% and 60% by volume; the other cell types are in case of softwoods, ray parenchyma (5–10 vol%), ray tracheid (<5 vol%), and epithelial parenchyma cells and in case of hardwoods, vessel elements (20–30 vol%) as well as ray parenchyma (10–15 vol%) and longitudinal parenchyma cells [1]. The main non-fiber cells of non-wood materials are parenchyma cells, epidermal cells, and vessels [9]. Of these cells, the content of parenchyma cells is much higher than that of the others. As the non-fiber cells are typically much shorter than the normal fibers, their ability to form a paper network is relatively poor. Most of them act as filler in the paper network and have a rather negative effect on the papermaking process and on the dry and wet properties of the paper. On the other hand, besides many problems related to the large fines fraction of non-wood fiber pulps, the unique morphological characteristics of non-wood fibers (i.e., short and slender) allow papermaking with low energy consumption in beating, good formation, a smooth and closed surface, and good printability.

    The ultrastructure of non-wood fibers is typically quite different from that of wood fibers [17]. In case of all technically important wood fibers (tracheids) (Figure 2.3), the cell wall is basically composed of two layers including the relatively thin primary wall (P) and the thick secondary wall (S) (Table 2.3) [1]. Based on differences in the microfibrillar orientation, the latter layer is divided into three sublayers that are termed as

Table 2.2: Average fiber dimensions of wood and non-wood plant fibers
[9, 11, 17–19].

Raw material	Length, mm	Diameter, μm
Hardwoods (in the temperate zone)	0.7–1.6	30
Softwoods (in the temperate zone)	2.7–4.6	38
Bamboo	2.3–2.7	15
Rice straw	0.9–1.2	18
Wheat straw	1.2–1.5	14
Kenaf (bark)	2.1–2.5	22
Miscanthus	0.9–1.2	14
Switchgrass	1.1–1.3	13
Cotton	0.7–0.9	20
Bagasse[a]	1.5–1.7	21

[a]The remaining fibrous dry material after crushing sugarcane stalks to recover their juice.

follows: *i*) the outer layer of the secondary wall (S_1), *ii*) the middle layer of the secondary wall (S_2), and *iii*) the inner layer of the secondary wall (S_3), which is sometimes also referred to as the tertiary wall (T). In certain cases (*e.g.*, conifer tracheids and some hardwood cells), the inside of S_3 layer is covered with a thin membrane called the "warty layer" (W). All these layers differ from one another in the different orientation of their cellulose-containing microfibrils and, to some extent, in their chemical composition.

A wood fiber cell consists mainly of celluloses, hemicelluloses, and lignin. A simplified picture is that cellulose forms a skeleton (*i.e.*, microfibril-containing lamellas) that is surrounded by other substances functioning as matrix (hemicelluloses) and encrusting material (lignin). The microfibrils wind around the cell axis with varying microfibrillar angles in different directions either to the right (Z helix) or to the left (S helix). The central cavity of the hollow fiber is termed the "lumen" (L). The middle lamella (ML) is located between the P walls of adjacent cells and serves the function of binding the cells together. As it is difficult to distinguish ML from the two P walls on either side, the term "compound middle lamella" (CML) is generally used to designate the combination of ML with the two adjacent P walls. The term "compound secondary wall" is sometimes used for describing the $S_1 + S_2 + S_3$ layer.

One illustrative ultrastructure example of non-wood lignocellulosic biomass materials is a diverse group of evergreen perennial flowering plants, bamboos [21]. More than 1,450 bamboo species from 70 genera can be found in variable climates, from cold mountains to hot tropical regions [11, 22–27]. As a fast-growing plant with excellent mechanical properties bamboo is one of the most important non-wood material resources. It is widely used in papermaking, furniture, textiles, and as a component of advanced industrial materials. Bamboo has a lignin content and hemicelluloses content rather like those in woods and has also a low SiO_2 content. In terms of tissue

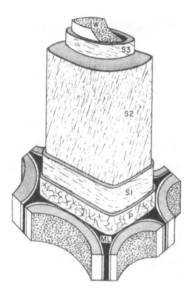

Figure 2.3: Schematic drawing of a wood fiber showing microfibril orientation and relative size of the different layers [20]. Key: ML = middle lamella, P = primary wall, S_1 = outer layer of the secondary wall, S_2 = middle layer of the secondary wall, S_3 = inner layer of the secondary wall, and W = warty layer.

Table 2.3: Average thickness of different wood cell wall layers, number of microfibrillar layers (lamellae), and microfibrillar angle within the layers [1].

Wall layer[a]	Thickness, μm	Number of microfibrillar layers	Average angle of microfibrils, degrees
P	0.05–0.1	–[b]	–[b]
S_1	0.1–0.3	3–6	50–70
S_2	1–8[c]	30–150[c]	5–30[d]
S_3	<0.1	<6	60–90
ML[e]	0.2–1.0	–	–

[a]P refers to primary wall, S_1 to outer layer of the secondary wall, S_2 to middle layer of the secondary wall, S_3 to inner layer of the secondary wall, and ML to middle lamella.
[b]Cellulose microfibrils form mainly an irregular network.
[c]Varies greatly between earlywood (1–4 μm) and latewood (3–8 μm).
[d]The microfibrillar angle varies between 20–30° (earlywood) and 5–10° (latewood).
[e]An intercellular layer bonding the cells together. Contains mainly nonfibrillar material.

types, the bamboo culm (or stem) consists of fibers, parenchyma cells, and conducting tissues, which constitute on average up to 32%, 59%, and 9% of the total volume, respectively. Figure 2.4 shows a structure model of Moso bamboo (*Phyllostachys heterocycla* f. *pubescens*) fiber cell wall also indicating microfibril angles with respect to the direction of cell axis.

The structural wood constituents are not uniformly distributed in wood cells, and their relative mass proportions can vary significantly, depending on the morphological region and age of the wood [1]. In practice, this means that differences are typically large between reaction wood and normal wood as well as between the various

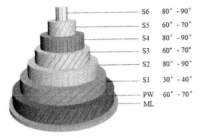

S6	80° - 90°
S5	60° - 70°
S4	80° - 90°
S3	60° - 70°
S2	80° - 90°
S1	30° - 40°
PW	60° - 70°
ML	

Figure 2.4: Structure model of Moso bamboo (*Phyllostachys heterocycla* f. *pubescens*) fiber cell wall, also showing microfibril angles with respect to the direction of cell axis [27]. Key = ML middle lamella, PW = primary wall, and S$_{1-6}$ = secondary walls.

cell types. The detailed data on the distribution of the major constituents within the cell wall layers are of great importance for a better understanding of the cell wall arrangement, but also for explaining the physical and chemical properties of wood fiber as a natural composite material. In general, the separation of the cell wall layers in a pure state is difficult due to the small dimensions of these layers. Hence, for example, limited carbohydrate analyses of microtomed wall layer fractions have been conducted. However, in most cases, several indirect methods have been used for the quantitative or semiquantitative determination of the cell wall components. Since the chemical composition data on the cell wall layers can vary greatly depending on the origin of the fibers, only some principal features are discussed in the following.

Table 2.4 shows an approximate chemical composition of the cell wall of softwood tracheids. These values have been calculated for a given average thickness of the cell wall layers by assuming a certain distribution profile of the constituents (Table 2.5). Thus, for example, the lignin content of the compound middle lamella (ML + P) is high but, because the layer is thin, only a minor fraction of the total lignin is in this compound layer. In softwoods, the lignin content of the ML is about 70% of the total material, and even higher values (10–30% higher) have been reported for the cell corner middle lamella (ML$_{CC}$) associated with fibers and vessels. In hardwoods, the lignin concentration in the ML regions is lower than in softwoods. Additionally, both the ray parenchyma cells, and the vessels have higher lignin content (30–35%) than the fibers (20–25%). The data in these tables also indicate that the compound secondary wall (S$_1$ + S$_2$ + S$_3$) has the highest polysaccharide (*i.e.*, cellulose and hemicelluloses: glucomannan and xylan) content; almost all the polysaccharides are in this layer.

The quantities and chemical composition of extractives vary significantly depending on wood species [1]. The extractives also occupy certain morphological sites in the wood structure and do not belong to the cell wall substances. For example, the resin acids are found in the resin canals of conifers, whereas the fats and waxes are located in the ray parenchyma cells of both softwoods and hardwoods. Heartwood has, besides many low-molar-mass as well as high-molar-mass phenols, also other aromatic-like compounds, which are not usually found in sapwood and which give the heartwoods of many species their typical dark color and resistance to decay. Although extractives are generally more abundant in heartwood than in sapwood, some variations in the radial and horizontal distribution of certain components in sapwood have also been observed.

Table 2.4: Relative mass proportions of the main chemical constituents in the cell wall of softwood tracheids (% of the total dry solids of each layer) [1].

Constituent	Morphological region[a]	
	(ML + P)	$(S_1 + S_2 + S_3)$
Polysaccharides	35	75
Cellulose	12	45
Glucomannan	3	20
Xylan	5	10
Others[b]	15	<1
Lignin	65	25

[a]ML is the middle lamella, P the primary wall, and S_1 the outer, S_2 the middle, and S_3 the inner layers of the secondary wall.
(ML + P) refers to the compound middle lamella and $(S_1 + S_2 + S_3)$ to the compound secondary wall.
[b]Consisting mainly of pectic substances.

Table 2.5: Distribution of the main chemical constituents in the cell wall of softwood tracheids (% of the total amount of each constituent) [1].

Constituent	Morphological region[a]	
	(ML + P)	$(S_1 + S_2 + S_3)$
Polysaccharides	5	95
Cellulose	3	97
Glucomannan	2	98
Xylan	5	95
Others[b]	75	25
Lignin	21	79

[a]ML is the middle lamella, P the primary wall, and S_1 the outer, S_2 the middle, and S_3 the inner layers of the secondary wall.
(ML + P) refers to the compound middle lamella and $(S_1 + S_2 + S_3)$ to the compound secondary wall.
[b]Consisting mainly of pectic substances.

A tree contains only rather low amounts of inorganic components, and their content in needles, leaves, bark, branch wood, and root wood can be much higher than that of stemwood [1]. In general, a tree takes up inorganic salts mainly from the forest soil through its root system and transports them to the stem and crown by sap flow. Hence, the greatest concentration of inorganic elements occurs in the living parts of the tree. Both the total minerals content and the concentration of each element vary greatly within and between the species. Unlike the cell wall structural components, the content of inorganic constituents varies significantly with the environmental conditions under which the tree has grown. Additionally, there are indications that young trees tend to have a higher concentration of inorganics than mature trees do and that hardwoods contain somewhat more inorganics than softwoods do.

2.3 Structural constituents

The chemical components of lignocellulosic biomass materials, including polysaccharides (*i.e.*, cellulose and hemicelluloses), lignin, and extractives, differ chemically from each other and they also behave in characteristic ways during their conversion processes [1, 28–39]. Hence, it can be concluded that to optimize the certain conversion process of a particular feedstock material, a deep knowledge of the behavior of these components, especially high-molar-mass constituents, under varying conditions is needed. In contrast, extractives, as a low-molar-mass and minor component, form typically a side-stream in many conversion processes. It has also been indicated that cellulose, hemicelluloses, and lignin act independently during chemical treatments and only negligible interactions between these major components can be found. However, some interactions of these three main lignocellulosic biomass components may still occur by influencing slightly, for example, the reaction rates of the components when being closely together in the lignocellulosic biomass sample compared to those rates detected separately for the pure constituents. It is known that the close association between carbohydrate and lignin components in the wood matrix also results in the formation of chemical linkages between these constituents, although the number of these common chemical bonds related to the total mass of lignin and carbohydrates is relatively low.

Polysaccharides are complex carbohydrates that consist of more than nine monosaccharide moieties covalently bound together by glycosidic linkages [40]. Their structures may range from linear to highly branched polymers. These polymeric carbohydrates occur naturally as such or as structural parts, for example, of peptideglycans and lipopolysaccharides. Polysaccharides can be roughly divided into plant (*e.g.*, cellulose, hemicelluloses, starch, pectins, and gums) and animal (*e.g.*, glycogen and chitin) carbohydrates. Additionally, polysaccharides produced by algae, fungi, and microbes are often separately classified. The trivial names of polysaccharides typically indicate their origin.

The general term "glycan" or "homopolysaccharide" refers to polysaccharides that contain only one type of monosaccharide unit (*i.e.*, glycose) [40]. For example, a glucan means a glycan with a repeating unit of glucose. Similarly, established names are normally used for arabinan, xylan, galactan, and mannan. It is also possible that the accurate configuration of the monosaccharide unit as well as the numbers of the carbon atoms participating in the formation of a glycosidic bond, are shown; for example, cellulose $[\rightarrow4)\text{-}\beta\text{-D-Glc}p\text{-}(1\rightarrow]_n$, (*i.e.*, the monosaccharide β-D-glucopyranose, β-D-Glc$p$, is a repeating unit) is alternatively called "(1→4)-β-D-glucan" or "(1→4)-β-D-glucopyranan". There are also many similar specific names, such as that of pectin or (1→4)-α-D-galacturonan with repeating α-D-galacturonic acid units. Polysaccharides that contain more than one type of monosaccharide unit are generally called "heteroglycans" or "heteropolysaccharides". Hemicelluloses in lignocellulosic biomass materials are typical examples of these heterogeneous polysaccharides.

It has been estimated that over 90% of the carbohydrate mass in nature consists of polysaccharides [40]; they also form a significant part of the total carbohydrates used industrially. Commercially, the most important polysaccharides are cellulose (an H_2O-insoluble material), hemicelluloses (a partly H_2O-soluble material), starches (only swell in H_2O), and H_2O-soluble gums. Additionally, a wide range of industrial cellulose derivatives (*e.g.*, cellulose nitrate, cellulose acetate, and carboxymethylcellulose) are produced.

2.3.1 Cellulose

Cellulose is the world's most abundant natural biopolymer, which is distributed in its native form throughout the plant kingdom and thus, the major organic component in most lignocellulosic biomass materials [1, 40]. It has been estimated that almost half of the biomass formed in the photosynthesis consists of cellulose; globally about $1.5 \cdot 10^{12}$ tons of carbon dioxide (CO_2) are chemically bound by photosynthesis each year. Almost all cellulose is formed by photosynthesis (*i.e.*, plant cellulose), but there are also microbial extracellular carbohydrates, such as bio cellulose or bacterial cellulose (BC) that are synthesized by various bacteria. Plant cellulose and BC have the same chemical structure, but somewhat different physical properties. Based on detailed microscopy observations, the plant cell was discovered in 1663 by Robert Hook who named this biological unit for its resemblance to cells (in Latin, "cella" means "small room") inhabited by Christian monks in a monastery. Anselme Payen noted in 1837–1842 that most plant materials contain a relatively resistant fraction of the fibrous material with essentially the same elemental composition $C_6H_{10}O_5$ and named it "cellulose".

Cellulose is a polydispersed and completely linear homopolysaccharide that consists of repeating β-D-glucopyranose (β-D-Glc*p*) moieties (in a $^4C_1$ conformation) linked together by (1→4)-glycosidic (or in this case, glucosidic) bonds (Figure 2.5). In the $^4C_1$ conformation, all substituents (C_1-OR, C_2-OH, C_3-OH, C_4-OR, and C_5-CH$_2$OH) of the β-D-glucopyranose chain units are equatorially oriented, making the chain very stable due to the minimized interaction between the pyranose ring substituents. It should be noted that the cellulose chain has both reducing (C_1-OH) and non-reducing (C_4-OH) units in its molecular structure.

The degree of polymerization (DP) of native wood cellulose is of the order of about 10,000 and is lower than that of cotton cellulose (about 15,000) [40]. These DP values correspond, respectively, to molar masses of 1.6 million Da and 2.4 million Da and to molecular lengths of 5.15 μm and 7.73 μm. In chemical pulping, the DP of cellulose can decrease to 500–2,000, corresponding to molecular lengths of 0.26 μm and 1.03 μm, respectively. The polydispersity (M_w/M_n) of cellulose is rather low (<2), indicating that the weight average molar mass (M_w) and the number average molar mass (M_n) do not deviate much from each other.

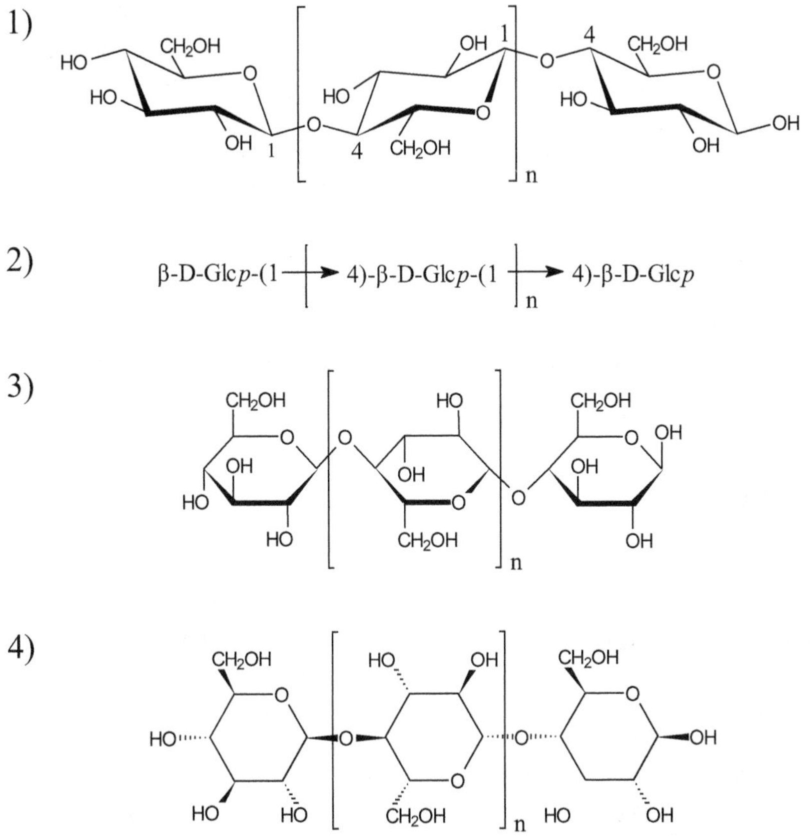

1)

2) β-D-Glc$p$-(1 $\longrightarrow$ 4)-β-D-Glc$p$-(1 $\longrightarrow$ 4)-β-D-Glc$p$

3)

4)

Figure 2.5: Structure of cellulose presented by stereochemical (1), abbreviated (2), Haworth perspective (3), and Mills (4) formulas [40].

Due to the strong tendency for intramolecular and intermolecular hydrogen bonding, bundles of cellulose molecules aggregate to microfibrils and further to fibrils (and finally lamellae in the cell wall) that contain both highly ordered crystalline (60–75% of the total cellulose and 50–150 nm in length) and less ordered (disordered) amorphous regions (25–50 nm in length) [40]. Several detailed crystalline structures of cellulose (*i.e.*, polymorphous lattices of cellulose) have also been determined: celluloses I (native cellulose), II (monoclinic cellulose), III, and IV. The unit cell of the native cellulose consists of four β-D-glucopyranose residues and, for example, in the chain direction the repeating unit is a cellobiose residue (1.03 nm in length). The form II is obtained whenever the lattice of cellulose I is destroyed, for example, on swelling with strong alkali (*e.g.*, during mercerization) or on dissolution of cellulose. The conversion of cellulose I to cellulose II is irreversible. Celluloses III and IV can be produced from celluloses I or II with specific chemical treatments.

Naturally occurring microbial respiration and decomposition of lignocellulosic bio-mass materials are significant reactions that release the carbon bound in various plants to the carbon cycle of the Earth [40]. Additionally, herbivores (*e.g.*, ruminants and ter-mites) can digest cellulose due to the microorganisms in their gastrointestinal tract. In contrast, humans are not able to degrade cellulose into sugars to utilize it as a source of energy. The reason for this is the fact that the glucosidic bonds (*i.e.*, β-(1→4)-bonds) in cellulose differ from those in starch (*i.e.*, α-(1→4)-bonds) and the starch-hydrolyzing en-zymes, amylases, present in the saliva of humans and some other mammals cannot af-fect cellulose. In general, cellulose is relatively inert against chemical treatments; it is also soluble only in a few solvents, although some of them can be used consequently to improve its reactivity and accessibility. Different solvents may cause interfibrillar or in-trafibrillar swelling or even dissolution of cellulose.

2.3.2 Hemicelluloses

Hemicelluloses are heteropolysaccharides that are not as well-defined as cellulose [1, 40–46]. Their chemical and thermal stabilities are also clearly lower than those of cellu-lose, presumably due to their lack of crystallinity and lower DP (100–200). Moreover, hemicelluloses differ from cellulose with respect to their solubility in alkalis. This char-acteristic property is usually utilized when fractionating various polysaccharides in lig-nin-free samples. Some hemicelluloses, such as certain fragments of hardwood xylan and arabinogalactan are partly or even totally H_2O-soluble. Therefore, in these cases, distinction between H_2O-soluble hemicelluloses, sugars (*i.e.*, mainly monosaccharides and disaccharides), and some extractives-derived compounds is sometimes difficult. The polysaccharide chain of hemicelluloses is usually linear, but it can be also branched and contains side groups or short side chains.

The term "hemicellulose" was introduced by E. Schulze in 1891 when it was noted that specific polysaccharides extracted from plant tissues with diluted alkali could be more readily hydrolyzed with acids than cellulose [40]. In spite of its long use, this term still has no unique definition. These cellulose-resembling polysaccharides (*i.e.*, they differ from (1→4)- and (1→3)-glucans and their derivatives) were first mislead-ingly believed to represent intermediates of the biosynthesis of cellulose (note the pre-fix hemi-). Today, essential chemical data are obtained about the main hemicelluloses from many important origins. However, in this connection, only the essential features of the most common hemicelluloses are briefly discussed. Due to many current biore-finery concepts suggested, the interest toward the commercial utilization of hemicel-luloses is gradually increasing. Table 2.6 summarizes the main structural features of the major hemicelluloses appearing in both softwoods [47–50] and hardwoods [51, 52].

The building moieties of hemicelluloses are hexoses (D-glucose, D-mannose, and D-galactose), pentoses (D-xylose, L-arabinose, and D-arabinose), or deoxyhexoses (L-rhamnose or 6-deoxy-L-mannose and rare L-fucose or 6-deoxy-L-galactose). Small amounts of uronic

Table 2.6: The major hemicelluloses in softwoods and hardwoods [1, 43].

Hemicellulose type	Occurrence[a]	Amount, % of dry wood	Composition		
			Units	Molar ratios	Linkage
Glucomannan	SW	10–15	α-D-Galp	0.3	1→6
			β-D-Glcp	2	1→4
			β-D-Manp	7	1→4
			Acetyl	2	
(Galacto)glucomannan	SW	5–8	α-D-Galp	1	1→6
			β-D-Glcp	1	1→4
			β-D-Manp	3	1→4
			Acetyl	1	
Xylan	SW	6–11	α-L-Araf	1	1→3
			4-O-Me-α-D-GlcpU	2	1→2
			β-D-Xylp	8	1→4
Arabinogalactan	SW[b]	10–20	α-L-Araf	2	1→6
			β-L-Arap	1	1→3
			β-D-Galp	11	1→3, 1→6
Glucomannan	HW	1–4	β-D-Glcp	2	1→4
			β-D-Manp	3	1→4
Xylan	HW	20–30	4-O-Me-α-D-GlcpU	1	1→2
			β-D-Xylp	10	1→4
			Acetyl	7	

[a]SW refers to softwood and HW to hardwood.
[b]Occurs in the heartwood of larches, while its content in other softwoods is typically <1% of the dry wood solids.
[c]α-L-Araf is α-L-arabinofuranose, β-L-Arap β-L-arabinopyranose, α-D-Galp α-D-galactopyranose, β-D-Glcp β-D-glucopyranose, β-D-Manp β-D-mannopyranose, 4-O-Me-α-D-GlcpU 4-O-methyl-α-D-glucuronic acid, and β-D-Xylp β-D-xylopyranose (for their chemical structures, see Figure 2.6). Acetyl group corresponds to CH_3CO-.

acids (4-O-methyl-D-glucuronic acid, D-galacturonic acid, and D-glucuronic acid) are also present. These units mainly exist as six-membered (pyranose) structures in α or β forms (Figure 2.6). As shown in Table 2.6, softwoods and hardwoods differ not only in the total content of hemicelluloses, but also in the percentages of individual hemicellulose constituents (mainly glucomannan and xylan) and in the detailed composition of these constituents. In general, softwood hemicelluloses have more mannose and galactose units and less xylose units and acetylated hydroxyl groups than those from hardwoods.

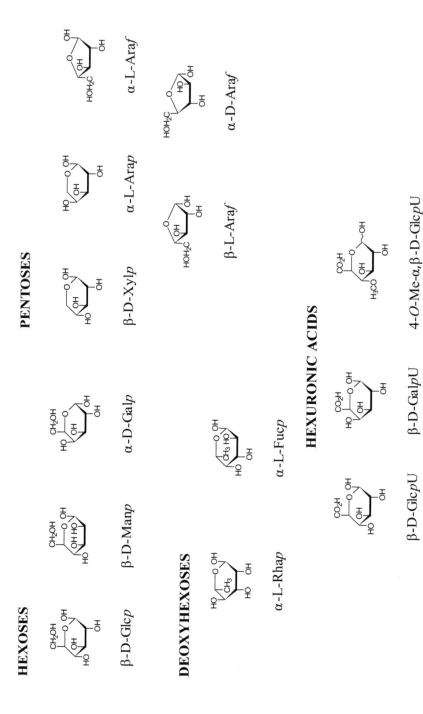

Figure 2.6: Common sugar units of wood hemicelluloses [40]. Note that U is used for uronic acid (*e.g.*, GlcU) instead of A (*e.g.*, GlcA), which is also typically used in the literature.

In softwoods, the primary hemicellulose component is glucomannan or (galacto)glucomannan or O-acetylgalactoglucomannan (Table 2.6). Its framework is built of a linear backbone of $(1\rightarrow4)$-linked β-D-glucopyranose (β-D-Glc$p$) and β-D-mannopyranose (β-D-Man$p$) units. The framework moieties are partly acetylated at C_2-OH and C_3-OH; the acetyl group content is about 6% of the total glucomannan corresponding to on the average one acetyl group per 3–4 hexose units. Additionally, it is substituted by $(1\rightarrow6)$-linked α-D-galactopyranose (α-D-Gal$p$) units. This hemicellulose component can be roughly classified into two fractions with different galactose contents: i) the galactose-poor fraction (glucomannan, two-thirds of the total amount) and ii) the galactose-rich fraction ((galacto)glucomannan, one-third of the total). Hence, for the total fraction of glucomannan, the typical overall molar ratio of the galactose:glucose:mannose is 0.5:1.0:3.5. The partial chemical structure of glucomannan from softwoods can be presented as follows:

$$\rightarrow4)\text{-β-D-Glc}p\text{-}(1\rightarrow4)\text{-β-D-Man}p\text{-}(1\left[\rightarrow4)\text{-β-D-Man}p\text{-}(1\right]_2$$

$$\overset{\displaystyle 6}{\uparrow}$$
$$\underset{\displaystyle 1)}{\big|}$$
$$\text{α-D-Gal}p$$

Softwoods contain (6–11% of the dry wood mass) xylan or arabinoglucuronoxylan or L-arabino(4-O-methylglucurono)xylan (Table 2.6). They are composed of a practically linear framework of $(1\rightarrow4)$-linked β-D-xylopyranose (β-D-Xyl$p$) units containing branches of both $(1\rightarrow2)$-linked pyranoid 4-O-methyl-α-D-glucuronic acid (4-O-Me-α-D-Glc$p$U) and $(1\rightarrow3)$-linked α-L-arabinofuranose (α-L-Ara$f$, containing a five-membered furanose structure) groups. In contrast to hardwood xylan, no acetyl groups are present in softwood xylan. The partial chemical structure of xylan from softwoods can be presented as follows:

$$\rightarrow4)\text{-β-D-Xyl}p\text{-}(1\left[\rightarrow4)\text{-β-D-Xyl}p\text{-}(1\right]\rightarrow4)\text{-β-D-Xyl}p\text{-}(1\left[\rightarrow4)\text{-β-D-Xyl}p\text{-}(1\right]_4$$

4-O-Me-α-D-Glc$p$U $\quad\quad$ α-L-Ara$f$

The content of glucomannan in hardwoods is clearly lower (1–4% of the dry wood mass) than that in softwoods and, unlike softwood glucomannan, hardwood glucomannan is not acetylated (Table 2.6). It has a practically linear backbone with a partial chemical structure as follows:

$$\rightarrow4)\text{-}β\text{-D-Glc}p\text{-}(1\rightarrow4)\text{-}β\text{-D-Man}p\text{-}(1\rightarrow4)\text{-}β\text{-D-Man}p\text{-}(1\rightarrow$$

In hardwoods, the content of xylan or O-acetyl-(4-O-methylglucurono)xylan is 15–30% of the dry wood mass (Table 2.6). It is composed of the same framework (i.e., $(1\rightarrow4)$-

linked β-D-Xylp units) as the softwood xylan, but it contains much fewer (1→2)-linked pyr-anoid uronic acid substituents that are not evenly distributed within the xylan chain. The framework moieties are also partly acetylated at C_2-OH and C_3-OH. The acetyl group content varies in the range of 8–17% of the total xylan, corresponding to 3.5–7.0 acetyl groups per 10 xylose units. In addition to these structural units, hardwood xylan has been reported [52] to contain small amounts of L-rhamnose (α-L-Rhap) and galacturonic acid (α-D-GalpU) in the structural sequence at the reducing end of the xylan molecule. The partial chemical structure of xylan from hardwoods can be presented as follows:

→4)-β-D-Xylp-(1→4)-β-D-Xylp-(1→ 4)-β-D-Xylp-...
        ↑(2→1)   9

4-O-Me-α-D-GlcpU

...→4)-β-D-Xylp-(1→3)-α-L-Rhap-(1→2)-α-D-GalpU-(1→4)-β-D-Xylp

As indicated in Table 2.6 arabinogalactan may occur in significant proportions (10–20% of the dry wood mass) in the heartwood of larches (*Larix sibirica/L. decidua*), while its content in other softwoods is generally less than 1% of the dry wood mass. It consists of a backbone of (1→3)-linked β-D-galactopyranose (β-D-Galp) residues most of which carry a side group or chain attached to their C_6 position. The side chains consist of (1→6)-linked β-D-Galp) chains of variable length and arabinose substituents (α-L-Ara*f* and β-L-Ara*p*). Unlike all the other wood hemicelluloses (*i.e.*, they are matrix substances), larch arabinogalactan is extracellular and it can be extracted almost quantitatively from the heartwood with H_2O. An example of a possible partial chemical structure of arabinogalactan from hardwoods is (it may also contain β-D-GlcpU units) as follows:

→3)-β-D-Galp-(1→3)-β-D-Galp-(1→3)-β-D-Galp-(1→3)-β-D-Galp-(1→3)-β-D-Galp-(1→

| α-L-Araf | β-D-Galp | β-D-Galp | α-L-Araf | (β-D-Galp)₃ |

β-L-Arap   β-D-Galp   β-D-Galp

Additionally, different acidic galactans (about 10% of the dry wood mass) are mainly present in the reaction wood (*i.e.*, compression wood in softwoods and tension wood in hardwoods) [40]. For example, an acidic galactan, built up of (1→4)-linked β-D-Galp units substituted at C_6 mainly with a single α-D-GalpU unit (α-D-GlcpU units are also present in small amounts), is a major hemicellulose in compression wood:

$$\rightarrow 4)\text{-}\beta\text{-D-Gal}p\text{-}(1\rightarrow 4)\text{-}\beta\text{-D-Gal}p\text{-}(1\overbrace{\rightarrow 4)\text{-}\beta\text{-D-Gal}p\text{-}(1}^{}\Big\rightarrow_{40}$$

β-D-GalpU   β-D-GalpU

A small amount of rhamnoarabinogalactan is also present in hardwoods [40]. It consists of a slightly branched backbone of (1→3)-linked β-D-Galp units. For example, in sugar maple (*Acer saccharum*) the molar ratio of galactose:arabinose:rhamnose is 1.7:1.0:0.2. The arabinose and rhamnose components are, respectively, α-L-Ara*f* and α-L-Rha*p*.

The hemicellulose compositions in non-wood lignocellulosic biomass materials typically resemble more those in hardwoods than in softwoods. For example, more than 90% of the bamboo hemicelluloses consist of a linear chain of (1→4)-linked β-D-xylopyranose units with attachments of side groups of L-arabinose and 4-*O*-methyl-D-glucuronic acid [11, 53-59]. Based on the molar ratio of L-arabinose:4-*O*-methyl-D-glucuronic acid:D-xylose (1.2:1.0:24.5) (Table 2.7), this hemicellulose has a characteristic structure within the family of *Gramineae* (*i.e.*, a large family of monocotyledonous flowering plants, grasses) and is an intermediate between hardwood and softwood xylans. Its content of acetyl groups is 6-7% of the total xylan. It has been also shown that grass arabinoglucuronoxylan contains ferulic acid ester linked to C_5 hydroxyl of arabinose moieties [62, 63]. Some of the adjacent arabinoglucuronoxylan chains are interlinked *via* dehydrodiferulates, which may comprise 0.01-0.05% of grass dry mass and render the grasses more rigid [64-66].

Table 2.7: Structural examples of the major hemicellulose components (arabinoglucuronoxylans) in non-wood feedstocks [9].

Raw material	Monosaccharide units[a] and molar ratios
Bamboo [53, 54]	Ara:GlcU:Xyl / 1.2:1.0:24.5
Wheat straw [57]	Ara:GlcU:Xyl / 1.6:1.0:16.6
Kenaf [60]	Ara:GlcU:Xyl / 0.2:1.0:4.1
Reed canary grass [61]	Ara:GlcU:Xyl / 2.2:1.0:13.1

[a]The monosaccharide units are L-arabinose (Ara), 4-*O*-methyl-D-glucuronic acid (GlcU), and D-xylose (Xyl).

Although not generally classified as hemicelluloses, other miscellaneous polysaccharides in woods and non-wood lignocellulosic biomass materials are various (1→3)- and (1→4)-glucans present in small amounts [40, 63, 67]; for example, besides starch ((1→4)-α-glucan (see below), callose ((1→4)-β-glucan), laricinan ((1→4)-β-glucan), xyloglucan (with a backbone of β-(1→4)-linked glucose residues), fucoxyloglucan, and rhamnoarabinogalactan.

Pectic substances form a heterogeneous group of carbohydrates[40]; in wood chemistry, they are considered to include polysaccharides that contain acidic groups, such as galacturonans and galactans ("acidic galactans"), but also non-acidic arabinans. While there is no agreed-upon definition of these substances, they are conventionally connected only with pectic acids, which are galacturonoglycans or poly(α-D-galactopyranosyluronic acids). They consist of linear backbones of (1→4)-linked α-D-GalpU residues that are normally in the form of methyl esters or calcium salts (pectinates). Pectins are present in most primary cell walls and in the non-woody parts of terrestrial plants. They may also contain many other monosaccharide residues, such as arabinose and xylose.

Starch is composed of linear amylose and branched amylopectin parts:

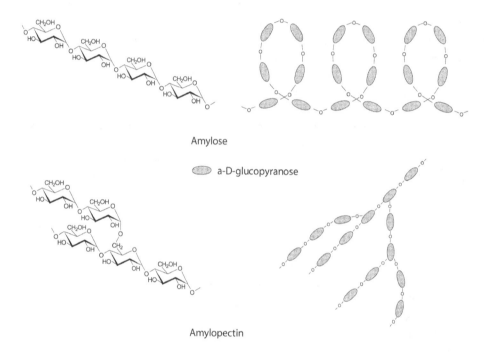

Amylose

a-D-glucopyranose

Amylopectin

Amylose is defined as a linear homoglycan, (1→4)-α-D-glucan, where α-D-glucopyranose units are linked to each other by (1→4)-glycosidic bonds [40]. However, there are also indications that some molecules are slightly branched by α-(1→6)-linkages. The molar masses of amylose typically vary in the range of $3 \cdot 10^5 - 9 \cdot 10^6$ Da. In contrast to amylose, amylopectin is the highly branched component of starch. It consists of hundreds of short (1→4)-α-D-glucan chains that are interlinked (about 5%) by (1→6)-α-D-glucan chains with an average DP of 20–30. Typical molar masses of amylopectin vary in the range of $10^7 - 10^9$ Da. Additionally, it seems that the crystalline regions are predominantly made up of amylopectin polymers where the outer branches are hydrogen bonded to each other to form crystallites.

2.3.3 Lignin

Lignocellulosic biomass materials essentially consist of a closely associated network of large-molar-mass polysaccharides (*i.e.*, cellulose and hemicelluloses) and lignin. Lignin is an amorphous and complex polymer with a chemical structure that distinctly differs from the other macromolecular constituents of wood [1, 3, 28, 68–74]. Furthermore, it is known that unlike wood carbohydrates, the chemical structure of lignin is characteristically irregular in the sense that different structural elements (*i.e.*, phenylpropane units or monolignols) are not linked to each other in any systematic order. The term "lignin" is derived from the Latin word used for wood, lignum.

In general, native lignins are roughly classified into three major groups: [1, 3, 28, 68–74], *i*) softwood lignin, *ii*) hardwood lignin, and *iii*) grass lignin. Besides the native lignins, which are typically separated from the wood in the form of milled wood lignin (MWL), dioxane lignin, cellulolytic enzyme lignin (CEL, enzymically liberated lignin), or enzymatic mild acidolysis lignin (EMAL), there are several industrial-based technical lignins that are mainly by-products of the chemical pulping. Kraft lignin (or sulfate lignin), alkali lignin (or soda lignin), and lignosulfonates are derived, respectively, from kraft, soda-anthraquinone (AQ), and sulfite pulping. Additionally, organosolv lignins, obtained from the pulping with organic solvents (primarily alcohols) and acid hydrolysis lignins, obtained from the acid hydrolysis processes of lignocellulosic biomass materials, are well-known, although their current production is limited. So-called "Klason lignin" can be obtained after hydrolyzing the polysaccharides of lignocellulosic biomass materials with 72% sulfuric acid (H_2SO_4). It is highly condensed and does not represent the lignin in its native state in the feedstock. It is also the fact that there are many characteristic differences between all these lignins.

According to a widely accepted concept confirmed originally by numerous detailed studies, lignin can be defined as a polyphenolic material arising primarily from enzymic dehydrogenative polymerization of three phenylpropanoid units (*p*-hydroxycinnamyl alcohols) (Figure 2.7). This biosynthesis reaction chain comprises various oxidative coupling reactions of the resonance-stabilized phenoxy radicals obtained from these α,β-unsaturated C_6C_3 precursors leading to the formation of a randomly cross-linked macromolecule [1]. Although the precursors of the *p*-hydroxycinnamyl alcohol type are practically the only building units of all kinds of lignins, in native lignins there is also a small proportion of other types of building units that cannot conceivably have been produced directly by oxidative radical coupling of the common precursors.

The proportions of the precursors in lignins vary with their botanical origin [1, 70, 75]. Normal softwood lignins are usually referred to as "guaiacyl (G) lignins", because the structural elements are derived principally from *trans*-coniferyl alcohol (more than 90%), with the remainder consisting mainly of *trans*-*p*-coumaryl alcohol. In contrast, hardwood lignins, generally termed "guaiacyl-syringyl (GS) lignins", are mainly composed of *trans*-coniferyl alcohol and *trans*-sinapyl alcohol-type units in varying ratios (*e.g.*, about 50% of *trans*-coniferyl alcohol and about 50% of *trans*-

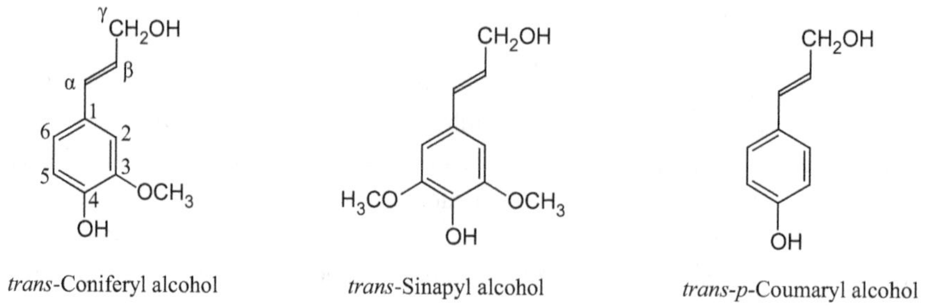

trans-Coniferyl alcohol   trans-Sinapyl alcohol   trans-p-Coumaryl alcohol

Figure 2.7: Phenylpropanoid units (C_6C_3 precursors) of lignin [1].

sinapyl alcohol). Grass lignins are also generally classified as GS lignins, although they additionally contain significant amounts of structural elements derived from *trans-p-coumaryl* alcohol (*i.e.*, p-hydroxyphenyl (H) lignin) and some aromatic acid residues (*e.g.*, about 45% of *trans*-coniferyl alcohol, about 45% of *trans*-sinapyl alcohol, and about 10% other precursors) [76–79]. Hence, grass lignins are often classified also as GHS lignins.

The structural building blocks of lignin are joined together by ether linkages C-O-C and carbon-carbon bonds C–C [1, 70, 71, 80–87]. Of these interunit linkages, the C-O-C ones dominate (more than two-thirds are of this type), with the most prominent type being the β-O-4 structure (Figure 2.8). Mainly due to the improved spectroscopic methods, versatile data on the frequency of the different types of linkages have substantially increased during the last 30 years and quite firmly established for the common lignins. Detailed knowledge about the characteristics of these linkages is of great theoretical and practical interest, for example, for a better understanding of the degradation reactions of lignin in many technical processes, such as delignification. Table 2.8 summarizes the dominating bond types and frequencies. Additionally, numerous miscellaneous linkages and minor structures have been identified. It is also known that the frequency of the structural groups can vary to some extent according to the morphological location of lignin.

The content of functional groups in lignin varies among the wood species as well as within the cell walls and hence, only approximate values for the frequencies of these groups can be given (Table 2.9). As its precursors, the lignin polymer contains characteristic methoxyl groups and phenolic hydroxyl groups, but also aliphatic hydroxyl groups (*i.e.*, are introduced into the lignin polymer during its biosynthesis) and some terminal aldehyde groups in the side chain. However, only relatively few of the phenolic hydroxyl groups are free because most of them form linkages to the neighboring phenylpropane units. In some types of native lignins, substantial amounts of the alcoholic hydroxyl groups are esterified with p-hydroxybenzoic acid (in aspen lignin) or p-coumaric acid (in grass lignins) [1, 89]. The approximative elemental mass ratios C:H:O for softwood lignin and hardwood lignin are 64:6:30 and 59:6:35, respectively.

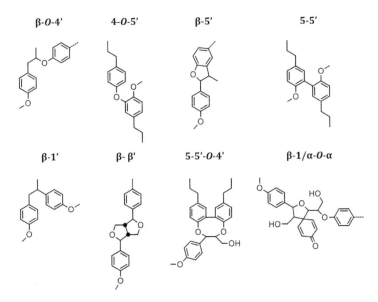

β-O-4'   **4-O-5'**   **β-5'**   **5-5'**

β-1'   **β-β'**   **5-5'-O-4'**   **β-1/α-O-α**

Figure 2.8: Major structures of the inter-unitary linkages in native softwood and hardwood lignins [83–86, 88]. For their frequencies, see Table 2.8.

Table 2.8: Average frequencies of the major inter-unitary linkages in native softwood and hardwood lignins (% of the total linkages) [1, 70, 86]. For the corresponding structures, see Figure 2.8.

Dimer structure[a]	Linkage type	Softwood	Hardwood
β-Aryl ether	β-O-4'	35–50	50–60
Diaryl ether	4-O-5'	5	7
Phenylcoumaran[b]	β-5'	8–12	5–9
Biphenyl	5-5'	10–15	5
1,2-Diarylpropane	β-1'	5	5
Linked through side chains	β-β'	2–3	3–4
Dibenzodioxocin	5-5'-O-4'	4–5	<1
Spiro-dienone	β-1/α-O-α	1–3	2–3

[a]The linkage α-O-4' (5 – 10%) as well as the linkages γ-O-4', γ-O-α', and β-6' exist in small amounts. In poplar (*Populus*) trees there are also α-esters (<5%) and γ-esters (<5%).
[b]Both ring and open structures.

Based on the detailed information obtained from studies of biosynthesis as well as versatile analyses of various linkage types and functional groups, several hypothetical structural formulas for lignins have been suggested; for example, for softwood lignin [81, 90–92], hardwood lignin [93–95], and non-wood lignin [79, 96]. Although these formulas are in good agreement with the analytical reality of isolated lignin preparations, it is obvious that it is not possible to isolate lignin purely from lignocellulosic biomass materials without partial degradation [97]. For this reason, also be-

Table 2.9: Functional groups of native softwood and hardwood lignins (per 100 C_6C_3 units) [1].

Functional group	Softwood	Hardwood
Aliphatic hydroxyl[a]	115–120	110–115
Phenolic hydroxyl	20–30	10–20
Methoxyl	90–95	140–160
Carbonyl	20	15

[a]Total sum of the primary and secondary hydroxyl groups.

sides challenges (*e.g.*, calibration with lignin-type standards) in the analytical determination of molar masses, the real molar mass of lignin in intact wood is difficult to obtain. However, different methods for measuring the M_n of softwood MWL suggest values in the range of 15,000–20,000 Da (DP 75–100), whereas slightly lower values for hardwood MWL have been reported [1, 98]. The M_w/M_n of softwood MWL is relatively high (2.3–3.5) when compared to that of cellulose and its derivatives. During alkaline delignification processes the average M_w of lignin decreases significantly resulting in the dissolution of lignin (Table 2.10).

Table 2.10: The weight average molar mass (M_w) of soluble lignin from the different phases of alkaline cooks (Da) [99].

Cooking method	Initial	End	Maximum
Softwood kraft	1,900	3,100	4,100
Birch kraft	2,200	2,600	3,400
Aspen kraft	1,500	2,700	2,950
Wheat kraft	5,050	3,400	5,050
Wheat soda-AQ[a]	5,900	4,900	5,900

[a]AQ refers to anthraquinone.

Electron microscopical observations on analytical lignin preparations have revealed mostly spherical particles of different sizes (10–100 nm) in addition to deformed structures [1]. With respect to the polymeric properties, lignin can be considered as a thermoplastic high-molar-mass material, which serves the dual purpose of acting as a binder between wood cells and imparting rigidity to the cell walls. Although native lignins behave as an insoluble, three-dimensional network, isolated lignins exhibit solubility in solvents, such as dioxane, acetone, ethylene glycol monomethyl ether (EGME or methyl cellosolve), tetrahydrofuran (THF), dimethylformamide (DMF), and dimethyl sulfoxide (DMSO).

2.3.4 Lignin-hemicellulose bonds

The close association between lignin and carbohydrate components in lignocellulosic biomass materials strongly suggests the existence of chemical linkages between these constituents [1]. Hence, for example, in some of the partial structural formulas proposed for lignin, chemical bonds between lignin and carbohydrates are also indicated. It has been obvious for a long time that physical and chemical interactions (*i.e.*, hydrogen bonds, van der Waals' forces, and chemical bonding) occur between lignin and carbohydrates, but, in spite of many intensive studies, it has been difficult to definitively verify the precise type and amount of chemical linkages. However, numerous experimental data obtained clearly suggest that the number of chemical bonds related to the mass of lignin and carbohydrates is relatively low. In general, the data on the association between lignin and carbohydrates in the wood cell wall are of great technical interest, especially when considering the need to separate lignin from polysaccharides as selectively as possible.

It is now generally accepted that lignin is chemically linked particularly with the main hemicellulose constituents, although there are also indications of lignin and cellulose bonds [1]. The terms "lignin-carbohydrate complex" (LCC) or "lignin-polysaccharide complex" (LPC) are generally used for describing the covalently bonded aggregates of lignin and hemicelluloses, but the more specific term "lignin-hemicellulose complex" (LHC) can be used as well. The chemical stability of such bonds and their resistance to acidic or alkaline treatments primarily depends on the type of linkage involved.

Examples of the most frequently suggested types of lignin-hemicellulose bonds include benzyl ether, benzyl ester, and phenyl glycoside linkages (Figure 2.9). The hemicelluloses side groups L-arabinose, D-galactose, and 4-*O*-methyl-D-glucuronic acid, as well as the hemicellulose chains' end groups D-xylose in xylan and D-mannose (and D-glucose) in glucomannan, are most commonly able to form connecting links to lignin [1]. This is mainly due to their sterically favored positions. The participation of these monosaccharides in the formation of LCCs also means that both lignin-xylan and lignin-glucomannan complexes are abundant. Additionally, it has been noted that especially the side group monosaccharide residues are generally enriched in many preparations of native lignins.

The α-carbon (C_α) of the phenyl propane units (*i.e.*, a benzylic carbon atom) is the most probable connection point between lignin and the hemicellulose blocks [1]. An ester linkage to xylan through 4-*O*-methyl-D-glucuronic acid as a bridging group is easily cleaved by alkali. However, more common and also much more alkali- and acid-stable than the ester bonds are the ether linkages formed through C_α and C_3 (or C_2) of L-arabinose units or through C_α and C_3 of D-galactose units.

Lignin/phenolic-carbohydrate complexes (LPCCs) present particularly in grass lignins are another type of linkages between lignin and hemicelluloses [89, 98]. They contain a hydroxycinnamic acid that is linked to lignin and may also be linked to a carbohydrate [100]. The two common hydrocinnamic acids, *p*-coumaric acid and

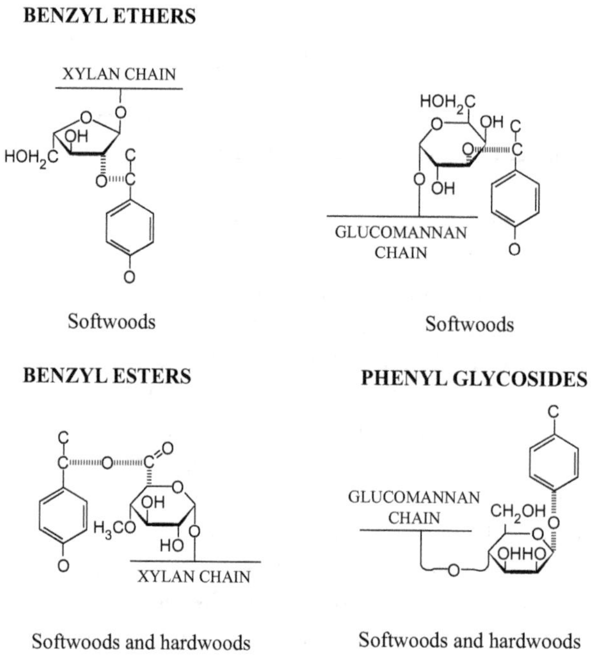
BENZYL ETHERS
XYLAN CHAIN
HOH2C
OH
GLUCOMANNAN CHAIN
Softwoods
Softwoods
BENZYL ESTERS
PHENYL GLYCOSIDES
H3CO
HO
XYLAN CHAIN
GLUCOMANNAN CHAIN
CH2OH
Softwoods and hardwoods
Softwoods and hardwoods

Figure 2.9: Typical lignin-hemicellulose linkages [1].

ferulic acid have mainly been detected, but in some non-wood plants, also sinapic acid forms these linkages [101–103]. Hydroxycinnamic acids form LPPCs as monomers or as dimers, and especially dimers of ferulic acid and sinapic acid have been found. It has been shown that ferulic acid is typically esterified to a hemicellulose and etherified to a core lignin forming either a ferulic acid bridge or a diferulic acid bridge [89, 100–102, 104]. In contrast, *p*-coumaric acid is mainly esterified to lignin without forming a bridge structure. The ester bonds are alkali-labile and are readily released even at mild alkali extraction [105].

There are also indications that lignin in the middle lamella and primary wall of the cell wall is associated with the pectic polysaccharides (galactan and arabinan) through ether linkages [1]. In these cases, C_6 in D-galactose units and C_5 in L-arabinose units seem to participate in the bridging. The glycosidic linkages can be formed by the reaction of the reducing end groups of hemicellulose chains and the phenolic hydroxyl groups (or the benzylic alcohol groups) of lignin. These linkages are easily cleaved by acids.

2.4 Nonstructural constituents

2.4.1 Major groups of extractives

In addition to the major structural constituents (*i.e.*, cellulose, hemicelluloses, and lignin), lignocellulosic biomass materials, as complex feedstocks, also contain a minor amount of nonstructural extractives, mainly with low molar masses [1]. Extractives comprise an extraordinarily large diversity of individual compounds (*i.e.*, several thousands), and depending on the morphological location of various extractives-based compounds, they have different defense and physiological functions in the lignocellulosic biomass materials [28, 106–115]. Bark (*i.e.*, the inner and outer bark layers) and the inner part of mature wood (xylem) in a stem, heartwood, usually contain higher amounts of extractives than the outer part of a stem, sapwood [1, 111, 115–117], as also indicated specifically for softwoods [118–124] and hardwoods [125–129]. Additionally, roots [130] as well as embedded branch knots in the stem [131, 132] contain a high concentration of extractives.

By a broad definition, extractives are either soluble in neutral organic solvents (*e.g.*, diethyl ether, methyl *tert*-butyl ether, dichloromethane, acetone, ethanol, methanol, hexane, toluene, and THF) or H_2O [1, 110, 112, 115]. Hence, they can be simply classified according to their chemical character (*i.e.*, based on their polarity); they are typically *i*) lipophilic (or hydrophobic) compounds, but some of them are *ii*) hydrophilic (or lipophobic). The term "resin" is often used as a collective name for the lipophilic extractives (with the exception of phenolic substances), which can be easily recovered from a wood sample by nonpolar organic solvents, but are insoluble in H_2O. The extractives typically impart color, odor, and taste to wood, and some of them (*e.g.*, fats and waxes) can be the energy source of the biological functions in the wood cells. Most resin components protect the wood against microbiological damage or insect attacks.

The chemical composition of the fraction of extractives in wood varies widely from species to species, and the total amount of extractives in a certain species primarily depends on growth conditions [1]. For example, the typical content of extractives in Norway spruce (*Picea abies*), Scots pine (*Pinus sylvestris*), and silver birch (*Betula pendula*) is, respectively, in the range of 1.0–2.0%, 2.5–4.5%, and 1.0–3.5% of the wood dry solids.

Table 2.11 shows the general classification of wood extractives. There are also some H_2O-soluble wood polysaccharides collectively called "gums" or "industrial gums", which can be classified by shape into different groups, such as linear, branched (*i.e.*, short branches on an essential linear backbone), or branch-on-branch structures [3]. Certain tropical trees spontaneously form exudate gums, which are exuded as viscous fluids at sites of injury. These gums have typically branch-on-branch structures, with illustrative examples being gum arabics, gum karaya, gum tragacanth, and gum ghatti [133–135]. In general, gums are polymeric substances that in an appropriate solvent or swelling agent form at low dry solids content highly viscous dispersions or gels. Gums are tasteless, colorless, odorless, and nontoxic, and are subjected to microbiological attack. Additionally, some polymeric extractives (*e.g.*, natural rubber) have been traditionally tapped from many resin-rich trees for several decades, and after processing they can be used effectively in different ways [136–138].

Table 2.11: Classification of wood extractives [3].

Aliphatic compounds	Phenolic compounds	Other compounds
Terpenes and terpenoids	Simple phenols	Sugars
(including, *e.g.*, resin acids	Stilbenes	Cyclitols
and steroids)	Lignans	Tropolones
Esters of fatty acids	Flavonoids	Amino acids
(fats and waxes)	Isoflavones	Alkaloids
Fatty acids	Condensed tannins	Coumarins
Alkanes	Hydrolyzable tannins	Quinones

After felling of the tree, the total content of lipophilic extractives begins to decrease and the chemical composition of the fraction of extractives is gradually changed [3, 123, 124, 139, 140]. This phenomenon is of great importance when storing extractives and their feedstock resources. For example, exposure to air affects the double bonds ($>C=C<$) in extractives and initiates a chain reaction that generates free radicals, which, in turn, are particularly strong oxidants. Furthermore, extractives are oxidized by certain enzymes, and some enzymes also act as catalysts in the hydrolysis of the esterified components (*e.g.*, glycerides). All these chemical and biochemical reactions are largely influenced by the conditions prevailing during wood storage and are markedly faster when the wood is stored in the form of chips instead of logs. Additionally, it is known that the hydrolysis of glycerides leading to free fatty acids and glycerol proceeds faster when the conditions for wood storage are wet instead of dry [112]. This is particularly important during storage of wood logs in H_2O in the summer. Besides extractives, even wood polysaccharides can be biodegraded to some extent during long storage. In practice, this may result in pulping in clearly reduced pulp yield and low pulp quality.

Extractives have been utilized for a wide range of purposes; for example, as solvents and various additives as well as raw materials in the chemical industry and for pharmaceutical purposes and in perfumery [115, 141]. Some classes of extractives also play an important role in the pulping and papermaking processes [1, 5, 142]. For example, the southern pines have a notably higher content of extractives, which provides substantial amounts of crude turpentine and tall oil soap as by-products from the kraft pulping. As indicated above, the extractives content of wood chips decreases during storage being halved during the first month [109, 143]. Hence, before sulfite pulping, spruce has been traditionally stored in the form of chips for a certain time to reduce the "pitch problems". Another way of reducing the content of extractives of the spruce sulfite pulps to an acceptable level is the removal of parenchyma cells through mechanical fiber fractionation, since much of the resin remains encapsulated inside these cells [1]. However, in contrast, in the kraft pulping, fresh wood chips can be used and prolonged wood storage results only in decreased yields of the important extractives-derived by-products [142].

As already indicated in Section 2.1 "Content of the main chemical constituents", in woods from temperate zones, elements other than carbon, oxygen, hydrogen, and nitrogen make up between 0.1% and 0.5% of the wood dry solids, whereas those from tropical and subtropical regions make up even 5% [1]. The total amount of wood inorganics is practically measured as ash, which is the residue obtained after the proper combustion of the organic matter of a wood sample. The ash contains mainly different metal oxides and average values for the ash content of commercial softwoods and hardwoods are generally in the range of 0.3–1.5% of the wood dry solids. Additionally, the metal ion concentrations depend to some extent on the tree species [144, 145] as well as the morphological parts of the tree [146].

There is also a remarkable dependence of the ash content and composition on the environmental conditions (*e.g.*, site fertility and climate) under which the tree has grown and, on the other hand, on the location within the tree [1]. It is also possible that the total amount of ash residue from tropical pulp woods occasionally may be relatively high, since their heavy trunks typically with a hollow center collect sand and other inorganic contaminants during logging and transport, for example, when pulling across the ground. However, in most cases, ash originates both from a variety of salt deposits in the cell walls and lumina as well as from inorganic moieties bound to the cell wall components, such as -CO_2H groups of pectins and xylan. Typical deposits are various metal salts including, for example, carbonates, phosphates, silicates, oxalates, and sulfates.

Some of the inorganic elements present are essential for wood growth [1]. The mineral constituents are generally a nuisance when wood is pulped or used for the production of energy [142, 144]. In the recovery of cooking chemicals, the presence of SiO_2 in high concentrations promotes scale formation in the evaporators. In bleaching, trace amounts of some transition elements, such as manganese (Mn), iron (Fe), and copper (Cu) significantly accelerate the degradation of pulp carbohydrates during certain bleaching stages and also detrimentally affect the brightness of the final pulp. Hence, in the bleaching process, most metal ions are generally displaced and washed out from pulp by aqueous acids or chelating agents, such as ethylenediaminetetraacetic acid (EDTA) and diethylenetriaminepentaacetic acid (DTPA). In spite of this metal recovery, the final pulp usually contains inorganic impurities at least partly originating from the wood feedstock. Additionally, it should be pointed out that since the primary and trace elements present in tree biomass are necessary for the growth of the trees, their recycling is of great importance from a soil conservation and fertilization point of view.

Typically, alkali and alkaline earth elements, such as potassium (K), calcium (Ca), and magnesium (Mg) make up about 80% of the total inorganic elemental constituents of softwoods and hardwoods, although a wide variety of other elements is also present in detectable amounts [1]. Table 2.12 represents the approximate concentration levels of various elements other than carbon, oxygen, hydrogen, and nitrogen in softwoods and hardwoods. The concentration of chlorine (Cl) (10–100 ppm) may vary to a great extent. This is at least partly due to the known fact that it mainly occurs in the form of chloride ions ($Cl^{\ominus}$), which are part of soluble and easily transporting salts in the living tree.

Table 2.12: Approximative concentration levels of various elements[a] in the dry stemwood of softwoods and hardwoods (ppm) [1].

Range	Element									
400–1,000	K	Ca								
100–400	Mg	P								
10–100	Na	Ba	Mn	Fe	Zn	Si	S	F	Cl	
1–10	Rb	Sr	Y	Ti	Nb	Ru	Pd	Pt	Cu	B
	Al	Ge	Se	Te						
0.1–1	Cs	Ta	Cr	Os	Rh	Ni	Ag	Sn	Br	
<0.1	Li	Sc	Zr	Hf	V	Mo	W	Re	Co	Ir
	Au	Hg	Ga	In	Pb	As	Bi	Sb	I	

[a]Small amounts of lanthanides are also present mainly at levels of <1 ppm.

2.4.2 Chemical features of extractives

In the following text, the different chemical groups of extractives are briefly presented. The main aim is to give only a general picture of typical structural features in each case. Hence, only some illustrative compounds belonging to each group are included.

More than 4,000 terpenes and their derivatives have been isolated and identified and they comprise a broad class of compounds with widespread appearance in the plant kingdom [1, 28, 108, 115]. Their basic structural unit is isoprene or 2-methyl-1,3-butadiene ($H_2C=C(CH_3)CH=CH_2$ or C_5H_8) and they can be divided into subgroups according to the number of isoprene units linked in a terpene (Table 2.13). Of these subgroups, the monoterpenes and diterpenes are of major industrial importance followed by sesquiterpenes and triterpenes. In woody tissues, there are, although as rare components, also present hemiterpenes, sesterterpenes, and tetraterpenes. Additionally, some polyterpenes, such as cis-1,4-polyisoprene (i.e., rubber, the exudates of *Hevea species*, such as *H. braziliensis*) are of great industrial importance.

The isoprene units are linked according to the "isoprene rule", which practically means that the isoprene units are joined together in a regular head-to-tail way [1, 114].

Table 2.13: The main structural types of terpenes in woody tissues.

Group name	Number of $C_{10}H_{16}$ units	Molecular formula
Hemiterpenes	0.5	C_5H_8
Monoterpenes	1	$C_{10}H_{16}$
Sesquiterpenes	1.5	$C_{15}H_{24}$
Diterpenes	2	$C_{20}H_{32}$
Sesterterpenes	2.5	$C_{25}H_{40}$
Triterpenes	3	$C_{30}H_{48}$
Tetraterpenes	4	$C_{40}H_{64}$
Polyterpenes	>4	$>C_{40}H_{64}$

However, this rule is strictly followed only up to five isoprene units, and, for example, the structure of many triterpenes can be explained by a tail-to-tail coupling of two sesquiterpenes. Besides this classification, terpenes can be generally classified according to the number of rings within a structure, *i.e.*, acyclic, monocyclic, bicyclic, tricyclic, and tetracyclic terpenes. The term "terpenes" refers generally to pure hydrocarbons, whereas the compounds collectively called "terpenoids" bear one or more oxygen-containing functional groups, such as -CO_2H, hydroxyl (-OH), and carbonyl (>C=O) groups. Despite this common practice, for simplicity, the term "terpenoids" is also occasionally used as a general name for all terpene-based compounds. Figure 2.10 shows some typical examples of terpenes and terpenoids.

Monoterpenes and monoterpenoids are volatile compounds and contribute substantially to the odor of wood [1, 115]. The compounds in this group are also dominant in turpentine from kraft pulping. They can be divided into acyclic, monocyclic, bicyclic, and the rare tricyclic structural types. Most of the monoterpenoid hydrocarbons are alicyclic and fewer aromatic compounds. Monoterpenes represent, together with diterpenes and some fatty acids and their glycerides, one of the most important constituents of the resin canal extractives and exudates of softwoods. The most important monoterpenes are α-pinene and β-pinene. Occasionally found in significant amounts are 3-carene, limonene, myrcene, and β-phellandrene. Although monoterpenes and monoterpenoids are rare in common hardwood species, some of these compounds are minor constituents of the oleoresins of tropical hardwoods.

Sesquiterpenes and sesquiterpenoids represent a wide range of compounds, and more than 2,500 compounds belonging to this group have been separated and identified [1]. They can be found as components of canal resin as well as of the heartwood deposits of softwoods. Hence, they usually represent a minor portion of the volatile substances (gum turpentines) in the gum oleoresins (resin canal exudates) of certain pines. Sesquiterpenes and sesquiterpenoids are also found in many tropical hardwoods, but are rare components of hardwoods from temperate zones. However, because these compounds occur usually only in small amounts in wood, they are industrially less important. They are classified into acyclic, monocyclic, bicyclic, and

MONOTERPENES AND MONOTERPENOIDS

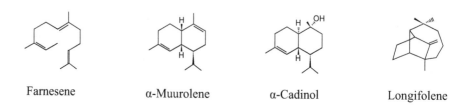

(-)-Limonene   α-Pinene   3-Carene   Camphene   Borneol

SESQUITERPENES AND SESQUITERPENOIDS

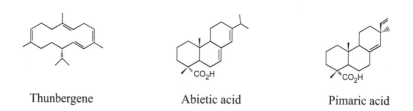

Farnesene   α-Muurolene   α-Cadinol   Longifolene

DITERPENES AND DITERPENOIDS

Thunbergene   Abietic acid   Pimaric acid

TRITERPENOIDS AND STEROIDS

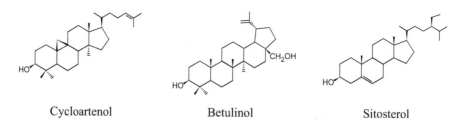

Cycloartenol   Betulinol   Sitosterol

Figure 2.10: Chemical structure of typical terpenes and terpenoids [1].

tricyclic compounds with different skeletal types. Typically, sesquiterpenes are more common compounds than sesquiterpenoids.

Diterpenes and diterpenoids constitute a major part of the canal extractives or oleoresin and are of great industrial importance [1, 28, 108, 112]. They seem to be restricted to softwood species mainly in the form of resin acids and apparently only some diterpenoids have been found in tropical hardwoods. The common resin acids are bicyclic, tricyclic, and teracyclic diterpenoids which can be classified into labdane, abietane, pimarane, and phyllocladene type derivatives. Of these resin acid types (tricyclic compounds), the abietane structures predominate, although the pimarane structures are also present in significant quantities. However, the resin acids of the abietane type with a conjugated dienoic structure are chemically less stable against isomerization and oxidation than those of the pimarane type. The substance group of extractives includes a great variety of acyclic, bicyclic, tricyclic, and macrocyclic derivatives.

Triterpenes and triterpenoids are widely distributed in the plant kingdom [1, 108, 112, 115, 126, 127]. They comprise mainly oxygenated derivatives and are traditionally treated as two classes of compounds: triterpenoids and steroids. These compound groups are both structurally and biogenetically closely related, and there is little reason for this distinction. Their biosynthesis proceeds from the acyclic squalene precursor according to almost identical pathways; the steroids differ from some of tetracyclic terpenoids only by the postcyclization loss of methyl groups. Thus, tetracyclic triterpenoids differ from steroids in that they have one or two methyl ($-CH_3$) groups at C_4. Therefore, they are also sometimes referred to as "methyl sterols" or "dimethyl sterols". Triterpenoids can be roughly divided into three subgroups: tetracyclic lanostane, pentacyclic lupane, and pentacyclic oleanane derivatives. Triterpenoids and steroids occur mainly as fatty acid esters and as glycosides, but also in the free form. Being sparingly soluble hydrophobic components, together with their degradation products, they can cause problems in pulping and papermaking processes.

Triterpenoids and steroids are common in softwoods, although generally in relatively small amounts [1, 108, 112, 115, 126, 127]. In the case of hardwoods, like softwoods, the most prominent compound is sitosterol, but many other compounds have been analyzed as well. In many hardwoods of tropical and temperate zones, a great variety of triterpenes and steroids are also present, although only in small amounts. The wood of the *Betula* species contains, besides sitosterol, the lupane triterpenoids (*i.e.*, betulinol and lupeol) and the crystalline betulinol is also mainly responsible for the white color of the birch bark. Both sitosterol and betulinol are potential raw materials for making wood-based chemicals. Additionally, several tropical woods contain glycosides of triterpenoids and steroids (called "saponins"), which produce lathering solutions in H_2O. The aglycones of the saponins are called "sapogenins".

Acyclic primary alcohols typically consisting of 6–9 isoprene units (*i.e.*, betulaprenols) and esterified with various saturated fatty acids are present in silver birch (*Betula pendula*) [1]. The >C=C< bonds have both *cis* and *trans* configurations. Various polyprenes with a high number of isoprene units are also present in certain wood species in the form of rubber (see above) and gutta percha. While the isoprene units in rubber are all arranged in *cis* configuration, those in gutta percha are all in *trans* configuration. In both polymers, the isoprene units are linked mainly by 1,4-bonds, and only a small proportion of 3,4-bonds are present.

Aliphatic extractives contain alkanes, fatty alcohols, fatty acids, fats, and waxes. Only small amounts of alkanes (the principal components are C_{22-30}-alkanes), free alcohols, and free fatty acids occur in woods [1]. The major parts of the fatty acids are esterified with glycerol (i.e., fats, mainly triglycerides) or with higher fatty alcohols (C_{18-22}) and terpenoids (i.e., waxes). More than 30 fatty acids or fatty acid moieties have been identified in softwoods and hardwoods (Figure 2.11); the most common fatty acid constituents belong both to the saturated and unsaturated compounds, mainly mono-, di-, and trienoic derivatives [108, 110, 112, 115]. Additionally, unsaturated tetraenoic fatty acids occur in small quantities. In softwoods, the parenchyma resin is mainly composed of fats, whereas in hardwoods, the parenchyma resin is virtually the only resin type and contains a significant proportion of waxes as well as fats. Fats and waxes are hydrolyzed during wood storage and kraft pulping.

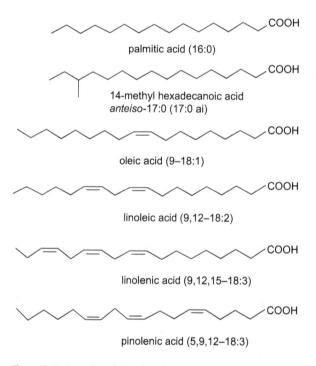

Figure 2.11: Examples of abundant fatty acids in trees [115].

Wood contains a large variety of aromatic extractives reaching from simple phenols to complex polyphenols and their related compounds (Table 2.11). It is characteristic for polyphenols that they tend to be species-specific and are often colored compounds, which are accumulated (i.e., primarily higher-molar-mass derivatives) abundantly in the heartwood, bark, and knots of many species, and only small amounts are present in sapwood (i.e., primarily lower-molar-mass derivatives) [1, 115]. Some of them are probably degradation products of compounds that can be hydrolyzed during extraction or steam distillation (*e.g.*, glycosides). This kind of extractives also has fungicidal properties and hence, protects the tree against microbiological attack. Certain phenolic compounds (*e.g.*, pinosylvin) present in the pine heartwood can generate harmful cross-links with lignin during acid sulfite pulping, thereby inhibiting the delignification. Additionally, for example, ellagic acid, which is the abundant component of hydrolyzable tannins and rich especially in certain *Eucalyptus* species, forms sparingly soluble salts that have negative effects (*e.g.*, scaling problems) on the alkaline pulping processes. However, tannins are, in most cases, commercially useful wood-derived products. They exhibit bioactive properties, including antioxidative, radical-scavenging, antimicrobial, antitumor, anticancer, and antiviral activities, which has expanded their use into nutritional properties pharmaceutical/medicine areas. Figure 2.12 shows selected examples of the chemical structures of the most important phenolic extractives.

STILBENES

LIGNANS

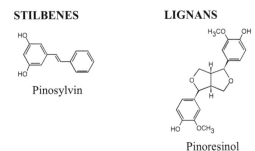

Pinosylvin

Pinoresinol

Constituents of HYDROLYZABLE TANNINS

Gallic acid

Ellagic acid

FLAVONOIDS

ISOFLAVONES

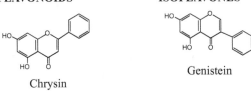

Chrysin

Genistein

Figure 2.12: Examples of typical phenolic extractives [1].

Stilbenes are derivatives of 1,2-diphenylethene [1, 115]. These compounds are mainly located in the heartwood of the *Pinus* species. In contrast, lignans (Figure 2.13) are distributed widely in the stemwood of both softwoods and hardwoods. They are basically formed by oxidative coupling of two C_6C_3 units and can be classified according to their chemical structures into several groups. In spruce, the lignan 7-hydroxymatairesinol (HMR) is the dominating polyphenol. Norlignans are the related compounds of lignans with one less carbon atom than lignans. In addition to certain dark-colored pigments, all these compounds principally contribute to the color of the heartwood.

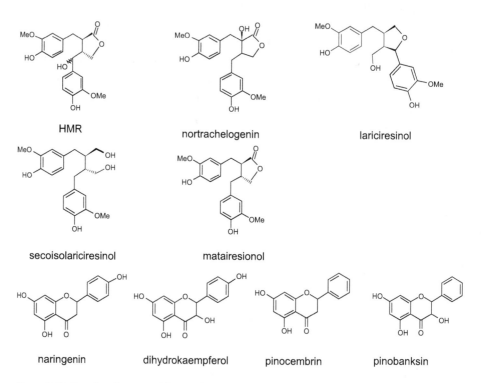

Figure 2.13: Abundant lignans and flavonoids in tree knots [115]. HMR refers to 7-hydroxymatairesinol.

Hydrolyzable tannins cover a broad range of esters of a sugar residue (usually D-glucose) with one or more polyphenol carboxylic acids, such as gallic acid and ellagic acid (Figure 2.14) [1, 115, 147]. Hence, they are divided into gallotannins and ellagitannins. The leaves and needles typically contain hydrolyzable tannins. Their main use is for leather tanning, where their performance (*i.e.*, in terms of color and light resistance) is excellent [147, 148]. The ester linkages in these structures are readily hydrolyzed by acids, alkalis, and enzymes (*e.g.*, tannase and takadiastase).

Flavonoids (Figure 2.13) have a typical diphenylpropane ($C_6C_3C_6$) skeletal structure [1]. These compounds are widely distributed in the stemwood of both softwoods and hardwoods. Condensed tannins or proanthocyanides are oligomers made of flavonoid units that can differ in their hydroxylation pattern [147, 149]. They typically consist of 3–8 flavonoid units. In European softwoods and hardwoods, condensed tannins are largely procyanidins with varying proportions of 3-flavanol units, catechin, and epica-

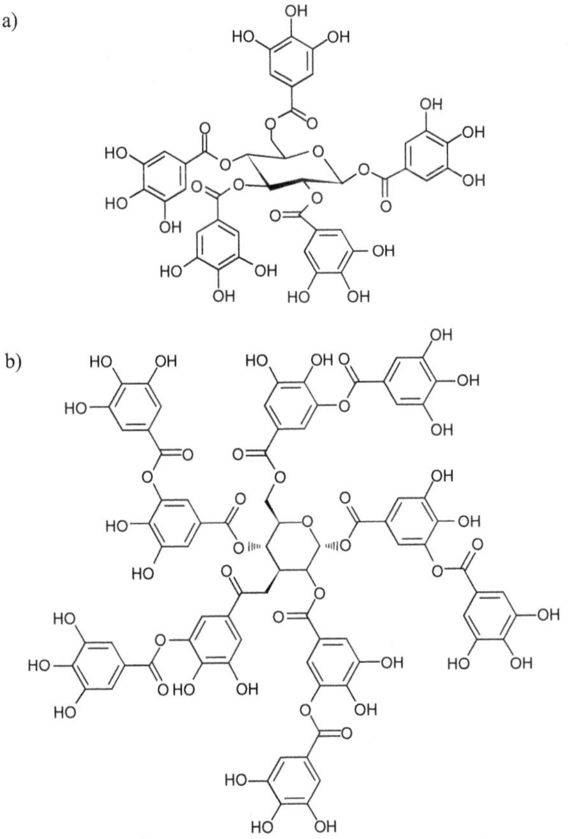

Figure 2.14: Molecular structure examples of pentagalloylglucose (a) and tannic acid (b) [147].

techin [115]. Nowadays, the major source of these substances is the bark of black wattle (*Acacia mearnsii*) and quebracho heatwood (*Schinopsis balansae* and *S. lorentzii*) [149]. Condensed tannins have been already extracted from oak in Europe in ancient times and used for tanning hides into leather. Leather tanning and wine making together with applications within the building material sector as well as a component in adhesives dominate and drive a growing international tannin market [147]. Isoflavones or isoflavonoids have a slightly different carbon skeleton from that of flavonoids (Figure 2.12).

Tropolones are 2-hydroxy-2,4,6-cycloheptatriene-1-one derivatives having a seven-membered ring and a close resemblance with phenols and terpenoids, *i.e.*, with mono-terpenoids and sesquiterpenoids [1]. The most abundant members are C_{10}-tropolones and C_{15}-tropolones (Figure 2.15). Tropolones are typical in the heartwood of many decay-resistant cedars, such as Western red cedar (*Thuja plicata*). Since they form strong metal complexes, they can cause corrosion problems in pulp digesters. Some quinones, cyclitols, coumarins, alkaloids, and amino acids have also occasionally been identified in certain wood species. Additionally, the typical monosaccharides, oligosacchardies, and polysaccharides can be found in woods. However, wood does not normally contain many H_2O-soluble organic substances. For example, the H_2O-soluble arabinogalactans (see Section 2.3.2 "Hemicelluloses") belong structurally to hemicellulose constituents and are not considered to be extractives.

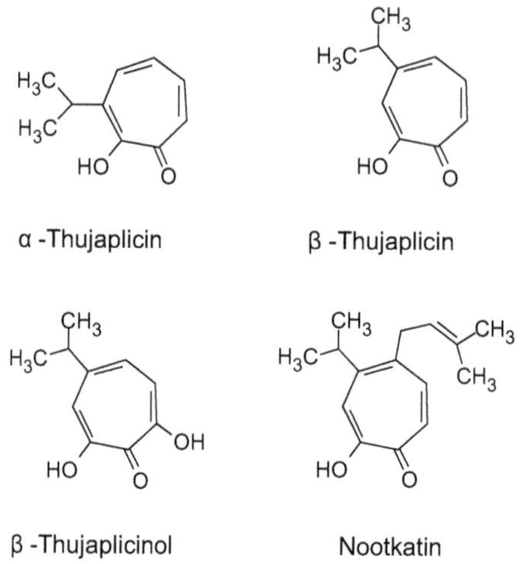

Figure 2.15: Examples of tropolenes [1].

2.4.3 Recovery of extractives from lignocellulosic materials

The principal emphasis in this chapter is paid to the selection of solvent since it has the most significant effect on extraction. The extraction of compounds from lignocellulosic biomass materials comprises many stages, beginning with the material pretreatment for extraction and ending up with the treatment of extract after recovery [150, 151]. Before a feedstock material is subjected to solvent extraction (i.e., solid-liquid extraction (SLE)), normally several preprocessing phases are applied [152]. The pretreatment of the initial lignocellulosic biomass material is generally recommended prior to the extraction phase, as the mass-transfer rate strongly depends on the structure of the feedstock material [152, 153]. In practice, this typically means the reduction of particle-size either by pressing, grinding, milling, or extrusion. The extraction efficiency is generally improved with reduced particle size as smaller particles can be more easily penetrated by the solvent and reducing the particle size leads to shorter diffusion pathways and hence, the reduction of the extraction time needed [151]. However, it is also possible that the size reduction of particles can be limited due to possible technical difficulties depending on the type of extraction equipment and operation [150]. Particles that are too fine may sometimes result in a high pressure drop, slow drainage, agglomeration, flow instabilities, channelling, formation of flakes, or entrainment of fines into the extract. Additionally, too fine particle size may also cause the excessive absorption of solute in the solid, which makes subsequent filtration difficult.

The raw material is usually pretreated for extraction, including drying to define moisture content [151–153]. To avoid reactions of sensitive compounds, it is advisable to use a mild drying technique [154]. For example, in case of tannins in the bark, the feedstock material is generally air-dried or dried at comparatively low temperatures (≤40 °C) to avoid the possible self-condensation of the extractives and the formation of bonds between the extractives and fiber or protein leading to low extractability of tannins [155]. Other possible preprocessing stages can be enzymatic pretreatment, micronization, pelletization, granulation, swelling, and prewetting. When choosing appropriate solvents for the extraction of desired compounds, selectivity, solubility, cost, and safety should be the primary concerns [150, 156]. Hence, in general, based on the first-principles theory of similarity and intermiscibility (*i.e.*, "like dissolves like"), solvents with a polarity value close to the polarity of the solute are likely to perform better and *vice versa* (Table 2.14).

Table 2.14: Solubilities of different extractives groups in organic solvents and H_2O [110, 156]. Key: +++ easily soluble, ++ soluble, + slightly soluble, and 0 insoluble.

Solvent	Resin acids, monoterpenoids, and other terpenoids	Fats, fatty acids, steryl esters, and sterols	Phenolic compounds	Glycosides, sugars, starch, and proteins
Alkanes (*e.g.*, hexane)	+++	+++	0	0
Diethyl ether	+++	+++	++	0
Dichloromethane	+++	+++	++	0
Acetone	+++	+++	+++	++
Ethanol	++	++	+++	+
Water	0	0	+	+++

↑ Decreasing polarity

Increasing polarity →

Nonpolar solvents are used to solubilize primarily lipophilic compounds (*e.g.*, alkanes, fatty acids, waxes, sterols, some terpenoids, and alkaloids), and the choice of solvents for these compounds ranges from those with a very low polarity, such as hexane, to those that are less nonpolar, like dichloromethane [153, 156–158]. The nonpolar solvents hexane, cyclohexane, benzene, toluene, diethyl ether, chloroform, and ethyl acetate are used to extract alkaloids, coumarins, fatty acids, flavonoids, and terpenoids. Medium-polarity solvents are used to extract compounds of intermediate polarity (*e.g.*, some alkaloids and flavonoids), whereas more polar solvents are used for polar compounds (*e.g.*, flavonoid glycosides, tannins, and some alkaloids). Hence, the choice of the solvent for hydrophilic compounds will fall on a polar solvent, which may be non-protic, such as acetone, or protic, such as ethanol, methanol, and even H_2O. Higher selectivity of solvents enables fewer steps to be used [153, 156]. If the feedstock material is a complex mixture, where multiple components need to be recovered, group selectivities become important. Selective extraction can also be carried out sequentially with solvents of increasing polarity [158]. This technique allows a preliminary separation of the compounds present in the material within distinct extracts and simplifies further isolation.

The choice of solvent needs to be considered also based on the extraction method [153]. Optimum solvents for traditional extractions may not be the best option for advanced extraction methods, such as ultrasound-assisted extraction (UAE) [157]. The selection of solvent is usually based on achieving high-molecular affinity between the solvent and solute, but in the case of UAE, factors affecting cavitation, such as solvent vapor pressure or surface tension, need to be considered as well. In this case, cheaper solvents, like H_2O mixtures, could work better than the traditional volatile and pure solvents used in conventional extraction methods.

2.4.4 General principles of extraction methods

In the SLE method, the feedstock material is placed in direct contact with a solvent [150, 151, 158]. The extraction process progresses through the different steps: *i*) a solvent diffuses into a solid feedstock matrix, *ii*) a solute (or a substrate) dissolves in a solvent to form a solution, *iii*) a solute is then diffused out of a solid feedstock matrix, and *iv*) the collection of extracted solutes. Extracting lignocellulosic biomass materials generally results in a wide range of removed compounds [152]. Hence, the extract must be often subjected to further treatments and refining before achieving the desired product in its final form. Typical required treatments after extraction are as follows: *i*) filtration to separate solids, *ii*) concentration of extracts *via* evaporation of solvents, *iii*) fractioning and enrichment of desired components (*e.g.*, by chromatographic methods, liquid-liquid extraction, membrane methods, and crystallization), *iv*) removal of impurities, such as heavy metals by complexation, and *v*) drying of products (*e.g.*, by spray dying) and applying product forming technologies (*e.g.*, by granulation).

Two major principles can be distinguished according to contact modes of solvent and solid material, *i.e.*, percolation and immersion (Table 2.15), and most available industrial equipment can be classified into these two contact types [151]. Additionally, laboratory-scale maceration can be differentiated from immersion, and Soxhlet extraction is mostly used for the determination of solubilities on the laboratory scale [152].

In the percolation method, the solid feedstock material is stacked as a fixed bed and the solvent passes through this bed driven by gravity from top to bottom [150–152]. The solvent can be recycled and passed through the fixed bed several times until sufficient mass transfer has been achieved. If the saturated solvent is continuously being replaced by fresh solvent, removal of the desired components can be essentially complete. In the immersion method, the solid feedstock particles are immersed in

Table 2.15: Various extraction methods for lignocellulosic biomass materials [151].

Method	Solvent	Temperature	Pressure	Time
Percolation [150, 152, 159]	H_2O, aqueous and nonaqueous solvents	Room temperature or under heat	Atmospheric	Long
Immersion [152]	H_2O, aqueous and nonaqueous solvents	Room temperature or under heat	Atmospheric	Long
Maceration [152]	H_2O, aqueous and nonaqueous solvents	Room temperature	Atmospheric	Long
Soxhlet extraction [152]	Organic solvents	Under heat	Atmospheric	Long
Ultrasound-assisted extraction (UAE) [152, 157]	H_2O, aqueous and nonaqueous solvents	Room temperature or under heat	Atmospheric	Short
Microwave-assisted extraction (MAE) [152, 160, 161]	H_2O, aqueous and nonaqueous solvents	Under heat	From atmospheric to high pressure	Short
Supercritical fluid extraction (SFE) [152, 160, 162–165]	Supercritical fluid (usually S-CO_2), often with a modifier	Near room temperature	High	Short
Pressurized liquid extraction (PLE) [152, 160]	H_2O, aqueous and nonaqueous solvents	Under heat	High	Short
Hydrodistillation and steam distillation [152]	H_2O	Under heat	From atmospheric to moderate pressure	Long

stirred solvent, which ensures very intense contact between the two phases. The mechanical stress on the particles is relatively high and the filtering of the extract must be separately achieved. Additionally, at most, only an equilibrium between solid and solvent can be obtained, which may leave, depending on the solubility properties, a significant amount of the desired components in the solid. Maceration is a very simple extraction method with the disadvantage of long extraction time and low extraction efficiency. The clear difference between the maceration and immersion methods is that in the latter case, the liquid and the solid are in motion with each other, while in the former case, the solid is just contacted with the liquid without motion of liquid or solid for a defined, usually quite long residence time. The maceration method can be used, for example, for the extraction of thermolabile components.

In Soxhlet extraction [150–152] the straightforward principle is to continuously recycle the extraction solvent through the solid feedstock matrix, i.e., by circulating a small amount of the same solvent after repeating condensation and boiling steps several times with the desired component being gradually collected in the main condensed solvent. It represents an extraction method with high extraction efficiency that requires less solvent consumption than the percolation or maceration methods. However, the disadvantage is the high temperature and long extraction time, which may increase the possibilities of thermal degradation of the products.

The simplest application of SLE is the operation in a single batch mode [152, 166]. This process can be performed, for example, as immersion extraction in a stirred vessel or a centrifugal extractor or as percolation extraction in a vessel with a filtering bottom. Another mode of batch extraction is the application of the Soxhlet method. An improved multistage operation is the combination of several batch extractors, such as a series of batch percolators.

To extract as much solute as possible using a limited quantity of solvent (*i.e.*, to obtain as concentrated extract as possible), multistage extraction processes are needed [165]. A great number of options exist with respect to the movement of the solid and the liquid streams with respect to each other. One improvement of industrial extractions with respect to extraction efficiency, solvent consumption, and energy consumption as well as convenience in operation is obtained by the application of continuous fully countercurrent operating processes [152, 166]. The countercurrent mode practically means that the feedstock material and the solvent run countercurrently from stage to stage through the whole equipment (Figure 2.16). The advantage of this procedure is that the fresh solvent is brought into contact with the most depleted feedstock material, while the fresh material is contacted with already enriched miscella (*i.e.*, a solution containing an extracted product). This arrangement ensures that a high concentration difference of the extract components between the solid material and the liquid is always obtained. The industrial countercurrently working equipment can be classified as either immersion-type or percolation-type systems.

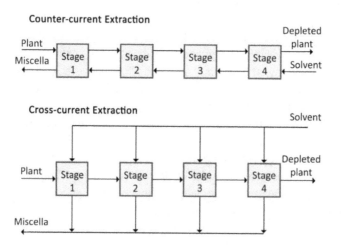

Figure 2.16: Countercurrent and cross-current solid-liquid extractions [152, 166].

The continuous extraction in cross-current mode means that the solids are transferred from stage to stage through the whole equipment, while fresh solvent is introduced into each extraction stage (Figure 2.16) [152, 166]. This operation mode favors complete extraction of the feedstock material since the highest possible concentration difference is always maintained. The disadvantage is that the required amount of solvent is higher than that needed in the countercurrent extraction. In most commercial equipment, both operation procedures are combined in such a way that in certain sections of the extractor, solids and liquids run countercurrently to enrich the miscella, while in another section, the solids are washed with fresh solvent in cross-current mode, to obtain a complete extraction of the desired compounds.

Extraction is a mature technology, where most of the practices used today were developed almost 100 ago [152, 166]. Classical extraction techniques are typically based on the extraction power of different solvents and the application of heat and/or mixing [167]. In recent years, the interest in the use of various extractives has grown due to the demand for nature-derived ingredients to replace synthetic chemicals, and, on the other hand, due to the growing markets of natural health products [168]. This general trend as well as the search for green and sustainable extraction methods has initiated new research and developments in the field of biomass extraction [152, 169].

On the industrial scale, conventional SLE processes have some major drawbacks, such as insufficient recovery of extracts, low extraction selectivity, extensive extraction duration, and intensive heating and/or mixing leading to high energy consumption and possible thermal decomposition of thermolabile compounds [157, 162, 163]. They also require often large volumes of high purity and costly organic solvents, which results in the need to evaporate vast amounts of solvent. Hence, recent trends in extraction techniques have largely focused on finding solutions that minimize the use of solvents, while at the same time enabling process intensification and a cost-effective production of high-quality extracts [169]. Additionally, more environmentally friendly solvents, such as alcohol-H_2O mixtures, often increase the costs and lower the extraction yield. Examples of the most promising new techniques include UAE, MAE, SFE, and PLE (or PHWE, *i.e.*, pressurized hot-water extraction) methods (Table 2.15) and the use of alternative solvents [157, 167, 170–175]. These technologies may also be used in combination with each other; for example, UAE may be combined with MAE [176]. The new techniques enable, besides the reduction of organic solvent consumption, shorter extraction times, efficient extractions, and automation [177].

In general, extraction can be facilitated and higher extraction yields obtained by increasing the temperature and solvent-to-solid ratio to favor solubilization and diffusion [150, 158, 166]. However, temperatures that are too high may cause loss in solvents, resulting in the extracts of undesirable impurities, and the decomposition of thermolabile components. For example, too high extraction temperatures can degrade the condensed tannins [178]. On the other hand, too high solvent-to-solid ratio is also not economical, because it will lead to excessive solvent consumption and requires a long time for concentration. Hence, when choosing the right extraction method, it is important to optimize all the relevant parameters influencing the extraction efficiency, not only the solvent composition, but also extraction temperature and time, particle size of the material to be extracted, liquid-to-solid ratio, and pH value [152, 179]. Additionally, it is important to take into consideration the properties of the extracted compounds to avoid chemical modification of these compounds during extraction by hydrolysis, oxidation, and isomerization reactions [180].

2.4.5 Extractives-derived products from stemwood

The extractives from stemwood are typically recovered either *via* extraction or as by-products from kraft pulping. The main products can be classified to the following groups [181–183]:
- terpenes and terpenoids-based products
    - turpentine
    - resin acids
- fatty acids-based products
- polyphenols-based products from knot lignans
- pine tar

In this subchapter, only the products from knot lignans as well as their applications are briefly described. They can be separated by direct extraction and other extractives-based products discussed in the separate chapters (see Sections 4.2 "Kraft pulping" and 4.3 "Sulfite pulping") dealing with the by-products of kraft pulping (*i.e.*, turpentine, resin acids, and fatty acids) or wood pyrolysis (pine tar). Figure 2.17 represents general possibilities for recovering extractives-based products from pine stemwood.

It has been reported [115, 132, 184] that Norway spruce (*Picea abies*) knots (*i.e.*, branch bases inside tree stems) constitute an extraordinarily rich source of lignans. Knots contain significantly higher concentrations of lignans and oligolignans than the adjacent stemwood; the lignan content may even be several hundred times higher in knots [131, 185, 186]. The amount of lignans in Norway spruce knots varies in the range of 6–24% of the wood dry solids, the average value being 10% of the wood dry solids [115, 131]. Additionally, knots also contain 2–6% (w/w) oligolignans [187].

Lignans have attracted much interest due to their broad range of biological activity [181, 188]. Several lignans could be produced on large scale from knots either directly by separation or by semi-synthesis starting from the dominating lignan in spruce, HMR (see Section 2.4.2 "Chemical features of extractives"), which comprises 65–85% of the lignans in knots [184]. When tree stems are industrially chipped, most of the wood knots produce thick chips (*i.e.*, oversized chips), which can be enriched through chip screening [181]. Knotwood chips are undesired material in chemical or mechanical pulping and they can be removed from other chips before pulping by a proven technology called "ChipSep".

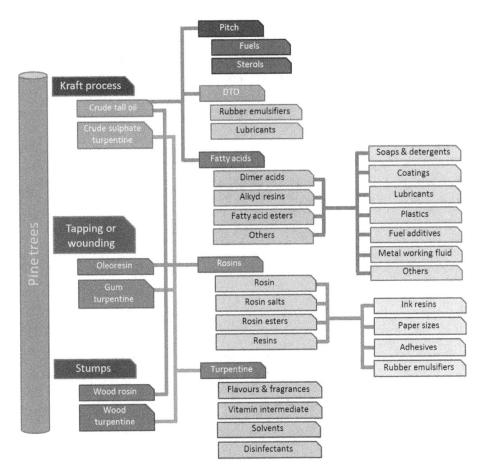

Figure 2.17: The pine extractives-derived products, which are the most significant revenue generators [181]. DTO refers to distilled tall oil.

Lignans can help maintain good cardiovascular health and moderate other estrogen-dependent health problems, such as menopause and osteoporosis [115]. In general, lignans and oligolignans are strong antioxidants and radical scavengers. Some lignans, such as HMR, attract much attention for their risk-reducing activity of cancer and cardiovascular disease [115, 189]. HMR has also been stated to have a positive influence on inhibiting the development of breast, prostate, and colon cancers. In 2006, the HMRlignan™ product (*i.e.*, dietary supplement) came onto the market in the form of capsules. Additionally, HMR is valuable as an optically pure chemical, which can be utilized in high-value applications, for example, as a platform chemical for organic synthesis.

The bottom of the root neck of Norway spruce is another rich source of lignans [130, 181, 190]. For example, an increase in HMR concentration in the heartwood close to the lowermost part of the root neck has been detected. Additionally, some sapwood samples close to the bottom contained relatively high amounts of HMR. The highest concentration of HMR found from the heartwood sample of root neck has been more than 10%. However, spruce stumps are not currently utilized commercially as a source of lignan products.

References

[1]   Alén, R. 2000. Structure and chemical composition of wood. In: Stenius, P. (Ed.). *Forest Products Chemistry*. Fapet, Helsinki, Finland. Pp. 11–57.

[2]   Alén, R. 2018. *Carbohydrate Chemistry – Fundamentals and Applications*. World Scientific, Singapore. Pp. 472–496.

[3]   Alén, R. 2011. Structure and chemical composition of biomass feedstocks. In: Alén, R. (Ed.). *Biorefining of Forest Resources*. Paper Engineers' Association, Helsinki, Finland. Pp. 17–54.

[4]   Goldstein, I.S. 1981. Composition of biomass. In: Goldstein, I.S. (Ed.). *Organic Chemicals from Biomass*. CRC Press, Boca Raton, FL, USA. Pp. 9–18.

[5]   Alén, R. 2011. Principles of biorefining. In: Alén, R. (Ed.). *Biorefining of Forest Resources*. Paper Engineers' Association, Helsinki, Finland. Pp. 55–114.

[6]   Piskorz, J. 2002. Fundamentals, mechanisms and science of pyrolysis. In: Bridgwater, A.V. (Ed.). *Fast Pyrolysis of Biomass: A Handbook, Volume 2*. CPL Press, Newbury, England. Pp. 103–125.

[7]   Murwanashyaka, J.N., Pakdel, H., and Roy, C. 2002. Fractional vacuum pyrolysis of biomass and separation of phenolic compounds by steam distillation. In: Bridgwater, A.V. (Ed.). *Fast Pyrolysis of Biomass: A Handbook, Volume 2*. CPL Press, Newbury, England. Pp. 407–418.

[8]   Cao, N., Darmstadt, H., and Roy, C. 2001. Activated carbon produced from charcoal obtained by vacuum pyrolysis of softwood bark residues. *Energy & Fuels* 15:1263–1269.

[9]   Feng, Z. 2001. *Alkaline Pulping of Non-Wood Feedstocks and Characterization of Black Liquors*. Doctoral Thesis. University of Jyväskylä, Laboratory of Applied Chemistry, Jyväskylä, Finland. 54 p.

[10] Saijonkari-Pahkala, K. 2001. *Non-Wood Plants as Raw Material for Pulp and Paper*. Doctoral Thesis. University of Helsinki, the Faculty of Agriculture and Forestry, Helsinki, Finland. 101 p.

[11] Vu, M.T.H. 2004. *Alkaline Pulping and the Subsequent Elemental Chlorine-Free Bleaching of Bamboo (Bambusa procera)*. Doctoral Thesis. University of Jyväskylä, Laboratory of Applied Chemistry, Jyväskylä, Finland. 69 p.

[12] Madakadze, I.C., Masamvu, T.M., Radiotis, T., Li, J., and Smith, D.L. 2010. Evaluation of pulp and paper making characteristics of elephant grass (*Pennisetum purpureum* Schum) and switchgrass (*Panicum virgatum* L.). *African Journal of Environmental Science and Technology* 4(7):465–470.

[13] Brosse, N., Dufour, A., Meng, X., Sun, Q., and Ragauskas, A. 2012. *Miscanthus*: a fast-growing crop for biofuels and chemicals production. *Biofuels, Bioproducts and Biorefining* 6(5): (https://doi.org/10.1002/bbb.1353).

[14] Dungani, R., Khalil, H.P.S.A., Sumardi, I., Suhaya, Y., Sulistyawati, E., Islam, M.N., Suraya, N.L.M., and Aprilia, N.A.S. 2014. Non-wood renewable materials: properties improvement and its application. In: Hakeem, K.R., Jawaid, M., and Rashid, U. (Eds.). *Biomass and Bioenergy: Applications*. Springer International Publishing, Cham, Switzerland. Pp. 1–29.

[15] Dhyani, V., and Bhaskar, T. 2018. A comprehensive review on the pyrolysis of lignocellulosic biomass. *Renewable Energy* 129(Part B):695–716.

[16] Salami, A. 2021. *Biorefining of Lignocellulosic Biomass and Chemical Characterization of Slow Pyrolysis Distillates*. Doctoral Thesis. University of Eastern Finland, Kuopio, Finland. 143 p.

[17] Hua, W.J., and Xi, H.N. 1988. Morphology and ultrastructure of some non-woody papermaking materials. *Appita Journal* 41(5):365–374.

[18] Atchison, J.E. 1990. Data on non-wood plant fibers. In: Hamilton, F., and Leopold, B. (Eds.). *Pulp and Paper Manufacture, Vol. 3, Secondary Fibers and Non-Wood Pulping*. The Joint Textbook Committee of the Paper Industry (CPPA and TAPPI), Atlanta, GA, USA. Pp. 4–21.

[19] Ververis, C., Georghiou, K., Christodoulakis, N., Santas, P., and Santas, R. 2004. Fiber dimensions, lignin and cellulose content of various plant materials and their suitability for paper production. *Industrial Crops and Products* 19:245–254.

[20] Côté, H.A. Jr. 1967. *Wood Ultrastructure*. University of Washington Press, Seattle, WA, USA.

[21] Farrelly, D. 1984. *The Book of Bamboo*. Sierra Club Books, San Francisco, CA, USA. 332 p.

[22] Grosser, D., and Liese, W. 1971. On the anatomy of Asian bamboos, with special reference to their vascular bundles. *Wood Science and Technology* 5:290–312.

[23] Gratani, L., Crescente, M.F., Varone, L., Fabrini, G., and Digiulio, E. 2008. Growth pattern and photosynthetic activity of different bamboo species growing in the botanical garden of Rome. *Flora* 203:77–84.

[24] Jiang, J.X., Yang, Z.K., Zhu, L.W., Shi, L.M., and Yan, L.J. 2008. Structure and property of bamboo fiber. *Chinese Forestry Science and Technology* 30:128–132.

[25] Khalil, H.P.S.A., Bhat, I.U.H., Jawaid, M., Zaidon, A., Hermawan, D., and Haidi, Y.S. 2012. Bamboo fibre reinforced biocomposites: a review. *Materials & Design* 42:353–368.

[26] Liu, D., Song, J., Anderson, D.P., Chang, P.R., and Hua, Y. 2012. Bamboo fiber and its reinforced composites: structure and properties. *Cellulose* 19:1449–1480.

[27] Hu, K., Huang, Y., Fei, B., Yao, C., and Zhao, C. 2017. Investigation of the multilayered structure and microfibril angle of different types of bamboo cell walls at the micro/nano level using a LC-Polscope imaging system. *Cellulose* 24:4611–4625.

[28] Herrick, F.W., and Hergert, H.L. 1977. Utilization of chemicals from wood: retrospect and prospect. In: Loewus, F.A., and Runecles, V.C. (Eds.). *The Structure, Biosynthesis, and Degradation of Wood, Recent Advances in Phytochemistry, Volume 11*. Plenium Press, New York, NY, USA. Pp. 443–515.

[29] Goldstein, I.S. (Ed.). 1981. *Organic Chemicals from Biomass*. CRC Press, Boca Raton, FL, USA. 310 p.

[30] Sjöström, E. 1993. *Wood Chemistry – Fundamentals and Applications*. 2nd edition. Academic Press, San Diego, CA, USA. 293 p.

[31] Hon, D.N.-S., and Shiraishi, N. (Eds.). 2001. *Wood and Cellulosic Chemistry*. 2nd edition. Marcel Dekker, New York, NY, USA. 914 p.

[32] Hu, T.Q. (Ed.). 2008. *Characterization of Lignocellulosic Materials*. Wiley-Blackwell, Hoboken, NJ, USA. 392 p.

[33] Ek, M., Gellerstedt, G., and Henrikson, G. 2009. *Wood Chemistry and Wood Biotechnology*. Walter de Gruyter, Berlin, Germany. 308 p.

[34] Heitner, C., Dimmel, R.R., and Schmidt, J.A. (Eds.). 2010. *Lignin and Lignans: Advances in Chemistry*. CRC Press, Boca Raton, FL, USA, 683 p.

[35] Alén, R. (Ed.). *Biorefining of Forest Resources*. Paper Engineers' Association, Helsinki, Finland. 381 p.

[36] Fengel, D., and Wegener, G. 2011. *Wood: Chemistry, Ultrastructure, Reactions*. Walter de Gruyter, Berlin, Germany. 626 p.

[37] Rowell, R.B. (Ed.). 2012. *Handbook of Wood Chemistry and Wood Composites*. 2nd edition. CRC Press, Boca Raton, FL, USA, 703 p.

[38] Chen, H. 2014. *Biotechnology of Lignocellulose*. Springer, Heidelberg, Germany. 510 p.

[39] Alén, R. 2018. *Carbohydrate Chemistry – Fundamentals and Applications*. World Scientific, Singapore. 586 p.

[40] Alén, R. 2018. *Carbohydrate Chemistry – Fundamentals and Applications*. World Scientific, Singapore. Pp. 280–341.

[41]   Timell, T.E. 1967. Recent progress in the chemistry of wood hemicelluloses. *Wood Science and Technology* 1:45–70.

[42]   Wilkie, K.C.B. 1979. The hemicelluloses of grasses and cereals. *Advances in Carbohydrate Chemistry and Biochemistry* 36:215–264.

[43]   Sjöström, E. 1993. *Wood Chemistry – Fundamentals and Applications*. 2nd edition. Academic Press, San Diego, CA, USA. Pp. 63–70.

[44]   Thompson, N.S. 1995. Hemicellulose. In: *Kirk-Othmer – Encyclopedia of Chemical Technology, Volume 13*. 4th edition. John Wiley & Sons, Chichester, United Kingdom. Pp. 54–72.

[45]   Laine, C. 2005. *Structures of Hemicelluloses and Pectins in Wood and Pulp*. Doctoral Thesis. Helsinki University of Technology, Laboratory of Organic Chemistry, Espoo, Finland. 63 p.

[46]   Zhou, X., Li, W., Mabon, R., and Broadbelt, L.J. 2017. A critical review on hemicellulose pyrolysis. *Energy Technology* 5:52–79.

[47]   Xu, C., Leppänen, A.-S., Eklund, P., Holmlund, P., Sjöholm, R., Sundberg, K., and Willför, S. 2010. Acetylation and characterization of spruce (*Picea abies*) galactoglucomannans. *Carbohydrate Research* 345:810–816.

[48]   Willför, S. 2002. *Water-Soluble Polysaccharides and Phenolic Compounds in Norway Spruce and Scots Pine Stemwood and Knots*. Doctoral Thesis. Åbo Akademi University, Laboratory of Forest Products Chemistry, Åbo, Finland. 70 p.

[49]   Willför, S., and Holmbom, B. 2004. Isolation and characterization of water soluble polysaccharides from Norway spruce and scots pine. *Wood Science and Technology* 38:173–179.

[50]   Xu, C. 2008. *Physicochemical properties of water-soluble spruce galactoglucomannans*. Doctoral Thesis. Åbo Akademi University, Laboratory of Wood and Paper Chemistry, Åbo, Finland. 84 p.

[51]   Ericsson, T., Petersson, G., and Samuelson, O. 1977. Galacturonic acid groups in birch xylan. *Wood Science and Technology* 11(3):219–223.

[52]   Johansson, M.H., and Samuelson, O. 1977. Reducing end groups in birch xylan and their alkaline degradation. *Wood Science and Technology* 11(4):251–263.

[53]   Higuchi, T. 1980. Chemistry and biochemistry; bamboo for pulp and bamboo for paper. In: Lessard, G., and Chouinard, A. (Eds.). *Bamboo Research in Asia*. The International Development Research Centre (IDRC), Ottawa, ON, Canada. Pp. 51–56.

[54]   Liese, W. 1985. Anatomy and properties of bamboo. In: *Proceedings of the International Bamboo Workshop*. Hangzhou, China. Pp. 196–208.

[55]   Liese, W. 1987. Research on bamboo. *Wood Science and Technology* 21(3):189–209.

[56]   Gascoigne, J.A. 1988. Bamboo grow and utilization. In: *Proceedings of Tropical Wood Pulping Symposium '88*. Singapore. Pp. 183–193.

[57]   Li, Z. 1988. Further discussion on the basic behavior of the pulping of grasses. *China Pulp & Paper* 7(5):53–59.

[58]   Nakamura, S., Wakabayashi, K., Nakai, R., Aono, R., and Horikoshi, K. 1993. Purification and some properties of an alkaline xylanase from alkaliphilic *basillus* sp. strain 41M-1. *Applied and Environmental Microbiology* 59:2311–2316.

[59]   Ebringerová, A., Hromádková, Z., and Heinze, T. 2005. Hemicellulose. *Advances in Polymer Science* 186:1–67.

[60]   Shi, S. 1993. Structural characteristic of lignin, hemicelluloses, and cellulose. In: Nie, S., and Fang, S. (Eds.). *Practical Handbook of Alkaline Pulping of Frequently-Used Non-Wood Fibers*. China Publishing House of Light Industry, Beijing. P.R. China. Pp. 86–103.

[61]   Holmbom, B. 1996. Research on production and use of agro-fibers in Finland. In: *Proceedings of the International Symposium on Vegetal (Non-Wood) Biomass as a Source of Fibrous Materials and Organic Products*, October 21–25, 1996, Guangzhou, P.R., China. Pp. 30–35.

[62]   Grabber, J.H., Hatfield, R.D., Ralph, J., Zón, J., and Amrhein, N. 1995. Ferulate cross-linking in cell walls isolated from maize cell suspensions. *Phytochemistry* 40:1077–1082.

[63] Sipponen, M. 2015. *Effect of Lignin Structure on Enzymatic Hydrolysis of Plant Residues*. Doctoral Thesis. Aalto University, Department of Biotechnology and Chemical Technology, Espoo, Finland. 62 p.

[64] Ishii, T. 1991. Isolation and characterization of a diferuloyl arabinoxylan hexasaccharide from bamboo shoot cell walls. *Carbohydrate Research* 219:15–22.

[65] Iiyama, K., Lam, T.B.-T., and Stone, B.A. 1994. Covalent cross-links in the cell wall. *Plant Physiology* 104:315–320.

[66] Ralph, J., Quideau, S., Grabber, J.H., and Hatfield, R.D. 1994. Identification and synthesis of new ferulic acid dehydrodimers present in grass cell walls. *Journal of the Chemical Society, Perkin Transactions* 1:3485–3498.

[67] Fengel, D., and Wegener, G. 1989. *Wood: Chemistry, Ultrastructure, Reactions*. Walter de Gruyter, Berlin, Germany. Pp. 106–131.

[68] Sarkanen, K.V., and Ludwig, C.H. (Eds.). 1971. *Lignins: Occurrence, Formation, Structure and Reactions*. Wiley-Interscience, New York, NY, USA. 916 p.

[69] Fengel, D., and Wegener, G. 1989. *Wood: Chemistry, Ultrastructure, Reactions*. Walter de Gruyter, Berlin, Germany. Pp. 132–181.

[70] Sjöström, E. 1993. *Wood Chemistry – Fundamentals and Applications*. 2nd edition. Academic Press, San Diego, CA, USA. Pp. 71–89.

[71] Sakakibara, A., and Sano, Y. 2001. Chemistry of lignin. In: Hon, D.N.-S., and Shiraishi, N. (Eds.). *Wood and Cellulosic Chemistry*. 2nd edition. Marcel Dekker, New York, NY, USA. Pp. 109–174.

[72] Glasser, W.G., and Sarkanen, S. 2008. *Lignin: Properties and Materials*. American Chemical Society (ACS), Cellulose, Paper, and Textile Division, Washington, DC, USA. 545 p.

[73] Heitner, C., Dimmel, D.R., and Schmidt, J.A. (Eds.). 2010. *Lignin and Lignans: Advances in Chemistry*. CRC Press, Boca Raton, FL, USA. 683 p.

[74] Bauer, S., Sorek, H., Mitchell, V.D., Ibáñez, A.B., and Wemmer, D.E. 2012. Characterization of *Miscanthus giganteus* lignin isolated by ethanol organosolv process under reflux condition. *Journal of Agricultural and Food Chemistry* 60:8203–8212.

[75] Dence, C.W., and Lin, S.Y. 1992. Introduction. In: Dence, C.W., and Lin, S.Y. (Eds.). *Methods in Lignin Chemistry*. Springer-Verlag, Berlin, Germany. Pp. 3–19.

[76] Rolando, C., Monties, B., and Lapierre, C. 1992. Thioacidolysis. In: Dence, C.W., and Lin, S.Y. (Eds.). *Methods in Lignin Chemistry*. Springer-Verlag, Berlin, Germany. Pp. 334–349.

[77] Fidalgo, M.L., Terrón, M.C., Martinez, A.T., González, A.E., González-Vila, F.J., and Galetti, G.C. 1993. Comparative study of fractions from alkaline extraction of wheat straw through chemical degradation, analytical pyrolysis, and spectroscopic techniques. *Journal of Agricultural and Food Chemistry* 41:1621–1626.

[78] Bocchini, P., Galletti, G.C., Camarero, S., and Martinez, A.T. 1997. Absolute quantitation of lignin pyrolysis products using an internal standard. *Journal of Chromatography A* 773:227–232.

[79] Yelle, D.J., Kaparaju, P, Hunt, C.G., Hirth, K., Kim, H., Ralph, J., and Felby, C. 2013. Two-dimensional NMR evidence for cleavage of lignin and xylan substituents in wheat straw through hydrothermal pretreatment and enzymatic hydrolysis. *BioEnergy Research* 6:211–221.

[80] Higuchi, T., Tanahashi, M., and Nakatsubo, F. 1972. Acidolysis of bamboo lignin. III. Estimation of arylglycerol-β-aryl ether groups in lignin. *Wood Research* 54:9–18.

[81] Adler, E. 1977. Lignin chemistry – past, present and future. *Wood Science and Technology* 11:169–218.

[82] Crestini, C., and Argyropoulos, D.S. 1997. Structural analysis of wheat straw lignin by quantitative 31P and 2D NMR spectroscopy. the occurrence of ester bonds and α-O-4 substructures. *Journal of Agricultural and Food Chemistry* 45:1212–1219.

[83] Zhang, L., and Gellerstedt, G. 2001. NMR observation of a new lignin structure, a spiro-dienone. *Chemical Communication* 24:2744–2745.

[84] Chakar, F.S., and Ragauskas, A.J. 2004. Review of current and future softwood kraft lignin process chemistry. *Industrial Crops and Products* 20(2):131–141.

[85] Gellerstedt, G., and Henriksson, G. 2008. Lignin: major sources, structures and properties. In: Belgacem, M.N., and Gandini, A. (Eds.). *Monomers, Polymers and Composites from Renewable Resources*. Elsevier, Oxford, United Kingdom. Pp. 201–224.

[86] Brodin, I. 2009. *Chemical Properties and Thermal Behaviour of Kraft Lignins*. Licentiate Thesis. KTH Royal Institute of Technology, School of Chemical Science and Engineering (CHE), Stockholm, Sweden. 47 p.

[87] Dimmel, D.R. 2010. Overview. In: Heitner, C., Dimmel, D.R., and Schmidt, J.A. (Eds.). *Lignin and Lignans: Advances in Chemistry*. CRC Press, Boca Raton, FL, USA. Pp. 1–10.

[88] Huttunen, M. 2015. *Pyrolysis of Softwood and Hardwood Lignins*. Master's Thesis. University of Jyväskylä, Laboratory of Applied Chemistry, Jyväskylä, Finland. 92 p.

[89] Berg Miller, M.E., Brulc, J.M., Bayer, E.A., Lamed, R., Flint, H.J., and White, B.A. 2010. Advanced technologies for biomass hydrolysis and saccharification using novel enzymes. In: Vertès, A.A., Qureshi, N., Blaschek, H.P., and Yukawa, H. (Eds.). *Biomass to Biofuels – Strategies for Global Industries*. Wiley, Chichester, United Kingdom. Pp. 199–212.

[90] Glasser, W.G., Glasser, H.R., and Morohoshi, N. 1981. Simulation of reactions with lignin by computer (Simrel). VI. Interpretation of primary experimental analysis data ("Analysis Program"). *Macromolecules* 14(2): 253–262.

[91] Brunow, G., Kilpeläinen, I., Sipilä, J., Syrjänen, K., Karhunen, P., Setälä, H., and Rummakko, P. 1998. Oxidative coupling of phenols and the biosynthesis of lignin. In: Lewis, N.G., and Sarkanen, S. (Eds.). *Lignin and Lignan Biosynthesis*. ACS Symposium Series, Washington, DC, USA. Pp. 131–147.

[92] Ralph, J., Brunow, G., and Boerjan, W. 2007. Lignins. In: *Encyclopedia of Life Sciences*. Wiley, Chichester, United Kingdom. Pp. 1–10.

[93] Nimz, H. 1974. Beech lignin – proposal of a constitutional scheme. *Angewandte Chemie* 13:313–321.

[94] Ralph, J., Brunow, G., Harris, P.J., Dixon, R.A., Schatz, P.F., and Boerjan, W. 2008. Lignification: are lignins biosynthesized via simple combinatorial chemistry via proteinaceous control and template replication? In: Daayf, F., and Lattanzio, V. (Eds.). *Recent Advances in Polyphenol Research*. Wiley-Blackwell, Oxford, United Kingdom. Pp. 36–66.

[95] Yáñez-S, M., Matsuhiro, B., Nuñez, C, Pan, S., Hubbell, C.A., Sannigrahi, P., and Ragauskas, A.J. 2014. Physicochemical characterization of ethanol organosolv lignin (EOL) from *Eucalyptus globulus*: effect of extraction conditions on the molecular structure. *Polymer Degradation and Stability* 110:184–194.

[96] Del Rio, J.C., Rencoret, J., Prinsen, P., Martinez, A.T., Ralph, J., and Gutierrez, A. 2012. Structural characterization of wheat straw lignin as revealed by analytical pyrolysis, 2D-NMR, and reductive cleavage methods. *Journal of Agricultural and Food Chemistry* 60:5922–5935.

[97] Brunow, G., Lundquist, K., and Gellerstedt, G. 1999. Lignin. In: Sjöström, E., and Alén, R. (Eds.). *Analytical Methods in Wood Chemistry, Pulping, and Papermaking*. Springer, Heidelberg, Germany. Pp. 77–124.

[98] Pakkanen, H. 2012. *Characterization of Organic Material Dissolved during Alkaline Pulping of Wood and Non-Wood Feedstocks*. Doctoral Thesis. University of Jyväskylä, Laboratory of Applied Chemistry, Jyväskylä, Finland. 76 p.

[99] Pakkanen, H., and Alén, R. 2012. Molecular mass distribution of lignin from the alkaline pulping of hardwood, softwood, and wheat straw. *Journal of Wood Chemistry and Technology* 32:279–293.

[100] Buranov, A.U., and Mazza, G. 2008. Lignin in straw of herbaceous crops. *Industrial Crops and Products* 28:237–259.

[101] Atsushi, K., Azuma, J.I., and Koshijima, T. 1984. Lignin-carbohydrate complexes and phenolic-acids in bagasse. *Holzforschung* 38:141–149.

[102] Scalbert, A., Monties, B., Lallemand, J.Y., Guittet, E., and Rolando, C. 1985. Ether linkage between phenolic-acids and lignin fractions from wheat straw. *Phytochemistry* 24:1359–1362.

[103] Grabber, J.H., Ralph, J., Lapierre, C., and Barrière, Y. 2004. Genetic and molecular basis of grass cell-wall degradability. I. Lignin-cell wall matrix interactions. *Comptes Rendus-Biologies* 347:455–465.

[104] Sun, R.-C., Sun, X.-F., and Zhang, S.-H. 2001. Quantitative determination of hydroxycinnamic acids in wheat, rice, rye, and barley straws, maize stems, oil palm frond fiber, and fast-growing poplar wood. *Journal of Agricultural and Food Chemistry* 49:5122–5129.

[105] Sun, R.-C., Lawther, J.M., and Banks, W.B. 1997. A tentative chemical structure of wheat straw lignin. *Industrial Crops and Products* 6:1–8.

[106] Rowe, J., and Conner, A.H. 1979. Extractives in eastern hardwoods – a review. *General Technical Report FPL 18*. Forest Products Laboratory, Forest Service, U.S. Department of Agriculture, Madison, WI, USA. 66 p.

[107] Ekman, R. 1980. *Wood Extractives of Norway Spruce – A Study on Nonvolatile Constituents and their Effects on Fomes annosus*. Doctoral Thesis. Åbo Akademi, Åbo, Finland. 42 p.

[108] Fengel, D., and Wegener, G. 1989. *Wood: Chemistry, Ultrastructure, Reactions*. Walter de Gruyter, Berlin, Germany. Pp. 182–226.

[109] Sjöström, E. 1993. *Wood Chemistry – Fundamentals and Applications*. 2nd edition. Academic Press, San Diego, CA, USA. Pp. 90–108.

[110] Holmbom, B. 1999. Extractives. In: Sjöström, E., and Alén, R. (Eds.). *Analytical Methods in Wood Chemistry, Pulping, and Papermaking*. Springer, Heidelberg, Germany. Pp. 125–148.

[111] Back, E.L. 2000. The locations and morphology of resin components in the wood. In: Back, E.L., and Allen, L.H. (Eds.). *Pitch Control, Wood Resin and Deresination*. TAPPI Press, Atlanta, GA, USA. Pp. 1–35.

[112] Ekman, R., and Holmbom, B. 2000. The chemistry of wood resin. In: Back, E.L., and Allen, L.H. (Eds.). *Pitch Control, Wood Resin and Deresination*. TAPPI Press, Atlanta, GA, USA. Pp. 37–76.

[113] Dorado, J., Van Beek, T.A., Claassen, F.W., and Sierra-Alvarez, R. 2001. Degradation of lipophilic wood extractive constituents in *Pinus sylvestris* by the white-rot fungi *Bjerkandera* sp. and *Trametes versicolor*. *Wood Science and Technology* 35(1–2):117–125.

[114] Umezawa, T. 2001. Chemistry of extractives. In: Hon, D.N.-S., and Shiraishi, N. (Eds.). *Wood and Cellulosic Chemistry*. 2nd edition. Marcel Dekker, New York, NY, USA. Pp. 213–241.

[115] Holmbom, B. 2011. Extraction and utilisation of non-structural wood and bark components. In: Alén, R. (Ed.). *Biorefining of Forest Resources*. Paper Engineers' Association, Helsinki, Finland. Pp. 178–224.

[116] Fengel, D., and Wegener, G. 1989. *Wood: Chemistry, Ultrastructure, Reactions*. Walter de Gruyter, Berlin, Germany. Pp. 240–267.

[117] Kokki, S. 2001. Chemistry of bark. In: Hon, D.N.-S., and Shiraishi, N. (Eds.). *Wood and Cellulosic Chemistry*. 2nd edition. Marcel Dekker, New York, NY, USA. Pp. 243–273.

[118] Ånäs, E., Ekman, R., and Holmbom, B. 1983. Composition of nonpolar extractives in bark of Norway spruce and Scots pine. *Journal of Wood Chemistry and Technology* 3(3):119–130.

[119] Yesil-Celiktas, O., Ganzera, M., Akgun, I., Sevimli, C., Korkmaz, K.S., and Bedir, E. 2009. Determination of polyphenolic constituents and biological activities of bark extracts from different *Pinus* species. *Journal of the Science of Food and Agriculture* 89:1339–1345.

[120] Krogell, J., Holmbom, B., Pranovich, A., Hemming, J., and Willför, S. 2012. Extraction and chemical characterization of Norway spruce inner and outer bark. *Nordic Pulp and Paper Research Journal* 27(1):6–17.

[121] Ruuskanen, M. 2017. *The Influence of the Origin and Treatment History of Spruce and Pine Bark on the Extraction of Tannin*. Master's Thesis. University of Helsinki, Department of Forest Sciences, Helsinki, Finland. 59 p.

[122] Tamminen, T., Ruuskanen, M., and Grönqvist, S. 2017. The influence of softwood bark origin on tannin recovery by hot-water extraction. In: *Proceedings of the International Symposium on Wood, Fibre and Pulping Chemistry*, August 28 – September 1, 2017, Porto Seguro, BA, Brazil. 5 p.

[123] Halmemies, E.S., Brännström, H.E., Nurmi, J., Läspä, O., and Alén, R. 2021. Effect of seasonal storage on single-stem bark extractives of Norway spruce (*Picea abies*). *Forests* 12(6):736.

[124] Halmemies, E.S., Alén, R., Hellström, J., Läspä, O., Nurmi, J., Hujala, M., and Brännström, H.E. 2022. Behaviour of extractives in Norway spruce (*Picea abies*) bark during pile storage. *Molecules* 27(4):1186.

[125] Suokas, E. 1977. *Studies on Chemical Transformations in the Lupane Series*. Doctoral Thesis. Helsinki University of Technology, Department of Chemistry, Espoo, Finland. 46 p.

[126] Ekman, R. 1983. The suberin monomers and triterpenoids from the outer bark of *Betula verrucosa* Ehrh. *Holzforschung* 37(4):205–211.

[127] Cole, B.J.W., Bentley, M.D., and Hua, Y. 1991. Triterpenoid extractives in the outer bark of *Betula lenta* (black birch). *Holzforschung* 45(4):265–268.

[128] Krasutsky, P.A. 2006. Birch bark research and development. *Natural Product Reports* 23:919–942.

[129] Gominho, J., Figueira, J., Rodrigues, J.C., and Pereira, H. 2007. Within-tree variation of heartwood, extractives and wood density in the eucalypt hybrid *urograndis* (*Eucalyptus grandis × E. urophylla*). *Wood Fiber Science* 33(1):3–8.

[130] Latva-Mäenpää, H. 2017. *Bioactive and Protective Polyphenolics from Roots and Stumps of Conifer Trees (Norway Spruce and Scots Pine)*. Doctoral Thesis. University of Helsinki, Faculty of Science, Chemistry, Helsinki, Finland. 74 p.

[131] Willför, S., Hemming, J., Reunanen, M., Eckerman, C., and Holmbom, B. 2003. Lignans and lipophilic extractives in Norway spruce knots and stemwood. *Holzforschung* 57:27–36.

[132] Nisula, L. 2018. *Wood Extractives in Conifers – a Study of Stemwood and Knots of Industrially Important Species*. Doctoral Thesis. Åbo Akademi, Laboratory of Wood and Paper Chemistry, Åbo, Finland. 253 p.

[133] Ali, B.H., Ziada, A., and Blunden, G. 2009. Biological effects of gum arabic: a review of some recent research. *Food and Chemical Toxicology* 47(1):1–8.

[134] Patel, S., and Goyal, A. 2015. Applications of natural polymer gum arabic: a review. *International Journal of Food Properties* 18(5):986–998.

[135] Sanchez, C., Nigen, M., Tamayo, V.M., Doco, T., Williams, P., Amine, C., and Renard, D. 2018. Acacia gum: history of the future. *Food Hydrocolloids* 78:140–160.

[136] Tanaka, Y., Mori, M., Takei, A., Boochathum, P., and Sato, Y. 1990. Structural characterisation of naturally occurring *trans*-polyisoprenes. *Journal of Natural Rubber Research (Malaysia)* 5(4):241–245.

[137] Rose, K., and Steinbuchel, A. 2005. Biodegradation of natural rubber and related compounds: recent insights into a hardly understood catabolic capability of microorganisms. *Applied and Environmental Microbiology* 71(6):2803–2812.

[138] Yikmis, M., and Steinbuchel, A. 2012. Historical and recent achievements in the field of microbial degradation of natural and synthetic rubber. *Applied and Environmental Microbiology* 78(13):4543–4551.

[139] Lappi, H. 2012. *Production of Hydrocarbon-Rich Biofuels from Extractives-Derived Materials*. Doctoral Thesis. University of Jyväskylä, Laboratory of Applied Chemistry, Jyväskylä, Finland. 111 p.

[140] Halmemies, E., and Brännström, H. 2021. Factors contributing to the loss of extractives. In: Alén, R. (Ed.). *Pulping and Biorefining – ForestBioFacts, Digital Learning Environment*. Paperi ja Puu Ltd., Helsinki, Finland.

[141] Back, E.L., and Allen, L.H. (Eds.). 2000. *Pitch Control, Wood Resin and Deresination*. TAPPI Press, Atlanta, GA, USA. 392 p.

[142] Alén, R. 2000. Basic chemistry of wood delignification. In: Stenius, P. (Ed.). *Forest Products Chemistry*. Fapet, Helsinki, Finland. Pp. 58–104.

[143] Ekman, R. 2000. Resin during storage and biological treatment. In: Back, E.L., and Allen, L.H. (Eds.). *Pitch Control, Wood Resin and Deresination*. TAPPI Press, Atlanta, GA, USA. Pp. 185–204.

[144] Werkelin, J. 2002. *Distribution of Ash-Forming Elements in Four Trees of Different Species*. Master's Thesis. Åbo Akademi, Laboratory of Inorganic Chemistry, Åbo, Finland. 65 p.

[145] Werkelin, J., Skrifvars, B.-J., and Hupa, M. 2005. Ash-forming elements in four Scandinavian wood species. part I: summer harvest. *Biomass and Bioenergy* 29:451–466.

[146] Berglund, A. 1999. *Morphological Investigation of Metal Ions in Spruce Wood*. Licentiate Thesis. Chalmers University of Technology, Department of Forest Products and Chemical Engineering, Göteborg, Sweden. 49 p.

[147] Brännström, H., and Halmemies, E. 2021. Bark extractives – polyphenols and others. In: Alén, R. (Ed.). *Pulping and Biorefining – ForestBioFacts, Digital Learning Environment*. Paperi ja Puu Ltd., Helsinki, Finland.

[148] Pizzi, A. 2008. Tannins: major sources, properties and applications. In: Belgacem, M.N., and Gandini, A. (Eds.). *Monomers, Polymers and Composites from Renewable Resources*. Elsevier, Oxford, United Kingdom. Pp. 179–199.

[149] Bianchi, S. 2017. *Extraction and Characterization of Bark Tannins from Domestic Softwood Species*. Doctoral Thesis. University of Hamburg, Hamburg, Germany. 95 p.

[150] Zhang, Q., Lin, L., and Ye, W. 2018. Techniques for extraction and isolation of natural products: a comprehensive review. *Chinese Medicine* 13(1):20.

[151] Brännström, H., and Halmemies, E. 2021. Separation of valuable extractives from trees. In: Alén, R. (Ed.). *Pulping and Biorefining – ForestBioFacts, Digital Learning Environment*. Paperi ja Puu Ltd., Helsinki, Finland.

[152] Pfennig, A., Delinski, D., Johannisbauer, W., and Josten, H. 2011. Extraction technology. In: Bart, H., and Pilz, S. (Eds.). *Industrial Scale Natural Products Extraction*. Wiley-VCH Verlag, Weinheim, Germany. Pp.181–220.

[153] Bart, H. 2011. Extraction of natural products from plants – an introduction. In: Bart, H., and Pilz, S. (Eds.). *Industrial Scale Natural Products Extraction*. Wiley-VCH Verlag, Weinheim, Germany. Pp.1–25.

[154] Willför, S., Smeds, A., and Holmbom, B. 2006. Chromatographic analysis of lignans. *Journal of Chromatography A* 1112(1):64–77.

[155] Feng, S., Cheng, S., Yuan, Z., Leitch, M., and Xu, C. 2013. Valorization of bark for chemicals and materials: a review. *Renewable and Sustainable, Energy Reviews* 26:560–578.

[156] Brännström, H., and Halmemies, E. 2021. Choosing the right solvent – hydrophobic or hydrophilic? In: Alén, R. (Ed.). *Pulping and Biorefining – ForestBioFacts, Digital Learning Environment*. Paperi ja Puu Ltd., Helsinki, Finland.

[157] Esclapez, M., García-Pérez, J.V., Mulet, A., and Cárcel, J. 2011. Ultrasound-assisted extraction of natural products. *Food Engineering Reviews* 3(2):108–120.

[158] Seidel, V. 2012. Initial and bulk extraction of natural products isolation. In: Sarker, S., and Nahar, L. (Eds.). *Natural Products Isolation. Methods in Molecular Biology (Methods and Protocols)*. Vol 864. Humana Press, Totowa, NJ, USA. Pp. 27–41.

[159] Li, Y., Radoiu, M., Fabiano-Tixier, A., and Chemat, F. 2012. From laboratory to industry: scale-up, quality, and safety consideration for microwave-assisted extraction. In: Chemat, F., and Cravotto, G. (Eds.). *Microwave-Assisted Extraction for Bioactive Compounds*. Springer, Boston, MA, USA. Pp. 207–229.

[160] Turner, C. 2006. Overview of modern extraction techniques for food and agricultural samples. In: *Modern Extraction Techniques*. ACS Symposium Series, Washington, DC, USA, 926:3–19.

[161] Destandau, E., Michel, T., and Elfakir, C. 2013. Microwave-assisted extraction. In: Rostagno, M. A., and Prado, J. M. (Eds.). *Natural Product Extraction: Principles and Applications*. Royal Society of Chemistry Cambridge, United Kingdom. Pp. 113–156.

[162] Rombaut, N., Tixier, A., Bily, A., and Chemat, F. 2014. Green extraction processes of natural products as tools for biorefinery. *Biofuels, Bioproducts and Biorefining* 8(4):530–544.

[163] Selvamuthukumaran, M., and Shi, J. 2017. Recent advances in extraction of antioxidants from plant by-products processing industries. *Food Quality and Safety* 1(1):61–81.

[164] Gandhi, K., Arora, S., and Kumar, A. 2017. Industrial applications of supercritical fluid extraction: a review. *International Journal of Chemical Studies* 5:336–340.

[165] Berk, Z. 2018. *Food Process Engineering and Technology*. 3rd edition. Academic Press, London, United Kingdom. 742 p.

[166] Brännström, H., and Halmemies, E. 2021. Operation modes and procedures in industrial extraction processes. In: Alén, R. (Ed.). *Pulping and Biorefining – ForestBioFacts, Digital Learning Environment.* Paperi ja Puu Ltd., Helsinki, Finland.

[167] Yasri, A., Nabouls, I., Aboulmouhajir, A., Kouisni, L., and Bekkaoui, F. 2018. Plants extracts and secondary metabolites, their extraction methods and use in agriculture for controlling crop stresses and improving productivity: a review. *Academia Journal of Medicinal Plants* 6(8):223–240.

[168] Royer, M., Houde, R., Viano, Y., and Stevanovic, T. 2012. Non-wood forest products based on extractives – a new opportunity for the Canadian forest industry. part 1: hardwood forest species. *Journal of Food Research* 1(3):8–45.

[169] Chemat, F., Vian, M.A., and Cravotto, G. 2012. Green extraction of natural products: concept and principles. *International Journal of Molecular Sciences* 13(7):8615–8627.

[170] Herrero, M., Cifuentes, A., and Ibañez, E. 2006. Sub- and supercritical fluid extraction of functional ingredients from different natural sources: plants, food-by-products, algae and microalgae: a review. *Food Chemistry* 98(1):136–148.

[171] Mustafa, A., and Turner, C. 2011. Pressurized liquid extraction as a green approach in food and herbal plants extraction: a review. *Analytica Chimica Acta* 703(1):8–18.

[172] Co, M., and Turner, C. 2012. Pressurized fluid extraction and analysis of bioactive compounds in birch bark. In: Carrier, J., Ramaswamy, S., and Bergeron, C. (Eds.). *Biorefinery Co-Products: Phytochemicals, Primary Metabolites and Value-Added Biomass Processing Carrier.* Wiley, Hoboken, NJ, USA. Pp. 259–285.

[173] Destandau, E., Michel, T., and Elfakir, C. 2013. Microwave-assisted extraction. In: Rostagno, M.A., and Prado, J.M. (Eds.). *Natural Product Extraction: Principles and Applications.* Royal Society of Chemistry Cambridge, United Kingdom. Pp. 113–156.

[174] Plaza, M., and Turner, C. 2015. Pressurized hot water extraction of bioactives. *Trends in Analytical Chemistry* 71:39–54.

[175] Chemat, F., Rombaut, N., Sicaire, A., Meullemiestre, A., Fabiano-Tixier, A., and Abert-Vian, M. 2017. Ultrasound assisted extraction of food and natural products. mechanisms, techniques, combinations, protocols and applications. a review. *Ultrasonics Sonochemistry* 34:540–560.

[176] Vinatoru, M., Mason, T., and Calinescu, I. 2017. Ultrasonically assisted extraction (UAE) and microwave assisted extraction (MAE) of functional compounds from plant materials. *Trends in Analytical Chemistry* 97:159–178.

[177] Njila, M.N., Mahdi, E., Massoma-Lembe, D., Zacharie, N., and Nyonseu, D. 2017. Review on extraction and isolation of plant secondary metabolites. In: *Proceedings of the 7th International Conference on Agricultural, Chemical, Biological and Environmental Sciences*, Kuala Lumpur, Malaysia.

[178] Naima, R., Oumam, M., Hannache, H., Sesbou, A., Charrier, B., Pizzi, A., and Charrier-El Bouhtoury, F. 2015. Comparison of the impact of different extraction methods on polyphenols yields and tannins extracted from moroccan *Acacia Mollissima* barks. *Industrial Crops and Products* 70:245–252.

[179] De Monte, C., Carradori, S., Granese, A., Di Pierro, G.B., Leonardo, C., and De Nunzio, C. 2014. Modern extraction techniques and their impact on the pharmacological profile of serenoa repens extracts for the treatment of lower urinary tract symptoms. *BMC Urology* 14(1):63.

[180] Tura, D., and Robards, K. 2002. Sample handling strategies for the determination of biophenols in food and plants. *Journal of Chromatography A* 975(1):71–93.

[181] Brännström, H., and Halmemies, E. 2021. Stemwood extractives-based products. In: Alén, R. (Ed.). *Pulping and Biorefining – ForestBioFacts, Digital Learning Environment.* Paperi ja Puu Ltd., Helsinki, Finland.

[182] Rajendran, V., Breitkreuz, K., Kraft, A., Maga, D., and Brucart, M. 2016. *Analysis of the European Crude Tall Oil Industry – Environmental Impact, Socio-Economic Value & Downstream Potential.* Fraunhofer Institute for Environmental, Safety and Energy Technology UMSICHT, Oberhausen, Germany. Retrieved from http://www.harrpa.eu/images/Publications/EU_CTO_Added_Value_Study_Fin.pdf.

[183] Anon. 2016. *Global Impact of the Modern Pine Chemical Industry.* Pine Chemical Association, Fernandina Beach, FL, USA. Retrieved from https://cdn.ymaws.com/www.pinechemicals.org/re source/resmgr/Studies/PCA-_Global_Impact_of_the_Mo.pdf.

[184] Holmbom, B., Eckerman, C., Eklund, P., Hemming, J., Nisula, L., Reunanen, M., Sjöholm, R., Sundberg, A., Sundberg, K., and Willför, S. 2003. Knots in trees – a new rich source of lignans. *Phytochemistry Reviews* 2(3):331–340.

[185] Willför, S., Nisula, L., Hemming, J., Reunanen, M., and Holmbom B. 2004. Bioactive phenolic substances in industrially important tree species. part 1: knots and stemwood of different spruce species. *Holzforschung* 58(4):335–344.

[186] Willför, S., Sundberg, A., Rehn, P., Holmbom, B., and Saranpää, P. 2005. Distribution of lignans in knots and adjacent stemwood of *picea abies*. *Holz als Roh- und Werkstoff* 63(5):353–357.

[187] Willför, S., Reunanen, M., Eklund, P., Sjöholm, R., Kronberg, L., Fardim, P., Pietarinen, S., and Holmbom, B. 2004. Oligolignans in Norway spruce and Scots pine knots and Norway spruce stemwood. *Holzforschung* 58(4):345–354.

[188] Eklund, P.C., Willför, S.M., Smeds, A.I., Sundell, F.J., Sjöholm, R.E., and Holmbom, B.R. 2004. A new lariciresinol-type butyrolactone lignan derived from hydroxymatairesinol and its identification in spruce wood. *Journaal of Natural Products* 67(6):927–931.

[189] Masuda, T., Akiyama, J., Fujimoto, A., Yamauchi, S., Maekawa, T., and Sone, Y. 2010. Antioxidation reaction mechanism studies of phenolic lignans, identification of antioxidation products of secoisolariciresinol from lipid oxidation. *Food Chemistry* 123(2):442–450.

[190] Latva-Mäenpää, H., Laakso, T., Sarjala, T., Wähälä, K., and Saranpää, P. 2014. Root neck of Norway spruce as a source of bioactive lignans and stilbenes. *Holzforschung* 68(1):1–7.

3 Chemical and biochemical conversion

3.1 Acid-catalyzed hydrolysis

Today, considerable quantities of chemicals and various materials together with energy based on lignocellulosic biomass feedstocks in their different forms are commonly used as shown in the background data in Chapter 1 "Introduction to biorefining". Lignocellulosic biomass feedstocks are chemically highly complex raw materials and hence, they also offer a huge number of possibilities to be extensively used as described in a wide range of excellent books and other publications [1–22]. In these kinds of approaches, the life cycle assessment (LCA) ranging from the extraction or production of raw materials to final disposal is an important tool for quantifying the environmental impact of bio-based materials [23]. The main chemical components of the renewable biomaterials are polymeric carbohydrates (i.e., cellulose and hemicelluloses), and one of the most obvious ways of using such carbohydrates as chemicals is via selective acid or enzymatic hydrolysis (i.e., saccharification) to low-molar-mass carbohydrates, mainly monosaccharides and disaccharides, which then can be converted into a multitude of value-added products. This straightforward approach is discussed in short in this chapter showing only the main principles and products. The use of other chemical constituents of lignocellulosic biomass materials as well as polymeric carbohydrates as such is included in other chapters of this book.

Acid hydrolysis of lignocellulosic biomass materials has a long history, and it has been known, for example, for over 200 years that cellulose can be converted into its monomeric moieties, glucose units by this method [24–29]. Traditionally, wood-derived cellulose was typically first treated with strong sulfuric acid (H_2SO_4) at low temperatures of 4–20 °C followed by heating with dilute acids. In principle, acid hydrolysis can be accomplished either by concentrated mineral acids at low temperatures with long reaction times or by dilute mineral acids (i.e., the typical acid concentration is in the range of 2–5%) at higher temperatures (160–230 °C) and pressures (about 1 MPa) with shorter reaction times in the range of seconds to minutes [28, 30–40]. Additionally, hemicelluloses can be readily hydrolyzed by dilute acids under moderate conditions, whereas much more extreme conditions are needed for cellulose (i.e., a higher conversion of cellulose to glucose). In general, larger amounts of degradation products of monosaccharides (e.g., furans) are formed at higher temperatures. Acid hydrolysis of carbohydrates can be catalyzed by a wide range of acids, including H_2SO_4, hydrochloric acid (HCl), orthophosphoric acid (H_3PO_4), hydrofluoric acid (HF), sulfurous acid (H_2SO_3), nitric acid (HNO_3), and trifluoroacetic acid (F_3CO_2H (TFA)) as well as various combinations of some of these acids [11, 28, 30, 33, 35–45].

The commercial hydrolysis techniques have been of interest for more than 100 years [39]. The first wood saccharification processes were developed by Alexander Classen in 1901 employing concentrated H_2SO_4 and in 1909 employing dilute H_2SO_4 by

https://doi.org/10.1515/9783110608366-003

Malcom F. Ewen and George H. Tomlinson. The saccharification processes performed by mineral acids have a much longer history than the enzymatic process (see Section 3.2 "Enzymatic hydrolysis"), and early attempts to commercialize the mineral acids-based techniques also laid a firm foundation for modern development aimed at the production of a wide range of marketable chemicals by fermentation [1, 3, 11, 16, 17, 26, 40]. Due to this development during its early stage, several facilities were built for hydrolysis of wood and already at that time, the monosaccharides produced were fermented, for example, to ethanol (CH_3CH_2OH), which was and still is an acceptable motor fuel (see Section 3.3 "Use of carbohydrates in hydrolyzates") [28]. In the hydrolysis-based conversion of lignocellulosics to useful products *via* water (H_2O)-soluble sugars (*i.e.*, mainly glucose and other monosaccharides), the most common and key step is the pretreatment of various feedstocks (see Section 3.2 "Enzymatic hydrolysis") [39, 40]; hydrolysis can then be accomplished by chemical (with mineral acids) or biochemical (with enzymes) treatments (Figure 3.1).

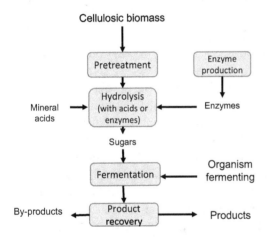

Figure 3.1: A simplified process scheme for the conversion of lignocellulosic biomass materials-derived carbohydrates to various products by hydrolysis either with mineral acids or enzymes [39]. The sugars formed can be utilized, besides by fermentation, also by different chemical treatments.

Two main carbohydrate reactions occur during the acid hydrolysis process: [24, 27, 29, 30, 39, 46–49], *i*) depolymerization of polysaccharides to their monosaccharide moieties and in part to oligosaccharide residues (*i.e.*, the cleavage of glycosidic linkages between the monosaccharide units) and *ii*) formation of monosaccharides-derived products (*e.g.*, furans) that can inhibit the subsequent fermentation [11, 39, 50]. Hence, in the most prominent reaction, the acid-catalyzed conversion of high-molar-mass carbohydrates in lignocellulosic biomass materials to hexoses and pentoses takes place. However, there is only a limited demand for glucose and other monosaccharides from hydrolysis unless they are further refined. For example, many factors influence the reactivity and digestibility of the cellulose fraction of these raw materials [30, 51]; the factors include the content of hemicelluloses and lignin, the crystallinity of cellulose, and the porosity of the feedstock materials. Additionally, the effective isolation of individual components often requires complicated separation techniques and thus, the most important and the simplest method for utilization of the released sugars is fermentation. It enables their use in the production of food, chemicals, and energy (*e.g.*, *via* CH_3CH_2OH and 1-butanol ($CH_3CH_2CH_2CH_2OH$)).

Hemicelluloses are relatively easily hydrolyzed and the composition of the resulting sugar mixture (*i.e.*, hydrolyzate) essentially depends on the used raw material [38]. The major hydrolysis products from hemicelluloses (mainly glucomannan and xylan) [52] are, besides glucose, other aldohexoses (mannose and galactose) as well as aldopentoses (xylose and arabinose) [30, 33, 42, 53, 54]. In contrast, cellulose is hydrolyzed clearly more slowly, and the glucose concentration increases toward the end of hydrolysis. The slow hydrolysis of cellulose also permits the secondary degradation reactions of the glucose produced to become important [11, 26]. The evident degradation products of glucose are 5-(hydroxymethyl)furfural (HMF) and levulinic acid. To minimize contamination of glucose, which in most cases is the desired product, a two-stage acid hydrolysis is often recommended [39]. It first comprises a mild treatment where hemicelluloses are preferentially hydrolyzed, followed by a second step under harsher conditions to recover mainly glucose from cellulose. In general, a complete hydrolysis of cellulose is difficult. The kinetics of cellulose [29, 55], hemicelluloses [42, 53, 54], and wood [31–33, 56] during acid hydrolysis has been studied in detail.

As the harsh process conditions of acid hydrolysis cause corrosion risks and generate, besides a product mixture, large amounts of acid and solid (*e.g.*, sulfur-containing lignin) wastes, these hydrolysis methods can be costly and require special equipment [40, 57, 58]. Additionally, the recycling of mineral acid is still a challenge. As indicated above, an additional drawback is the further degradation of monosaccharides (*i.e.*, to limit the reaction to hydrolysis) to a wide range of harmful by-products. It can be concluded that wood acid hydrolysis processes for producing biochemicals have mainly been hampered by incomplete simplification of the manufacturing procedure and the high energy consumption arising from the need to recycle the waste acid. Hence, development of new economical "green processes" for the conversion of cellulose into glucose under mild conditions with a high selectivity would still be essential. There have also been investigations, for example, dealing with hydrolysis of cellulose in supercritical H_2O [59, 60], cellulose hydrolysis catalyzed by highly acidic lignin-derived carbonaceous catalyst [61], as well as fractionation of sugarcane straw by the sulfur dioxide (SO_2)- CH_3CH_2OH-H_2O method [62–64].

The gradual reversion to fermentation as the preferred process has created a need for a cheap and widely available source of fermentable sugars [28]. A variety of lignocellulosic sources, such as agricultural crop residues and wood residues, appear to meet these requirements. The product yields, especially from various carbohydrates derived from common agricultural resources are relatively high, thus also generally leading to lower conversion costs [40]. In contrast, forest-derived biomass, typically with a relatively high content of branches, tops, needles or foliage, and bark, has a lower content of carbohydrates compared to stemwood material [38]. Hence, for softwood-derived raw materials, acid hydrolysis has been claimed to be more a competitive method for producing fermentable sugars than enzymatic hydrolysis [40]. In contrast, hardwood-derived residues would be a more potential choice when considering the costs, processability, and yields in carbohydrate conversion by acid hydrolysis. However, it is obvious that the easily hydrolyzed hemicelluloses comprise a major part of the wood components, which also typically can be easily removed [26]. Normally, monosaccharides from softwoods (mainly aldohexoses) are more fermentable than those (mainly aldopentoses) from hardwoods and non-wood materials. Additionally, several process side streams from the pulp and paper industry (see Sections 4.3.3 "Spent liquor" and 5.2.2 "Acidic pretreatments") may also represent interesting sources of raw materials. These side streams offer carbohydrate sources, which are already adequately pretreated for economic bioconversion. Due to the limited volumes, many potential side streams from the food sector seem to be insignificant, although in each case, accurate calculations should be carried out to determine the cost relationships between the raw material price, availability, processability, and yield.

Since its early days in 1945 [31], the research effort on the kinetics of dilute acid hydrolysis of wood cellulose at elevated temperatures has continued, and a great number of new results, mainly dealing with a broader range of reaction conditions, have been obtained [53, 65]. During recent decades, biomass hydrolysis processes have economically been attractive only in a few cases. However, recent studies on reactor design and simulation have suggested that acid hydrolysis technology may again be a viable alternative among feasible biomass saccharification processes. In several acid hydrolysis processes, as also in enzymatic hydrolysis processes, economic (*e.g.*, capital investment) and even technical obstacles have hampered their common utilization. During the development work many different reactor configurations have been investigated including, for example, plug flow [66], progressive batch [67], counter-current [68], and percolation [69] reactors, as well as single-state batch and bed-shrinking flow-through reactors [35] or a twin-screw extruder [32]. Although, for many reasons, HCl is the most efficient hydrolyzing agent, the dilute H_2SO_4 method has traditionally been developed and maintained as an industrial-scale application.

One of the application examples dealing with the dilute acid hydrolysis processes of pure cellulose is a continuous-flow reactor using 1% H_2SO_4 with a residence time of 0.22 min and a temperature of 237 °C providing a sugar yield of about 50% [40]. In this case, the most important advantage is the fast reaction rate, which facilitates continuous processing. However, due to enhanced carbohydrate degradation, the biggest disadvantage is the relatively low sugar yield.

Mild acidic prehydrolysis of wood using 0.5–1.0% H_2SO_4 at 120–130 °C was commercially used first in Germany during the 1940s in the two-stage modification of kraft pulping processes (see Section 4.2 "Kraft pulping") [65]. The hydrolyzates obtained were rich in monosaccharides in varying proportions depending on the wood raw material, and were useful, for example, for aerobic fermentation by *Torula* or fodder yeast [1]. Such pretreatment stages prior to kraft pulping have later been used in several kraft mills and they have also recently restudied as a potential integrated biorefinery concept by many researchers (see Section 5.2.2 "Acidic pretreatments"). Additionally, the spent liquors from the acid sulfite pulping contain large amounts of sugars (see Section 4.3.3 "Spent liquor"); this fraction can traditionally be fermented to products, such as CH_3CH_2OH and feed protein, or their main monosaccharides, xylose and mannose, can be isolated and used in other applications [40, 70]. Furthermore, a number of commercial-scale wood-based sugar plants producing CH_3CH_2OH and other products, such as furfural and yeast cells have been in operation and these kinds of processes have been a part of the sugar industry in many other countries as well [71]. In recent years, the acidic treatment of lignocellulosic biomass materials with dilute H_2SO_4 has been primarily used as a means of pretreatment for enzymatic hydrolysis of celluloses (see Section 3.2 "Enzymatic hydrolysis") [34, 38].

One of the principal ideas behind strong acid hydrolysis is its ability to disrupt the crystalline structure of cellulose by solution or swelling in the acid [26, 48, 55]. The subsequent hydrolysis of cellulose can then be accomplished at temperatures low enough to avoid degradation affording almost quantitative glucose yields. Hence, it is possible to avoid in strong acid hydrolysis at lower temperatures, the yield limitations taking place in dilute acid hydrolysis of the crystalline cellulose at higher temperatures. Among the cellulose hydrolyzing agents are three strong acids: 41% HCl, 93% H_2SO_4, and 85–100% H_3PO_4. The action of these acids is to dissolve the cellulose completely at low temperatures of 4–20 °C. Furthermore, it has been shown that minimum acid concentrations needed for dissolving the cellulose are critical and the kinetics can vary significantly with only small changes in acid concentration. For example, at room temperature in 16 h 41% HCl has shown to cause a hydrolysis yield of 100%, whereas 40%, 39%, and 38% HCl resulted, respectively, in a hydrolysis yield of 73%, 45%, and 22%.

One of the recent concentrated acid hydrolysis process, the Arkenol process, typically works by adding 70–77% H_2SO_4 to biomass dried to a moisture content of 10% at 50 °C, the ratio of acid to biomass being 1.25 [40, 65]. H_2O is then added to dilute the acid to a concentration of 20–30%, and the mixture is heated to 100 °C for one hour. H_2SO_4 is separated for recycling from the sugar mixture by a chromatographic column. In acid hydrolysis, in addition to H_2SO_4, concentrated HCl has also attained some commercial significance. However, despite improvements in corrosion-resistant materials and acid recovery techniques, industrial installations and pilot plants still use almost totally dilute H_2SO_4 as the hydrolysis catalyst.

Besides strong mineral acids, aqueous weak organic acids, such as formic acid (HCO_2H), acetic acid (CH_3CO_2H), and oxalic acid (HO_2CCO_2H) have been used either for obtaining directly different sugar fractions or as a pretreatment stage to improve the accessibility of lignocellulosic biomass materials for subsequent treatments, like enzymatic hydrolysis [72–79]. As a significant advantage, weak organic acids are, compared to mineral acids, noncorrosive, and, for example, lower-molar-mass organic acids (*e.g.*, HCO_2H and CH_3CO_2H) are volatile compounds, which can be easily recovered by distillation for reuse. Additionally, mineral acids cause the formation of large amounts of further-degraded products, such as furfural and HMF, whereas in case of milder organic acids, the formation of these furan derivatives is not as critical. In practice, this means that there is less possibility for the formation of inhibitors to fermentation in the subsequent process stage. As an additional value, high-quality lignin is also possible to be recovered from these kinds of hydrolyzates.

Acid-catalyzed autohydrolysis of lignocellulosic biomass materials is a process without the addition of any external acid [39]. This is possible since a thermal process will cause the splitting of thermally labile acetyl groups ($-COCH_3$) in the native hemicelluloses [52], and the CH_3CO_2H formed catalyzes the hydrolysis. However, a further addition of CH_3CO_2H may still improve the hydrolysis efficiency. For example, the pretreatment of corn stover with diluted CH_3CO_2H toward enhancing acidogenic fermentation indicates the optimal conditions of hydrolysis (0.25% CH_3CO_2H at 191 °C) when reaching the highest fermentation level [76].

The use of the simplest carboxylic acid, HCO_2H with a relatively high acidity enables an effective fractionation of lignocellulosic biomass materials [80]. The hydrolysis process with HCO_2H not only degrades hemicelluloses into monosaccharides and oligosaccharides, but, due to cleavage of the lignin β-*O*-4 bonds, also dissolves lignin into the hydrolyzate, thus remaining cellulose is a residue solid (see also Section 4.4.2 "Organosolv pulping") [81]. Lignin can then be precipitated from the hydrolyzate by adding H_2O. More recently, a combination of autohydrolysis and HCO_2H hydrolysis has been shown to be efficient in fractionation of the three major components of lignocellulosic materials [82]. In one illustrative example, the highest cellulose yield of about 70% was obtained from eastern cottonwood at a $HCO_2H/CH_3CO_2H/H_2O$ ratio of 30:50:20 (v/v/v) with a solid-to-liquor ratio of 1:12 (g/mL) at 105 °C for 90 min [83]. In general, HCO_2H hydrolysis is often used as a pretreatment approach prior to enzymatic hydrolysis with or without the addition of small amounts of mineral acid catalyst.

Due to its high efficiency in producing sugars and the good compatibility with downstream operations, aqueous HO_2CCO_2H has also been found as a promising alternative to mineral acids for biomass hydrolysis [72, 74, 84]. In the biorefinery process where CH_3CH_2OH is produced as the main product, HO_2CCO_2H holds a few advantages; it is less toxic to yeasts and does not inhibit glycolysis. The recovery of HO_2CCO_2H can be made by conventional techniques, such as electrodialysis, ion exchange, and adsorption. Among these methods, especially electrodialysis has proven effective in the recovery and reuse of HO_2CCO_2H [75]. The amorphous regions of cellulose fibers can also be hydrolyzed by HO_2CCO_2H [85]. Hence, a mild hydrolysis with HO_2CCO_2H has been applied to extract cellulose nanocrystals from the bleached wood pulp.

3.2 Enzymatic hydrolysis

Due to the microbial decomposition of lignocellulosic biomass materials, enzymatic hydrolysis of cellulose to glucose has been known to be as an integral part of the carbon cycle since its inception [26]. However, the detailed development of commercial conversion plans was not possible, until the detailed mechanism of microbial degradation of cellulose was gradually clarified, especially in the 1970s. Hydrolysis of wood and other lignocellulosic biomass materials by enzymes to fermentable sugars and further to their conversion products has been intensively studied in recent decades [65].

In this hydrolysis method, cellulases (*i.e.*, specific for cellulose) together with various hemicellulases (*i.e.*, specific for hemicelluloses), are chosen as enzymes for preferentially cleaving β-(1→4)-glycosidic linkages in polysaccharide chains, ultimately aiming at almost complete degradation of carbohydrates-containing raw materials to monosaccharides, aldohexoses, and aldopentoses [86–97]. Hence, the enzymatic hydrolysis of lignocellulosic biomass materials to fermentable sugars is a common step in several conversion processes where sugars are used as intermediates to produce value-added chemicals or fuels. During the past decade, especially the production of CH_3CH_2OH has been the driver for developing more efficient conversion processes. In practice, this has meant that in many countries, commercial-scale wood-based sugar plants producing CH_3CH_2OH and other products, such as furfural and yeast cells, are in operation as a part of the sugar industry; [28, 71, 98–107], for example, traditional Brazil's numerous fuel alcohol plants (*i.e.*, mainly producing CH_3CH_2OH for automobiles) are fed by the by-products of their large sugar industry. Various steps of the technologies involved, including pretreatment, enzyme production, hydrolysis, and to some extent also the process configuration, can be considered common to all the biotechnical production processes. However, in each case, the fermentation process is designed for the final product.

Enzymatic hydrolysis, compared to conventional acid hydrolysis, is considered as a more promising technology for converting biomass into H_2O-soluble sugars to be used as raw materials for the biotechnical production of different bulk products [28]. This is primarily due to higher yields (*i.e.*, the selectivity is good), the minimized formation of harmful by-products, and the generally improved outlook for improvements in biotechnology [65, 108]. Additionally, enzymatic hydrolysis proceeds under mild and noncorrosive conditions.

Despite these clear benefits, enzymatic hydrolysis has not yet definitely broken through on full scale and the current technology seems to be fully applicable only to less-lignified plants, such as straws and reeds, rather than woods with a higher content of lignin [40]. On the other hand, after proper pretreatment (*i.e.*, the distribution of the naturally resistant structure of feedstocks to create reactive sites to biological processes), various lignocellulosic biomass materials as well as their carbohydrate components can be readily hydrolyzed into monosaccharides, using cellulolytic and hemicellulolytic enzymes besides acids. It is also obvious that new and more economical processes are likely to emerge from the versatile research now in progress. Since glucose is the main monosaccharide component produced by hydrolysis from almost all lignocellulosic biomass materials, it also represents the most important carbon source for different biotechnical conversion processes.

Besides substrate characteristics, the nature of the enzyme complex has a significant effect on how effectively a pretreated cellulosic biomass matrix is hydrolyzed with respect to the saccharification of cellulose and hemicelluloses [65]. In general, lignocellulosic raw materials are rather resistant to enzymatic hydrolysis, mainly because of the crystalline structure of cellulose and the presence of lignin, but the other important factors affecting the reactivity and digestibility of these materials include, for example, the content of hemicelluloses and the porosity of the feedstock material (Table 3.1) [40, 111–113]. The use of enzymatic hydrolysis is also highly dependent on the cost of the selective enzymes required. Additionally, due to the heterogeneity of the lignocellulosic feedstock, the low conversion rate may cause difficulties. Enzymatic degradation is also relatively slow compared to acid hydrolysis.

Table 3.1: Characteristics of lignocellulosic feedstocks influencing their digestibility [40, 109, 110].

Chemical properties	Physical properties
Cellulose and hemicellulose content	Degree of polymerization of carbohydrates
Lignin content and distribution	Crystallinity of cellulose
Interlinkages of hemicelluloses and lignin	Pore volume
Acetyl content of hemicelluloses	Accessible surface area
Ash (silica) content	Particle size

Since many structural and compositional factors hinder enzymatic digestion of carbohydrates in lignocellulosic biomass materials, various pretreatment methods are usually needed, prior to the enzymatic process, and they can significantly enhance, especially the hydrolysis of cellulose [109, 113–115]. Most pretreatment technologies (Table 3.2) have been developed for the conversion of lignocellulosic feedstocks into CH_3CH_2OH, but they are in principle applicable for any hexose- or pentose-utilizing bioconversion process as well. The primary goal of such pretreatment technologies is simply to render biomass materials more accessible to enzymatic hydrolysis (i.e., to improve the rate of enzymatic hydrolysis) for efficient product generation. The accessible surface area, cellulose crystallinity, and lignin content (Table 3.1) have received most attention, and according to many studies, they have the greatest impact on biomass digestibility [40, 112, 113, 116–119]. Increased accessibility of cellulose enables more enzymes to bind to cellulose fiber surfaces, whereas decreased crystallinity increases the reactivity of cellulose, i.e., the rate at which the bound enzyme hydrolyzes glycosidic linkages. According to the overall principles the successful pretreatment should [113, 120]:

- maximize the enzymatic convertibility
- minimize the loss of carbohydrates
- maximize the production of valuable by-products (*e.g.*, lignin)
- not require the addition of chemicals toxic to the enzymes or the fermenting microorganisms
- minimize the use of chemicals, energy, and capital equipment
- be scalable to the full-scale process

Table 3.2: The main pretreatment processes of lignocellulosic biomass materials.

Physical pretreatments[a]	Selected references
Grinding and milling (various milling procedures)	Refs. [73, 77, 90, 99, 110, 121, 122]
Irradiation (γ-ray, electron-beam, and microwave irradiation)	Refs. [77, 99, 110, 121, 123]
Physicochemical pretreatments	
Hydrothermal (steam explosion (SE or STEX) and liquid hot water (LHW) or autohydrolysis)	Refs. [77, 99, 105–107, 121, 124–129]
Ammonia fiber explosion (AFEX)	Refs. [77,99, 105–107, 1, 121, 128, 130]
Ammonia recycle percolation (ARP)	Refs. [38, 128, 130, 131].
Carbon dioxide (CO_2) explosion	Refs. [38, 77, 105, 107, 121, 128].
Chemical pretreatments	
Acid pretreatment (dilute acids, concentrated acids, and organic acids)	Refs. [33, 34, 38, 73, 77, 90, 99, 106, 107, 110] Refs. [130, 133, 135, 138, 139]
Alkali pretreatment	Refs [38, 73, 77, 90, 99, 106, 107, 110] Refs. [130, 133, 135, 138, 139, 141–146]
Solvent extract (organosolv process, ionic liquids, and sulfite pretreatment to overcome the recalcitrance of lignocellulose (SPORL))	Refs. [38, 77, 90, 99, 105–107, 110] Refs. [120, 123, 128, 130, 147–149]
Oxidation (ozonolysis, wet oxidation (WO), and H_2O_2)	Refs. [38, 73, 90, 99, 105–107, 150]
Biological pretreatments	
Fungi (white-rot, brown-rot, and soft-rot fungi)	Refs. [38, 73, 77, 105–107, 121, 128, 151]
Actinomycetes[b]	Refs. [77, 99, 107, 121, 130].

[a]Other physical pretreatment processes are high pressure steaming, extrusion, expansion, sonication, pulsed-electric-field treatment, and pyrolysis [38, 77, 90, 99, 110, 121].
[b]Gram-positive bacteria producing a number of enzymes (*e.g.*, ligninases, peroxidases, laccases, and hydrolases) that help degrade lignocellulosic material.

Most pretreatment methods, especially those belonging to physical pretreatments involve expensive instruments or equipment that may have high energy requirements [38]. Physical pretreatments usually with no addition of chemicals increase the accessible surface area and the size of the pores as well as decrease the degree of polymerization (DP) of cellulose and the crystallinity of cellulose [73, 99, 121]. Most of them cause the partial degradation of hemicelluloses and lignin. Physicochemical and chemical pretreatments have typically rapid treatment rates under harsh conditions and there are requirements for external chemicals. These methods also increase in the accessible surface area, due to swelling, and the size of the pores (*i.e.*, increase porosity as well) and decrease the DP of cellulose and the cellulose crystallinity. They cause the partial or complete hydrolysis of hemicelluloses and the enhanced removal of lignin is possible. The maximum digestibility typically coincides with the complete removal of hemicelluloses. Hence the dilute acid treatment of biomass (together with steam explosion) aimed at hemicellulose hydrolysis has become a widely accepted pretreatment method for enzymatic hydrolysis, and physicochemical and chemical pretreatments are among the most effective and the most promising processes for industrial applications. Biological pretreatments as non-energy-intensive processes do not require high energy for lignin removal from lignocellulosic biomass materials, despite extensive lignin degradation [38, 106, 128, 130]. However, these pretreatments have, compared to other pretreatment methods, a very low treatment rate, which tends to be too slow for industrial purposes.

Agricultural residues and other non-wood feedstocks are normally more easily treated than hardwood materials [65]. It is also generally agreed that softwood substrates are inherently more recalcitrant to hydrolysis than hardwood substrates. This fact is attributed to factors, such as the lignin content and structure. As shown above, most pretreatment processes exploit a variety of mechanisms to render the carbohydrate components of lignocellulosic biomass materials more susceptible to enzymatic hydrolysis. Hence, most successful pretreatments rely on removing either hemicelluloses or lignin to create a material with sufficient porosity to allow significant access for enzymatic attack.

In lignocellulosic biomass materials, especially the physical state of cellulose as such or after suitable pretreatments makes them challenging substrates for enzymes in the conversion to sugars, which then can be, for example, fermented to various products by bacteria, yeast, or fungi [87, 88, 97, 104, 152, 153]. In the lignocellulosic matrix, single glucose polymers are packed onto each other to form highly crystalline microfibrils in which the individual cellulose chains are held together by hydrogen bonds [52]. Cellulose microfibrils also contain some amorphous regions. In wood fibers, the winding direction of cellulose microfibrils varies in different cell wall layers, giving the fiber its unique strength and flexibility (see Section 2.2 "Fiber dimensions and ultrastructure").

Efficient degradation of cellulose requires different cellulases (*i.e.*, the combination of enzymes) acting sequentially or in concert [40, 95, 154]. One long-term goal of the cellulase research has long been to engineer mixtures of cellulases that have higher specific activities, greater thermostability, and improved recyclability in the hydrolysis of pretreated biomass than the cellulase mixtures known before [92, 95]. However, to design better cellulases, it is necessary to determine the rate limiting step in the hydrolysis of crystalline cellulose. Hence, the target of increasing the rate of this step, most effectively also increases the rate of hydrolysis.

Cellulases have traditionally been categorized into following classes [40, 94, 104, 113, 155]:

- endo-1,4-β-glucanases (EG) (EC 3.2.1.4), which randomly hydrolyze internal β-(1→4)-glucosidic bonds in the cellulose chains leading to shorter cellulose chains with reducing ends
- exo-1,4-β-glucanases or cellobiohydrolases (CBH) (EC 3.2.1.91), which include *i*) 1,4-β-D-glucanglucohydrolase (releases glucose moieties from the reducing end of cellulose chains and hydrolyzes slowly cellobiose formed by another enzyme) and *ii*) 1,4-β-D-glucancellobiohydrolase, which releases cellobiose moieties from the reducing end of cellulose chains
- β-D-glucosidases (BGL) (EC 3.2.1.21) and β-D-glucosideglucohydrolases or cellobiases, which degrade short cellooligosaccharides and cellobiose to glucose

It can be simply concluded that when the enzymatic cellulase system acts *in vitro* on an insoluble cellulose substrate, three important processes take simultaneously place [104]: *i*) physical and chemical changes in the cellulosic fraction, *ii*) primary hydrolysis leading to soluble sugars from the surface of cellulose, and *iii*) secondary hydrolysis, which involves hydrolysis of soluble sugars to lower-molar-mass sugars, and finally to glucose. However, in practice, the straightforward classification of cellulases is rather complicated as some enzymes also have activity on other polysaccharides, such as xylan [40]. Hence, it is often used in a recent classification of these enzymes as well as of other glycosyl hydrolases; the different glycosyl hydrolases are categorized according to the structures of their catalytic domains into more than 90 families. On the other hand, this system requires data on the amino acid sequence, which is not always available. For example, the cellulolytic system of *Trichoderma reesei* has been studied in most detail, and most industrial cellulases are produced with this fungus. Additionally, cellulases have been shown to act synergistically, *i.e.*, their combined hydrolytic activity is higher than the sum of their individual activities.

A wide range of species of anaeorbic bacteria, such as *Bacteroides*, *Clostridium*, *Clostrium*, and *Ruminococcus* and aerobic bacteria, such as *Actinomycetes*, *Bacillus*, *Cellulomonas*, *Pseudomonas*, and *Streptomyces* together with filamentous fungi, such as *Aspergillus niger*, *Fusarium graminearum*, *Penicillium chrysogenum*, *Phanerochaete chrysosporium*, and *Trichoderma reesei* are able to produce cellulases [97, 104].

Most cellulases have a multidomain structure which contains a core domain separated from a cellulose-binding domain (CBD) by a linker peptide [40, 94, 96]. The core domain has the active site, whereas the CBD interacts with cellulose by binding the enzyme to it. The CBDs are particularly significant in the hydrolysis of crystalline cellulose. It also seems that the ability of exo-1,4-β-glucanases to degrade crystalline cellulose clearly decreases when the CBD is absent. Furthermore, it has been claimed that the CBD enhances the enzymatic activity merely by increasing the effective enzyme concentration on the surface of cellulose and/or by loosening single cellulose chains from the cellulose surface. On the other hand, the CBDs are also mostly responsible for the nonspecific binding of cellulases onto lignin, which decreases their efficiency and potential reuse.

The two most common hemicelluloses are heterogeneous xylan (mainly in hardwoods and non-woods) and glucomannan (mainly in softwoods) with various side groups and hence, the hemocellulolytic systems (hemocellulases) needed for their hydrolysis are complex [52, 113]. Hemicellulases have generally been categorized into following classes [104]:

- endohemicellulases, which act within the hemicellulose chains and have only limited activity on short-chain oligomers
- exohemicellulases, which act progressively outside the hemicellulose chains
- hemicellulases, which hydrolyze hemicellulose side groups (*e.g.*, acetyl xylan esterases (EC 3.1.1.72) and *p*-coumaric acid esterases (EC 3.1.1.73))

A major part of the published work on hemicellulases deals with the production, properties, mode of action, and applications of xylanases [40, 113, 155]. For example, xylanases 1,4-β-D-xylanases (EC 3.2.1.8) catalyze the random hydrolysis of (1→4)-β-D-xylosidic linkages in xylans, whereas 1,4-β-D-xylosidases (EC 3.2.1.37) attack xylooligosaccharides from the non-reducing end and liberate xylose. Most of the xylanases characterized are able to hydrolyze different types of xylan showing only differences in the spectrum of end products [156]. The main products formed from the hydrolysis of xylan are xylobiose, xylotriose, and higher oligomers and the distribution of the products depend on the mode of action of the individual xylanases. In practice, the most important characteristics of xylanases are their high specific activity as well as high temperature stability and activity. For example, the microorganisms *Trichoderma longibrachiatum*, *Basillus pumilus*, *Thermoanaerobacter ethanolicus*, *Schizophyllum commune*, and *Aspergillus nidulans* have a high activitity for xylanases [104, 157, 158].

Mannanases 1,4-β-D-mannanases (EC 3.2.1.78) catalyze the random hydrolysis of D-mannopyranosyl β-(1→4)-linkages within the main chain of mannan and various, mainly mannose-containing polysaccharides, such as glucomannan, galactomannan, and galactoglucomannan, whereas 1,4-β-D-mannosidases (EC 3.2.1.25) cleave mannooligosaccharides to mannose [40, 91, 113, 155]. It also seems that mannanases form a more heterogeneous group of enzymes than xylanases. Additionally, some mannanases have been found to have a multidomain structure similar to that of several cellulolytic enzymes; *i.e.*, the protein contains a catalytic core domain, which is separated by a linker from a CBD. The main enzymatic hydrolysis products from galactomannan and glucomannan are mannobiose, mannotriose, and mixtures of higher oligosaccharides. The hydrolysis yield of mannan is dependent on the degree of substitution as well as on the distribution of the substituents, although the ratio of glucose to mannose in gluco-mannans also affects their hydrolysis [159]. For example, the microorganisms *Basillus subtilis*, *Sclerotium rolfsii*, *Aspergus niger*, and *Pyrococcus furiosus* have a high activitity for xylanases [104, 157, 158].

The side groups connected to xylan and glucomannan main chains are removed by accessory enzymes, such as α-glucuronidase (EC 3.2.1.131), α-arabinosidase (EC 3.2.1.55), and α-D-galactosidase (EC 3.2.1.22) [40]. Acetyl substituents of xylans are re-moved by esterases (EC 3.1.1.72). However, there is a great variety of carbohydrate side group-cleaving enzymes differing in their specificity and protein properties. Such enzymes are needed for reaching a high degree of monosaccharides in the total hy-drolysis of soluble or insoluble hemicelluloses. The synergism between various hemi-cellulolytic enzymes can be also observed as the accelerated action of endoglucanases in the presence of accessory enzymes.

In enzymatic hydrolysis, compared to acid hydrolysis, high substrate conversion yields are possible, although the hydrolytic processes are quite slow, and the costs of enzymes are typically relatively high for being purchased or produced on-site [40, 160, 161]. In general, hydrolysis yields depend on the type and pretreatment of the substrate, the type and dosage of the enzyme, and the hydrolysis time. The main parameters affect-ing the hydrolysis are shown in Table 3.3. An increasing number of commercial prepara-tions designed for the hydrolysis of lignocellulosic biomass materials are also available. For example, the high cost of cellulases has traditionally been considered to be the main bottleneck in commercializing lignocellulosics-based CH_3CH_2OH production processes, although many significant improvements have been achieved from the 1990s [162].

Table 3.3: The main parameters having influence on enzymatic hydrolysis [40].

Enzyme-related parameters	Process-related parameters
Composition of enzyme mixtures and specific activity	Raw material and pretreatment
Thermal stability and activity of enzymes	Enzyme loading and agitation
Synergism between different enzymes	Raw material consistency
End-product inhibition	Hydrolysis temperature
Adsorption (*i.e.*, irreversibilty and unspecific adsorption of enzymes on lignin)	Residence time

The type of raw material affects most significantly the yields of enzymatic hydrolysis [40]. The highest conversion is typically obtained with extensively delignified lignocel-lulosic biomass materials, such as kraft pulp or with hydrothermally pretreated agri-cultural residues with a low content of hemicelluloses. The process conditions to be used are determined by the optimum conditions of the enzymes, which may limit the process options; most cellulases have a maximum activity at 50 °C and at pH 4–5. However, for example, novel thermostable enzymes have been developed, which allow more flexibility in the process configurations [163]. The cellulase loadings usually range from 5 FPU to 10 FPU/g of substrate (or cellulose). FPU is a standard activity unit measured against filter paper, and β-glucosidase is usually supplemented for a separate hydrolysis.

3.3 Use of carbohydrates in hydrolyzates

Traditional organic chemical industry uses various physicochemical processes to convert common fossil raw materials into more valuable products that often are more oxidized than the raw materials [164–167]. However, in the use of lignocellulosic biomass raw materials, one of the advantageous features, compared to fossil feedstocks, is that these raw materials characteristically contain reactive oxygen-containing functional groups: hydroxyl (-OH), aldehyde (-CHO), carbonyl (>C=O), and carboxyl (-CO$_2$H) groups [1–3, 9, 10, 39, 65, 168, 169].

The global chemical industry is highly competitive and new processes in the utilization of lignocellulosic biomass resources are steadily introduced and hence, the situation is gradually changing [65]. In the past, the chemical industry has, for example, not considered biotechnology as an appropriate field for diversification, or as an effective tool for feedstock conversion to industrial chemicals. However, today, it can be with good reason claimed that in several cases, lignocellulosic biomass feedstocks processed by biotechnical and/or chemical conversion technologies have, due to extensive research, already acquired a firm, but somewhat limited position in large-scale production [40].

In the present new era, the Green Industrial Revolution, when designing novel processes, products, and materials that are based on renewable resources, some basic principles must be generally considered [65, 170]. These principles focus on evaluating and improving current procedures together with developing new ones that will have a limited impact on the environment and are in the long run, more economical and sustainable. The key issues of green chemistry include:

- the use of renewable feedstocks (*i.e.*, safer chemicals and auxiliaries as well as selective catalysts) rather than depleting one
- the use of well designed and straightforward synthesis with minimum energy requirements and environmental impacts
- the development of real-time (on-line) analyses for the production
- the manufacture of degradative (biodegradable) in the environment
- the prevention of possible chemical accidents in the production

The additional principles concerned with green technology primarily comprise, besides the use of renewable material and energy inputs, the maximization of process efficiency (*i.e.*, with respect to feedstock and chemical requirements, process conditions, and energy consumption) and, on the other hand, the minimization of waste formation.

There have been indications about some important building-block chemicals (*i.e.*, platform chemicals) that can be produced from lignocellulosics-derived carbohydrates [6]. All of these platform chemicals are derived from sugars on a biochemical biorefinery platform and are supposed to be competitive with existing petrochemicals. The suggested chemicals comprise alditols (glycerol, sorbitol, xylitol, and arabitol), monoacids (3-hydroxypropanoic acid and 3-hydroxybutyrolactone), and diacids (fumaric, malic, succinic, itaconic, 2,5-furandicarboxylic, glucaric, levulinic, aspartic, and glutamic acids). Additionally, a similar report dealing with the use of lignin has been published [13].

Biochemical transformations of lignocellulosic biomass materials *via* hydrolysis to sugars followed by fermentation to chemicals account for many important routes, but chemical transformations predominate in the conversion of these key chemicals into a wide range of derivatives and intermediates [40, 65, 170]. However, several current technologies for the biological conversion of renewable resources are still at an experimental stage. With the rapid advance of biotechnology, better biocatalysts can most likely be developed to overcome some of the shortcomings of the existing ones. Additionally, better bioreactors can be designed for enhancing productivity, which also in many cases especially requires more efficient product recovery systems; *i.e.*, separation and purification of products from the broth.

This subchapter describes only some illustrative examples of the production of chemicals and their significant derivatives from lignocellulosic biomass after hydrolysis. However, it is obvious that an enormous number of chemicals can be manufactured from lignocellulosic-derived sugars, not only by traditional chemical conversion but also by biochemical conversion, which also seems to be gaining increasing importance. Hence, the following description is mainly intended to be tutorial in nature with the objective of showing only a broad perspective on the versatile possibilities rather than giving any detailed data, for example, on the manufacturing processes. Furthermore, in many cases, the price of carbohydrates based on different agricultural resources or residues from the wood industry has often been prohibitive for the economic production of several chemicals. This practically means that so far, some common chemicals have not been able to compete commercially with those produced synthetically in the petrochemical industry.

There are two types of processes in the production of a great number of primary glucose-based chemicals by fermentation (Figure 3.2), representing a substantial volume of today's chemical industry [39, 65]: *i*) one type consists of well-established anaerobic or aerobic production of simple chemicals, such as various alcohols and carboxylic acids and *ii*) the other type of fermentation is aiming at more complex chemicals, such as antibiotics, enzymes, and hormones. For clarity, glucose is selected in Figure 3.2 as an example of feedstock monosaccharides, but the analogous or the same products can be produced from other monosaccharides as well. Additionally, an example of typical biochemical modification routes is enzymatic isomerization in the production of starch-based sweeteners, so-called "high fructose syrups". These sugar mixtures are replacing sucrose in such applications where it is possible to for the sweetener to be added in the form of an aqueous solution.

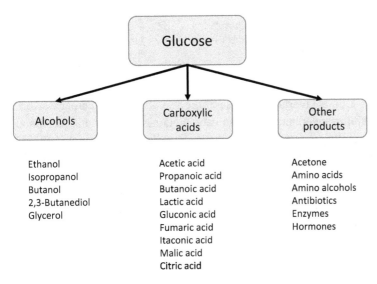

Figure 3.2: Examples of important fermentation products from glucose [39].

In principle, the conversion of cellulose *via* monomers to polymers with properties totally different from those of the initial polysaccharides is possible (Table 3.4). These platform chemicals are employed, besides as monomers for polymer synthesis, also as precursors for production of other chemicals, for example, by adding oxygen- or nitrogen-containing functional groups [171]. One of the obvious chemicals, gluconic acid, has been excluded because it is mainly used as such in the food industry and in many other applications as its sodium salt.

In the chemical industry, synthetic CH_3CH_2OH has typically been produced from the petrochemical feedstock by the acid-catalyzed hydration of ethylene over a H_3PO_4 at about 300 °C with high-pressure steam [105, 166]:

$$H_2C{=}CH_2 + H_2O \rightarrow CH_3CH_2OH \tag{3.1}$$

However, today, the vast majority of CH_3CH_2OH is based on the lignocellulosic biomass resources. There are several drivers, such as the increasing demand of oil, the decreasing reserves of fossil-based carbon sources, and global warming, to produce CH_3CH_2OH (in this chapter the term "bioethanol" is systematically used throughout the text below) and other energy carriers for transport [40]. For example, the greenhouse gas (GHG) balance of sugarcane (*i.e.*, the main production raw material) bioe-

Table 3.4: Examples of plastic industrial monomers and their polymers based on glucose fermentation products [39].

Fermentation products	Monomers	Polymers
Ethanol	Ethylene, butadiene	Polyethylene, polybutadiene
Butanol	Vinyl chloride, chloroprene	Polyvinyl chloride, neoprene
2,3-Butanediol	Vinyl acetate	Polyvinyl acetate
Acetic acid	Acrylic acid, acrylonitrile	Acrylic resins, polyacrylonitrile
Lactic acid	Styrene, methyl methacrylate	Polystyrene
Acetone	Bisphenol A	Epoxy and polycarbonate resins

thanol [71, 100, 172–175], is favorable because of the use of the residual bagasse (*i.e.*, both sucrose-containing sugarcane juice and cane molasses are used for bioethanol production) for generating energy. Similarly, a plant producing bioethanol from typical other lignocellulosic biomass substrates, such as agricultural crop residues [73, 101, 104–107, 176–180] could meet almost all its energy requirements, including process heat and electricity, by burning process residues, mostly lignin. In many cases, several starch-containing agricultural crops are also directly suitable for bioethanol production but due to their primary value of food and feed, these crops are unable to meet the global demand for bioethanol production [175]. Additionally, effects of integrating a bioethanol production process into a kraft pulp mill has been studied [181–183].

The processes for production of bioethanol from cellulosic materials consist of several steps [40, 175]. They include *i*) collection and storage of raw feedstock, *ii*) size reduction and pretreatment or fractionation of biomass components to make the feedstock material amenable to hydrolysis, *iii*) enzyme production, *iv*) hydrolysis (also acid hydrolysis), *v*) separation of the sugar solution from the residual materials (*e.g.*, lignin), *vi*) microbial fermentation of the sugar solution, *vii*) product recovery (*i.e.*, distillation to produce roughly 95% pure bioethanol and possible dehydration by molecular sieves to bring the bioethanol concentration to over 99.5%), *viii*) energy and steam generation and recycling, *ix*) waste treatment, and *x*) recovery and utilization of by-products. These steps are in principle the same for producing any other biotechnical product from lignocellulosic biomass materials.

Bioethanol production as such is relatively inexpensive [40, 101, 172]. When considering the entire process, the most important characteristics of the biological conversion system to be selected are the type of substrates that can be used, the bioethanol yield (typically below 15%), and the achieved bioethanol concentration. However, the technologies for producing bioethanol from sugar juices (from cane, beet, or even sulfite spent liquors) differ totally from those used to convert lignocellulosic raw materials. Batch process systems currently used for producing bioethanol from sugar juices typically have a volumetric bioethanol productivity of 2–3 g/L in one hour. In contrast, continuous stirred-tank reactors have somewhat higher productivity, and the systems employing immobilized cells or bioethanol removal can achieve a productivity of 100 g/L in one hour. A high final bioethanol concentration has been a major factor in reducing distillation costs in purification. In lignocellulosic biomass materials-based processes, the high solids content in the production is normally the main bottleneck limiting the feasibility and economics of the process. Nowadays, new and more attractive separation techniques, for example, such as vacuum distillation, solvent extraction, membrane reactors, and gas stripping, have been developed for removing bioethanol during or after fermentation.

The bioethanol-producing organism should be able to ferment all the monosaccharides present and furthermore, withstand the potential inhibitors in the hydrolysate [40, 65]. The fact is that conventional yeast strains are able to produce bioethanol only from hexoses; *i.e.*, glucose, mannose, and galactose. However, some pentose-fermenting microorganisms are found in Nature among bacteria, yeasts, and fungi but the bioethanol yields and production rates are often not fully comparable to those from hexoses [102, 178, 184–187]. Hence, by means of modern genetic engineering, new fermenting organisms (yeasts or bacteria), such as engineered robust *Saccharomyces cerevisiae* yeast strains and bacterial strains of *Escherichia coli, Klebsiella oxytoca*, and *Zymomonas mobilis* have been constructed to achieve more efficient conversion of the sugars present in biomass-derived hydrolyzates. Due to the obstacles to efficient fermentation of pentoses to bioethanol, other uses of the pentose fractions have also been considered.

The materials from the acidic pretreatment process contain various degradation products, mostly consisting of the breakdown products of lignin and sugars, including furans as well as HCO_2H and CH_3CO_2H [39, 40]. Most of these components are toxic to microorganisms and slow down the subsequent fermentation process. However, efficient and tolerant yeast strains based on adaptation and selection have been introduced. In most cases, this detoxification is a significant cost factor and a number of different detoxification methods have also been developed, such as treatments with lime or charcoal [187].

There are typically three major types of process used to convert solid, pretreated lignocellulosic biomass materials to sugars from cellulose and hemicelluloses, and further to bioethanol (or other compounds) [40, 101, 102, 172, 178]: separate hydrolysis and fermentation (SHF), simultaneous saccharification and fermentation (SSF), simultaneous saccharification and co-fermentation (SSCF), and consolidated bioprocessing (CBP) or direct microbial conversion (DMC), although additional process modifications are frequent. The main advantage of the SHF technology is that its each step is performed at the optimum temperature; 45–50 °C for currently available enzymes and 30–35 °C for the fermenting organism. In the SSF technology, both cellulose hydrolysis and fermentation of glucose are carried out in a single step at 37–38 °C. It seems to be the most promising approach to convert biotechnically lignocellulosic biomass materials into bioethanol, and its main benefits comprise the enhanced rate of cellulose hydrolysis, due to the continuous removal of the product sugars, which inhibit the cellulase activity. This practice results in decreased enzyme loading and higher product yields. In the SSCF technology, the pretreated lignocellulosic biomass material is neutralized and directly exposed to different enzymes and microorganisms, which are capable of hydrolyzing cellulose and hemicelluloses to fermentable sugars as well as ferment them to bioethanol. The SSCF technology is slightly superior to the SSF technology in terms of cost effectiveness, better yields, and shorter processing time. The CBP technology includes all the three processes, $i.e.$, enzyme production, hydrolysis, and fermentation, in a single process step and in one reactor. The main disadvantages are the relatively low capability of the organisms to produce enzymes under the anaerobic fermentation conditions as well as the low bioethanol tolerance and production rate of many organisms.

The first biochemical synthesis of $CH_3CH_2CH_2CH_2OH$ (1-butanol or n-butanol, in this chapter the term "biobutanol" is systematically used throughout the text below) was already made by Louis Pasteur in 1861 when he isolated a butyric-acid-forming bacterium and named it $Vibrion$ $butyrique$ [65, 188–194]. In 1915 in England Chaim Weizmann patented the process producing acetone (CH_3COCH_3) and biobutanol by the bacterium $Clostridium$ $acetobutylicum$ and this process was later known as "ABE (acetone-butanol-ethanol) fermentation". At that time CH_3COCH_3 was needed for the manufacture of the smokeless gunpowder, cordite, but World War I meant that England was unable to import large amounts of calcium acetate, which was the feedstock for the production of CH_3COCH_3. The traditional ABE fermentation with $C.$ $acetobutylicum$ or $C.$ $beijerinckii$ is among the first industrial-scale fermentation processes, and at the beginning of the 20th century, it was already one of the most important fermentation processes after the bioethanol fermentation by yeast. A further modification of the ABE fermentation process is to produce 2-propanol or isopropanol ($CH_3CH(OH)CH_3$) instead of CH_3COCH_3 together with biobutanol [195, 196].

Anhydrous bioethanol is an effective octane booster for spark ignition internal combustion engines (gasoline engines), and in blends of 5–30% no engine modification requirements are needed [105–107, 174, 189]. It also provides additional oxygen in combustion, and when blended with gasoline it burns, compared to pure gasoline, relatively more completely with lower carbon monoxide (CO) and hydrocarbon emissions. Additionally, the lower boiling point of bioethanol also leads to better combustion efficiency. However, compared to gasoline, bioethanol has lower energy density and vapor pressure affecting the fuel consumption. In contrast, due to its low cetane number, bioethanol does not burn efficiently by compression ignition and is moreover not easily miscible with diesel fuel. On the other hand, biobutanol is also an excellent fuel extender being, for example, superior to bioethanol. It contains only 22% oxygen (in bioethanol 35%), is able to tolerate H_2O contamination, and is easy to blend with gasoline in higher concentrations without harm to engines [65]. Besides its favorable properties when used currently as a fuel with many advantageous properties, biobutanol is suitable as a solvent, as such or as an ester ($e.g.$, as butyl acetate ($CH_3CH_2CH_3CH_2OCOCH_3$)) [171, 191].

Besides bioethanol and biobutanol, there is a great number of other alcohols (mainly polyalcohols) that can be produced from carbohydrates-containing raw materials either by biotechnical or chemical methods [40, 65, 170, 188, 197, 198]. Examples of these major alcohols and polyalcohols include methanol (CH_3OH), 1,3-propanediol, 2,3-butanediol, glycerol or 1,2,3-propanetriol, xylitol, L-arabitol, D-glucitol or D-sorbitol, and D-mannitol (Figure 3.3). Petroleum-derived poly(ethylene glycol) (PEG) $\left(HO[CH_2CH_2\text{-}O\text{-}]_nH\right)$ and poly(propylene glycol) (PPG) $\left(HO[CH(CH_3)CH_2\text{-}O\text{-}]_nH\right)$ are polyether substances with a wide range of applications.

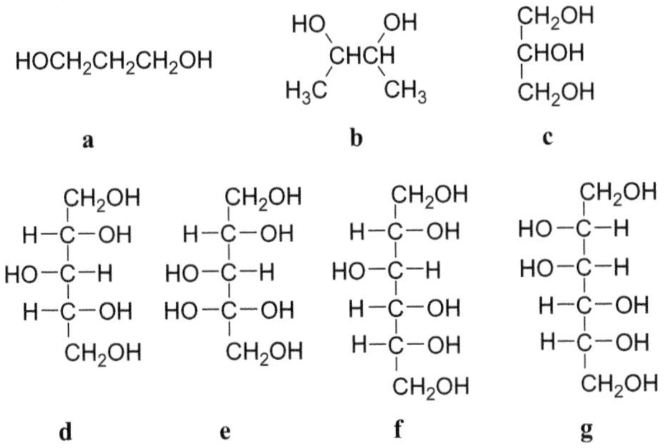

Figure 3.3: Examples of the major alcohols (besides methanol, bioethanol, 2-propanol or isopropanol, and biobutanol, see the text) that can be produced from lignocellulosic biomass. 1,3-Propanediol (a), 2,3-butanediol (b), glycerol or 1,2,3-propanetriol (c), xylitol (d), L-arabitol (e), D-glucitol or D-sorbitol (f), and D-mannitol (g).

Glycerol is mainly produced by the saponification of various fats (*i.e.*, esters of glycerol and fatty acids) in vegetable oils, representing an inevitable by-product when manufacturing free fatty acids or biodiesel [39, 52], but it is also formed in certain aerobic fermentations [199, 200], and can be used as an attractive substrate for further fermentations. It has a wide range of applications in the foodstuffs, pharmaceutical, and chemical industries. Additionally, it would be possible to manufacture, for example, glyceric acid, 1,3-propanediol, propylene glycol, and diglyceraldehyde from glycerol [6, 201].

Several aliphatic carboxylic acids can be produced, besides by petrochemical methods, by different biotechnical methods, representing a natural route to their manufacturing from renewable raw materials [40, 65]. However, in many cases, the acid produced inhibits its biotechnical production and should be continuously removed or neutralized. Common aliphatic acids have a great number of industrial applications and, for example, certain multifunctional acids (*e.g.*, low-molar-mass hydroxy carboxylic acids) in kraft black liquors (see Section 4.2.2 "Reactions of the wood chemical constituents") could be utilized as feedstocks for "green chemicals" and biodegradable polymers. Figure 3.4

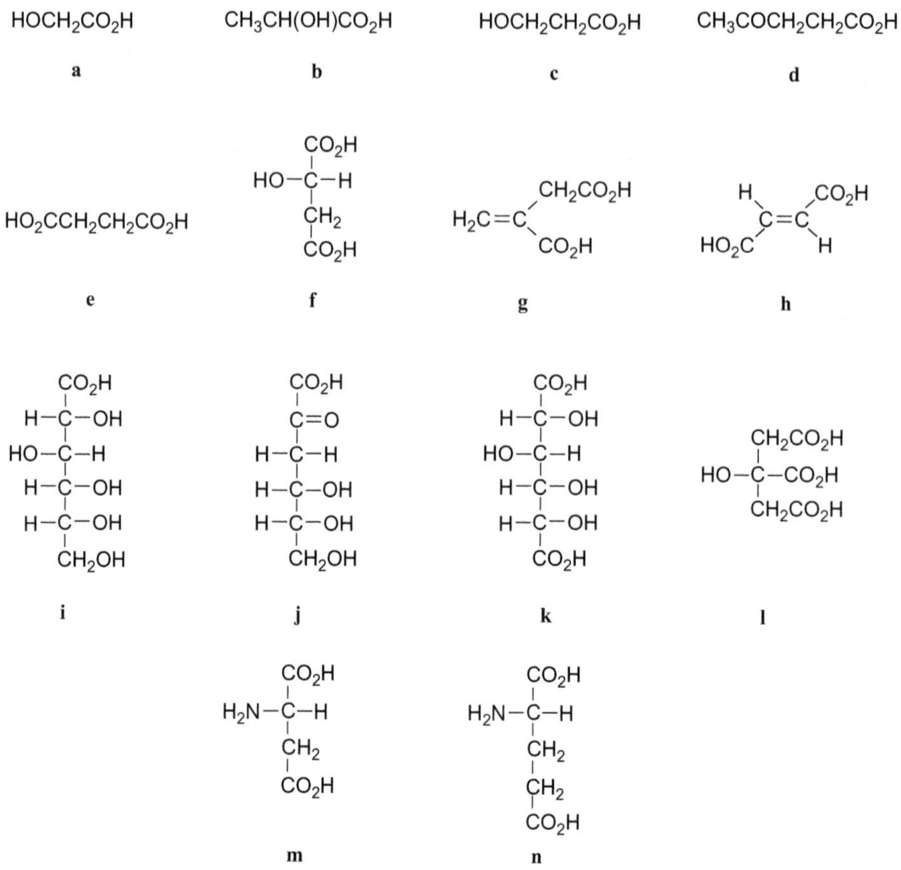

$HOCH_2CO_2H$ (a) $CH_3CH(OH)CO_2H$ (b) $HOCH_2CH_2CO_2H$ (c) $CH_3COCH_2CH_2CO_2H$ (d)

$HO_2CCH_2CH_2CO_2H$ (e)

(f)
$$
\begin{array}{c}
CO_2H \\
HO-C-H \\
CH_2 \\
CO_2H
\end{array}
$$

(g)
$$
H_2C=C\begin{array}{l} CH_2CO_2H \\ CO_2H \end{array}
$$

(h)
$$
\begin{array}{c}
H \quad\ CO_2H \\
C=C \\
HO_2C \quad\ H
\end{array}
$$

(i)
$$
\begin{array}{c}
CO_2H \\
H-C-OH \\
HO-C-H \\
H-C-OH \\
H-C-OH \\
CH_2OH
\end{array}
$$

(j)
$$
\begin{array}{c}
CO_2H \\
C=O \\
H-C-H \\
H-C-OH \\
H-C-OH \\
CH_2OH
\end{array}
$$

(k)
$$
\begin{array}{c}
CO_2H \\
H-C-OH \\
HO-C-H \\
H-C-OH \\
H-C-OH \\
CO_2H
\end{array}
$$

(l)
$$
\begin{array}{c}
CH_2CO_2H \\
HO-C-CO_2H \\
CH_2CO_2H
\end{array}
$$

(m)
$$
\begin{array}{c}
CO_2H \\
H_2N-C-H \\
CH_2 \\
CO_2H
\end{array}
$$

(n)
$$
\begin{array}{c}
CO_2H \\
H_2N-C-H \\
CH_2 \\
CH_2 \\
CO_2H
\end{array}
$$

Figure 3.4: Examples of the major aliphatic carboxylic acids (besides acetic acid or ethanoic acid, propionic acid or propanoic acid, and butyric acid or butanoic acid, see the text) that can be produced from lignocellulosic biomass. Glycolic acid or hydroxyethanoic acid (a), lactic acid or 2-hydroxypropanoic acid (b), 3-hydroxypropanoic acid or β-lactic acid (c), levulinic acid or 4-oxopentanoic acid (d), succinic acid or butanedioic acid (e), L-malic acid or L-hydroxysuccinic acid (f), itaconic acid or methylidenesuccinic acid (g), fumaric acid or *trans*-butenedioic acid (h), D-gluconic acid (i), 2-oxogluconic acid or 2-keto-D-gluconic acid (j), D-glucaric acid or saccharic acid (k), citric acid or 2-hydroxy-1,2,3-propanetricarboxylic acid (l), L-aspartic acid (m), and L-glutamic acid (n).

shows examples of the typical acids; CH_3CO_2H, propionic acid or propanoic acid $(CH_3CH_2CO_2H)$, and butyric acid or butanoic acid $(CH_3CH_2CH_2CO_2H)$ are not included.

The production of CH_3CO_2H as a secondary fermentation product *via* the oxidation of the primary fermentation product, bioethanol, by bacteria, such as *Acetobacter* species (e.g., *A. suboxydans*) under aerobic conditions is probably the oldest organic acid realized by fermentation [3, 65, 171, 199, 202]. The present biotechnical production of CH_3CO_2H as vinegar by *A. aceti* (can make vinegar up to 14% CH_3CO_2H) from various substrates accounts for only 10% of world production. Additionally, specific anaerobic bacteria (*e.g., Clostridium thermoaceticum* and *C. lentocellum*) can convert sugars directly into this acid, without using bioethanol as an intermediate [203]. CH_3CO_2H is used as a solvent but is also a chemical feedstock when producing vinyl acetate, acetic acid anhydride, and monochloroacetic acid (MCA) as well as various salts and esters, including ethyl, *n*-butyl, isobutyl, and propyl acetates [65]. The trivial name "glacial acetic acid" is used for H_2O-free CH_3CO_2H.

α-Hydroxy acids (AHAs) are a class of chemical compounds that contain a carboxylic acid ($R-CO_2H$) group substituted with a -OH group on the adjacent carbon [204, 205]. Lactic acid (Figure 3.4) is the most important AHA [9, 65, 171, 206–208]. It can be found in many processed food products naturally (*e.g.*, in tomato juice) or as a product of *in situ* microbial fermentation, particularly in sour milk products, and it is also responsible for the sour flavor of sourdough breads. It was first discovered in 1780 by Carl Wilhelm Scheele who found in milk an acid ("milk acid") that could be purified by crystallizing its calcium salt. Lactic acid can be manufactured on an industrial scale by chemical synthesis or carbohydrate fermentation. It exists in three different forms: *i*) optically active D-(-)-lactic acid (*R*-lactic acid), *ii*) optically active L-(+)-lactic acid (*S*-lactic acid), and *iii*) optically inactive (a racemic mixture) D,L-lactic acid (Figure 3.5).

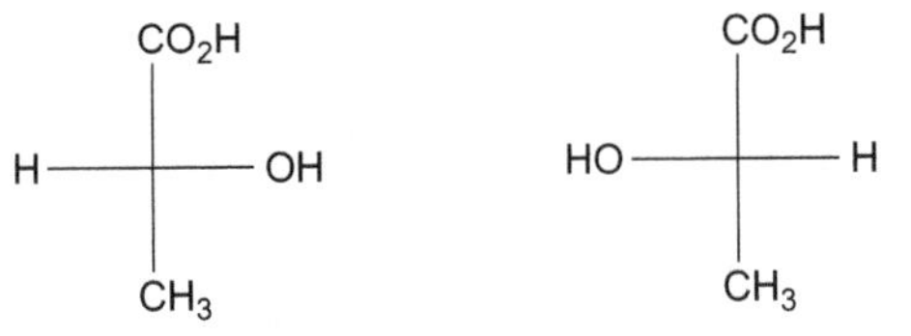

Figure 3.5: Optical isomers of lactic acid. D-(-)-Lactic acid or (*R*)-(-)-lactic acid and L-(+)-lactic acid or (*S*)-(+)-lactic acid.

The most commonly used synthetic method for producing D,L-lactic acid is the hydrolysis of lactonitrile derived from acetaldehyde and hydrogen cyanide [207, 209]. Currently, the manufacture of lactic acid is, however, almost exclusively based on carbohydrate fermentation in the form of L- or D-lactic acids; *i.e.*, the fermented medium contains one lactic acid stereoisomer as pure (or its salt, lactate) or is a mixture of both stereoisomers (or their lactates) in different proportions. The commercial production processes mainly use microorganisms, such as *Lactobacillus delbrueckii*, *L. amylophilus*, *L. bulgaricus*, *L. leichmanii*, *L. helveticus*, and *L. paracasei* [65, 206, 207, 209–217]. The choice of organism depends primarily on the carbohydrate feedstock to be fermented. Several techniques can be used for separating the lactic acid product from the fermented medium, for example, including extraction by solvents and separation with ion-exchange resins and membranes as well as separation by electrodialysis, adsorption, or vacuum distillation (also as ethyl lactate). The production of lactic acid from sulfite spent liquor by fermentation has also been studied [218].

Lactic acid is traditionally used in food and food-related applications as an acidulant, a food preservative, and a feedstock for the manufacture of various esters of lactate salts with fatty acids (*e.g.*, calcium and sodium stearoyl-2-lactylate, glyceryl lactostearate, and glyceryl lactopalmitate) suitable for emulsifying agents in baking foods [9, 40, 65, 206–209, 219–222]. Lactic acid and ethyl lactate also have several uses in pharmaceutical and cosmetic formulations (*e.g.*, in topical ointments, lotions, and parenteral solutions) and are used in detergents as well. Environmentally friendly "green solvents" based on lactic acid esters with low-molar-mass alcohols, such as bioethanol, propanol, and biobutanol, are a potential growth area. Technical lactic acid grades are generally used as an acidulant in vegetable and leather tanning industries. Additionally, because of its advantageous chemical properties, lactic acid has the potential to become a large-volume chemical intermediate in producing different "oxygenated chemicals". Examples of such chemicals derived from lactic acid are acetaldehyde or ethanal, acrylic acid or propenoic acid, propionic acid or propanoic acid, propylene oxide or 1,2-epoxypropane, propylene glycol or 1,2-propanediol, and acetylacetone or 2,3-pentanedione. One industrially relevant microbial product is an isomer of lactic acid, 3-hydroxypropanoic acid [223], which can be used for producing acrylic acid, methyl acrylate, acrylamide, and 1,3-propanediol [6].

In general, "biodegradable plastics" are considered plastics that can be decomposed within reasonable time by the action of living organisms, usually microbes, into CO_2, H_2O, and some organic residues [224–227]. These plastics are typically produced with renewable feedstock materials, microorganisms, petrochemicals, or their combinations. Hence, there might be no intrinsic difference between the biodegradability of bio-based polymers and those produced synthetically; for example, natural rubber, *cis*-poly(isoprene) from the rubber tree and the same synthetic polymer manufactured from petrochemical feedstocks. Other illustrative examples include the well-known biodegradable polyesters poly(caprolactone) (PCL, from ε-caprolactone *via* the ring opening polymerization) and poly(butylene succinate) (PBS, from succinic acid and 1,4-butanediol *via* the direct esterification). On the other hand, the general term "bioplastic" is also somewhat unspecified for the product originated from natural raw materials (*e.g.*, from lignocellulosic biomass materials) and misleading because it suggests that any polymer derived from the biomass is environmentally friendly. Thus, a better term for these products might be "bio-based plastics". In contrast, the term "biopolymer" refers only to natural polymers, such as cellulose, hemicelluloses, and proteins.

Poly(lactic acid) (PLA) $(CH_3CH(OH)CO[-OCH(CH_3)CO-]_nOCH(CH_3)CO_2H)$ is a rigid bio-based thermoplastic polymer (*i.e.*, is remoldable with heat and pressure) that can be semicrystalline or totally amorphous depending on the stereopurity of the polymer backbone [228–239]. It was the first polymer produced from renewable resources and behaves like the most common petrochemicals-derived thermoplastic resin of the polyester family, poly(ethylene terephthalate) (PET) $(H[-O_2C-Bz-CO_2-CH_2CH_2-]_nOH$, where Bz is a benzene ring), but also performs a lot like the polyolefin poly(propylene)

(PP) ($[-CH(CH_3)CH_2-]_n$). However, despite many favorable properties and a high consumption volume, perhaps it cannot be considered fully a commodity polymer, because of its certain physical and processing shortcomings.

As lactic acid, like glycolic and 2-hydroxybutanoic acids, has both -OH and -CO$_2$H functional groups, on heating it undergoes condensation reactions by removal of H$_2$O, which in the first stage of this reaction, leads readily to linear (dimers and trimers) and cyclic (dimers) structures [240]. Despite this characteristic property, there are still many different reaction routes to produce the high-molar-mass PLA (M$_w$ > 100,000 g/mol) and which depend on the accurate reaction conditions selected (*i.e.*, temperature, pressure, and pH) [229, 238]. Three generally applied routes are as follows: *i*) in the first phase, various PLA prepolymers with a low molar mass (M$_w$ 2,000–10,000 g/mol) are produced by conventional condensation and then these short chains are linked with each other using coupling agents (*e.g.*, 1,6-hexamethylene diisocyanate) [238], *ii*) the high-molar-mass PLA is directly produced by azeotropic dehydrative condensation [236], and *iii*) first, several PLA prepolymers with a low molar mass (M$_w$ 1,000–5,000 g/mol, DP <100) are produced by condensation and then these prepolymers are depolymerized to a cyclic dimer (lactide or 3,6-dimethyl-1,4-dioxane-2,5-dione, sometimes called also as "dilactide") *via* transesterification by backbiting reaction; this lactide structure is a monomer for the production of the final PLA when its ring structure is catalytically (usually with tin(II) 2-ethylhexanoate) opened and the opened moieties are quickly joined together (*i.e.*, by a ring-opening polymerization) now forming long PLA chains [229, 231, 235, 236, 238]. Of these alternatives the last reaction route *iii*) is the most used on industrial scale, and by which the molar mass of PLA can also be controlled in the most reliable way.

A lactide ring formed in the reaction route *iii*) has, however, depending on the stereochemical structure of lactic acid (Figure 3.5), two chiral centers and the formation of three optical alternatives is possible (Figure 3.6). The structure of the monomer lactides essentially affects the structure of the PLA chain. If the lactide intermediate is the mixture of D- and L-lactides, it is called "racemic lactide" (or rac-lactide or D,L-lactide) [238]. On the other hand, if the lactides are pure D-type or L-type, the PLA formed is called "PDLA" and "PLLA", respectively. In case of rac-lactides, the product is called "PDLLA". In practice, in most cases, the PLLA and PDLLA products are normally used [234, 241].

PDLA and PLLA have different characteristics and their mutual ratio (*i.e.*, the used ratio of D- and L-lactides in the production) determines the properties of this product, which also allows PLA to be tailored for specific applications [234, 235]. In general, PLA plastics (*i.e.*, containing D- and L-lactide units in varying proportions) are biodegradable, transparent, and have many characteristics similar to those of the thermoplastics now used in packaging consumer goods, and can be also used in a variety of common products [207, 232, 233, 242, 243]. In contrast, the L-isomer of lactic acid is the preferred monomer to produce lactide since it provides a high lactide yield and the high-molar-mass polymer PLLA with a high degree of crystallinity (about 38%) and tensile strength [207, 244]. PLLA has also the highest melting point of different PLA grades, and by add-

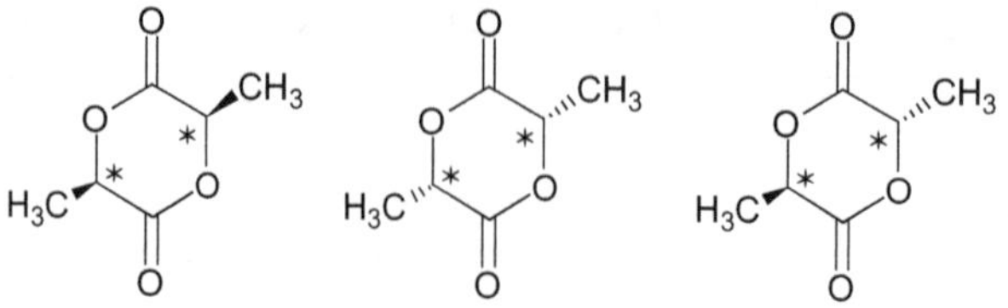

Figure 3.6: The three diastereomeric structures of lactide (3,6-dimethyl-1,4-dioxane-2,5-dione). D-Lactide or (R,R)-lactide (a), L-lactide or (S,S)-lactide (b), and *meso*-lactide or (R,S)-lactide (c). The asymmetric carbon atom is marked with an asterisk*.

ing the D-lactide monomer to its production, the melting point of the PLA polymer formed can be decreased; this also leads to a lower crystallinity, and when the PDLA content is higher than 12–15%, the end product turns amorphous [235, 245]. The PLLA biodegradable polymers can be used especially for medical applications, such as surgical sutures (a stitch or stitches), additives in controlled drug delivery, and surgical implants (prostheses); formerly, titanium and other suitable metals have been normally used in different orthopedic screws and plates, and because they are not degradable, they stay in the body and to remove them, another surgery is needed [239, 242, 246–248]. Furthermore, a fairly wide range of properties of PLA can be obtained by cooperation with monomers, such as glycolide (i.e., poly(lactide-co-glycolide) (PLGA), see below), ε-caprolactone (i.e., poly(lactide-co-ε-caprolactone) (PLCL)), and polyether polyols [207, 238, 242]. Other ways to alternate the properties of PLA are to change its structure by branching the molecule chains, adding nanoparticles to form nanocomposites, and coating with high barrier materials [234, 236].

Glycolic acid (Figure 3.4) is the smallest AHA and occurs naturally, for example, as a component of sugarcane and sugar beet juices [65, 171, 191, 205]. It is mainly obtained either by the acid-catalyzed reaction between methanal and CO and H_2O in the presence of acid catalysts at high pressure (>30 MPa) and temperature (eq. (3.2)) or by hydrolyzing molten chloroacetic acid with 50% aqueous NaOH at 90 °C (eq. (3.3)).

$$HCHO + CO + H_2O \rightarrow HOCH_2CO_2H \tag{3.2}$$

$$ClCH_2CO_2H + NaOH \rightarrow HOCH_2CO_2H + NaCl \tag{3.3}$$

Glycolic acid can also be produced from renewable resources by fermentation [249–251] as well as from ethylene glycol by *Gluconobacter oxydans* [252]. It is used in adhesives, metal cleaning, textile and leather processing, and as a component in personal-care products, but it also is an important intermediate for organic synthesis (i.e., oxidation, reduction, and esterification) [171, 205].

If only cyclic glycolide (1,4-dioxane-2,5-dione) [240] monomers are used in the condensation polymerization (see above for PLA and PLGA) of glycolic acid, the poly(glycolic acid) polymer (PGA) $\left(HOCH_2CO[-OCH_2CO-]_n OCH_2CO_2H \right)$ is formed [244]. Due to the excellent gas-barrier properties of PGA, it is as a lightweight material used, for example, as a coating material for PET bottles and other food and beverage packaging applications, and its (together with PLA) low inflammatory response with animal tests has also been confirmed [253]. PLA and PGA are biodegradable and their degradation in the environment takes place by ester bond hydrolysis [231, 238, 254]. Additionally, under industrial compost conditions, the natural enzymes (esterases, lipases, and proteases) excreted by various microorganisms participate in this biodegradation [236].

Polyhydroxyalkanoates (PHAs) are either thermoplastic or elastomeric linear polyesters, which are produced in Nature by several microorganisms and are of interest as biodegradable plastics [255–257]. The most common type of PHAs is the poly(hydroxybutyrate) (PHB) poly(3-hydroxybutyrate) (P3HB) $(H[-OCH(CH_3)CH_2CO-]_nOH)$, and examples of other polymers belonging to PHAs are poly(4-hydroxybutyrate) (P4HB) $((H[-OCH_2CH_2CH_2CO-]_nOH)$, poly(hydroxyvalerate) (PHV), poly(hydroxyhexanoate) (PHH), poly(hydroxyoctanoate) (PHO), and their copolymers. PHAs have many advantageous properties and potential application possibilities.

Succinic acid or amber acid (Figure 3.4) is probably the most common natural dicarboxylic acid, since it is an intermediate of the citric acid cycle or the tricarboxylic acid (TCA) cycle and one of the fermentation end-products of anaerobic metabolism, thus synthesized in almost all microbial, plant, and animal cells [40, 65, 191, 258, 259]. It can be manufactured by several alternative petrochemical-based processes [260]. However, it should preferably be produced by a fermentation process to obtain a high-purity product, which is essential, for example, when it is used as a raw material for polymer manufacture [261]. Succinic acid is one of many organic acids produced from glucose by anaerobic rumen bacteria. Examples of the most investigated microorganisms are *Actinobacillus succinogenes*, *Anaerobiospirillum succiniciproducens*, *Mannheimia succiniciproducens*, and recombinant *Escherichia coli* [40, 259, 260, 262, 263].

Succinic acid can be used as a platform chemical of several important specialty chemicals in the food, chemical, and pharmaceutical industries [40, 65, 171, 208, 259]. For example, it is a flavor-enhancing organic acid used in dairy products and fermented beverages and is also used for manufacturing surfactants and detergents, green solvents, and plastics. Due to its many different uses, the conventional production methods are likely to be gradually replaced by fermentable production systems. Examples of chemicals derived from succinic acid are 1,4-butanediol, 1,4-diaminobutane, succindiamine or succinamide, succinonitrile, tetrahydrofuran (THF), γ-butyrolactone (GBL) or 4-butanolide, 2-pyrrolidone or 4-butanelactam, and N-methyl-2-pyrrolidone or 1-methyl-2-pyrrolidone.

Itaconic acid (Figure 3.4) was conventionally obtained by the distillation of citric acid, but currently, it is produced by fermentation [40, 65, 206, 208, 264–266]. Itaconic acid anhydride is mainly produced by sugar fermentation using *Aspergillus terreus*, although many other microorganisms, including yeast strains belonging to the genus *Candida* and *Rhodotorula* as well as the smut fungus *Ustilago maydis* are also suitable for this purpose [40, 206, 266–270]. The chemical property that makes this dicarboxylic acid an important intermediate is the conjunction of its two -CO_2H groups and one methylidiene (=CH_2) group. Hence, for example, the =CH_2 group can engage in addition polymerization, leading to polymers with several free -CO_2H groups that give rise to advantageous properties. Itaconic acid is a versatile feedstock for the synthesis of polymers and its other significant reactions are esterification and the formation of N-substituted pyrrolidinones with amines. The latter products are used in detergents, shampoos, and other products in which their surface activity is needed. Examples of chemicals from itaconic acid include 2-methyl-1,4-butanediol, 2-methyl-1,4-butanediamine, itaconic acid diamide, 3-methyltetrahydrofuran, 3-methyl-γ-butyrolactone, 4-methyl-γ-butyrolactone, 3-methylpyrrolidine, 3-methyl-N-substituted-2-pyrrolidone, and 4-methyl-N-substituted-2-pyrrolidone.

Fumaric acid (Figure 3.4) occurs naturally and is normally produced from petroleum-based maleic acid by a catalytic isomerization process, but as the intermediate of the TCA cycle, it is often also found as a metabolic product produced by microorganisms and some mycelial fungi; particularly those belonging to *Rhizopus* are known to produce it from glucose under special cultural conditions [171, 206]. Due to one double bond (>C=C<) and two -CO_2H groups, it is a good starting material for chemical synthesis. Fumaric acid is used for making polyesters and other types of synthetic polymers and has several potential industrial applications.

Like fumaric acid, L-malic acid or (S)-malic acid (Figure 3.4) occurs naturally and is also an intermediate in the TCA cycle [171, 205, 206]. It has traditionally been recovered from apple juice in which it occurs in small quantities. In contrast, the large-scale production of the racemic malic acid (D,L-malic acid or (R,S)-malic acid) is based on the hydration of either maleic acid or fumaric acid. However, L-malic acid can be specifically obtained by enzymatic hydration of fumaric acid. Since L-malic acid (as well as D,L-malic acid) closely resembles both citric and tartaric acids in its physical and chemical properties together with a more neutral flavor than citric acid, it is used in the food industry as an acidulant (*i.e.*, as a citric acid replacement). Additionally, it is suitable to produce various ester and amide derivatives and has potential use as a raw material to manufacture biodegradable polymers.

Table 3.5 illustrates some straightforward examples of processes in making useful chemicals from D-glucose by well-known chemical processes [39, 65]. Hence, D-glucose can be oxidized either under acidic or alkaline conditions to mono- and dicarboxylic acids, or reduced to various polyhydroxy compounds by means of hydrogenation or hydrogenolysis [170, 271–274]. Other possibilities for its conversion include dehydration and isomerization [5, 24, 170, 275]. The production of furans (Figure 3.7), such as furfural and HMF, is also of importance. Levulinic acid (LEVA) (Figure 3.4) can be obtained, for example, from HMF *via* the liberation of HCO_2H during heating. In contrast, for example, the direct oxidation of cellulose by sodium metaperiodate ($NaIO_4$) leads to 2,3-dialdehyde cellulose (*i.e.*, the -CHO groups are formed from the carbon atoms C_2 and C_3 in an anhydroglucose unit) [276, 277].

Table 3.5: Examples of chemical products from D-glucose by common chemical treatments [39].

Process	Products
Oxidation	D-Gluconic acid, D-glucaric acid, and 2-keto-D-gluconic acid
Alkaline oxidation	Formic acid, glycolic acid, D-glyceric acid, D-erythronic acid, and D-arabinonic acid
Hydrogenation	D-Glucitol or D-sorbitol
Hydrogenolysis	Ethylene glycol, propylene glycol, and glycerol
Dehydration	5-(Hydroxymethyl)furfural[a]
Isomerization	D-Fructose

[a]Degrades into levulinic acid and formic acid on heating.

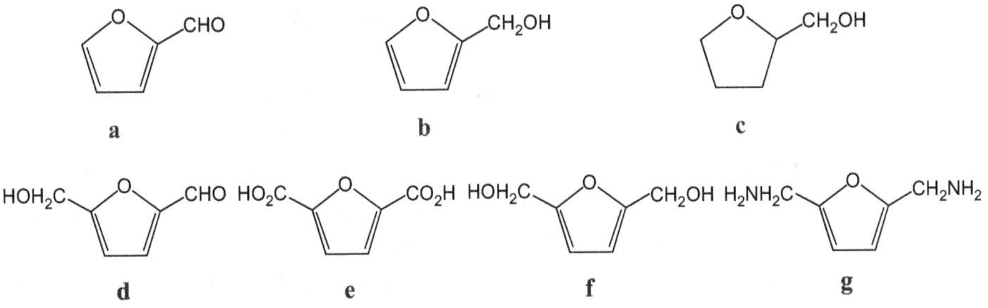

Figure 3.7: Examples of furan derivatives. Furfural (a), furfuryl alcohol or 2-furylmethanol (b), tetrahydrofurfuryl alcohol or 2-tetrahydrofurylmethanol (c), 5-(hydroxymethyl)furfural or 5-hydroxymethyl-2-furaldehyde (HMF) (d), 2,5-furandicarboxylic acid or furan-2,5-dicarboxylic acid (e), 2,5-bis(hydroxymethyl)furan (f), and 2,5-bis(aminomethyl)furan (g).

In recent years, furfural (Figure 3.7) has been recognized as an important bio-based platform chemical, and therefore, its valorization has been attracting increasing attention [278]. There is no petrochemical route for the economic production of furfural and its full-scale production is principally based on either in a single-stage process in which both the acid-catalyzed hydrolysis of xylan (i.e., from agricultural residues and hardwoods) and the dehydration of the liberated xylose take place at the same time in a single-stage process often at high temperatures (>160 °C) or in a two-stage process in which xylan is first hydrolyzed to xylose and then the isolated xylose is converted to furfural [279–287]. It is an important intermediate that finds wide applications in oil refining, plastics, pharmaceutical, and agrochemical industries. However, the most important use of furfural is focused on its catalytic hydrogenation to furfuryl alcohol (FFA) (Figure 3.7), which also has a wide range of applications [278, 288–297]. Examples of other chemicals derived from furfural include tetrahydrofurfuryl alcohol (THFFA) (Figure 3.7), 2-methylfuran, 2-methyltetrahydrofuran (MTHF), furan, tetrahydrofuran (THF), dihydropyran (DHP), tetrahydropyran (THP), cyclopentanone, furoic acid, maleic acid anhydride or 2,5-furandione, fumaric acid, furfurylamine, and polytetrahydrofuran or poly(tetramethylene ether)glycol $(HO[-CH_2CH_2CH_2CH_2O-]_nH)$ (PTMEG) from THF [5, 10, 39, 171, 278, 290, 294–302].

Hexosans or hexose sugars can be converted into HMF in an analogous way as furfural from pentosans or pentose sugars and it is also formed as by-product in the acid hydrolysis of lignocellulosic feedstocks [24, 40, 65, 275, 303–307]. Especially, fructose (a hexulose) derived either from sucrose (or isomerization of glucose) or polyfructans like inulin is the most suitable raw material for its industrial-scale production [5, 306, 308–311]. HMF is a versatile and important intermediate for many useful chemicals, and many of its derivatives can be used as fuel, fuel additives, and polymers [10, 312–314]. Examples of chemicals derived from HMF are 2,5-furandicarboxylic acid (FDCA) (Figure 3.7), 2,5-bis(hydroxymethyl)furan (Figure 3.7), 2,5-bis(aminomethyl)furan (Figure 3.7), LEVA (Figure 3.4, see the text below), and ethers of HMF [5, 10, 39, 171, 314].

Among the green chemical substances produced from lignocellulosic biomass materials, FDCA is one of the promising monomers, which is applicable to the plastic industries [315–318]. The copolymerization of ethylene glycol $(HOCH_2CH_2OH)$ and FDCA yields poly(ethylene furanoate) $(PEF)(H[-O_2C-Fr-CO_2-CH_2CH_2-]_nOH,$ where Fr is a furan ring), which is analogous to the reaction of $HOCH_2CH_2OH$ and the petroleum-derived terephthalic acid leading to the commodity PET $(H[-O_2C-Bz-CO_2-CH_2CH_2-]_nOH,$ where Bz is a benzene ring) (see the text above). PEF, like PET and poly(butylene terephthalate) (PBT), has many advantageous properties (*e.g.*, PEF acts as an effective gas- and H_2O-vapor barrier) especially prevalent in packaging [319–322]. Other interesting FDCA-based polyesters include, for example, poly(propylene furanoate) [323] (PPF) and poly(butylene furanoate) [324] (PBF).

LEVA has traditionally been prepared by treating lignocellulosic biomass with mineral acids, leading to low yields and a great number of unwanted side products, although in recent years, some improved methods have also emerged [10, 24, 39, 40, 65, 275, 325–331]. In principle, the process proceeds from the hexose sugars via the intermediate HMF, which, on prolonged heating, is decomposed to LEVA and HCO_2H. LEVA can be further cyclized to α- and β-angelica lactones [325, 330]. It is a potential petrol additive, but it is also considered as an important platform chemical, since it can be further converted to a variety of other potential chemicals, some of which are direct substitutes for petrochemicals [4, 47–50, 332, 333]. Hence, LEVA is a versatile precursor, for example, to synthetic rubbers, nylon-like polymers, and other plastics. Examples of potentially interesting derivatives of LEVA are acrylic acid or propenoic acid $(H_2C=CHCO_2H)$, β-acetylacrylic acid, 1,4-pentanediol, α-angelica lactone, γ-valerolactone or 4-pentanolide, and 2-methyltetrahydrofuran together with several esters of LEVA.

References

[1]   Herrick, F.W., and Hergert, H.L. 1977. Utilization of chemicals from wood: retrospect and prospect. In: Loewus, F.A., and Runecles, V.C. (Eds.). *The Structure, Biosynthesis, and Degradation of Wood, Recent Advances in Phytochemistry, Volume 11*. Plenium Press, New York, NY, USA. Pp. 443–515.

[2]   Goldstein, I.S. (Ed.). 1981. *Chemicals from Cellulose*. CRC Press, Boca Raton, FL, USA. 310 p.

[3]   Sinsky, A.J. 1983. Organic chemicals from biomass: an overview. In: Wise, D.L. (Ed.). *Organic Chemicals from Biomass*. The Benjamin/Cummins Publishing Company, London, England. Pp. 1–67.

[4]   Hon, D.N.-S. (Ed.). 1996. *Chemical Modification of Lignocellulosic Materials*. Marcel Dekker, New York, NY, USA. 370 p.

[5]   Lichtenthaler, F.W., and Peters, S. 2004. Carbohydrates as green raw materials for the chemical industry. *Comptes Rendus Chimie* 7:65–90.

[6]   Werpy, T., Petersen, G., Aden, A., Bozell, J., Holladay, J., White, J., Manheim, A., Elliot, D., Lasure, L., Jones, S., Gerber, M., Ibsen, K., Lumberg, L., and Kelley, S. 2004. *Top Value Added Chemicals from Biomass – Volume I: Results of Screening for Potential Candidates from Sugars and Synthesis Gas*. U.S. Department of Energy, Oak Ridge, TN, USA. 69 p.

[7]   Clements, L.D., and Van Dyne, D.L. 2006. The lignocellulosic biorefinery – a strategy for returning to a sustainable source of fuels and industrial organic chemicals. In: Kamm, B., Gruber, P.R., and Kamm, M. (Eds.). *Biorefineries – Industrial Processes and Products, Volume 1*. Wiley-VCH, Weinheim, Germany. Pp. 115–128.

[8]   Hill, C.A.S. 2006. *Wood Modification – Chemical, Thermal and Other Processes*. John Wiley & Sons, Chichester, England. 239 p.

[9]   Kamm, B., Kamm, M., Gruber, P.R., and Kromus, S. 2006. Biorefinery systems – an overview. In: Kamm, B., Gruber, P.R., and Kamm, M. (Eds.). *Biorefineries – Industrial Processes And Products, Volume 1*. Wiley-VCH, Weinheim, Germany. Pp. 3–40.

[10]   Kamm, B., Kamm, M., Schmidt, M., Hirth, T., and Schulze, M. 2006. Lignocellulose- based chemical products and product family trees. In: Kamm, B., Gruber, P.R., and Kamm, M. (Eds.). *Biorefineries – Industrial Processes and Products, Volume 2*. Wiley-VCH, Weinheim, Germany. Pp. 97–149.

[11]   Katzen, R., and Schell, D.J. 2006. Lignocellulosic feedstock biorefinery: history and plant development for biomass hydrolysis. In: Kamm, B., Gruber, P.R., and Kamm, M. (Eds.). *Biorefineries – Industrial Processes and Products, Volume 1*. Wiley-VCH, Weinheim, Germany. Pp. 129–138.

[12]   Argyropoulos, D.S. (Ed.). 2007. *Materials, Chemicals, and Energy from Forest Biomass*. In: *ACS Symposium Series 954*. American Chemical Society, Washington, DC, USA. 591 p.

[13]   Bozell, J., Holladay, J., Johnson, D., and White, J. 2007. *Top Value Added Chemicals from Biomass – Volume II: Results of Screening for Potential Candidates from Biorefinery Lignin*. U.S. Department of Energy, Oak Ridge, TN, USA. 79 p.

[14]   Rosillo-Calle, F., de Groot, P., Hemstock, S.L., and Woods, J. (Eds.). 2007. *The Biomass Assessment Handbook – Bioenergy for a Sustainable Environment*. Earthscan, London, United Kingdom. 269 p.

[15]   Clark, J., and Deswarte, F. 2008. *Introduction to Chemicals from Biomass*. John Wiley & Sons, Chichester, England. 184 p.

[16]   Singh nee' Nigam, and Pandey, A. (Eds.). 2009. *Biotechnology for Agro-Industrial Residues Utilisation*. Springer-Verlag, Heidelberg, Germany. 630 p.

[17]   Demirbas, A. 2010. *Biorefineries – For Biomass Upgrading Facilities*. Springer-Verlag, Heidelberg, Germany. 240 p.

[18]   Ramaswamy, S., Huang, H.-J., and Ramarao, B.V. 2013. *Separation and Purification Technologies in Biorefineries*. John Wiley & Sons, Chichester, England. 258 p.

[19]   Bajpai, P. 2013. *Biorefinery in the Pulp and Paper Industry*. Elsevier, Amsterdam, The Netherlands. 103 p.

[20]   Ragauskas, A.J. (Ed.). 2014. *Materials for Biofuels*. World Scientific, Singapore. 340 p.

[21]   Pandey, A., Höfer, R., Taherzadeh, M., Nampoothiri, K.M., and Larroche, C. (Eds.). 2015. *Industrial Biorefineries & White Biotechnology*. Elsevier, Amsterdam, The Netherlands. 710 p.

[22]   Rastegari, A.A., Yadav, A.N., and Gupta, A. (Eds.). 2019. *Prospects of Renewable Bioprocessing in Future Energy Systems*. Springer Nature Switzerland, Chan, Switzerland. 537 p.

[23]   Dommett, L. 2008. An introduction to life cycle assessment. *Biofuels, Bioproducts and Biorefining* 2:385–388.

[24]   Harris, J.F. 1975. Acid hydrolysis and dehydration reactions for utilizing plant carbohydrates. *Applied Polymer Symposium* 28:131–144.

[25]   Goldstein, I.S. 1980. The hydrolysis of wood. *TAPPI* 63(9):141–143.

[26]   Goldstein, I.S. 1981. Chemicals from cellulose. In: Goldstein, I.S. (Ed.). *Organic Chemicals from Biomass*. CRC Press, Boca Raton, FL, USA. Pp. 101–124.

[27]   Goldstein, I.S. 1983. Acid processes for cellulose hydrolysis and their mechanisms. In: Soltes, E.J. (Ed.). *Wood and Agricultural Residues – Research on Use for Feed, Fuels, and Chemicals*. Elsevier, Amsterdam, The Netherlands. Pp. 315–328.

[28]   Wayman, M. 1986. Comparative effectiveness of various acids for hydrolysis of cellulosics. In: Young, R.A., and Rowell, R.M. (Eds.). *Cellulose – Structure, Modification and Hydrolysis*. John Wiley & Sons, New York, NY, USA. Pp. 265–279.

[29]   Fan, L.T., Gharpuray, M.M., and Lee, Y.-H. 1987. *Cellulose Hydrolysis*. Springer-Verlag, Heidelberg, Germany. Pp. 121–148.

[30]   Katzen, R., and Othmer, D.F. 1942. Wood hydrolysis. a continuous process. *Industrial Engineering Chemistry* 34:314–322.

[31]   Saeman, J.F. 1945. Kinetics of wood hydrolysis – decomposition of sugars in dilute acid at high temperature. *Journal of Industrial Engineering Chemistry* 37:43–52.

[32]   Madelaine, J.L., Bouvier, J.M., and Gelus, M. 1990. HTST hydrolysis of wood: modelling of the kinetics and projections on reactor configuration. *Wood Science and Technology* 24:143–157.

[33]   Parajó, J.C., Vázquez, D., Alonso, J.L., Santos, V., and Dominguez, H. 1993. Prehydrolysis of *Eucalyptus* wood with dilute sulphuric acid: operation at atmospheric pressure. *Holz als Roh- und Werkstoff* 51:357–363.

[34]   Lee, Y.Y., Iyer, P., and Torget, R.W. 1999. Dilute-acid hydrolysis of lignocellulosic biomass. *Advances in Biochemical Engineering/Biotechnology* 65:93–115.

[35]   Kim, J.S., Lee, Y.Y., and Torget, R.W. 2001. Cellulose hydrolysis under extremely low sulfuric acid and high-temperature conditions. *Applied Biochemistry and Biotechnology* 91–93:331–340.

[36]   Ojumu, T.V., AttahDaniel, B.E., Betiku, E., and Solomon, B.O. 2003. Auto-hydrolysis of lignocellulosics under extremely low sulphur acid and high temperature conditions in batch reactor. *Biotechnology and Bioprocess Engineering* 8:291–293.

[37]   Marzialetti, T., Valenzuela Olarte, M.B., Sievers, C., Hoskins, T.J.C., Agrawal, P.K., and Jones, C.W. 2008. Dilute acid hydrolysis of loblolly pine: a comprehensive approach. *Industrial and Engineering Chemistry Research* 47:7131–7140.

[38]   Kumar, P., Barrett, D.M., Delwiche, M.J., and Stroeve, P. 2009. Methods for pretreatment of lignocellulosic biomass for efficient hydrolysis and biofuel production. *Industrial & Engineering Chemistry Research* 48:3713–3729.

[39]   Alén, R. 2011. Principles of biorefining. In: Alén, R. (Ed.). *Biorefining of Forest Resources*. Paper Engineers' Association, Helsinki, Finland. Pp. 55–114.

[40]   Viikari, L., and Alén, R. 2011. In: Alén, R. (Ed.). *Biorefining of forest resources*. In: *Paper Engineers' Association*, Helsinki, Finland. Pp. 225–261.

[41]   Defaye, J., Gadelle, A., Papadopoulos, J., and Pedersen, C. 1983. Hydrogen fluoride saccharification of cellulose and lignocellulosic materials. *Journal of Applied Polymer Science: Applied Polymer Symposium* 37:653–670.

[42]   Maloney, M.T., Chapman, T.W., and Baker, A.J. 1985. Dilute acid hydrolysis of paper birch: kinetics studies of xylan and acetyl-group hydrolysis. *Biotechnology & Bioengineering* 27:355–361.

[43]   Hawley, M.C., Downey, K.W., Selke, S.M., and Lamport, D.T.A. 1986. Hydrogen fluoride saccharification of lignocellulosic materials. In: Young, R.A. and Rowell, R.M. (Eds.). *Cellulose – Structure, Modification and Hydrolysis*. John Wiley & Sons, New York, NY, USA. Pp. 297–321.

[44]   Fan, L.T., Gharpuray, M.M., and Lee, Y.-H. 1987. *Cellulose Hydrolysis*. Springer-Verlag, Heidelberg, Germany. Pp. 149–187.

[45]   Rowland, S.P., and Howley, P.S. 1989. Simplified hydrolysis of cellulose and substituted cellulose: observations on trifluoroacetic acid hydrolyses. *Journal of Applied Polymer Science* 37:2371–2382.

[46]   Bunton, C.A., Lewis, D.R., Llewellyn, D.R., and Vernon, C.A. 1955. Mechanisms of reaction in the sugar series. part I. the acid-catalysed hydrolysis of α- and β-methyl and α- and β-phenyl D-glucopyranosides. *Journal of the Chemical Society* 1955, 4419–4423.

[47]   Edward, J.T. 1955. Stability of glycosides to acid hydrolysis. a conformational analysis. *Chemistry and Industry* 1955, 1102–1104.

[48]   BeMiller, J.V. 1967. Acid-catalyzed hydrolysis of glycosides. In: Wolfrom, M.L., and Tipson, R.S. (Eds.). *Advances in Carbohydrate Chemistry, Volume 22*. Academic Press, New York, NY, USA. Pp. 25–108.

[49]   Sjöström, E. 1993. *Wood Chemistry – Fundamentals and Applications*. 2nd edition. Academic Press, San Diego, CA, USA. Pp. 38–50.

[50]   Larsson, S., Palmqvist, E., Hahn-Hägerdal, B., Tengborg, C., Stenberg, K., Zacchi, G., and Nilvebrant, N.-O. 1999. The generation of fermentation inhibitors during dilute acid hydrolysis of softwood. *Enzyme & Microbial Technology* 24:151–159.

[51]   Chang, M., and Tsao, G.T. 1981. The effect of structure on hydrolysis of cellulose. *Cellulose Chemistry and Technology* 15(4):383–395.

[52]   Alén, R. 2000. Structure and Chemical Composition of Wood. In: Stenius, P. (Ed.). *Forest Products Chemistry*. Fapet, Helsinki, Finland. Pp. 11–57.

[53]   Lee, Y.Y., and McCaskey, T.A. 1983. Hemicellulose hydrolysis and fermentation of resulting pentoses to ethanol. *Tappi Journal* 66(5):102–107.

[54]   Xu, C., Pranovich, A., Vähäsalo, L., Hemming, J., Holmbom, B., Schols, H.A., and Willför, S. 2008. Kinetics of acid hydrolysis of water-soluble spruce o-acetyl galactoglucomannans. *Journal of Agricultural and Food Chemistry* 56:2429–2435.

[55]   Lenz, J., Esterbauer, H., Sattler, W., Schurz, J., and Wrentschur, E. 1990. Changes of structure and morphology of regenerated cellulose caused by acid and enzymatic hydrolysis. *Journal of Applied Polymer Science* 41:1315–1326.

[56]   Conner, A.H., Wood, B.F., Hill, C.G. Jr., and Harris, J.F. 1986. In: Young, R.A., and Rowell, R.M. (Eds.). *Cellulose – Structure, Modification and Hydrolysis*. John Wiley & Sons, New York, NY, USA. Pp. 281–296.

[57]   Herman, B.G., and Patel, M. 2007. Today's and tomorrow's bio-based bulk chemicals from white biotechnology. *Applied Biochemistry and Biotechnology* 136:361–388.

[58]   Tian, J., Wang, J., Zhao, S., Jiang, C., Zhang, X., and Wang, X. 2010. Hydrolysis of cellulose by the heteropoly acid $H_3PW_{12}O_{40}$. *Cellulose* 17:587–594.

[59]   Rogalinski, T., Liu, K., Albrecht, T., and Brunner, G. 2008. Hydrolysis kinetics of biopolymers in subcritical water. *The Journal of Supercritical Fluids* 46:335–341.

[60]   Martínez, C.M., Cantero, D.A., Bermejo, M.D., and Cocero, M.J. 2015. Hydrolysis of cellulose in supercritical water: reagent concentration as a selectivity factor. *Cellulose* 22:2231–2243.

[61]   Gan, L., Zhu, J., and Lv, L. 2017. Cellulose hydrolysis catalyzed by highly acidic lignin-derived carbonaceous catalyst synthesized via hydrothermal carbonization. *Cellulose* 24:5327–5339.

[62]   You, X., van Heiningen, A., Sixta, H., and Iakovlev, M. 2016. Kinetics of SO₂-ethanol-water (AVAP®) fractionation of sugarcane straw. *Bioresource Technology* 212:111–119.

[63]   You, X., van Heiningen, A., Sixta, H., and Iakovlev, M. 2017. Sulfur balance of sulfur dioxide-ethanol-water fractionation of sugarcane straw. *Bioresource Technology* 241:998–1002.

[64]   You, X., van Heiningen, A., Sixta, H., and Iakovlev, M. 2017. Lignin and ash balances of sulfur dioxide-ethanol-water fractionation of sugarcane straw. *Bioresource Technology* 244:1111–1120.

[65]   Alén, R. 2018. *Carbohydrate Chemistry – Fundamentals and Applications*. World Scientific, Singapore. 586 p.

[66]   Thompson, D.R., and Grethlein, H.E. 1979. Design and evaluation of a plug flow reactor for acid hydrolysis of cellulose. *Industrial & Engineering Chemistry Product Research and Development* 18(3):166–169.

[67]   Wright, J.D., Bergeron, P.W., and Werdene, P.J. 1987. Progressing bath hydrolysis reactor. *Industrial & Engineering Chemistry Research* 26:699–705.

[68]   Song, S.K., and Lee, Y.Y. 1982. Countercurrent reactor in acid catalyzed cellulose hydrolysis. *Chemical Engineering Communications* 17(1–6):23–30.

[69]   Cahela, D.R., Lee, Y.Y., and Chambers, R.P. 1983. Modelling of percolation process in hemicellulose hydrolysis. *Biotechnology and Bioengineering* 25(1):3–17.

[70]   Yu, S., and Wayman, M. 1986. Fermentation of spent sulphite liquor to butanol and ethanol. *Journal of Pulp and Paper Science* 12(3):J72–J77.

[71]   Bon, E.P.S., and Ferrara, M.A. 2007. Bioethanol production via enzymatic hydrolysis of cellulosic biomass. In: *Proceedings of the FAO Seminar on The Role of Agricultural Biotechnologies for Production of Bioenergy in Developing Countries*, October 12, 2007, Rome, Italy.

[72]   Kootstra, A.M.J., Beeftink, H.H., Scott, E.L., and Sanders, J.P.M. 2009. Comparison of dilute mineral and organic acid pretreatment for enzymatic hydrolysis of wheat straw. *Biochemical Engineering Journal* 46:126–131.

[73]   Talebnia, F., Karakashev, D., and Angelidaki, I. 2010. Production of bioethanol from wheat straw: an overview on pretreatment, hydrolysis and fermentation. *Bioresource Technology* 101:4744–4753.

[74]   Qin, L., Liu, Z.H., Li, B.Z., Dale, B., and Yuan, Y.J. 2012. Mass balance and transformation of corn stover by pretreatment with different dilute organic acids. *Bioresource Technology* 112:319–326.

[75]   Lee, H.J., Seo, Y.J., and Lee, J.W. 2013. Characterization of oxalic acid pretreatment on lignocellulosic biomass using oxalic acid recovered by electrodialysis. *Bioresource Technology* 133:87–91.

[76]   Zhao, X., Wang, L., Lu, X., and Zhang, S. 2014. Pretreatment of corn stover with diluted acetic acid for enhancement of acidogenic fermentation. *Bioresource Technology* 158(4):12–18.

[77]   Silveira, M.H.L., Morais, A.R.C., da Costa Lopes, A.M., Olekszyszen, D.N., Bogel-Łukasik, R., Andreaus, J., and Ramos, L.P. 2015. Current pretreatment technologies for the development of cellulosic ethanol and biorefineries. *ChemSusChem* 8:3366–3390.

[78]   Lin, Q., Li, H., Ren, J., Deng, A., Li, W., Liu, C., and Sun, R. 2017. Production of xylooligosaccharides by microwave-induced, organic acid-catalyzed hydrolysis of different xylan-type hemicelluloses: optimization by response surface methodology. *Carbohydrate Polymers* 157:214–225.

[79]   Xu, C. 2021. Acid-catalyzed hydrolysis with organic acids. In: Alén, R. (Ed.). *Pulping and Biorefining – ForestBioFacts, Digital Learning Environment*. Paperi ja Puu Ltd., Helsinki, Finland.

[80]   Goldmann, W.M., Ahola, J., Mikola, M., and Tanskanen, J. 2017. Formic acid aided hot water extraction of hemicellulose from European silver birch (*Betula pendula*) sawdust. *Bioresource Technology* 232:176–182.

[81]   Zhang, M., Qi, W., Liu, R., Su, R., Wu, S., and He, Z. 2010. Fractionating lignocellulose by formic acid: characterization of major components. *Biomass and Bioenergy* 34(4):525–532.

[82]   Zhang, Y., Qin, M., Xu, W., Fu, Y., Wang, Z., Li, Z., Willför, S., Xu, C., and Hou, Q. 2018. Structural changes of bamboo-derived lignin in an integrated process of autohydrolysis and formic acid inducing rapid delignification. *Industrial Crops and Products* 115:194–201.

[83]   Zhuang, J., Lin, L., Liu, J., Luo, X., Pang, C., and Ouyang, P. 2009. Preparation of xylose and kraft pulp from poplar based on formic/acetic acid/water system hydrolysis. *BioResources* 4(3):1147–1157.

[84]   Lee, J.W., and Jeffries, T.W. 2011. Efficiencies of acid catalysts in the hydrolysis of lignocellulosic biomass over a range of combined severity factors. *Bioresource Technology* 102(10):5884–5890.

[85] Xu, W., Grénman, H., Liu, J., Kronlund, D., Li, B., Backman, P., Peltonen, J., Willför, S., Sundberg, A., and Xu, C. 2017. Mild oxalic-acid-catalyzed hydrolysis as a novel approach to prepare cellulose nanocrystals. *ChemNanoMat* 3(2):109–119.

[86] Su, T.-M., and Paulavicius, I. 1975. Enzymatic saccharification of cellulose by thermophilic actinomyces. *Applied Polymer Symposium* 28:221–236.

[87] Esterbauer, H., Hayn, M., Jungschaffer, G., Taufratzhofer, E., and Schurz, J. 1983. Enzymatic conversion of lignocellulosic materials to sugars. *Journal of Wood Chemistry and Technology* 3(3):261–287.

[88] Rollings, J. 1985. Enzymatic depolymerization of polysaccharides. *Carbohydrate Polymers* 5:37–82.

[89] Murray, W.D. 1986. Cellulose hydrolysis by *bacteroides cellulosolvens*. *Biomass* 10:47–57.

[90] Fan, L.T., Gharpuray, M.M., and Lee, Y.-H. 1987. *Cellulose Hydrolysis*. Springer-Verlag, Heidelberg, Germany. Pp. 21–119.

[91] Harjunpää, V. 1998. *Enzymes Hydrolysing Wood Polysaccharides – A Progress Curve Study of Oligosaccharide Hydrolysis by two Cellobiohydrolases and three β-Mannanases*. Doctoral Thesis. University of Helsinki, Laboratory of Organic Chemistry, Helsinki, Finland. Published in *VTT Publications 372*, Espoo, Finland, 76 p.

[92] Cen, P., and Xia, L. 1999. Production of cellulase by solid-state fermentation. *Advances in Biochemical Engineering/Biotechnology* 65:69–92.

[93] Itävaara, M., Siika-aho, M., and Viikari, L. 1999. Enzymatic degradation of cellulose-based materials. *Journal of Environmental Polymer Degradation* 7(2):67–73.

[94] Mosier, N.S., Hall, P., Ladisch, C.M., and Ladisch, M.R. 1999. Reaction kinetics, molecular action, and mechanisms of cellulolytic proteins. *Advances in Biochemical Engineering/Biotechnology* 65:23–40.

[95] Tolan, J.S., and Foody, B. 1999. Cellulase from submerged fermentation. *Advances in Biochemical Engineering/Biotechnology* 65:41–67.

[96] Wilson, D.B., and Irwin, D.C. 1999. Genetics and properties of cellulases. *Advances in Biochemical Engineering/Biotechnology* 65:1–21.

[97] Yang, B., Dai, Z., Ding, S.-Y., and Wyman, C.E. 2011. Enzymatic hydrolysis of cellulosic biomass. *Biofuels* 2(4):421–450.

[98] Wright, J.D. 1988. Ethanol from biomass by enzymatic hydrolysis. *Chemical Engineering Progress* 84(8):62–74.

[99] Taherzadeh, M.J., and Karimi, K. 2007. Enzyme-based hydrolysis processes for ethanol from lignocellulosic materials: a review. *BioResources* 2(4):707–738.

[100] Drapcho, C.M., Nhuan, N.P., and Walker, T.H. 2008. *Biofuels Engineering Process Technology*. McGraw-Hill, New York, NY, USA. Pp. 105–121.

[101] Kumar, S., Singh, S.P., Mishra, I.M., and Adhikari, D.K. 2009. Recent advances in production of bioethanol from lignocellulosic biomass. *Chemical Engineering and Technology* 32(4):517–526.

[102] Gírio, F.M., Carvalheiro, F.F., Duarte, L.C., and Bogel-Łukasik, M.R. 2010. Hemicelluloses for fuel ethanol: a review. *Bioresource Technology* 101:4775–4800.

[103] Soccol, C.R., Vandenberghe, L.P.S., Medeiros, A.B.P., Karp, S.G., Buckeridge, M., Ramos, L.P., Pitarelo, A.P., Ferreira-Leitão, V., Gottschalk, L.M.F., Ferrara, M.A., Bon, E.P.S., de Moraes, L.M.P., Araújo, J.A., and Torres, F.A.G. 2010. Bioethanol from lignocelluloses: status and perspectives in Brazil. *Bioresource Technology* 101:4820–4825.

[104] Buruiana, C.-T., Garrote, G., and Vizireanu, C. 2013. Bioethanol production from residual lignocellulosic materials: a review – part 2. *The Annals of the University Dunarea De Jos of Galati (AUDJG) – Food Technology* 37(1):25–38.

[105] Gashaw, A., and Getachew, T. 2014. Fermentation of crop residues and fruit wastes for production of ethanol and its pretreatment methods: a review. *International Journal Research (IJR)* 1(11):543–555.

[106] Sebayang, A.H., Masjuki, H.H., Ong, H.C., Dharma, S., Silitonga, A.S., Mahlia, T.M.I., and Aditiya, H.B. 2016. A perspective on bioethanol production from biomass as alternative fuel for spark ignition engine. *RSC Advances* 6:14964–14992.

[107] Yusuf, A.A., and Inambao, F.L. 2019. Bioethanol production techniques from lignocellulosic biomass as alternative fuel: a review. *International Journal of Mechanical Engineering and Technology (IJMET)* 10(6):34–71.

[108] Liu, Z.L., and Blaschek, H.P. 2010. Biomass conversion inhibitors and in situ detoxification. In: Vertès, A.A., Qureshi, N., Blaschek, H.P., and Yukawa, H. (Eds.). *Biomass to Biofuels – Strategies for Global Industries*. John Wiley & Sons, New York, NY, USA. Pp. 233–259.

[109] Alvira, P., Tomás-Pejó, E., Ballesteros, M., and Negro, M.J. 2010. Pretreatment technologies for an efficient bioethanol production process based on enzymatic hydrolysis: a review. *Bioresource Technology* 101:4851–4861.

[110] Karimi, K., Shafiei, M., and Kumar, R. 2013. Progress in physical and chemical pretreatment of lignocellulosic biomass. In: Gupta, V.K., and Tuohy, M.G. (Eds.). *Biofuel Technologies*. Springer-Verlag, Heidelberg, Germany. Pp. 53–96.

[111] Moriyama, S., and Saida, T. 1986. Continuous pretreatment and enzymatic saccharification of lignocellulosics. In: Young, R.A., and Rowell, R.M. (Eds.). *Cellulose – Structure, Modification and Hydrolysis*. John Wiley & Sons, New York, NY, USA. Pp. 323–336.

[112] Chang, V.S., and Holtzapple, M.T. 2000. Fundamental factors affecting biomass enzymatic reactivity. *Applied Biochemistry and Biotechnology* 84/86:5–37.

[113] Jørgenssen, H., Kristensen, J.B., and Felby, C. 2007. Enzymatic conversion of lignocellulose into fermentable sugars: challenges and opportunities. *Biofuels, Bioproducts and Biorefining* 1:119–134.

[114] Sun, Y., and Cheng, J. 2002. Hydrolysis of lignocellulosic materials for ethanol production: a review. *Bioresource Technology* 83:1–11.

[115] Mosier, N., Wyman, C., Dale, B., Elander, R., Lee, Y.Y., Holtzapple, M., and Ladisch, M. 2005. Features of promising technologies for pretreatment of lignocellulosic biomass. *Bioresource Technology* 96:673–686.

[116] Driemeier, C., Pimenta, M.T.B., Rocha, G.J.M., Oliveira, M.M., Mello, D.B., Maziero, P., and Conçalves, A.R. 2011. Evolution of cellulose crystals during prehydrolysis and soda delignification of sugarcane lignocellulose. *Cellulose* 18:1509–1519.

[117] Hidayat, B.J., Felby, C., Johansen, K.S., and Thygesen, L.G. 2012. Cellulose is not just cellulose: a review of dislocations as reactive sites in the enzymatic hydrolysis of cellulose microfibrils. *Cellulose* 19:1481–1493.

[118] Junior, C.S., Milagres, A.M.F., Ferraz, A., and Carvalho, W. 2013. The effects of lignin removal and drying on the porosity and enzymatic hydrolysis of sugarcane bagasse. *Cellulose* 10:3165–3177.

[119] Sun, Q., Foston, M., Sawada, D., Pingali, S.V., O'Neill, H.M., Li, H., Wyman, C.E., Langan, P., Pu, Y., and Ragauskas, A.J. 2014. Comparison of changes in cellulose ultrastructure during pretreatments of poplar. *Cellulose* 21:2419–2431.

[120] Holtzapple, M.T., and Humphrey, A.E. 1984. The effect of organosolv pretreatment on the enzymatic hydrolysis of poplar. *Biotechnology and Bioengineering* 26(7):670–676.

[121] Taherzadeh, M.J., and Karimi, K. 2008. Pretreatment of lignocellulosic wastes to improve ethanol and biogas production: a review. *International Journal of Molecular Sciences* 9:1621–1651.

[122] Huang, P., Wu, M., Kuga, S., and Huang, Y. 2013. Aqueous pretreatment for reactive ball milling of cellulose. *Cellulose* 20:2175–2178.

[123] Wang, Y., and Liu, S. 2012. Pretreatment technologies for biological and chemical conversion of woody biomass. *TAPPI Journal* 11(1):9–16.

[124] Vallander, L., and Eriksson, K.-E. 1985. Enzymic saccharification of pretreated wheat straw. *Biotechnology and Bioengineering* 27:650–659.

[125] Pou-Ilinas, J., and Canellas, J. 1990. Steam pretreatment of almond shells for xylose production. *Carbohydrate Research* 207:126–130.

[126] Kaar, W.E., Gutierrez, C.V., and Kinoshita, C.M. 1998. Steam explosion of sugarcane bagasse as a pretreatment for conversion to ethanol. *Biomass and Bioenergy* 14(3):277–287.

[127] Datar, R., Huang, J., Maness, P.-C., Mohagheghi, A., Czernik, S., and Chornet, E. 2007. Hydrogen production from the fermentation of corn stover biomass pretreated with steam-explosion process. *International Journal of Hydrogen Energy* 32:932–939.

[128] Agbor, V.B., Cicek, N., Sparling, R., Berlin, A., and Levin, D.B. 2011. Biomass pretreatment: fundamentals toward application. *Biotechnology Advances* 29:675–685.

[129] Dong, Z., Hou, X., Sun, F., Zhang, L., and Yang, Y. 2014. Textile grade long natural cellulose fibers from bark of cotton using steam explosion as a pretreatment. *Cellulose* 21:3851–3860.

[130] da Costa Sousa, L., Chundawat, S.P.C., Balan, V., and Dale, B.E. 2009. "Cradle-to-grave" assessment of existing lignocellulose pretreatment technologies. *Current Opinion in Biotechnology* 20:1–9.

[131] Yoon, H.H. 1998. Pretreatment of lignocellulosic biomass by autohydrolysis and aqueous ammonia percolation. *Korean Journal of Chemical Engineering* 15(6):631–636.

[132] Ucar, U., and Fengel, D. 1988. Characterization of the acid pretreatment for the enzymatic hydrolysis of wood. *Holzforschung* 42(3):141–148.

[133] Uçar, G. 1990. Pretreatment of poplar by acid and alkali for enzymatic hydrolysis. *Wood Science and Technology* 24:171–180.

[134] Allen, S.G., Schulman, D., Lichwa, J., and Antal, M.J. Jr. 2001. A comparison of aqueous and dilute-acid single-temperature pretreatment of yellow poplar sawdust. *Industrial & Engineering Chemistry Research* 40:2352–2361.

[135] Carvalheiro, F., Duarte, L., and Gírio, F.M. 2008. Hemicellulose biorefineries: a review on biomass pretreatments. *Journal of Scientific & Industrial Research* 67:849–864.

[136] Aden, A., and Foust, T. 2009. Technoeconomic analysis of the dilute sulfuric acid and enzymatic hydrolysis process for the conversion of corn stover to ethanol. *Cellulose* 16:535–545.

[137] Zhang, C., Zhuang, X., Wang, Z.J., Matt, F., St. John, F., and Zhu, J.Y. 2013. Xylanase supplementation on enzymatic saccharification of dilute acid pretreated poplars at different severities. *Cellulose* 20:1937–1946.

[138] Afifi, M.M.I., Massoud, O.N., and El-Akasher, Y.S. 2015. Bioethanol production by simultaneous saccharification and fermentation using pretreated rice straw. *Middle East Journal of Applied Sciences* 5(3):769–776.

[139] Loow, Y.-L., Wu, T.Y., Jahim, J.Md., Mohammad, A.W., and Teoh, W.H. 2016. Typical conversion of lignocellulosic biomass into reducing sugars using dilute acid hydrolysis and alkaline pretreatment. *Cellulose* 23:1491–1520.

[140] Min, D., Wei, L., Zhao, T., Li, M., Jia, Z., Wan, G., Zhang, Q., Qin, C., and Wang, S. 2018. Combination of hydrothermal pretreatment and sodium hydroxide post-treatment applied on wheat straw for enhancing its enzymatic hydrolysis. *Cellulose* 25:1197–1206.

[141] Simionescu, CR.I., Popa, V.I., Rusan, V., and Rusan, M. 1985. The influence of the enzymatic hydrolysis conditions on the transformation degree of different lignocellulose materials. *Cellulose Chemistry and Technology* 19:525–530.

[142] Takai, M., Shimizu, Y.-I., Tanno, K., and Hayashi, J. 1985. A basic research on enzymatic saccharification of cellulosic biomass (I). Chemical and physical pretreatments for pure cellulose. *Cellulose Chemistry and Technology* 19:217–229.

[143] Simionescu, CR.I., Popa, V.I., Popa, M., and Maxim, S. 1990. On the possibilities of immobilization and utilization of some cellulase enzymes. *Journal of Applied Polymer Science* 39:1837–1846.

[144] Schimper, C.B., Ibanescu, C., and Bechtold, T. 2009. Effect of alkali pre-treatment on hydrolysis of regenerated cellulose fibers (part 1: viscose) by cellulases. *Cellulose* 16:1057–1068.

[145] Mirahmadi, K., Kabir, M.M., Jeihanipour, A., Karimi, K., and Taherzadeh, Md.J. 2010. Alkaline pretreatment of spruce and birch to improve bioethanol and biogas production. *BioResources* 5(2):928–938.

[146] Meng, X., Geng, W., Ren, H., Jin, Y., Chang, H.-m., and Jameel, H. 2014. Enhancement of enzymatic saccharification of poplar by green liquor pretreatment. *BioResources* 9(2):3236–3247.

[147] Majdanac, L. and Jakševac, J.R. 1983. The influence of some pretreatment conditions of hardwood cellulose on its enzyme degradability. *Cellulose Chemistry and Technology* 17:315–322.

[148] Kangas, H., Tamminen, T., Liitiä, T., Hakala, T.K., Vorwerg, W., and Poppius-Levlin, K. 2014. Lignofibre (LGF) process – a flexible biorefinery for lignocellulosics. *Cellulose Chemistry and Technology* 48(9–10):765–771.

[149] Elgharbawy, A.A., Alam, Md.Z., Moniruzzaman, M., and Goto, M. 2016. Ionic liquid pretreatment as emerging approaches for enhanced enzymatic hydrolysis of lignocellulosic biomass. *Biochemical Engineering Journal* 109:252–267.

[150] Karagöz, P., Rocha, I.V., Özkan, M., and Angelidaki, I. 2012. Alkaline peroxide pretreatment of rapeseed straw for enhancing bioethanol production by same vessel saccharification and co-fermentation. *Bioresource Technology* 104:349–357.

[151] Hatakka, A.I. 1983. Pretreatment of wheat straw by white-rot fungi for enzymatic saccharification of cellulose. *Applied Microbiology and Biotechnology* 18:350–357.

[152] Koshijima, T., Yaku, F., Muraki, E., Tanaka, R., and Azuma, J. 1983. Wood saccharification by enzyme systems without prior delignification. *Journal of Applied Polymer Science: Applied Polymer Symposium* 37:671–683.

[153] Hamid, S.B.A., Islam, M.M., and Das, R. 2015. Cellulase biocatalysis: key influencing factors and mode of action. *Cellulose* 22:2157–2182.

[154] Himmel, M.E., Ding, S.Y., Johnson, D.K., Adney, W.S., Nimlos, M.R., Brady, J.W., and Foust, T.D. 2007. Biomass recalcitrance: engineering plants and enzymes for biofuels production. *Science* 315:804–807.

[155] Ulaganathan, K., Goud, B.S., Reddy, M.M., Kumar, V.P., Balsingh, J., and Radhakrishna, S. 2015. Proteins for breaking barriers in lignocellulosic bioethanol production. *Current Protein and Peptide Science* 16(1):100–134.

[156] Collins, T., Gerday, C., and Feller, G. 2005. Xylanases, xylanase families and extremophilic xylanases. *FEMS Microbiology Reviews* 29:3–23.

[157] Howard, R.L., Abotsi, E., Jansen van Rensburg, E.L., and Howard, S. 2003. Lignocellulose biotechnology: issue of bioconversion and enzyme production. *African Journal of Biotechnology* 2(12):602–619.

[158] Menon, V., and Rao, M. 2012. Trends in bioconversion of lignocellulose: biofuels, platform chemicals & biorefinery concept. *Progress in Energy and Combustion Science* 38:522–550.

[159] Moreira, L.R.S., and Filho, E.X.F. 2008. An overview of mannan structure and mannan-degrading enzyme systems. *Applied Microbiology and Biotechnology* 79:165–178.

[160] Kaylen, M., Van Dyne, D.L., Choi, Y.-S., and Blase, M. 2000. Economic feasibility of producing ethanol from lignocellulosic feedstocks. *Bioresearch Technology* 72:19–32.

[161] Wilson, D. 2009. Cellulases and biofuels. *Current Opinion in Biotechnology* 20:295–299.

[162] Himmel, M., Ruth, M., and Wyman, C. 1999. Cellulase for commodity products from cellulosic biomass. *Current Opinion in Biotechnology* 10:358–364.

[163] Viikari, L., Alapuranen, M., Puranen, T., Vehmaanperä, J., and Siika-aho, M. 2007. Thermostable enzymes in lignocellulose hydrolysis. *Advances in Biochemical Engineering/Biotechnology* 108:121–145.

[164] Wiseman, P. 1983. *An Introduction to Industrial Organic Chemistry.* 2nd edition. Applied Science Publishers, Essex, England. 366 p.

[165] Basta, N. 1999. *Shreve's Chemical Process Industries.* 5th edition. McGraw-Hill, New York, NY, USA. 878 p.

[166] Kent, J.A. (Ed.). 2003. *Riegel's Handbook of Industrial Chemistry.* 10th edition. Springer, New York, NY, USA, 1374 p.

[167] Maloney, G.T. 1978. *Chemicals from Pulp and Wood Waste*. Noyes Data Corporation, Park Ridge, NJ, USA. 289 p.

[168] Vertès, A.A., Qureshi, N., Blaschek, H.P., and Yukawa, N. (Eds.). 2010. *Biomass to Biofuels – Strategies for Global Industries*. Wiley, Chichester, United Kingdom. 559 p.

[169] Kerton, F.M. 2008. Green Chemical Technologies. In: Clark, J.H., and Deswarte, E.I. (Eds.). *Introduction to Chemicals from Biomass*. John Wiley & Sons, New York, NY, USA. Pp. 47–76.

[170] Yuan, Z., Dai, W., Zhang, S., Wang, F., Jian, J., Zeng, J., and Zhou, H. 2022. Heterogeneous strategies for selective conversion of lignocellulosic polysaccharides. *Cellulose* 29:3059–3077.

[171] Alén, R. 2009. *Encyclopedia of Organic Chemicals – Properties and Utilization*. Consalen Consulting, Helsinki, Finland. 1370 p. (In Finnish).

[172] Joshi, B., Bhat, M.R., Sharma, D., Joshi, J., Malla, R., and Sreerama, L. 2011. Lignocellulosic ethanol production: current practices and recent developments. *Biotechnology and Molecular Biology Review* 6(8):172–182.

[173] Canilha, L., Chandel, A.K., Milessi, T.S.S., Antunes, F.A.F., Freitas, W.L.C., Felipe, M.G.A., and Silva, S.S. 2012. Bioconversion of sugarcane biomass into ethanol: an overview about composition, pretreatment methods, detoxification of hydrolysates, enzymatic saccharification, and ethanol fermentation. *Journal of Biomedicine & Biotechnology*, article ID 989572.

[174] Kang, Q., Appels, L., Tan, T., and Dewil, R. 2014. Bioethanol from lignocellulosic biomass: current findings determine research priorities. *The Scientific World Journal*, article ID 298153.

[175] Gupta, A., and Verma, J.P. 2015. Sustainable bio-ethanol production from agro-residues: a review. *Renewable and Sustainable, Energy Reviews* 41:550–567.

[176] Wilke, C.R., Yang, R.D., Sciamanna, A.F., and Freitas, R.P. 1981. Raw materials evaluation and process development studies for conversion of biomass to sugars and ethanol. *Biotechnology and Bioengineering* 26:163–183.

[177] Sharma, D.K. 1989. Two-step process for the selective production of fermentable sugars and ethanol from biomass residues (agricultural wastes). *Cellulose and Technology* 23:45–51.

[178] Gong, C.S., Cao, N.J., Du, J., and Tsao, G.T. 1999. Ethanol production from renewable resources. *Advances in Biochemical Engineering/Biotechnology* 65:207–241.

[179] Saimi, J.K., Saini, R., and Tewari, L. 2015. Lignocellulosic agriculture wastes as biomass feedstocks for second-generation bioethanol production: concepts and recent developments. *3 Biotech* 5(4):337–353.

[180] Mohanty, B., and Abdullahi, I.I. 2016. Bioethanol production from lignocellulosic waste – a review. *Biosciences Biotechnology Research Asia* 13(2):1153–1161.

[181] Frederick, W.J. Jr., Lien, S.J., Courchene, C.E., DeMartini, N.A., Ragauskas, A.J., and Iisa, K. 2008. Co-production of ethanol and cellulose fiber from Southern Pine: a technical and economic assessment. *Biomass and Bioenergy* 32:1293–1302.

[182] Hytönen, E., and Stuart, P.R. 2009. Integrating bioethanol production into an integrated kraft pulp and paper mill: techno-economic assessment. *Pulp & Paper Canada* 110(5–6):25–32.

[183] Kautto, J., Henricson, K., Sixta, H., Trogen, M., and Alén, R. 2010. Effects of integrating a bioethanol production process to a kraft pulp mill. *Nordic Pulp and Paper Research Journal* 25(2):233–242.

[184] Chiang, L.C., Hsiao, H.Y., Flickinger, M.C., Chen, L.F., and Tsao, G.T. 1982. Ethanol production from pentoses by immobilized microorganisms. *Enzyme and Microbial Technology* 4(4):93–95.

[185] Ho, N.W.Y., Chen, Z., Brainard, A., and Sedlak, M. 1999. Successful design and development of genetically engineered *saccharomyces* yeasts for effective cofermentation of glucose and xylose from cellulosic biomass to ethanol. *Advances in Biochemical Engineering/Biotechnology* 65:163–192.

[186] Jeffries, T.W., and Shi, N.-Q. 1999. Genetic engineering for improved xylose fermentation by yeasts. *Advances in Biochemical Engineering/Biotechnology* 65:116–161.

[187] Margeot, A., Hahn-Hägerdahl, B., Edlund, M., Slade, R., and Monot, F. 2009. New improvements for lignocellulosic ethanol. *Current Opinion in Biotechnology* 20:372–380.

[188] Dürre, P. 1998. New insight and novel developments in clostridial acetone/ butanol/isopropanol fermentation. *Applied Microbiology and Biotechnology* 49:639–648.

[189] Ezeji, T.C., Qureshi, N., Karcher, B., and Blaschek, H.P. 2006. Production of butanol from corn. In: Minteer, S. (Ed.). *Alcoholic Fuels*. CRC Taylor & Francis, Boca Raton, FL, NY, USA. Pp. 99–122.

[190] Ezeji, T.C., Qureshi, N., and Blaschek, H.P. 2007. Bioproduction of butanol from biomass: from genes to bioreactors. *Current Opinion in Biotechnology* 18:220–224.

[191] Qureshi, N., and Blaschek, H.P. 2010. *Clostridia* and Process Engineering for Energy Generation. In: Vertès, A.A., Qureshi, N., Blaschek, H.P., and Yukawa, H. (Eds.). *Biomass to Biofuels – Strategies for Global Industries*. John Wiley & Sons, New York, NY, USA. Pp. 347–358.

[192] Cousin Saint Remi, J., Rémy, T., Van Hunskerken, V., van de Perre, S., Duerinck, T., Maes, M., De Vos, D., Gobechiya, E., Kirschhock, C.E.A., Baron, G.V., and Denayer. J.F.M. 2011. Biobutanol separation with the metal-organic framework ZIF-8. *ChemSusChem* 4:1074–1077.

[193] Jurgens, G., Survase, S., Berezina, O., Sklavounos, E., Linnekoski, J., Kurkijärvi, A., Väkevä, M., van Heiningen, A., and Granström, T. 2012. Butanol production from lignocellulosics. *Biotechnology Letters* 34(8):1415–1434.

[194] Morone, A., and Pandey, R.A. 2014. Lignocellulosic biobutanol production: gridlocks and potential remedies. *Renewable and Sustainable Energy Reviews* 37:21–35.

[195] Matsumura, M., Takehara, S., and Kataoka, H. 1992. Continuous butanol/isopropanol fermentation in down-flow column reactor coupled with pervaporation using supported liquid membrane. *Biotechnology and Bioengineering* 39:148–156.

[196] Jojima, T., Inui, M., and Yukawa, H. 2008. Production of isopropanol by metabolically engineered *escherichia coli*. *Applied Microbiology and Biotechnology* 77:1219–1224.

[197] Herrick, F.W., Casebier, R.L., Hamilton, J.K., and Wilson, J.D. 1975. Mannose chemicals. *Applied Polymer Symposium* 28:93–108.

[198] Cohen, S., Marcus, Y., Migron, Y., Dikstein, S., and Shafran, A. 1993. Water sorption, binding and solubility of polyols. *Journal of the Chemical Society, Faraday Transactions* 89(17):3271–3275.

[199] Detroy, R.W., and St Julian, G. 1983. Biomass conversion: fermentation chemicals and fuels. *Critical Reviews in Microbiology* 10:203–228.

[200] Wang, Z.X., Zhuge, J., Fang, H., and Prior, B.A. 2001. Glycerol production by microbial fermentation: a review. *Biotechnology Advances* 19:201–223.

[201] Biebl, H., and Marten, S. 1995. Fermentation of glycerol to 1,3-propanediol: use of cosubstrates. *Applied Microbiology and Biotechnology* 44:15–19.

[202] Jin, F., Zhou, Z., Moriya, T., Kishida, H., Higashijima, H., and Enomoto, H. 2005. Controlling hydrothermal reaction pathways to improve acetic acid production from carbohydrate biomass. *Environmental Science Technology* 39:1893–1902.

[203] Ravinder, T., Ramesh, B., Seenayya, G., and Reddy, G. 2000. Fermentative production of acetic acid from various pure and natural cellulosic materials by *clostridium lentocellum* SG6. *World Journal of Microbiology & Biotechnology* 16:507–512.

[204] Coppola, C.M., and Schuster, H.F. 1997. *α-Hydroxy Acids in Enantioselective Syntheses*. Wiley-VCH, Weinheim, Germany. 513 p.

[205] Miltenberger, K. 2012. Hydroxycarboxylic acids, aliphatic. In: *Ullmann's Encyclopedia of Industrial Chemistry*. Wiley-VCH, Weinheim, Germany. Pp. 481–492.

[206] Tsao, G.T., Cao, N.J., Du, J., and Gong, C.S. 1999. Production of multifunctional organic acids from renewable resources. *Advances in Biochemical Engineering/Biotechnology* 65:243–280.

[207] Datta, R., and Henry, M. 2006. Lactic acid: recent advances in products, processes and technologies – a review. *Journal of Chemical Technology & Biotechnology* 81:1119–1129.

[208] Corma, A., Iborra, S., and Velty, A. 2007. Chemical routes for the transformation of biomass into chemicals. *Chemical Reviews* 107:2411–2502.

[209] Narayanan, N., Roychoudhury, P.K., and Srivastava, A. 2004. L (+) lactic acid fermentation and its product polymerization. *Electronic Journal of Biotechnology* 7(2):167–179.

[210] Borsook, H., Huffman, H.M., and Liu, Y.-P. 1933. The preparation of crystalline lactic acid. *Journal of Biological Chemistry* 102:449–460.

[211] Ratchford, W.P., Harris, E.H. Jr., Fisher, C.H., and Willits, C.O. 1961. Extraction of lactic acid from water solution by amine-solvent mixtures. *Industrial & Engineering Chemistry* 43(3):778–781.

[212] Tay, A., and Yang, S.-T. 2002. Production of L-(+)-lactic acid from glucose and starch by immobilized cells of *Rhizopus oryzae* in a rotating fibrous bed bioreactor. *Biotechnology and Bioengineering* 80(1):1–12.

[213] Joglekar, H.G., Rahman, I., Babu, S., Kulkarni, B.D., and Joshi, A. 2006. Comparative assessment of downstream processing options for lactic acid – a review. *Separation and Purification Technology* 52:1–17.

[214] Venus, J. 2006. Utilization of renewables for lactic acid fermentation. *Biotechnology Journal* 1:1428–1432.

[215] Xiaboo, X., Jianping, L., and Peilin, C. 2006. Advances in the research and development of acrylic acid production from biomass. *Chinese Journal of Chemical Engineering* 14:419–427.

[216] John, R.P., and Nampoothiri, K.M. 2007. Fermentative production of lactic acid from biomass: an overview on process developments and future perspectives. *Applied Microbiology and Biotechnology* 74:524–534.

[217] Gavilà, L., Constantí, M., and Medina, F. 2015. D-Lactic acid production from cellulose: dilute acid treatment of cellulose assisted by microwave followed by microbial fermentation. *Cellulose* 22:3089–3098.

[218] Leonard, R.H., Peterson, W.H., and Johnson, M.J. 1948. Lactic acid from fermentation of sulfite waste liquor. *Industrial and Engineering Chemistry* 40(1):57–67.

[219] Smith, L.T., and Claborn, H.V. 1939. Utilization of lactic acid. *Industrial & Engineering Chemistry, News Edition* 17:370–371.

[220] Xu, X., Lin, J., and Cen, P. 2006. Advances in the research and development of acrylic acid production from biomass. *Chinese Journal of Chemical Engineering* 14:419–427.

[221] De Vuyst, L., and Leroy, F. 2007. Bacteriocins from lactic acid bacteria; production, purification, and food applications. *Journal of Molecular Microbiology and Biotechnology* 13:194–197.

[222] Esmaeili, N., Bakare, F.O., Skrifvars, M., Afshar, S.J., and Åkesson, D. 2015. Mechanical properties for bio-based thermoset composites made from lactic acid, glycerol and viscose fibers. *Cellulose* 22:603–613.

[223] van Maris, A.J.A., Konings, W.N., van Dijken, J.P., and Pronk, J.T. 2004. Microbial export of lactic acid and 3-hydroxypropanoic acid: implications for industrial fermentation processes. *Metabolic Engineering* 6:245–255.

[224] Sott, G. 2000. 'Green' polymers – an invited review. *Polymer Degradation and Stability* 68:1–7.

[225] Ammala, A., Bateman, S., Dean, K., Petinakis, E., Sangwan, P., Wong, S., Yuan, Q., Yu, L., Patrick, C., and Leong, K.H. 2011. An overview of degradable and biodegradable polyolefins. *Progress in Polymer Science* 36(8):1015–1049.

[226] Hernández, N., Williams, R.C., and Cochran, E.W. 2014. The battle for the "green" polymer. different approaches for biopolymer synthesis: bioadvantaged vs. bioreplacement. *Organic and Biomolecular Chemistry* 12(18):2834–2849.

[227] Kubowicz, S., and Booth, A.M. 2017. Biodegradability of plastics: challenges and misconceptions. *Environmental Science & Technology* 51(21):12058–12060.

[228] Amass, W., Amass, A., and Tighe, B. 1998. A review of biodegradable polymers: uses, current developments in the synthesis and characterization of biodegradable polyesters, blends of biodegradable polymers and recent advances in biodegradable studies. *Polymer International* 47:89–144.

[229] Lunt, J. 1998. Large-scale production, properties and commercial applications of polylactic acid polymers. *Polymer Degradation and Stability* 59:145–152.

[230] Garlotta, D. 2001. A literature review of poly(lactic acid). *Journal of Polymers and the Environment* 9:63–84.

[231] Södergård, A., and Stolt, M. 2002. Properties of lactic acid based polymers and their correlation with composition. *Progress in Polymer Science* 27:1123–1163.

[232] Henton, D.E., Gruber, P., Lunt, J., and Randall, J. 2005. Polylactic acid technology. In: Mohanty, A.K., Misra, M., and Drzal, L.T. (Eds.). *Natural Fibers, Biopolymers, and Biocomposites*. CRC Taylor & Francis, Boca Raton, FL, NY, USA. Pp. 527–577.

[233] Avérous, L. 2008. Polylactic acid: synthesis, properties and applications. In: Belgacem, M.N., and Gandini, A. (Eds.). Monomers, Polymers and Composites from Renewable Resources. Elsevier, Amsterdam, The Netherlands. Pp. 433–450.

[234] Lim, L.-T., Auras, R., and Rubino, M. 2008. Processing technologies for poly(lactic acid). *Progress in Polymer Science* 33:820–852.

[235] Groot, W., van Krieken, J., Sliekersl, O., and de Vos, S. 2010. Production and purification of lactic acid and lactide. In: Auras, T., Lim, L.-T., Selke, S.E.M., and Tsuji, H. (Eds.). *Poly(lactic acid): Synthesis, Structures, Properties, Processing, and Applications*. John Wiley & Sons, New York, NY, USA. Pp. 3–18.

[236] Madhavan Nampoothiri, K., Nair, N.R., and John, R.P. 2010. An overview of the recent developments in polylactide (PLA) research. *Bioresource Technology* 101:8493–8501.

[237] Södergård, A., and Stolt, M. 2010. Industrial production of high molecular weight poly(lactic acid). In: Auras, T., Lim, L.-T., Selke, S.E.M., and Tsuji, H. (Eds.). *Poly(lactic acid): Synthesis, Structures, Properties, Processing, and Applications*. John Wiley & Sons, New York, NY, USA. Pp. 27–41.

[238] Lasprilla, A.J.R., Martinez, G.A.R., Lunelli, B.H., Jardini, A.L., and Filho, R.M. 2012. Poly-lactic acid synthesis for application in biomedical devices – a review. *Biotechnology Advances* 30:321–328.

[239] Ramot, Y., Haim-Zada, M., Domb, A.J., and Nyska, A. 2016. Biocompatibility and safety of PLA and its copolymers. *Advanced Drug Delivery Reviews* 107:153–162.

[240] Alén, R., and Sjöström, E. 1980. Condensation of glycolic, lactic and 2-hydroxybutanoic acids during heating and identification of the condensation products by GLC-MS. *Acta Chemica Scandinavica* 34:633–636.

[241] Hiljanen-Vainio, M., Karjalainen, T., and Seppälä, J. 1996. Biodegradable lactone copolymers. I. Characterization and mechanical behavior of ε-caprolactone and lactide copolymers. *Journal of Applied Polymer Science* 59:1281–1288.

[242] Rasal, R.M., Janorkar, A.V., and Hirt, D.E. 2010. Poly(lactic acid) modifications. *Progress in Polymer Science* 35:338–356.

[243] Fang, C.-c., Zhang, Y., Qi, S.-y., Li, Y.-y., and Wang, P. 2020. Characterization and analyses of degradable composites made with needle-punched jute nonwoven and polylactic acid (PLA) membrane. *Cellulose* 27:5971–5980.

[244] Middleton, J.C., and Tipton, A.J. 2000. Synthetic biodegradable polymers as orthopedic devices. *Biomaterials* 21:2335–2346.

[245] Siracusa, V., Rocculi, P., Romani, S., and Rosa, M.D. 2008. Biodegradable polymers for food packaging: a review. *Trends in Food Science and Technology* 19:634–643.

[246] Conn, R.E., Kolstad, J.J., Borzelleca, J.F., Dixler, D.S., Filer, L.J., Ladu, B.N., and Pariza, M.W. 1995. Safety assessment of polylactide (PLA) for use as a food-contact polymer. *Food and Chemical Toxicology* 33:273–283.

[247] Chen, G., Ushida, T., and Tateishi, T. 2002. Scaffold design for tissue engineering. *Macromolecular Bioscience* 2:67–77.

[248] Anderson, M.J., and Shive, M.S. 2012. Biodegradation and biocompatibility of PLA and PLGA microspheres. *Advanced Drug Delivery Reviews* 64:72–82.

[249] Soucaille, P. 2007. Glycolic acid production by fermentation from renewable resources. Patent WO/2007/140816.

[250] He, Y.-C., Xu, J.-H., Su, J.H., and Zhou, L. 2010. Bioproduction of glycolic acid from glycolonitrile with a new bacterial isolate of *Alcaligens* sp. ECU401. *Applied Biochemistry and Biotechnology* 160:1428–1440.

[251] Yunhai, S., Houyong, S., Haiyong, C., Deming, L., and Qinghua, L. 2011. Synergistic extraction of glycolic acid from glycolonitrile hydrolysate. *Industrial & Engineering Chemistry Research* 50:8216–8224.

[252] Wei, G., Yang, X., Gan, T., Zhou, W., Lin, J., and Wei, D. 2009. High cell density fermentation of *gluconobacter oxydans* DSM 2003 for glycolic acid production. *Journal of Industrial Microbiology and Biotechnology* 36:1029–1034.

[253] Athanasiou, K.A., Niederauer, G.G., and Agarwal, C.M. 1996. Sterilization, toxicity, biocompatibility and clinical applications pf polylactic acid/polyglycolic acid copolymers. *Biomaterials* 17:93–102.

[254] Malin, M., Hiljanen-Vainio, M., Karjalainen, T., and Seppälä, J. 1996. Biodegradable lactone copolymers. II. hydrolytic study of ε-caprolactone and lactide copolymers. *Journal of Applied Polymer Science* 59:1289–1298.

[255] Doi, Y., and Steinbüchel, A. (Eds.). 2002. *Biopolymers, Polyesters I – Biological Systems and Biotechnological Production*. Wiley-VCH, Weinheim, Germany. 472 p.

[256] Lu, J., Tappel, R.C., and Nomura, C.T. 2009. Mini-review: biosynthesis of poly(hydroroxyalkanoates). *Polymer Reviews* 49(3):226–248.

[257] Haddadi, M.H., Asadolahi, R., and Negahdari, B. 2019. The bioextraction of bioplastics with focus on polyhydroxybutyrate: a review. *International Journal of Environmental Science Technology* 16:3935–3948.

[258] Moran, L.A., Scrimgeour, K.G., Horton, H.R., Ochs, R.S., and Rawn, J.D. 1994. *Biochemistry*. 2nd edition. Prentice Hall, Upper Saddle River, NJ, USA. Pp. 16-1–16-28.

[259] Song, H., and Lee, S.Y. 2006. Production of succinic acid by bacterial fermentation. *Enzyme and Microbial Technology* 39:352–361.

[260] Bechthold, I., Bretz, K., Kabasci, S., Kopitzky, R., and Springer A. 2008. Succinic acid: a new platform chemical for biobased polymers from renewable resources. *Chemical Engineering and Technology* 31:647–654.

[261] Zeikus J.G., Jain, M.K., and Elankovan, P. 1999. Biotechnology of succinic acid production and markets for derived industrial products. *Applied Microbiology and Biotechnology* 51:545–552.

[262] Huh, Y.S., Jun, Y.-S., Hong, Y.K., Song, H., Lee, S.Y., and Hong, W. H. 2006. Effective purification of succinic acid from fermentation broth produced by *Mannheimia succiniciproducens*. *Process Biochemistry* 41:1461–1465.

[263] Zheng, P., Dong, J.-J., Sun, Z.-H., Ni, Y., and Fang, L. 2009. Fermentative production of succinic acid from straw hydrolysate by *Actinobacillus succinogenes*. *Bioresource Technology* 100:2425–2429.

[264] Steiger, M.G., Blumhoff, M.L., Mattanovich, D., and Sauer, M. 2013. Biochemistry of microbial itaconic acid production. *Frontiers in Microbiology* 4:article 23.

[265] Sheldon, R.A. 2014. Green and sustainable manufacture of chemicals from biomass: state of art. *Green Chemistry* 16(3):950–963.

[266] Hajian, H., and Yusoff, W.M.W. 2015. Itaconic acid production by microorganisms: a review. *Current Research Journal of Biological Sciences* 7(2):37–42.

[267] Tabuchi, T., Sugisawa, T., Ishidori, T., Nakahara, T., and Sugiyama, J. 1981. Itaconic acid fermentation by a yeast belonging to the genus *Candida*. *Agricultural and Biological Chemistry* 45:475–479.

[268] Wilke, T., and Vorlop, K.-D. 2001. Biotechnological production of itaconic acid. *Applied Microbiology and Biotechnology* 56:289–295.

[269] Reddy, C.S.K., and Singh, R.P. 2002. Enhanced production of itaconic acid from corn starch and market refuse fruits by genetically manipulated *Aspergillus terreus* SKR10. *Bioresearch Technology* 85:69–71.

[270] Meena, V., Sumanjali, A., Dwarka, K., Subburathinam, K.M., and Sambasiva Rao, K.R.S. 2010. Production of itaconic acid through submerged fermentation employing different species of *aspergillus*. *Rasayan Journal of Chemistry* 3(1):100–109.

[271] Kieboom, A.P.G., and van Bekkum, H. 1984. Aspects of the chemical conversion of glucose. *Recueil, Journal of the Royal Netherlands Chemical Society* 103(1):1–12.

[272] Röper, H. 1990. Selective oxidation of D-glucose: chiral intermediates for industrial utilization. *Starch* 42(9):336–341.

[273] Vuorinen, T., Hyppänen, T., and Sjöström, E. 1991. Oxidation of D-glucose with oxygen in alkaline methanol-water mixtures: a convenient method of producing crystalline sodium D-arabinonate. *Starch* 43(5):194–198.

[274] Watanabe, M., Aizawa, Y., Iida, T., Aida, T.M., Levy, C., Sue, K., and Inomata, H. 2005. Glucose reactions with acid and base catalysts in hot compressed water at 473 K. *Carbohydrate Research* 340:1925–1930.

[275] Garves, K. 1981. Dehydration and oxidation of cellulose hydrolysis products in acidic solution. *Journal of Wood Chemistry and Technology* 1(2):223–235.

[276] Hearon, W.M., Lo, C.F., and Witte, J.F. 1975. Chemical from cellulose. *Applied Polymer Symposium* 28:77–84.

[277] Strong, E.B., Kirschbaum, C.W., Martinez, A.W., and Martinez, N.W. 2018. Paper miniaturization via periodate oxidation of cellulose. *Cellulose* 25:3211–3217.

[278] Wang, Y., Zhao, D., Rodriguez-Padrón, D., and Len, C. 2019. Recent advances in catalytic hydrogenation of furfural. *Catalysts* 9(10):796–829.

[279] Hitchcock, L.B., and Duffey, H.R. 1948. Commercial production of furfural in its twenty-fifth year. *Chemical Engineering Progress* 44(9):669–674.

[280] Root, D.F., Saeman, J.F., Harris, J.F., and Neill, W.K. 1959. Chemical conversion of wood residues, part II: kinetics of the acid-catalyzed conversion of xylose to furfural. *Forest Products Journal* 9(5):158–165.

[281] Kalninsh, A.Y., and Vedernikov, N.A. 1975. Utilization of hardwood as a chemical raw material in Latvian SSR. *Applied Polymer Symposium* 28:125–130.

[282] Harris, J.F. 1978. Process alternatives for furfural production. *TAPPI* 61(1):41–44.

[283] Mamman, A.S., Lee, J.-M., Kim, Y.-C., Hwang, I.T., Park, N.-J., Hwang, Y.K., Chang, J.-S., and Hwang, J.-S. 2008. Furfural: Hemicellulose/xylose-derived biochemical. *Biofuels, Bioproducts and Biorefining* 2:438–454.

[284] Choudhary, V., Pinar, A.B., Sandler, S.I., Vlachos, D.G., and Lobo, R.F. 2011. Xylose isomerization to xylulose and its dehydration to furfural in aqueous media. *ACS Catalysis* 1:1724–1728.

[285] Rong, C., Ding, X., Zhu, Y., Li, Y., Wang, L., Qu, Y., Ma, X., and Wang, Z. 2012. Production of furfural from xylose at atmospheric pressure by dilute sulfuric acid and inorganic salts. *Carbohydrate Research* 350:77–80.

[286] Yang, W., Li, P., Bo, D., Chang, H., Wang, X., and Zhu, T. 2013. Optimization of furfural production from D-xylose with formic acid as catalyst in a reactive extraction system. *Bioresource Technology* 133:361–369.

[287] Wang, Y., Delbecq, F., Kwapinski, W., and Len, C. 2017. Application of sulfonated carbon-based catalyst for the furfural production from D-xylose and xylan in a microwave-assisted biphasic reaction. *Molecular Catalysis* 438:167–172.

[288] Seo, G., and Chon, H. 1981. Hydrogenation of furfural over copper-containing catalysts. *Journal of Catalysis* 67:424–429.

[289] Liu, D., Zemlyanov, D., Wu, T., Lobo-Lapidus, R.J., Dumesic, J.A., Miller, J.T., and Marshall, C.L. 2013. Deactivation mechanistic studies of copper chromite catalyst for selective hydrogenation of 2-furfuraldehyde. *Journal of Catalysis* 299:336–345.

[290] Fulajtárova, K., Soták, T., Hronec, M., Vávra, I., Dobročka, E., and Omastová, M. 2015. Aqueous phase hydrogenation of furfural to furfuryl alcohol over Pd-Cu catalysts. *Applied Catalysts A: General* 502:78–85.

[291] Manikandan, M., Venugopal, A.K., Nagpure, A.S., Chilukuri, S., and Raja, T. 2016. Promotional effect of Fe on the performance of supported Cu catalyst for ambient pressure hydrogenation of furfural. *RSC Advances* 6:3888–3898.

[292] Romano, P.N., de Almeida, J.M.A.R., Carvalho, Y., Priecel, P., Sousa-Aquiar, E.F., and Lopez-Sanchez, J.A. 2016. Microwave-assisted selective hydrogenation of furfural to furfuryl alcohol employing a green and noble metal-free copper catalyst. *ChemSusChem* 9:3387–3392.

[293] Dohade, M.G., and Dhepe, P.L. 2017. Efficient hydrogenation of concentrated aqueous furfural solutions into furfuryl alcohol under ambient conditions in presence of PtCo bimetallic catalyst. *Green Chemistry* 19:1144–1154.

[294] Liu, X., Zhang, B., Fei, B., Chen, X., Zhang, J., and Mu, X. 2017. Tunable and selective hydrogenation of furfural to furfuryl alcohol and cyclopentanone over Pt supported on biomass-derived porous heteroatom doped carbon. *Faraday Discussions* 202:79–98.

[295] Gong, W., Chen, C., Wang, H., Fan, R., Zhang, H., Wang, G., and Zhao, H. 2018. Sulfonate group modified Ni catalyst for highly efficient liquid-phase selective hydrogenation of bio-derived furfural. *Chinese Chemical Letters* 29:1617–1620.

[296] Yang, X., Meng, Q., Ding, G., Wang, Y., Chen, H., Zhu, Y.I., and Li, Y.W. 2018. Construction of novel Cu/ZnO-Al$_2$O$_3$ composites for furfural hydrogenation: the role of al components. *Applied Catalysis A, General* 561:78–86.

[297] An, Z., and Li, J. 2022. Recent advances in the catalytic transfer hydrogenation of furfural to furfuryl alcohol over heterogeneous catalysts. *Green Chemistry* 24:1780–1808.

[298] Sawyer, R.L., and Andrus, D.W. 1955. 2,3-Dihydropyran. *Organic Synthesis, Collective* 3:276–277.

[299] Merat, N., Godawa, C., and Gaset, A. 1990. Hydrogenation selective de l'alcool furfurylique en alcool tetrahydrofurfurylique *Journal of Molecular Catalysis* 57:397–415.

[300] de Wild, P.J. 2011. *Biomass Pyrolysis for Chemicals.* Doctoral Thesis. University of Groningen, The Netherlands. 163 p.

[301] Biradar, N.S., Hengne, A.A., Birajdar, S.N., Swami, R., Rode, C.V. 2014. Tailoring the product distribution with batch and continuous process options in catalytic hydrogenation of furfural. *Organic Process Research & Development* 18:1434–1442.

[302] Zhang, F., Zhang, B., Wang, X., Huang, L., Ji, D., Du, S., Ma, L., and Lin, S. 2018. Synthesis of tetrahydropyran from tetrahydrofurfuryl alcohol over Cu-ZnO/Al$_2$O$_3$ under a gaseous-phase condition. *Catalysts* 8:105–115.

[303] McKibbins, S.W., Harris, J.F., Saeman, J.F., and Neill, W.K. 1962. Chemical conversion of wood residues, part V. Kinetics of the acid catalyzed conversion of glucose to 5-hydroxymethyl-2-furaldehyde and levulinic acid. *Forest Products Journal* 12(1):17–23.

[304] Su, Y., Brown, H.M., Huang, X., Zhou, X.-D., Amonette, J.E., and Zhang, Z.C. 2009. Single-step conversion of cellulose to 5-hydroxymethylfurfural (HMF), a versatile platform chemical. *Applied Catalysis A: General* 361:117–122.

[305] Abou-Yousef, H., Hassan, E.B., and Steele, P. 2013. Rapid conversion of cellulose to 5-hydroxymethylfurfural using single and combined metal chloride catalysts in ionic liquid. *Journal of Fuel Chemistry and Technology* 41(2):214–222.

[306] Kumar, A., Chauhan, A.S., Shaifali, and Das, P. 2021. Lignocellulosic biomass and carbohydrates as feedstock for scalable production of 5-hydroxymethylfurfural. *Cellulose* 28:3967–3980.

[307] Shi, N., Zhu, Y., Qin, B., Liu, Y., Zhang, H., Huang, H., and Liu, Y. 2022. Conversion of cellulose into 5-hydroxymethylfurfural in an H$_2$O/tetrahydrofuran/cyclohexane biphasic system with Al$_2$(SO$_4$)$_3$ as the catalyst. *Cellulose* 29:2257–2272.

[308] Kuster, B.F.M., and van der Steen, H.J.C. 1977. Preparation of 5-hydroxymethylfurfural. part I. dehydration of fructose in continuous stirrer tank reactor. *Die Stärke* 29(3):99–103.

[309] Román-Leshkov, Y., Chheda, J.N., and Dumesic, J.A. 2006. Phase modifiers promote efficient production of hydroxymethylfurfural from fructose. *Science* 312(5782):1933–1937.

[310] Zhao, H., Holladay, J.E., Brown, H., and Zhang, Z.C. 2007. Metal chlorides in ionic liquid solvents convert sugars to 5-hydroxymethylfurfural. *Science* 316(5831):1597–1600.

[311] Qu, Y.S., Song, Y.L., Huang, C.P., Zhang, J., and Chen, B.H. 2011. Dehydration of fructose to 5-hydroxymethylfurfural catalyzed by alkaline ionic liquid. *Advanced Materials Research* 287–290:1585–1590.

[312] Kuster, B.F.M. 1990. 5-Hydroxymethylfurfural (HMF). A review focusing on its manufacture. *Starch* 42:314–321.

[313] Huber, G.W., Iborra, S., and Corma, A. 2006. Synthesis of transportation fuels from biomass: chemistry, catalysts, and engineering. *Chemical Reviews* 106(9):4044–4098.

[314] Gruter, G.-J., and de Jong, E. 2009. Furanics: novel fuel options from carbohydrates. *Biofuels Technology* 1:11–17.

[315] Chen, G., van Straalen, N.M., and Roelofs, D. 2016. The ecotoxicogenomic assessment of soil toxicity associated with the production chain of 2,5-furandicarboxylic acid (FDCA), a candidate bio-based green chemical building block. *Green Chemistry* 18:4420–4431.

[316] Bello, S., Salim, I., Méndez-Trelles, P., Rodil, E., Feijoo, G., and Moreira, M.T. 2018. Environmental sustainability assessment of HMF and FDCA production from lignocellulosic biomass through life cycle assessment (LCA). *Holzforschung* 73(1):105–115.

[317] Kim, H., Lee, S., Ahn, Y., Lee, J., and Won, W. 2020. Sustainable production of bioplastics from lignocellulosic biomass: technoeconomic analysis and life-cycle assessment. *ACS Sustainable Chemistry & Engineering* 8:12419–12429.

[318] van Strien, N., Rautiainen, S., Asikainen, M., Thomas, D.A., Linnekoski, J., Niemelä, K., and Harlin, A. 2020. A unique pathway to platform chemicals: aldaric acids as stable intermediates for the synthesis of furandicarboxylic acid esters. *Green Chemistry* 22:8271–8277.

[319] Burgess, S.K., Leisen, J.E., Kraftschik, B.E., Mubarak, C.R., Kriegel, R.M., and Koros, W.J. 2014. Chain mobility, thermal, and mechanical properties of poly(ethylene furanoate) compared to poly(ethylene terephthalate). *Macromolecules* 47:1383–1391.

[320] Burgess, S.K., Kriegel, R.M., and Koros, W.J. 2015. Carbon dioxide sorption and transport in amorphous poly(ethylene furanoate). *Macromolecules* 48:2184–2193.

[321] Araujo, C.F., Nolasco, M.M., Ribeiro-Claro, P.J.A., Rudic, S., Silvestre, A.J.D., Vaz, P.D., and Sousa, A.F. 2018. Chain conformation and dynamics in crystalline and amorphous domains. *Macromolecules* 51:3515–3526.

[322] Kainulainen, T., Hukka, T.I., Özeren, H.D., Sirviö, J.A., Hedenqvist, M.S., and Heiskanen, J.P. 2020. Utilizing furfural-based bifuran diester as monomer and comonomer for high-performance bioplastics: properties of poly(butylene furanoate), poly(butylene bifuranoate), and their copolyesters. *Biomacromolecules* 21:743–752.

[323] Vannini, M., Marchese, P., Celli, A., and Lorenzetti, C. 2015. Fully biobased poly(propylene 2,5-furandicarboxylate) for packaging applications: excellent barrier properties as a function of crystallinity. *Green Chemistry* 17:4162–4166.

[324] Wang, J., Liu, X., and Jiang, Y. 2017. Copolyesters based on 2,5-furandicarboxylic acid (FDCA): effect of 2,2,4,4-tetramethyl-1,3-cyclobutanediol units on their properties. *Polymers* 9:305–319.

[325] Shilling, W.L. 1965. Levulinic acid from wood residues. *TAPPI* 48(10):105A–108A.

[326] Kitano, M., Tanimoto, F., and Okabayashi, M. 1975. Levulinic acid, a new chemical raw material – its chemistry and use –. *Chemical Economy & Engineering Review* 7(7):25–29.

[327] Himmel, M., Ruth, M., and Wyman, C. 1999. Cellulase for commodity products from cellulosic biomass. *Current Opinion in Biotechnology* 10:358–364.

[328] Timokhin, B.V., Baransky, V.A., and Eliseeva, G.D. 1999. Levulinic acid in organic synthesis. *Russian Chemical Reviews* 68(1):73–84.

[329] Hayes, D.J., Fitzpatric, S., Hayes, M.H., and Ross, J.R.H. 2006. The biofine process – production of levulinic acid, furfural, and formic acid from lignocellulosic feedstocks. In: Kamm, B., Gruber, P.R., and Kamm, M. (Eds.). *Biorefineries – Industrial Processes and Products, Volume 1*. Wiley-VCH, Weinheim, Germany. Pp. 139–164.

[330] Girisuta, B. 2007. *Levulinic Acid from Lignocellulosic Biomass*. Doctoral Thesis, the University of Groningen, Groningen, the Netherlands. 149 p.

[331] Balasubramanian, S., and Venkatachalam, P. 2022. Green synthesis of carbon solid acid catalysts using methane sulfonic acid and its application in the conversion of cellulose to platform chemicals. *Cellulose* 29:1509–1526.

[332] Wiggins, L.F. 1949. The utilization of sucrose. *Advanced Carbohydrate Chemistry* 4:293–336.

[333] Leonard, R.H. 1956. Levulinic acid as a basic chemical raw material. *Industrial and Engineering Chemistry* 48(8):1331–1341.

4 Chemical pulping-based methods

4.1 General aspects

The general term "pulping" refers to the different processes that are used to convert wood (*i.e.*, pulpwood) or other fibrous lignocellulosic biomass feedstocks into a product mass of liberated fibers by releasing the lignin that binds cellulosic fibers together [1–12]. This thermal conversion can be accomplished, according to a broad classification, either chemically (*i.e.*, by means of chemicals) or mechanically or by combining these two main types of treatment (Table 4.1). Hence, the term "pulp" is collectively used to denote products from all these defibration processes. Although pulps are predominantly used for producing paper and board (cardboard or paperboard) and thus used in their obvious applications (writing and printing), some pulps are processed into a great number of specialty papers, various cellulose derivatives (*e.g.*, cellulose esters and ethers), and regenerated cellulose (*e.g.*, viscose or rayon) (see Chapter 6 "Cellulose derivatives").

Table 4.1: Commercial pulping methods [8].

Method	Yield (% of dry wood)
Chemical pulping	35–60
Kraft, polysulfide kraft, and	
pre-hydrolysis kraft	
Soda-anthraquinone (AQ)	
Acid sulfite, bisulfite, and	
alkali sulfite-AQ	
Multistage sulfite	
Semichemical pulping	65–85
Neutral sulfite semichemical (NSSC)	
Soda	
Chemimechanical pulping	80–90
Chemithermomechanical (CTMP)	
Chemigroundwood (CGWP)	
Mechanical pulping	91–98
Thermomechanical (TMP)	
Refiner mechanical (RMP)	
Stone groundwood (SGWP)	
Pressure groundwood (PGWP)	

The average yield of chemical pulps is in the range of 45–55% and the yields of dissolving pulps (*i.e.*, pulps from acidic sulfite, multistage sulfite, and pre-hydrolysis kraft methods) are typically 35–40% [8, 13]. During chemical pulping, lignin dissolves

https://doi.org/10.1515/9783110608366-004

into the cooking liquor, and the process, delignification, results in almost lignin-free pulp. Chemical pulping accounts for about 70% of the total worldwide production and comprises a wide range of processes, from alkaline to acidic; it generally uses inorganic reagents in aqueous solution at elevated temperatures and pressures. Over 90% of the chemical pulps (*i.e.*, almost 150 million tons annually [14]) are currently produced in the dominant kraft (sulfate) process (see Section 4.2 "Kraft pulping"). In contrast, sulfite pulping (see Section 4.3 "Sulfite pulping") has clearly decreased during recent decades. The term "high-yield pulp" is normally used for different types of lignin-rich pulps (*i.e.*, mainly from neutral sulfite pulping) that require mechanical defibration. Additionally, about 150 million tons of recovered fibers (*i.e.*, deinked pulp (DIP)) are used for papermaking, and one of the major trends is its growing use.

There are several major reasons for the dominance of the kraft process over other delignification methods (*i.e.*, mainly sulfite pulping) [8, 13]. Kraft pulp has excellent pulp strength properties, and the process can use low-quality wood as fibrous feedstock with well-established recovery of cooking chemicals, energy, and by-products (see Sections 4.2.2 "Reactions of the wood chemical constituents" and 5.3.3 "Fractionation"). However, there are also clear disadvantages, including high investment costs, low pulp yields, formation of odorous sulfur-containing compounds, and the need for effective bleaching of pulp (see Section 4.5.2 "Delignifying bleaching"). The kraft process is principally the same method as that in its early days about 160 years ago, but it has evolved in many ways, leading to several process solutions and modifications. Besides the progress and modifications in the cooking stage, major development has occurred, especially in the bleaching of the product, where the present trend has been toward a strong reduction in the use of chlorine-containing compounds, to protect the environment. For example, many countries no longer use elemental chlorine (Cl_2) in the bleaching phase, and the use of chemicals containing oxygen (*e.g.*, oxygen (O_2), hydrogen peroxide (H_2O_2), and ozone (O_3)) is rapidly growing.

There are three basic process types of commercial kraft pulping: *i*) batch cooking, *ii*) continuous-flow cooking, and *iii*) conveyor cooking [1–12]. Most chemical pulping processes using wood chips are based on batch and continuous-flow cooking methods. Because of its limited capacity, only a small number of mills utilize the conveyor method, and it is mainly used to delignify sawdust and annual plants. The batch cooking process can be divided into *i*) conventional batch process and *ii*) displacement batch process. The most obvious drawback of the latter process, compared with the former process, is the relatively large space needed for the tank farm. Additionally, there are more instruments than in conventional batch cooking and especially, in continuous cooking. The continuous-flow cooking technology was also developed from its conventional applications with different modifications (*i.e.*, modified continuous cooking), such as those with temperature and alkali profiling and systems using washing sections. The process also provides several advantages over the batch process. In general, a continuous digester requires less reactor volume per unit of retention time compared to a batch digester. Other major benefits over batch-cooking systems include lower process energy requirements (steam for bringing the reactants to processing temperature), less problems in controlling environmental impacts (gaseous emissions and steam contamination), more efficient process energy recovery, and the ability to achieve an efficient first stage of brown stock washing within the reactor. However, there are also some drawbacks of the continuous digester process. Because of the constant movement of the wood chips during impregnation and cooking with continuous digesters, the mechanical impacts on the fibers are slightly greater, resulting in increased fiber damage and consequently, reduced pulp strength properties. Furthermore, the use of only one reactor makes the process more sensitive to production stops during process disturbances and maintenance.

Carbohydrates and their derivatives have a large potential as starting materials for manufacturing a wide range of useful low-molar-mass and/or high-molar-mass products [15]. One of the potential approaches is to separate a prominent part of carbohydrates as monosaccharides and oligosaccharides already, prior to kraft pulping (see Section 5.2 "Pretreatments of lignocellulosic feedstocks prior to pulping"), instead of recovering their degradation products, for example, from the acidic sulfite spent liquor. This kind of removal of organic material, prior to delignification under acidic conditions, has also traditionally been performed, especially for making dissolving pulps with a very low hemicellulose content (*cf.*, pre-hydrolysis kraft pulping) [3]. In the pre-hydrolysis methods developed in the 1940s, chips are pretreated with dilute acids or with steam, above atmospheric pressure.

Wood is the predominant source of chemical pulp, representing, at present, about 90% of the total, with the rest coming from various fibrous non-wood feedstocks [13]. The traditional delignification methods for non-wood feedstocks are soda and soda-AQ processes, although different sulfite and kraft processes are also applied. Pulping of non-wood feedstocks as well as wood raw materials with organic solvents (*i.e.*, organosolv methods) dates back to the beginning of the 1930s but was not seriously considered for practical applications until the 1980s (see Section 4.4.2 "Organosolv pulping"). The basic idea was to use suitable "lignin solvents" for this purpose and the processes were systematically investigated with a view to develop novel industrial pulping methods.

It can be concluded that according to the idea of biorefinery concept and circular economy (see Chapter 1 "Introduction to biorefining"), with respect to large-scale integrated production of present and new by-products, chemical pulping offers one of the most suitable alternatives for developing the wood-refinery concept into practice [8, 10, 16–23]. The wood material is primarily fractionated in the kraft process by means of the cooking liquor into pulp (*i.e.*, mainly cellulose), extractives (*i.e.*, turpentine and tall oil), and the most important by-product, black liquor dry matter, which consists, besides inorganic substances, preliminary degraded lignin and carbohydrates-derived (mostly from hemicelluloses) aliphatic carboxylic acids as well as some residual extractives. Today, the chemical pulp industry is an important branch of global industrial activities, based on vast and multidisciplinary technology.

4.2 Kraft pulping

4.2.1 Process description

Wood preparation in wood handling plants in a kraft mill consists of a series of operations, which cover processing of wood logs into a form (*i.e.*, bark-free chips) suitable for the subsequent pulping stage [11]. These operations also include the handling of bark, which is recognized as a valuable fuel source and normally burnt together with other wood waste and activated sludge in a bark-burning boiler. The quality of chips is generally measured by uniformity in size (*i.e.*, can contain oversized, pin chips, and fines) and by the relative absence of various contaminants (*i.e.*, can contain bark residues, burnt, and rotten wood as well as dirt and foreign materials).

The operations of the conventional woodroom flow sequence for logs is to present to pulping, the wood received from the forest in the desired form and purity and at a suitable rate (*i.e.*, to ensure a steady supply) [1]. Before debarking of wood logs, important operations include *i*) measurement of the incoming wood raw material, *ii*) wood storage arrangements, and *iii*) deicing (if needed), cutting, and feeding to debarking. The main post-digester operations consist of different screening systems for removal of knots and removal and disposition of undesirable fiber fractions (normally, after pulp washing) [24]. After these basic operations, the pulp is finally ready for further processing.

In kraft pulping, roughly half of the initial feedstock dry material degrades and dissolves in the cooking liquor [3–7, 9, 11, 25]. Figure 4.1 illustrates a simplified diagram of kraft pulping and cyclic nature of the chemical recovery process. Hence, the essential sequential steps in kraft pulping are, besides the delignification stage of debarked chips with white liquor, pulp (brown stock) washing, evaporation and combustion of black liquor, and causticizing (or recausticizing) of green liquor. It can be concluded that the kraft process is based on an efficient reuse and recirculation of inorganic cooking chemicals. Furthermore, it should be noted that an important function of the recovery furnace is also to reduce the oxidized inorganic sulfur compounds (mainly sodium sulfate (Na_2SO_4)) to sodium sulfide (Na_2S), and this operation is indicated as "reduction efficiency".

In the delignification process, white liquor (see Section 5.3 "Black liquor"), containing mainly the active cooking chemicals, sodium hydroxide (NaOH), and Na_2S in water (H_2O), is used for cooking the chips. After cooking or digestion, the spent cooking liquor (black liquor) is separated from the pulp and it, together with the pulp washing liquor, is concentrated to 65–80% solids content by the multiple-effect evaporators. The concentrated black liquor is then combusted in the recovery furnace for the recovery of cooking chemicals and generation of energy. The combustion produces an inorganic smelt of sodium carbonate (Na_2CO_3) and Na_2S, with a small amount of Na_2SO_4. The smelt is dissolved in H_2O to form green liquor, which is reacted in the causticizing stage with slaked lime (calcium hydroxide ($Ca(OH)_2$)) to convert Na_2CO_3 into NaOH and regenerate the original white liquor for a new cook. Due to the incomplete conversion reactions in the recovery cycle *i.e.*, $Na_2SO_4 \rightarrow Na_2S$ in the recovery furnace and $Na_2CO_3 \rightarrow 2NaOH$ in the causticizing stage

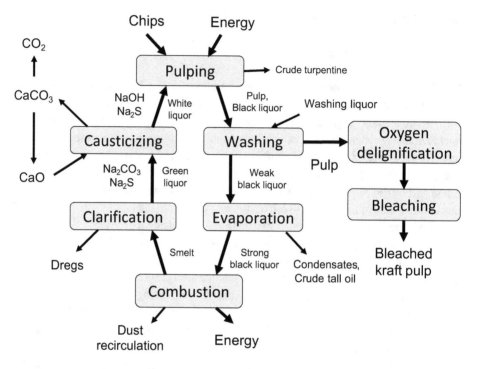

Figure 4.1: A simplified schematic illustration of kraft pulping and recovery cycle [26].

(both conversions are about 90%), the white liquor also contains some Na_2CO_3 and sodium salts of oxidized sulfur-containing anions as primary "dead-load chemicals".

According to the terminology, the active alkali ((AA): $[NaOH] + [Na_2S]$) and effective alkali ((EA): $[NaOH] + 1/2[Na_2S]$) are calculated as sodium equivalents (expressed as mass of NaOH or sodium oxide (Na_2O)) [4, 6]. Hence, the concentration of the liquor is expressed as grams of NaOH (predominantly in Europe) or Na_2O (predominantly in North America) per liter of solution. In modern pulping chemistry, molar units are also often used instead of mass units. It is also possible that both AA and EA are expressed as a percentage of dry-wood mass. The normal alkali requirement for kraft pulping of softwoods is 16–20% EA (as NaOH) on dry wood, while somewhat lower values are typical for hardwoods.

For example, if the concentrations of NaOH (molar mass 40.0 g/mol) and Na_2S (molar mass 78.0 g/mol) in white liquor are as follows [25]:

Chemical	Actual amount (g/L)	As NaOH (g/L)	As Na_2O (g/L)
NaOH	100	100.0	77.5
Na_2S	30	30.8	23.8

Then, AA is 130.8 g/L (100.0 g/L + 30.8 g/L) (as NaOH) or 101.3 g/l (77.5 g/L + 23.8 g/L) (as Na$_2$O) and EA is 115.4 g/L (100.0 + ½·30.8 g/L) (as NaOH) or 89.4 g/L (77.5 g/L + ½·23.8 g/L) (as Na$_2$S). Hence, AA > EA (as NaOH and Na$_2$O, respectively).

Sulfidity is defined as the ratio of Na$_2$S to AA (100 · ([Na$_2$S]/([NaOH] + [Na$_2$S])%), calculated as NaOH or Na$_2$O [4, 6]. The effects of Na$_2$S are quite dramatic up to a sulfidity of about 15%, but there appears to be only marginal additional benefit at higher sulfidity values. Most mills maintain white liquor sulfidity within the range of 25% to 35%. However, in today's kraft mills, sulfidity is gradually increased due to a more closed recovery cycle. In the example shown above, 100 × 30.8/130.8% = 23.5% (if all concentrations are expressed as NaOH) and as expected, the same 100 × 23.8/101.3% = 23.5% if all concentrations are expressed as Na$_2$O.

Besides the active cooking chemicals NaOH and Na$_2$S, the white liquor also contains appreciable amounts of other inorganic ions, and all these components are referred to collectively as dead load [3, 25]. These ineffective components include Na$_2$CO$_3$, Na$_2$SO$_4$, sodium sulfite (Na$_2$SO$_3$), and sodium thiosulfate(Na$_2$S$_2$O$_3$). Of them, as indicated above, the most prominent ones, Na$_2$CO$_3$ and Na$_2$SO$_4$, mainly result from incomplete causticizing and incomplete reduction in the recovery furnace, respectively. Although the dead-load chemicals do not take part in the pulping reactions, their high concentration levels in the white liquor are undesirable. They can cause scaling in the digester and especially in the evaporators, as well as increase generally the load of the recovery furnace. The combustion of black liquor in the recovery furnace produces an inorganic smelt of Na$_2$CO$_3$ and Na$_2$S with a small amount of Na$_2$SO$_4$ [6]. The smelt is dissolved in H$_2$O to form green liquor, which is reacted in the causticizing stage with lime (calcium oxide (CaO)) to convert Na$_2$CO$_3$ into NaOH and regenerate the original white liquor. The detailed chemical compositions of white and green liquors are shown in Section 5.3.1 "Chemical composition".

The term "liquor-to-wood ratio" describes the amount of total liquor per amount of dry wood charged to a digester [25]. This parameter is of importance, since the concentrations of the active cooking chemicals depends on it and thus, the actual effectiveness of these chemicals. Typical liquor-to-wood ratios vary between 3.5 and 5.0. For example, if a batch digester is loaded with 100 tons of chips with a moisture content of 25% and 275 tons of white liquor, the total amount of liquor is 100 × 0.25 tons + 275 tons = 300 tons and the total amount of dry wood is 100 × 0.75 tons = 75 tons. Hence, the liquor-to-wood ratio is 4.0 = 300 tons/75 tons. The unit can be also expressed as m^3/ton or L/kg.

The main purpose of kraft pulping is to liberate the fibers by dissolving them with the cooking chemicals, primarily the lignin-containing middle lamella, which binds the initial wood fibers together [6, 8]. However, the cooking yield is typically about 50% of the initial dry wood (the original total amount of lignin is 20–30% of the dry wood) and due to this nonselectivity of pulping, only about 90% of lignin is possible to be removed in a digester without a remarkable loss in pulp strength properties.

Technically, the digestion process can be conducted either as a batch or continuous process, having both typical functional features.

The pulp is separated after cooking from the black liquor in a controlled process stage known as brown stock washing [6, 25]. After this important stage, the pulp is introduced to more selective oxygen-alkali delignification or oxygen delignification (see Section 4.5.1 "Oxygen-alkali delignification"). The delignification degree in this phase is typically about 50%, if pulp quality is preserved. The final delignification stage is selective bleaching (see Section 4.5.2 "Delignifying bleaching"). However, the bleaching chemicals are rather expensive and cannot be recycled and reused. The chemical pulping and bleaching procedures are continuously modified, thus resulting in an increasing diversification of processes.

4.2.2 Reactions of the wood chemical constituents

Figure 4.2 indicates the typical material balance of wood organics over the kraft process, also including the subsequent oxygen-alkali delignification and the bleaching sequence, which produces kraft pulp with high brightness. As can be noted, about 90% of lignin, 60% of hemicelluloses, and 15% of cellulose are dissolved during the cook. Because of its crystalline nature and high degree of polymerization, cellulose suffers less material losses than the hemicelluloses [6]. Small amounts of polysaccharides that are removed are not completely degraded and they can be found as hemicellulose residues in the final black liquor [27–32]. In the conventional pine (*Pinus sylvestris*) and

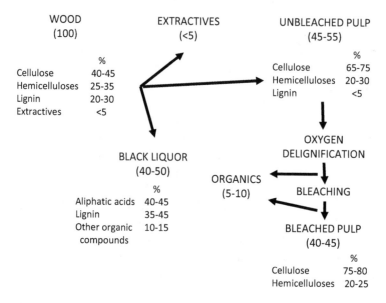

Figure 4.2: Typical material balance of wood organics in the kraft process for bleached pulp [8].

birch (*Betula pendula*) kraft pulping, the total cooking yield is 47% and 53% of the dry feedstock, respectively.

Table 4.2 outlines the basic reactions and phenomena between active alkali and the main wood constituents, resulting in the formation of various soluble fractions. In general, of the charged alkali (AA 18–24% of the wood dry solids), 70–75% is required for the neutralization of aliphatic carboxylic acids (*i.e.*, the degradation products of carbohydrates) and about 20% is consumed to neutralize the degraded fragments of lignin [6]. Aliphatic carboxylic acids in black liquor do not contain any sulfur, whereas the typical sulfur content of kraft lignin is 1–2% of the dry solids, corresponding to 10–20% of the initial charge. The approximate content of sodium in soluble aliphatic carboxylic acids and lignin is 20% and 8% of the dry solids, respectively.

Table 4.2: Main reactions and phenomena occurring in the feedstock chemical constituents during kraft pulping (for details, see text) [8].

LIGNIN
– Degradation (M_w decreases)
– Increase in hydrophilicity (liberation of phenolic groups)
$\rightarrow$ H_2O/alkali-soluble lignin fragments
HEMICELLULOSES, CELLULOSE
– Cleavage of acetyl groups
– Peeling reaction
– Stopping reaction
– Alkaline hydrolysis
$\rightarrow$ H_2O/alkali-soluble aliphatic carboxylic acids and hemicellulose fragments
EXTRACTIVES
– Hydrolysis of fatty acid esters (fats and waxes)
– Evaporation
$\rightarrow$ Tall oil soap and crude turpentine

Dissolution and degradation of lignin during kraft pulping produce a complex mixture of breakdown products with a wide molar mass distribution, ranging from simple low-molar-mass phenolic compounds to large macromolecules, and has been studied in detail [33–46]. On the other hand, the nature of the lignin remaining in the pulp at the end of the cook (*i.e.*, "residual lignin") has been characterized as well [47–52]. Due to the lack of selectivity for lignin in delignification, a substantial portion of polysaccharides is also degraded and converted, mainly to low-molar-mass hydroxy carboxylic acids that have also been explained in detail [4, 6, 8, 15, 43, 45]. The volatile extractives (*i.e.*, crude sulfate turpentine components) are chemically stable, but the nonvolatile fatty acid esters of the native wood are hydrolyzed (*i.e.*, saponification) almost completely and some of the unsaturated fatty acids and resin acids are also partly isomerized under alkaline pulping conditions [6, 8]. This nonvolatile part of extractives is recovered as tall oil soap from black liquor during its evaporation prior to combustion. The availability of these extractives-derived by-products in a mill strongly depends on the wood species used, the method and time of storing logs and chips, and the growth conditions of the trees; even among extractives-rich pine species, there are significant variations in the amounts of by-product extractives.

In general, it is known that hydrogen sulfide ions ($HS^\ominus$) primarily react with lignin, whereas carbohydrate reactions are only affected by alkalinity (*i.e.*, $HO^\ominus$ ions) [6]. This practically means that, as compared to soda pulping (*i.e.*, only active chemical is NaOH), kraft pulping is faster and provides a higher yield and a stronger pulp. In the conventional softwood kraft pulping where the cooking temperature has a relatively linear rise from 70 °C to 80 °C to the maximum value 160–170 °C (the temperature is then maintained constant) within a typical heating-up time of 90–120 min, the dissolution of lignin proceeds in three distinct phases [53]: *i*) extraction or initial phase, *ii*) bulk delignification, and *iii*) residual delignification. During the extraction phase, the selectivity of delignification is rather low, meaning that the lignin content is reduced only by 15–25% of the initial amount, as compared to about 40% of the hemicelluloses [2–4, 53]. As the temperature increases, delignification clearly accelerates, and above 140 °C, the rate of delignification is controlled by chemical reactions, and follows as a first-order reaction. The dissolution rate of lignin remains high during this bulk delignification phase until about 90% of the lignin is removed. During the residual delignification phase, delignification occurs at a much lower rate with an increasing loss of carbohydrates.

The overall effect of cooking time and temperature for kraft pulping can be estimated by a single numerical value, calculated according to the concept of the so-called "H-factor" [54]. In this method, the reaction rate at 100 °C is chosen as unity and reaction rates at all other temperatures are related to this standard. H-factor practically represents the area under a curve (*i.e.*, the integral of the corresponding function) in which the relative reaction rate is plotted against cooking time. Modern digester control systems automatically compute and accumulate this factor during the cook, thus monitoring the degree of delignification [6]. Depending on the mill, the final H-factor values typically vary in the range of 1,000 to 1,500. It is also known that the contribution of the heating-up time to H-factor is very small compared to that of cooking time at the maximum cooking temperature.

As indicated above, the selectivity of delignification is rather low, especially in the beginning of the cook, and leads to considerable carbohydrate losses [4, 6, 55]. In practice, this means that the wood polysaccharides are already attacked at the comparatively low temperatures when the chips come into contact with white liquor, or the recirculated black liquor is charged with white liquor to enhance impregnation prior to the actual cooking stage. The degradation of wood carbohydrates as well as the formation of several aliphatic carboxylic acids have also been studied in detail during kraft pulping [56–60]. Additionally, it has been found that there is a correlation between the degree of delignification and the formation of the selected carboxylic acids, analyzed by chromatographic methods [61–63]. One advantage of this approach is that instead of using absolute amounts, only the practical use of the concentration ratios of acids is needed. The same idea has also used for monitoring oxygen delignification [64] or predicting the combustion properties of black liquor [65] or even for monitoring kraft pulping with lignin monomers [66]. However, many new technologies have been adopted by the pulping industry since the 1980s [21, 67, 68], and most systems for the control of kraft pulping are normally based on the H-factor or other proper mathematical models [54, 69–75].

One of the most common volatile organic compounds (VOCs) formed (7–20 kg/ton of pulp) during kraft pulping is methanol (CH_3OH), which is the major alcohol in pulp mill process streams [76–81]. CH_3OH has been an environmental concern since, as a H_2O-soluble compound, it can increase the biochemical oxygen demand (BOD) and can also be released into the atmosphere at the process temperatures of kraft mill streams. CH_3OH is mainly formed (70–75% of the total) by the rapid alkali-catalyzed elimination from 4-O-methylglucuronic acid substituents in hemicelluloses, and a minor part of it can be obtained from the methoxyl (-OCH_3) groups of lignin [79, 81–83]. In general, it can be recovered from a mill foul condensate, formed by black liquor evaporation [84]. The recent CH_3OH purification process (consists of both CH_3OH cleaning and purification subprocess areas) [85] results in the cleaned CH_3OH, which contains about 99% less nitrogen and 90% less sulfur, than that present in the raw material. Hence, it is an ideal supporting biofuel, for example, for the recovery boiler or lime kiln without NO_X and SO_X emissions. The formation of volatile malodorous sulfur-containing compounds formed during kraft pulping is discussed below.

Comprehensive studies of metal profiles throughout the fiber lines and different process streams in the kraft mill have been performed [86–89]. Due to the trends toward more closure systems, the accumulation of non-process elements (NPEs) in kraft mill process streams is an increasingly important question. A major part of the intake of NPEs to the pulp mill is *via* the wood chips and a significant part of them dissolves into black liquor [88, 90]. NPEs can have many negative effects on kraft mill efficiency. High concentrations of NPEs may increase scaling on hot surfaces in heat exchangers and in black liquor evaporators [90]. Additionally, the accumulation of undesirable elements, such as chlorine and potassium (*e.g.*, the formation of KCl, K_2S, K_2CO_3, and K_2SO_4), may influence the recovery boiler corrosion and fouling rates caused by the enrichment of these elements in ash [88, 91, 92]. NPEs also reduce lime mud filtration efficiency and increase lime kiln fuel consumption [93–95]. In bleaching, calcium carbonate ($CaCO_3$) precipitates as well as sparingly soluble calcium oxalate, both lead to reduced production capacity [87, 90], and transition metals (*e.g.*, Mn, Fe, Co, and Cu) have negative impact on H_2O_2 bleaching of unbleached pulps (see Section 4.5.2 "Delignifying bleaching") [86, 88, 90, 96, 97]. Most such transition metals are precipitated from green liquor as dregs [88] and the removal of transition metals prior to bleaching can be made by chelating agents such as ethylenediaminetetraacetic acid (EDTA) or diethylenetriaminepentaacetic acid (DTPA) [6, 86]. The removal of metal ions from wood chips prior to pulping by acidic leaching has also been investigated [90, 98, 99].

As indicated in Section 5.3 "Black liquor", kraft black liquor as such is the most important by-product of kraft pulping. The entire liquor cycle comprises the generation of black liquor in the brownstock washers through evaporation, the recovery furnace, green and white liquor preparation, and lime kiln operation [100, 101]. Black liquor contains varying amounts of organic compounds derived from all the wood constituents, although the nature and amount of these compounds depend on the feedstock material used for the pulp production as well as on the pulping conditions applied [6, 8, 45, 102]. After the recovery of most extractives-based compounds (i.e., crude turpentine and tall oil soap), the remaining black liquor contains, besides inorganic substances, carbohydrate degradation products (i.e., aliphatic carboxylic acids and hemicellulose residues), lignin, and residual extractives. It is burned, after evaporation, in the recovery furnace to recover energy and cooking chemicals [100]. However, black liquor has many features, which when combined, make its combustion different from that of other fuels. The high content of inorganic material and H_2O (15–35%) decreases the heating value (12–15 MJ/kg dry solids) of the black liquor to a lower level than that of other common industrial fuels [103]. Additionally, black liquor generally swells substantially during combustion. In general, it can be concluded that each black liquor component plays a role in determining the thermochemical properties of black liquor.

The combustion behavior of black liquors has been studied in controlled laboratory conditions by burning single droplets or bigger quantities of liquor in an appropriate furnace with a challengeable gas atmosphere (either flowing or stagnant) [104, 105]. In these cases, the combustion parameters have included, for example, the duration of different burning stages (i.e., drying, pyrolysis, and char burning), droplet swelling (typically, by 10–60 times in volume [106–114]) during the pyrolysis stage and formation of different gaseous products during combustion as well as yields of pyrolysis and char burning. Besides these kinds of experiments, thermogravimetric (TG) investigations have also been used for clarifying the thermochemical properties of different black liquors. [100, 115–120].

The results obtained with the single-droplet technique have explained much about the relationships between the combustion behaviors of black liquors, which has allowed the combustion properties of various black liquors to be easily compared. For example, it has been generally observed by means of this technique that as a common trend in conventional laboratory-scale pulping [65, 121–123]: i) kraft black liquors from hardwood have slightly shorter burning times (pyrolysis time and char burning time) and swell more than those from softwood pulping, ii) large amounts of extractives reduce swelling, and iii) hemicelluloses residues increase swelling. Additionally, it has been found that burning time decreases and swelling increases with an increase in cooking time. However, in spite of these observations, so far there are only limited fundamental data available on the versatile mechanisms behind these overall effects.

Understanding the relationships between the main organic constituents and the different combustion stages of a black liquor droplet are of importance when considering the combustion properties of different black liquors. It has been assumed that aliphatic carboxylic acids have a significant influence on the drying rate of black liquor droplets because of the possibility of their forming intermolecular hydrogen bonds with H_2O molecules [8, 114]. On the other hand, compared to other constituents, these acids are relatively unstable on heating. Hence, it can be concluded that during the pyrolysis stage, the corresponding acid fraction is the main source of volatile degradation of products that is also necessary for swelling. In contrast, lignin has no such straightforward role in combustion (i.e., in pyrolysis and char burning stages). It is obvious that its aliphatic structural residues are readily converted into volatile compounds, whereas its aromatic moieties are important for char formation. There are also indications that the medium molar-mass lignin material (a fraction of 5–10 kDa), together with hemicellulose residues and aliphatic acids, plays an important role in hindering the escape of volatile products within a black liquor droplet. This requires that lignin polymers and other organic substances form a "highly elastic crust" in a certain temperature region during combustion. It has been also suggested [114] that the formation of a plastic state, essential for the swelling of black liquor, is partly a result of the melting of an array of the sodium salts of carboxylic acids and NaOH. Figure 4.3 shows some examples of the possible reaction routes occurring during the combustion of black liquor. These degra-

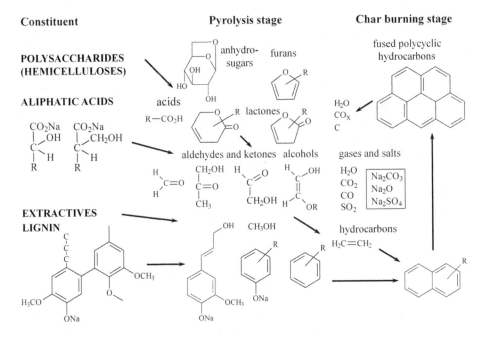

Figure 4.3: Examples of various thermochemical degradation reaction routes occurring during the combustion of black liquor in a recovery furnace [8].

dation mechanisms are similar to those detected in the characteristic pyrolysis of the main constituents of lignocellulosic biomass.

The operation of the kraft recovery furnace is initiated by spraying the concentrated liquor through one of several types of commercial nozzles [100]. Hence, the spraying characteristics of black liquor nozzles, together with liquor viscosity and nozzle pressure, are of practical importance [124, 125]. On the other hand, the fate of nitrogen and emissions of nitrogen-containing species (NO_x) from the black liquor recovery furnace [126–130], together with dust formation [131, 132], have also been studied in detail. In practice, in the modern mills, the sulfur emissions, such as sulfur dioxide (SO_2) and "total reduced sulfur" (TRS), from the black liquor recovery furnace are very low. Additionally, the combustion behavior of black liquors, for example, from heat treatment [133, 134], after partial lignin removal [135], from partial auto-caustization [136], and after the addition of totally chlorine-free (TCF) bleach plant filtrates [137] have been explained.

The recovery boilers that produce bioenergy have developed considerably in the past 80 years, culminating with units that are among the largest biofuel-fired boilers in the world [138]. The energy efficiencies of pulp and paper mills have improved, both in generation and consumption, and the newest pulp lines produce a surplus of energy without firing bark and wood residues or other auxiliary fuels [8]. Additionally, higher black liquor dry solids generate more steam. In spite of the increasing power generation (condensing power), electricity generation from black liquor has also become more significant.

The inorganic smelt from the recovery furnace is dissolved into weak liquor to form green liquor (see Section 5.3.1 "Chemical composition"), which contains mainly Na_2CO_3, Na_2S, and NaOH [11, 139]. The green liquor is clarified to remove "dregs" and in the causticizing stage, the Na_2CO_3 is converted to active NaOH (i.e., to white liquor, containing mainly NaOH and Na_2S) and various impurities introduced from the recovery furnace and lime kiln are removed. For this purpose, the clarified green liquor is reacted with $Ca(OH)_2$ and the white liquor formed is then clarified to remove precipitated "lime mud" (i.e., $CaCO_3$). The calcining ("reburning") of lime mud with external energy regenerates lime (i.e., to CaO), which reacts with H_2O to form $Ca(OH)_2$. The reactions are as follows:

$$Ca(OH)_2 + Na_2CO_3 \rightarrow 2NaOH + CaCO_3 \tag{4.1}$$

$$CaCO_3 \rightarrow CaO + CO_2 \tag{4.2}$$

$$CaO + H_2O \rightarrow Ca(OH)_2 + heat \tag{4.3}$$

The conventional causticization using CaO is today an integral part of the kraft process, although it also has some drawbacks in terms of high capital costs, external fuel consumption, low efficiency (80–90%), and energy economy (i.e., energy delivered to the system at high temperature and released at low temperature) [140]. Hence, these drawbacks have been the major drivers for the research and development of alternative processes [141–146]. According to the main concept of the non-conventional causticizing technologies (autocausticization), a H_2O-soluble amphoteric metal oxide or salt (*e.g.*, sodium metaborate $NaBO_2$ (or $Na_2O \cdot B_2O_3$), $Na_4P_2O_7$ (or $3Na_2O \cdot P_2O_5$), $Na_2Si_2O_5$ (or $Na_2O \cdot 2SiO_2$), or Al_2O_3) is added to the furnace that converts molten Na_2CO_3 directly to Na_2O and CO_2 under furnace conditions. The overall principle that takes place in the recovery furnace is between $NaBO_2$ and Na_2CO_3 in the molten smelt to form trisodium borate (Na_3BO_3), which reacts with H_2O in the smelt dissolving tank to form NaOH and regenerated $NaBO_2$:

$$NaBO_2 + Na_2CO_3 \rightarrow Na_3BO_3 + CO_2 \tag{4.4}$$

$$Na_3BO_3 + H_2O \rightarrow 2NaOH + NaBO_2 \tag{4.5}$$

In this system, however, a high amount of $NaBO_2$ that is carried through the pulping cycle and is inactive during delignification (*i.e.*, dead load), and the process has generally been found unattractive [140]. Hence, a concept of partial autocausticization has been developed. This practically means that only a minor part of the causticization occurs through the decarbonization agent and the remainder is causticized by conventional $Ca(OH)_2$. This kind of approach can be an attractive solution for mills that require incremental causticizing and lime kiln capacity. The system has also been tried in a full-scale application and, besides the reduction of lime requirement, the presence of $NaBO_2$ is been said to have only a little effect on the mill operations and pulp quality properties [147].

Lignin

During kraft pulping, the presence of $HS^\ominus$ ions greatly facilitates delignification because of their strong nucleophilicity, in comparison to $HO^\ominus$ ions [6]. Lignin undergoes more or less drastic degradation reactions and due to the liberation of phenolic hydroxyl (-OH) groups, the simultaneous increase in hydrophilicity of lignin fragments takes place. Most of the depolymerized fragments are H_2O/alkali-soluble and dissolved in the cooking liquor as sodium phenolates. The term "kraft lignin" (or in the case of soda pulping, "alkali lignin") is usually referred to as the soluble lignin degradation products in black liquor and their structures clearly differ from those of the native lignin and the residual pulp lignin [148–150].

In general, the nature of all the alkaline pulping reactions with lignin is nucleophilic [3, 4, 6, 7, 33, 35, 36]. In delignification, an essential aspect is the different behavior and stability of the various types of linkages between phenylpropane units in lignin. Depolymerization of lignin typically depends on the cleavage of all types of ether linkages (C_{aliph}-O-C_{arom}), whereas diaryl ethers (C_{arom}-O-C_{arom}) and the C-C bonds (especially C_{arom}-C_{arom}) are essentially stable. As α- and β-aryl ether linkages are the dominant types of linkages (50–70%) in both softwood and hardwood lignins, the cleavage of these linkages contributes essentially to lignin degradation [151]. However, during delignification, some condensation of lignin units (*i.e.*, the formation of new C-C bonds) also takes place, leading to fragments with an increased molar mass and a reduced solubility. The most common reactions are as follows (see Section 2.3.3 "Lignin"):

- cleavage of α-aryl ether linkages in free phenolic structures (nonetherified, free phenolic structures)
- cleavage of β-aryl ether linkages in free phenolic structures (nonetherified, free phenolic structures)
- cleavage of β-aryl ether linkages in nonphenolic structures (etherified phenolic structures)
- demethylation reactions
- condensation reactions

α-Aryl ether linkages in free phenolic structures are cleaved relatively easily during the initial phase of delignification (the first step of reaction route (a) in Figure 4.4) [6]. The corresponding linkages in pinoresinol and phenolic phenylcoumaran structures are also readily cleaved, usually followed by the release of formaldehyde (HCHO) or a proton (H⊕). However, the reaction results in the fragmentation of lignin only in the case of open α-ether structures. In contrast, the α-ether linkages are stable in nonphenolic structures.

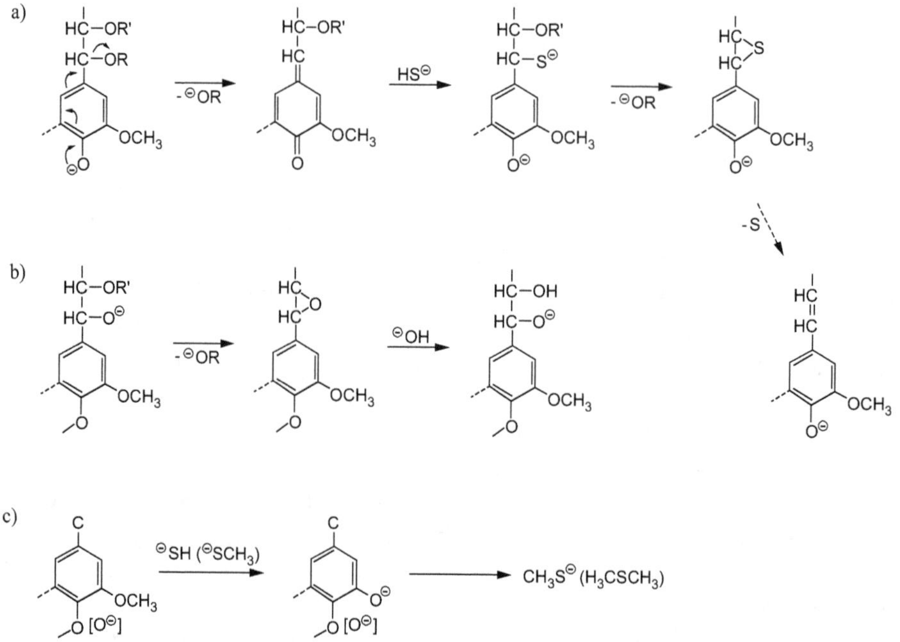

Figure 4.4: Degradation reactions of lignin during kraft pulping [6]. Cleavage of α- and β-aryl ether linkages in nonetherified phenolic structures (a), cleavage of β-aryl ether linkages in etherified structures (b), and cleavage of methyl aryl ether bonds (demethylation) (c).

β-Aryl ether linkages in free phenolic structures are also cleaved relatively easy during the initial phase of delignification (the reaction route (a) in Figure 4.4) [6]. The first step of the reaction involves after the elimination of the α-substituent (hydroxide, alkoxide or phenoxide ion) formation of a quinone-methide structure. The subsequent course of reactions depends on whether $HS^\ominus$ ions are present or not (cf., soda pulping without $HS^\ominus$ ions). In the case of kraft pulping, when $HS^\ominus$ ions are present, the $HS^\ominus$ ion reacts with the quinone methide to form a thiol derivative (a mercaptide structure), which is then converted into a thiirane intermediate, with the simultaneous cleavage of the β-ether bond.

The thiirane can be dimerized to a 1,4-dithiane structure, but this as well as other sulfur-containing intermediates are mainly decomposed, forming elemental sulfur and unsaturated styrene-type structures [6]. However, in the case of soda pulping, when only $HO^\ominus$ ions are present, the dominant reaction is the elimination of the hydroxymethyl group (i.e., hydroxylated γ-carbon atom) from the quinone-methide intermediate with the formation of HCHO and a styryl aryl ether structure (β-aroxy styrene unit), without cleavage of the β-ether bond.

β-Aryl ether linkages in nonphenolic structures are cleaved much more slowly, and the reaction is independent of the presence of the $HS^\ominus$ ions (the reaction route (b) in Figure 4.4) [6]. The reaction is promoted by anions at vicinal C-atoms and proceeds after generating a new free phenolic -OH group via an oxirane intermediate, which is subsequently opened with the formation of an α,β-glycol structure.

The -OCH_3 groups of lignin are predominantly cleaved by $HS^\ominus$ ions and only to a small extent by the less nucleophilic $HS^\ominus$ ions (i.e., the formation of CH_3OH, the reaction route (c) in Figure 4.4) [6, 152–154]. The decrease in the total amount of -OCH_3 groups in softwood lignins is about 10% during kraft pulping. In the case of $HS^\ominus$ ions, the formation of methyl mercaptan (CH_3SH (MM), b.p. 6.2 °C) takes place. This compound can further react in its ionized form with another -OCH_3 group, leading to dimethyl sulfide (CH_3SCH_3 (DMS), b.p. 37.3 °C). In the presence of oxygen, CH_3SH can be oxidized to dimethyl disulfide (CH_3SSCH_3 (DMDS), b.p. 109.7 °C). MM and DMS are highly volatile and extremely malodorous, causing an air pollution problem [155–157]. These gases, together with hydrogen sulfide (H_2S, b.p. −60.7 °C), are collectively referred to as TRS (see above), and the discharges are usually called "TRS emissions".

During kraft pulping, some condensation reactions can also occur [6, 34, 151, 158]. In these reactions, a phenolate unit is added to the quinone-methide intermediate, forming a new C-C bond by an irreversible release of a $H^\oplus$ (primary condensation). Since in the majority of cases, condensation takes place at the unoccupied C_5 position of phenolic units, an α-5 linkage is most frequently obtained. Additionally, it is possible that fragments with conjugated side-chain structures participate in a reaction, leading to a condensation product, but also liberating, simultaneously, one molecule of HCHO (secondary condensation). The released HCHO can further react with two phenolic units, resulting ultimately in a diarylmethane structure. However, a general result of all condensation reactions is that lignin dissolution is somewhat retarded.

Besides these kinds of reactions, certain unsaturated and highly conjugated lignin structures (*i.e.*, chromophores) and their precursors, leucochromophores, are introduced into pulps. These products are responsible for the typical dark color of unbleached pulps, which are generally called "brownstocks".

During kraft pulping, the degradation of lignin produces a variety of low-molar-mass compounds ("lignin monomers") as well [41, 42, 45]. The most extensive formation occurs in the bulk delignification phase and the total amount of these lignin-derived monomers increased continuously toward the end of the cook, being about 5% and 6% of the total lignin dissolved for birch and pine kraft pulping, respectively. In the case of birch pulping, the most prominent compounds (*e.g.*, syringol, acetosyringone, syringaldehyde, and syringic acid) originate from the syringyl units of lignin, whereas the corresponding products (*e.g.*, guaiacol, vanillin, and acetovanillone), derived from the guaiacyl units, are less abundant. In the case of pine pulping, the most prominent monomers are guaiacol, vanillin, acetovanillone, and dihydroconiferyl alcohol. The key intermediates in the formation of lignin monomers seem to be, at least, partly sinapyl alcohol (birch) and coniferyl alcohol (pine).

The weight average molar mass (M_w) of soluble lignin varies during alkaline pulping [159–161]. Typical values are presented in Table 2.10 in Section 2.3.3 "Lignin" for the beginning and the end of selected kraft cooks, carried out under conventional cooking conditions.

Carbohydrates

The degradation of carbohydrates in lignocellulosic materials under alkaline conditions, with or without oxygen, is a well-established phenomenon and it has been taken place clearly for more than a century in many industrial processes [6, 162–167]. The reactions proceed according to rather complex mechanistic pathways and several reaction mechanisms have been proposed over the years to describe the degradation process. However, the reactions under oxidative conditions, such as those occurring, for example, in the oxygen-alkali delignification of unbleached pulps (see Section 4.5.1 "Oxygen-alkali delignification") are analogous to the reactions under near complete absence of oxygen (*cf.*, kraft cooking), but with certain respects, the detailed mechanisms are somewhat different. In general, the behavior of cellulose under alkaline conditions is currently known in detail, including primary peeling and alkaline hydrolysis, with secondary peeling, together with stabilization reaction (stopping reaction) that competes with various peeling reactions.

Some exhaustive experiments dealing with the reactions of carbohydrates with alkali were performed 115–130 years ago [168–171], but the first generally accepted theory about alkaline carbohydrate degradation was not made until 1917 when Nef *et al.* [172]. described the two-step reaction of monosaccharides with $HO^{\ominus}$ ions; an isomerization to a α-dicarbonyl intermediate, followed by the formation of acidic degradation products through a benzilic acid-type rearrangement. Almost 30 years later,

Isbell [173] modified further the reaction mechanism that is generally referred to as the "Nef-Isbell mechanism" for alkaline carbohydrate degradation. The theory was clarified at that time in detail in many investigations [174–176] and has gradually become the present opinion. One decisive reason for this success was the gradual development of chromatography applications for determining more qualitatively and quantitatively various reaction products; a deeper understanding about the reaction mechanisms and kinetics could be gained. The degradation of polysaccharides, particularly in the context of kraft pulping, has been recently described in many publications [3, 6, 45, 177–184]. The data shown are generally based on the laboratory-scale experiments, such as those with cellulose [185–195] (under drastic conditions [196–198]), glucomannan [199–202], and xylan [203–206]. Additionally, some experiments, for example, with simple cellulose model compounds have been made [207, 208].

Under alkaline conditions, substantial mass losses of polysaccharides with a reducing end group (*i.e.*, a hemiacetal structure) occur [6]. Davidson [209] proposed for this phenomenon, the general name, "peeling reaction", which, however, is today often called more accurately, the "primary peeling reaction" (Figure 4.5) [6]. The stopping reaction (*i.e.*, a stabilization reaction) of polysaccharides competes with the peeling reaction and prevents further cleavage of the chains arising from the primary peeling process; it gives rise to the formation of carboxylic acid (-CO$_2$H) end groups that are stable under alkaline conditions [184]. The stopping reaction rate is also proportional to the concentration of reducing end groups [190]. The ratio between the peeling rate and the stopping rate is typically in the rough magnitude of 50–90:1 [6, 190, 210, 211]. Hence, the successive peeling degradation of cellulosic molecules in alkaline solvents does not proceed unto entire polymer dissolution. In general, amorphous hemicelluloses and lesser ordered amorphous cellulose regions degrade more readily than those of crystalline cellulose. The peeling rate decreases with an increasing time due to the reduced accessibility and the availability of reducing end group moieties. The peeling reaction rate can be even halted when the degrading chain reaches the chemically inaccessible crystalline region of the cellulose molecule, and the reducing group remains unattacked [45, 195]. This phenomenon is referred to as "the physical stopping reaction".

Toward the end of the heating-up period in a conventional cooking process (*i.e.*, a relatively linear rise of temperature from 70–80 °C to the maximum value 160–170 °C within 90–120 min), the random alkaline hydrolysis of glycosidic bonds of polysaccharides becomes energetically possible [6, 212, 213]. As this alkali-catalyzed hydrolysis happens, new reducing end groups are also simultaneously formed to the partially degraded polysaccharide chains or those already stabilized by the stopping reaction, thus making possible the secondary peeling reaction. Additionally, under the influence of the alkali, the acetyl (-COCH$_3$) groups of hemicelluloses are peeled almost completely in the beginning phase of the cooking (*i.e.*, deacetylation) stage, leading to the formation of soluble sodium acetate (CH$_3$CO$_2$Na).

As indicated above, the fundamental requirement for the peeling reaction is the presence of a reducing end group that acts in an alkaline environment. Figure 4.6 shows

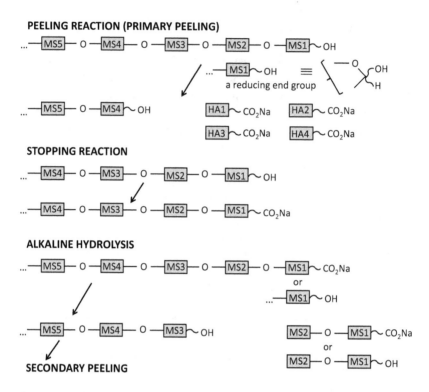

Figure 4.5: Main reactions of polysaccharides in alkaline pulping [6]. MS refers to a monosaccharide unit and HA to a hydroxy acid component.

the transformation between D-glucose, D-mannose, and D-fructose, which is generally called after its observers, the "Lobry de Bruyn-Alberda van Ekenstein transformation" [6, 184]. In this reaction, the aldose end group is isomerized *via* a 1,2-enediol intermediate into a 2-ketose end group. Three tautomeric forms of this 1,2-enediol are present: the initial aldose end group (D-glucose) and its 2-epimer (D-mannose), and a 2-ketose end group (D-fructose), which transforms further into 3-ketose end group and so on. Hence, the total reactions include isomerism, tautomerism, and epimerism.

After the isomerization of the aldose end group of the polysaccharide chain takes place, β-alkoxy elimination (Figure 4.7) happens, where the "end structural monosaccharide group" dissolves into the cooking liquor [6]. Hence, the remaining polysaccharide chain (an alkoxy (-OR) group, originally in the β position with respect to the carbonyl group (>C=O) of a ketose), is shortened by one monosaccharide unit, while exposing a new reducing end group open to further peeling. The peeling reaction involves the elimination of monosaccharide units that form various monocarboxylic and dicarboxylic acids; about one-and-half equivalents of acids per one monosaccharide unit are formed. The soluble monomeric compound, for its part, forms a dicarbonyl intermediate (*i.e.*, a 2,3-diulose structure), from which the main reaction route produces *via* a

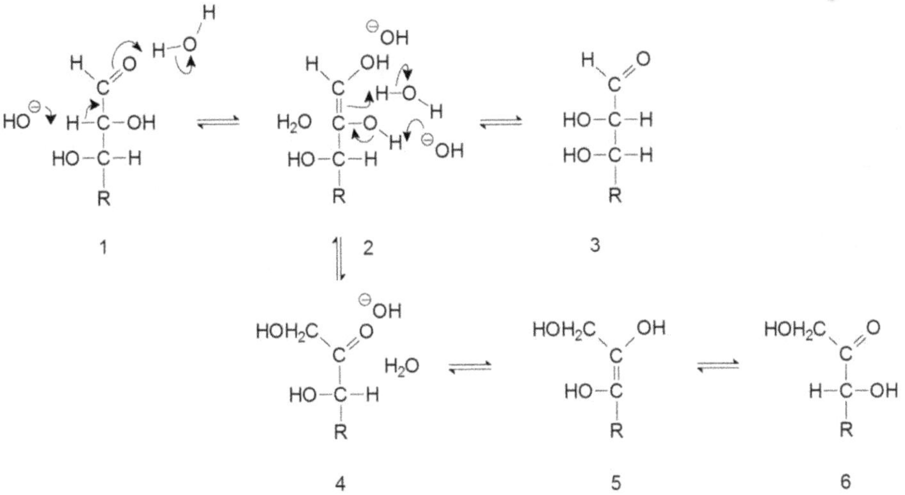

Figure 4.6: Lobry de Bruyn-van Ekenstein transformation of aldoses [184]. If the aldose is D-glucose (1), a 1,2-enediol (2), D-mannose (3), D-fructose (4), a 2,3-enediol (5), and D-allulose (6) are formed in the first stage of this transformation.

benzilic acid rearrangement-type reaction, either glucoisosaccharinic acid or xyloisosaccharinic acid. Several different peeling reactions lead, besides to "volatile" formic acid (HCO_2H) and acetic acid (CH_3CO_2H), to "nonvolatile" hydroxy monocarboxylic acids (*e.g.*, glycolic, lactic, 2-hydroxybutanoic, 3-deoxy-tetronic, 3,4-dideoxy-pentonic, 3-deoxy-pentonic, xyloisosaccharinic, and glucoisosaccharinic acids) as well as to minor amounts of hydroxy dicarboxylic acids [45]. The detailed composition of the acid fraction is shown in Section 5.3.1 "Chemical composition".

Of the main hydroxy acid components, 2-hydroxybutanoic and xyloisosaccharinic acids originate from pentosans (xylans), whereas 3,4-dideoxy-pentonic and glucoisosaccharinic acids (α- and β-forms) are formed from hexosans (glucomannan and cellulose) [6]. In contrast, the prominent glycolic, lactic, and 3-deoxy-pentonic acids, together with volatile HCO_2H, are formed from all polysaccharide constituents. Additionally, Figure 4.8 shows an example of a possible retro-aldol reaction that degrades the carbon chain of the monosaccharide structure.

In the benzilic acid rearrangement [214], the oxidation of benzoin (PhCH(OH)COPh, Ph is a phenyl group) leads to benzil (PhCOCOPh) with a 1,2-diketone structure that transforms in alkaline environments into an α-hydroxy acid anion ($Ph_2C(OH)CO_2^\ominus$), according to Figure 4.9. When applying the peeling reaction to carbohydrates (*cf.*, a 2,3-diulose structure), the reaction proceeds as shown in Figure 4.10. For clarity, the possible formation of the two alternative epimeric structures is not shown; the formation depends on the direction of approach of the $HO^\ominus$ ion in the beginning of the reaction [6].

In the stopping reaction of the carbohydrate chain, β-hydroxy elimination takes place directly from the terminal group aldose (Figure 4.11) and the peeling of the chain

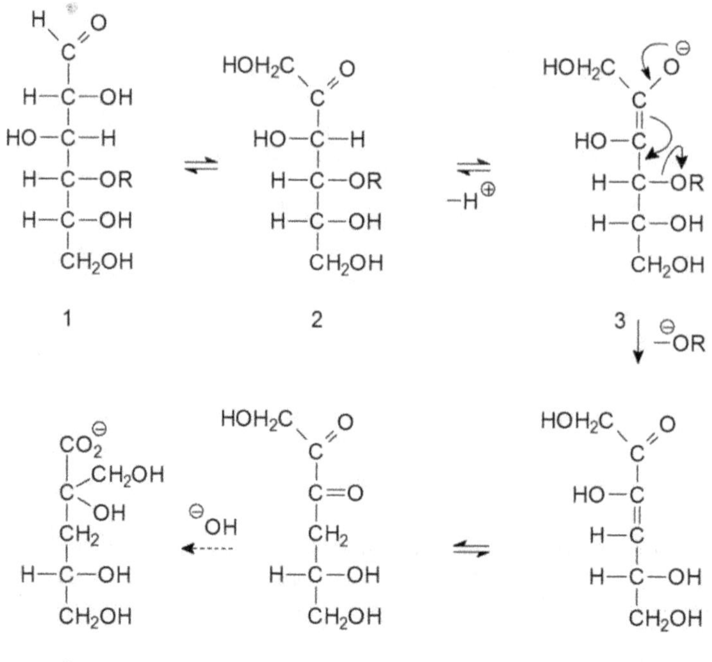

Figure 4.7: Peeling reaction of cellulose (R is an alkoxy group consisting of a glucan chain) [184]. Reaction steps include isomerization (1→2), formation of a 2,3-enediol structure (2→3), β-alkoxy elimination (3→4), tautomerization (4→5), and benzilic acid rearrangement-type reaction (5→6), during which an epimeric glucoisosaccharinic acid (3-deoxy-2-C-hydroxymethylpentonic acid) is formed.

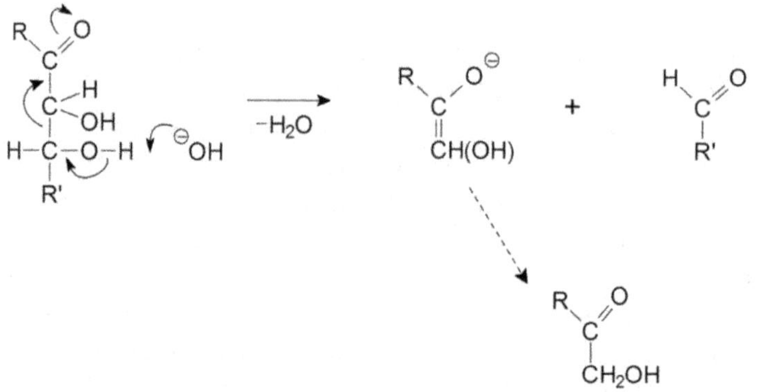

Figure 4.8: Example of an alkali-catalyzed reaction for cleaving a monosaccharide carbon chain [184].

is prevented; with respect to the original >C=O carbon, the β carbon now has a -OH group instead of an -OR group attached to it. Generally, this reaction leads to a terminal group where the functional unit is -CO$_2$H, and the isomerization into a 2-ketose is no

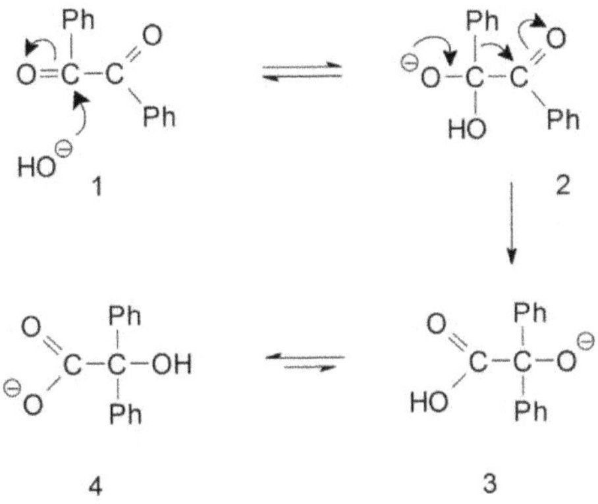

Figure 4.9: Alkali-catalyzed rearrangement reaction of benzil (1,2-diphenylethane-1,2-dione) (1) *via* intermediates (2) and (3) to benzilate anion (4) [184]. Ph is a phenyl group.

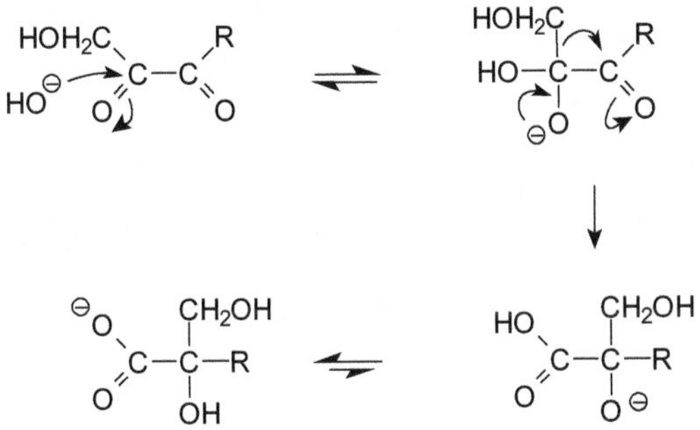

Figure 4.10: Benzilic acid rearrangement-type reaction of the compound residue dissolved during the peeling reaction (see Figure 4.7) [184].

longer possible. In this case, the dicarbonyl structure (a 1,2-diulose structure) that is also produced, undergoes a benzilic acid-type transformation (Figure 4.12), which primarily results in the formation of a metasaccharinic acid (3-deoxy-hexonic acid) end group into the polysaccharide chain [215]. Other possibilities are 2-C-methylglyceric acid and 2-C-methylribonic acid (glucosaccharinic acid) end groups and, in small quantities, also certain aldonic acid end groups (*i.e.*, mannonic, arabinonic, and erythronic acid groups), indicating the presence of oxygen in the reaction.

Deacetylated xylan, dissolved during alkaline delignification, is partly resorbed on the fibers, especially at the end of the conventional hardwood cook (*i.e.*, the alkalinity of the cooking liquor is decreased to some extent), which also effects the pulp properties [6, 30, 32, 177, 216, 217]. It is probable that the xylan fractions with a high uronic acid content are dissolved. Another reason for the significant decrease in the uronic acid content of the pulp at maximum cooking temperature of alkaline pulping is alkaline hydrolysis of these groups.

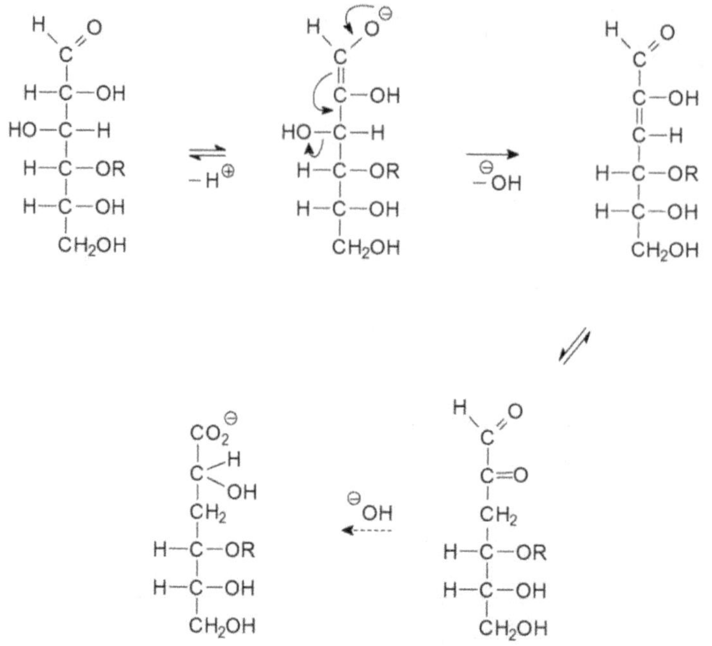

Figure 4.11: Stabilization of a carbohydrate chain in the stopping reaction (*i.e.*, termination of the peeling reaction) [184]. The reaction steps are analogous to those taking place in the peeling reaction (see Figure 4.7). However, in this case, due to β-hydroxy elimination (*i.e.*, β-alkoxy elimination is not occurring), no cleavage of the carbohydrate chain occurs.

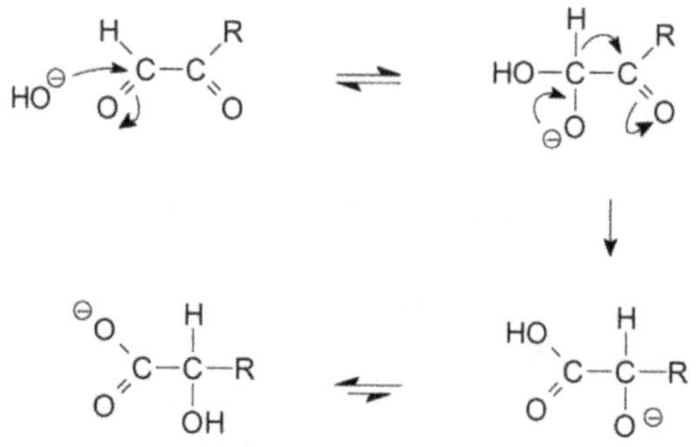

Figure 4.12: Benzilic acid rearrangement-type reaction of a carbohydrate chain end group during the stopping reaction (see Figure 4.11) [184].

It has been shown that, depending on cooking conditions, the 4-*O*-methylglucuronic acid groups of xylan in pulp are partly degraded and/or converted into 4-deoxy-4-hexenuronic (hexenuronic) acid groups (HexAs) by the β-elimination of CH_3OH during kraft pulping [6, 82, 218–223]. The HexAs are unreactive with O_2 and H_2O_2 in alkaline media, but can react, due to their ene functionality, with several other bleaching chemicals, such as chlorine dioxide (ClO_2), Cl_2, O_3, and peracids (RCO_3H), thus consuming these chemicals. Therefore, monitoring of the contents of HexAs is of importance, with potential benefits for the mill economy and management. Additionally, permanganate also attacks HexA, which, therefore, contributes greatly to the kappa number, especially in case of hardwood kraft pulps [224], and HexAs have long been recognized as a source of color generation in pulps [225–228]. However, HexAs can be converted into furan derivatives (*i.e.*, 2-furoic acid and 5-carboxy-2-furaldehyde), prior to bleaching by mild acid hydrolysis (*i.e.*, by an acidic washing treatment in an A stage of bleaching) [227, 229–231]. This kind of treatment essentially reduces the consumption of bleaching chemicals. There are also many sophisticated analytical methods for analyzing HexA and its conversion products [232–238].

Extractives
Extractives can be categorized into two groups that behave differently in the softwood kraft process, and form the main types of by-products [8, 239]: *i*) a volatile fraction (*i.e.*, crude turpentine or crude sulfate turpentine (CST)) and *ii*) a nonvolatile fraction (*i.e.*, tall oil soap or crude tall oil (CTO)). The yield of these by-product groups is strongly dependent on the wood species used for pulping, growth conditions of the tree, and the time and method of log and chip storage [10, 240–242]. CST is collected during digestion and purified in the distillation process [8]. CTO is composed of the sodium and calcium salts of fatty and resin acids, together with some neutral substances ("nonsaponifiables") [8, 243]. These soap constituents are suspended and/or dissolved in black liquor, and most of this material can be skimmed from the black liquor during the evaporation process. It should be noted that due to the saponification of esters (*i.e.*, fats and waxes, see Section 2.4.2 "Chemical features of extractives") with the simultaneous neutralization of aliphatic carboxylic acids, extractives also consume cooking chemicals. Although the extractives comprise a valuable fraction of by-products with a huge multitude of possible derivatives, they can form pitch problems, especially in alkaline chemical pulping and bleaching of hardwoods and non-wood feedstocks [244–254], and in the case of mechanical pulping, increased concentrations of dissolved and colloidal substances (DCS) can reduce production efficiency, affecting negatively the runnability and production capacity of the paper machine (*i.e.*, form pitch deposits) as well as the quality of the product [255–257]. Tall oil-based and turpentine-based products came to the market over one hundred years ago; the production of CTO started in 1899 in Sweden and in 1913, the first CTO distillation plant was started in Finland [258].

In principle, CST is recovered from the digester relief condensates [8, 259]. Batch digester recovery systems differ, however, substantially from continuous digester recovery systems; in these systems, the relief vapors are captured. The average yield of CST of pine species is 5–10 kg/ton of pulp, whereas the yields are somewhat lower for spruce species. CST is purified in the distillation process during which impurities, such as MM and DMS, as well as higher compounds and other impurities are removed. The main fractions are *i*) a monoterpene fraction, consisting primarily of α-pinene (50–80% of all compounds), β-pinene, and 3-carene (Δ^3-carene) and *ii*) a pine oil fraction, consisting primarily of hydroxylated monoterpenes (Table 4.3 and Figure 4.13). Distillation of CST yields "heads" that are volatile compounds, with almost no commercial value and isomeric α-pinene and β-pinene. Of these, β-pinene has the highest value for the chemical industry [260].

Table 4.3: Main turpentine compounds in commercial turpentines [259].

Component	Sulfate turpentine Nordic	Sulfate turpentine USA
α-Pinene	55–70	40–70
β-Pinene	2–6	15–35
3-Carene	7–30	2–10
Camphene	*ca.* 1	1–2
Dipentene	*ca.* 4	5–10

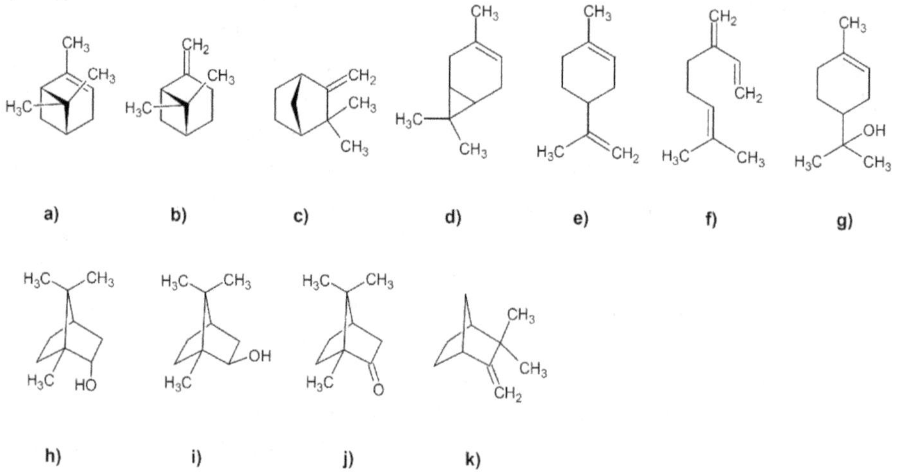

Figure 4.13.: Chemical structures of common monoterpenes in turpentine [239]. α-Pinene (a), β-pinene (b), camphene (c and k), 3-carene or Δ^3-carene (d), limonene (e), myrcene (f), terpienol (g), borneol (h), isoborneol (i), and camphor (j).

Monoterpene compounds of turpentine have traditionally been used as thinner for paints, varnishes, and lacquers or as a rubber solvent and a reclaiming agent, and as a vehicle carrying the solid components (*i.e.*, binders, pigments, and additives) of solvent-borne coatings, although it has been largely replaced in many applications by several mineral-based substitutes, such as white spirit [8, 10, 239–243]. Additionally, turpentine is still much employed in pharmaceutical purposes (antiseptic liniments) and in both general medical and veterinary practice as a rubefacient and vesicant. However, nowadays, it is mainly used for making different products for the chemical industry (*e.g.*, preparation of α- and β-pinene and crude 3-carene, and synthesis of camphor, menthol, and insecticides), and in perfumery. Another important use of turpentine and its fractions is to polymerize them into polyterpene resins. Such resins are used for various purposes, such as for the preparation of pressure-sensitive or as tackifiers in hot-melt-adhesives [243]. Additionally, terpene-based resins are also used in tires [261].

Of these monoterpenes, β-pinene is sold as such and α-pinene is further processed to make synthetic pine oil [8, 24, 239, 243, 262]. Pine oil (mainly α-terpineol) is prepared by the hydration of α-pinene with aqueous mineral acids, but it is also obtained by steam-distillation of needles, twigs, and cones from a variety of pine species. It is used like turpentine for a variety of purposes such as for applications requiring solvents with good emulsifying and dispersing properties and in the mining industry, for flotation of minerals. However, the main uses of pine oil include in household cleaning and as disinfectant products. α-Terpineol has a pleasant odor, which is like lilac, and is also a common ingredient in perfumes, cosmetics, and flavors.

Monoterpenes α- and β-pinenes are prone to a variety of isomerization and rearrangement reactions, resulting in an abundance of different flavors and fragrance chemicals [239, 243]. The main industrial use of β-pinene is its non-catalytic thermal isomerization reaction in the vapor phase to produce, besides limonene and p-menthadiene, β-myrcene; it is as an intermediate for commercial production of flavor and fragrance chemicals, such as terpenic alcohols (geraniol, nerol, and linalool) and synthetic aromas (citral, citronel, menthol, and myrcenol). Besides being processed to pine oil, α-pinene can be used to synthesize camphene, which is used in the preparation of fragrances, and as a food additive for flavoring. Borneol, isoborneol, camphor, and camphene are examples of monoterpenes, which are frequently used as such or for production of other fragrance and flavor compounds.

The resin and fatty acids in the removed tall-oil soap from the black liquor evaporation process are liberated from sodium by adding sulfuric acid (H_2SO_4) to yield CTO [8]. The average yield of CTO is in the range of 30–50 kg/ton of pulp, corresponding to 50–70% of the initial amount in the raw material used for pulping. CTO has the following main categories of uses [263]: *i*) distillation into a variety of products, *ii*) use as an aid component of petroleum extraction drilling fluid (can be used as such or after oxidation [243]) or for phosphate mining, *iii*) use as a process fuel in the pulp mill lime kiln, and *iv*) use as a feedstock for production of renewable fuels. The distillation of CTO from softwood pulping represents the largest end use, processing annually around 1.4 million tons. Due to the low distillable quality (*i.e.*, the enhanced presence of neutral components), CTO from hardwood pulp mills is mostly used for direct energy purposes.

CTO is purified and fractioned by vacuum distillation (3–30 mbar, 170–290 °C) [8, 243, 263]. The main commercial fractions, including their average mass proportions, are as follows: light oil ("tall-oil heads" (TOH), 10–15%% of total), tall-oil fatty acids (TOFA, 20–40% of total), and tall-oil resin or rosin (TOR, 25–35% of total), together with distilled tall oil (DTO, about 10% of total), and tall-oil pitch (TOP, 20–30% of total). Composition examples of commercial fractions of TOFA and TOR are presented in Table 4.4. Additionally, different extraction methods have been developed for removing the neutral components prior to distillation. One example of such extraction processes is the so-called "CSR process" by which the production of sitosterol is possible. Sitosterol can be used after reduction to sitostanol, followed by esterification; it can be used in foodstuff fats; it has been shown to lower the cholesterol level of human blood. Rosin generally means the natural resin of pine trees and is typically a mixture of resin acids (*i.e.*, 87–90%, after the removal of the most volatile components), neutral substances, and minor amounts of fatty acids [264].

Table 4.4: Typical chemical compositions of the commercial fractions of tall-oil fatty acids (TOF) and tall-oil resin (TOR) (% of total) [10, 259, 263].

Component	Nordic	American
TOFA		
Stearic acid ($C_{18:0}$)	1–2	ca. 2
Oleic acid (9–$C_{18:1}$)	25–28	45–55
Linoleic acid (9,12–$C_{18:2}$)	40–45	35–40
Pinolenic acid (5,9,12–$C_{18:3}$)	7–11	2–3
Minor fatty acids	10–15	10–15
Others	2–4	2–4
TOR		
Abietic acid	42–43	35–40
Palustric acid	10–12	8–12
Neoabietic acid	4–5	3–5
Dehydroabietic acid	21–22	20–25
Pimaric acid	4–5	2–4
Isopimaric acid	3–5	8–12
Minor resin acids	9–14	10–15
Others	4–7	4–7

From TOFA, many commercial fatty acid products, varying in both purity and composition are available. They are used in applications such as alkyds for coatings, dimer acids, and polyamides for coatings and adhesives, epoxy resins, surfactants, defoamers, emulsifiers, soaps, and cosmetics [261, 263, 265]. TOFA derivatives, with a wide range of properties, are suitable for different purposes. These derivatives include, for example, dimer acids, fatty acid esters, fatty acid soaps, fatty acid ethoxylates, and fatty acid amines. Fatty acid derivatives have also important applications in polyurethanes. One of the most common derivatives of TOFA is the group of esters that originate from the esterification of short chain alcohols, polyols (*i.e.*, glycerol, pentaerythritol, and trime-thylolpropane), and ethoxylates [266]. Additionally, the Diels-Alder and disproportion-ation (*i.e.*, an isomerization reaction of TOFA under heat, usually with a catalyst) reactions are commonly practiced methods for producing TOFA derivatives.

TOFA-based alkyds, mostly used in coatings, are made by reacting an unsaturated dibasic or anhydride with an unsaturated fatty acid ester that has been transesterified with polyols, such as glycerol [261, 263, 266]. The high content of unsaturated fatty acids favors the complete conversion (*i.e.*, dimerization) into well-defined dicarboxylic acids with twice the number of carbon atoms [265]. The dimer acids are the highest-molar-mass dibasic acids commercially available and their major use (about 76% of all dimer acid production) is non-nylon-type polyamide resins that have applications in inks, adhesives, and coatings [261].

The simple saponification reaction of TOFA comprises the formation of soaps, upon its treatment with bases [243, 263]. TOFA soaps, such as ammonium, potassium, and so-dium tallates are H_2O soluble. Due to the low content of saturated fatty acids, they can be utilized in many cosmetic and personal care applications, functioning as emulsifiers or surfactants. Hence, they can also be used in several industrial applications, such as bitumen emulsions that are commonly used in roadway construction and maintenance. The oil-soluble, but H_2O-insoluble TOFA salts of magnesium, aluminum, cobalt, or man-ganese are used in paint dryers and lubricants.

TOFA esters of monofunctional, low-molar-mass alcohols, such as butanol or 2-ethylhexanol, are principally used as solvents for coatings and printing inks, as coa-lescents in latex paints, and as lubricant esters for metal forming [243, 263]. Esterification of TOFA with CH_3OH using H_2SO_4 as a catalyst yields TOFA methyl esters. These com-pounds are comparable to rapeseed fatty acid methyl ester that is useful as biodiesel for compression-ignition engines. TOFA methyl esters may also be cooligomerized with eth-ene ($H_2C=CH_2$) or propene ($CH_3CH=CH_2$) to manufacture alkyl-branched chain fatty acid derivatives. The formed branched unsaturated fatty acids are typically hydrogenated, and the saturated branched fatty acids obtained are used as plasticizers, base fluids for lubricants, and for personal care formulations. Epoxidized TOFA esters, such as octyl tal-late, act as polyvinyl chloride (PVC) stabilizers and as plasticizers. The tallate esters of polyfuntional alcohols, such as pentaerythritol tetratallate are utilized as external lubri-cants and mold-release agents in the plastic industry or as industrial lubricating oils, gear oils, friction modifiers, and lubricating additives.

TOFA reacts also with ammonia, amines, and alkanolamines, yielding the corre-sponding TOFA amides or alkanolamides [263, 265]. Amidoamides of TOFA are produced *via* the reaction of TOFA with a polyalkylene polyamine, such as diethylenetriamine (DETA), triethylenetetramine (TETA), or tetraethylenepentamine (TEPA). The amido-amines have reactive amine sites that allow the cross-linking of epoxy resins in coatings, adhesives, and other applications. Additionally, they are used, for example, in asphalt to improve performance in road surfacing or as corrosion inhibitors in petroleum produc-tion operations.

TOR is primarily employed as chemical intermediaries, which can be further modified, for example, for adhesives, printing inks, emulsifiers, coatings, paint and lacquer vehicles, paper sizes, rosin soaps, rosin resin esters, sealants, tackifiers, and rubber processing aids as well as for many other specialty applications [243, 263, 265]. The chemical reactivity of TOR (*i.e.*, mainly resin acids) is mainly due to the presence of >C=C< structures and -CO$_2$H groups [262]. The former olefinic system is responsible for oxidation, reduction, hydrogenation, and dehydrogenation reactions, whereas the latter groups are involved in esterification, salt formation, decarboxylation, and anhydrides formation. With respect to industrial applications, Diels-Alder reaction (*i.e.*, the production of fortified sizing agents), salt formation, and esterification are the most relevant reactions of resin acids [243]. Other relevant reactions are those comprising the addition of resin acids to HCHO and/or phenols.

Papermaking fibers have a strong natural tendency to interact with H$_2$O [264, 266]. This behavior is needed to the development of significant interfiber hydrogen bonds during papermaking, although, due to this bond formation, paper also loses its strength when saturated with H$_2$O. H$_2$O adsorption is an advantage for certain paper grades (*e.g.*, towel and tissue) during use but for most paper grades, it is more typical to interact only moderately with H$_2$O. The processes that impart resistance to H$_2$O (or aqueous liquids) penetration in paper *via* an increase in the hydrophobicity of paper are called "sizing processes".

The most important use of rosin is paper sizing, which thus prevents printing inks from blurring and feathering and helps paper bags to retain their strength by resisting moisture absorption [246]. Rosin can be used in paper sizing in three forms [243, 263–268]: *i*) as sodium salts (known as neutral rosin or with multivalent ions, such as Al complexes), *ii*) free rosin acids, and *iii*) modified rosin products. To improve sizing efficiency, commercially used rosin can be also fortified [264]. In practice, this means that additional -CO$_2$H groups are added into the resin acids (*e.g.*, the reaction of levopimaric acid with fortifying agents, such as maleic acid anhydride or fumaric acid and itaconic acid) by means of the Diels-Alder reaction [269, 270]. Typical examples of such modified rosin adducts are maleopimaric acid anhydride and fumaropimaric acid. The traditional natural sizing system (*e.g.*, the alum-rosin sizing system) with rosin can only be utilized efficiently under acid pH conditions [267]. However, for many practical reasons, for example, including the use of CaCO$_3$ filler, several paper grades are currently produced under neutral or alkaline conditions. This trend has resulted in the use of synthetic sizing agents such as alkenyl succinic anhydride (ASA) and alkyl ketene dimer (AKD).

Alkali resinates have been used both as polymer latex stabilizers and as surfactants in emulsion polymerization from the early development of these techniques [233, 263]. However, the reactivity of the conjugated double bonds toward free radicals has made it more profitable to use hydrogenated or dehydrogenated rosins rather than their natural forms. Adhesive tackifiers that have different softening points can be produced by using rosin acids and varying polyols [261]. Polymerization is another approach to increase the molar mass rosin acids, followed by further derivatization (*e.g.*, to esters and amides). The polymerized rosin acids as well as their derivatives have been used as a binder or an additive, for example, in printing inks, paints, coatings, and tackifiers.

Rosin is used to modify the epoxy resins in photopolymerizable printing inks and coating compositions [261, 263]. Thermoplastic elastomers are produced by reacting rosin acids with acrylic acids, caprolactones or butyrolactones to form thermoplastic resins. These thermoplastic resins are used in varnishes and nanocomponents as well as in stimuli-responsive polymers that are suitable in areas such as drug delivery, tissue engineering, and sensors. Modification of rosin acids with ethylene oxides leads to polymers that are used as surface-active agents.

The use of rosin and rosin esters (*e.g.*, rosin and maleic rosin esters with glycerol, sorbitol, mannitol, and pentaerythritol) have been studied as coatings and microencapsulating materials for the controlled release of several pharmaceuticals, and as an implant matrix for the delivery of drugs [262, 263]. Other applications of rosin derivatives and polymers comprise emulsifiers for pharmaceuticals and cosmetics. There is also an example of utilizing the bioactive properties of TOR in animal feed supplement.

DTO is a mixture that mainly contains fatty acids and more than 10% resin acids (often between 25–35%) [263]. It can be used in metal working fluids, oil field chemicals, soaps, cleaners, and alkyd resins. The fraction of TOH consists of neutral components and short-chain fatty acids [266]. It has solvent properties and, for example, palmitic acid ($C_{16:0}$) is recovered from the TOH stream. This fraction is mostly used as a fuel oil and process fuel at CTO distiller, but it is also utilized as industrial oils and for rust protection [8]. TOP contains a high-molar-mass substance that is mainly cross-linked materials, esters of fatty acids and rosin acids, neutral and unsaponifiable materials, and many decarboxylated materials [261, 263]. One of the most significant uses of TOP has been its use as a feedstock for the extraction of sterols, suitable for cholesterol-reducing additives for food [259]. Another significant application of TOP is in asphalt; this area includes emulsions, pavement binders, rejuvenators, and plasticizers [271]. TOP can also be used for fuel purposes and in many specific applications.

4.3 Sulfite pulping

4.3.1 Process description

Unlike the alkaline kraft process, the sulfite process covers the whole range of pH (Table 4.5), which enables high flexibility in pulp yields and properties [6, 272–282]. The pulps extend from dissolving pulps for different end uses of chemical industry to high-yield neutral sulfite semichemical (NSSC) pulps. The latter pulps are produced by cooking the chips with Na_2SO_3/sodium bisulfite or sodium hydrogen sulfite ($NaHSO_3$) solutions, followed by the mechanical defibration of the partially delignified wood in disc refiners. The fiber properties of hardwood NSSC pulps make them suitable, especially for corrugating mediums. In contrast, most pulps produced by acid sulfite and bisulfite pulping are used for different paper grades, whereas alkaline sulfite-AQ methods generally produce kraft-type pulps [283–291]. However, due to possible health risks, it has been recommended by European Food Safety Authority (EFSA) and Confederation of European Paper Industries that the use of AQ should be decreased [292].

Table 4.5: Sulfite pulping processes [6].

Process	pH range	Base	Active reagent	Pulp type
Acid sulfite	1–2	$Ca^{2\oplus}$, $Mg^{2\oplus}$, $Na^{\oplus}$, $H_4N^{\oplus}$	$HSO_3^{\ominus}$, $H^{\oplus}$	Dissolving pulp Chemical pulp
Bisulfite	2–6	$Mg^{2\oplus}$, $Na^{\oplus}$, $H_4N^{\oplus}$	$HSO_3^{\ominus}$, $H^{\oplus}$	Chemical pulp High-yield pulp
Neutral sulfite (NSSC)[a]	6–9	$Na^{\oplus}$, $H_4N^{\oplus}$	$HSO_3^{\ominus}$, $SO_3^{2\ominus}$	High-yield pulp
Alkaline sulfite-AQ[b]	9–13	$Na^{\oplus}$	$SO_3^{2\ominus}$, $HO^{\ominus}$	Chemical pulp

[a]NSSC, neutral sulfite semichemical.
[b]AQ, anthraquinone.

Besides the basic processes shown in Table 4.5, sulfite pulping in two or even three stages with different pH regions has been developed as a means of improving pulp properties for different applications [6, 272, 277, 293]. One possibility of two-stage (or multistage) applications is to precook the wood chips first in a $Na_2SO_3/NaHSO_3$ solution at pH 6–7 and then use acid sulfite pulping conditions in a second cooking stage. In the first stage, lignin is sulfonated to a certain degree, but is mainly retained in the solid-wood phase. Delignification is accomplished in the second stage by charging liquid SO_2 to the digester.

The two-stage processes, compared to conventional acid sulfite pulping, greatly improve the uniformity of lignin sulfonation and can lead to an increase in pulp yield (maximally about 8% of dry wood) [272, 275]. This yield increase is mainly restricted to softwoods, and associated with an increased retention of glucomannan in the pulp. In contrast, compared with conventional pulping, two-stage pulping of hardwoods only moderately improves the yield of xylan, and is not used in the pulp industry. An additional advantage of the two-stage pulping is that it can utilize pine heartwood, which is not possible when applying the conventional acid sulfite process. In the first stage at pH 6–7, reactive groups of lignin are protected by their sulfonation, which blocks their condensation reactions with phenolic extractives (*e.g.*, taxifolin and pinosylvin) [277, 281].

In the two-stage application, the yield can be varied by adjusting the pH in the first stage so that optimum pulp properties are obtained for different application purposes [275]. On the other hand, the three-stage sodium base process, where the first stage at pH 6–8 is followed by an acid stage at pH 1–2 and finally by a weakly alkaline stage, is suitable for producing hardwood-based dissolving pulps with a high cellulose content.

The acid sulfite process has a long history. In 1866, a method, where the defibration of wood took place in a pressurized system of aqueous calcium hydrogen sulfite ($Ca(HSO_3)_2$) and SO_2 was patented in England by Benjamin C. Tilghman, although this acidic cooking process was not immediately realized on a large scale [281, 294, 295]. However, Carl Daniel Ekman conducted essential research that finally realized the practical execution of large-scale processes and the first sulfite pulp factory started operations in Sweden in 1874. Furthermore, independent of Ekman, Alexandre Mitscherlich in Germany also devised a similar large-scale production. Before the actual acid calcium-based sulfite cooking, Peter Claussen had, in 1851, proposed the possibility of making fiber suitable for papermaking from wood, with the help of NaOH and SO_2, which was the precursor for neutral sulfite cooking.

The active sulfur-containing species in the sulfite process are SO_2, hydrogen sulfite ions ($HSO_3^{\ominus}$), and sulfite ions ($SO_3^{2\ominus}$), together with active ions, $H^{\oplus}$ and $HO^{\ominus}$, in proportions that essentially depend on the pH of the cooking liquor [6, 281, 282]. According to the equilibrium, these species are present almost exclusively in the form of $HSO_3^{\ominus}$ in H_2O solutions of SO_2 (the total SO_2 charge is typically 20% of oven-dry wood feedstock) at pH around 4. Below and above this pH value, the concentrations of SO_2 and $SO_3^{2\ominus}$ ions increase, respectively.

According to the old terminology, the total amount of SO_2 (*i.e.*, "total SO_2") present in cooking acid or cooking liquor is divided into the so-called "free SO_2" and "combined SO_2" [6, 281]. They are usually expressed as grams of SO_2 per 100 mL of solution. A typical acid sulfite cooking liquor contains, respectively, about 5 g and 1 g of total SO_2 and free SO_2. The corresponding values for a bisulfite cooking liquor are 5 g and 2.5 g. However, the liquor is described more precisely by the actual concentration of "sulfurous acid" ("H_2SO_3", *i.e.*, SO_2 hydrate), $HSO_3^{\ominus}$, and $SO_3^{2\ominus}$, derivable from the standard titration curve. Therefore, depending on the different pulping reactions, the active cooking chemicals are consumed and inactivated during sulfite pulping.

The active base, normally $Na^{\oplus}$, $NH_4^{\oplus}$, $Mg^{2\oplus}$, or $Ca^{2\oplus}$, is the cation bound to the $HSO_3^{\ominus}$ and $SO_3^{2\ominus}$, and its concentration is usually expressed in grams of sodium oxide (Na_2O) per liter of solution [6, 281]. The conventional base, $Ca^{2\oplus}$, was mainly used due to its low cost (from limestone or $CaCO_3$) and because of the lack of stringent environmental quality regulations; there was no need to recover it. However, due to the limited solubility of calcium sulfite ($CaSO_3$), $Ca^{2\oplus}$ can be used only in acid sulfite pulping, where a large excess of SO_2 is available to prevent the formation of $CaSO_3$ from $Ca(HSO_3)_2$. When using the more soluble $Mg^{2\oplus}$ as a base, pH can be increased to about 5; above this range, magnesium sulfite ($MgSO_3$) starts to precipitate while $Mg^{2\oplus}$ precipitates as magnesium hydroxide ($Mg(OH)_2$) in the alkaline region. In contrast, Na_2SO_3 and $(NH_4)_2SO_3$ as well as NaOH and NH_4OH are easily soluble and there are no limitations caused by pH on their use in the cooking liquor. In modern sulfite cooking, mainly Na$^\oplus$ (and Mg$^{2\oplus}$ at pH <5) is used and inorganic chemicals are recovered and regenerated. In the recovery cycle, organic solids are burned to generate energy.

As indicated above, in the early days of sulfite pulping, the base used was almost exclusively Ca$^{2\oplus}$ from CaCO$_3$, and the recovery of the cooking chemicals was usually not integrated with the spent liquor combustion [6, 281]. Therefore, the main interest was only directed toward the relatively simple recovery of excess SO$_2$. Hence, all Na-based spent liquors can be burned in a kraft-type furnace and the cooking chemicals can be efficiently recovered. However, in the case of the combustion of Mg-based sulfite spent liquors, no smelt is obtained; instead, the base is completely recovered as magnesium oxide (MgO) in dust collectors. The sulfur escapes as SO$_2$ and is adsorbed from the combustion gases in scrubber towers. NH$_4$-based sulfite spent liquors can also be burned in the same type of furnace as the Ca-based liquors. This results in entirely volatile compounds and the problems with fly ash are eliminated. All sulfur escapes to the combustion gases as SO$_2$ which can be partly adsorbed in an ammonia-in-H$_2$O solution (NH$_3$(aq)).

The term "lignosulfonates" refers to neutralized lignin fragments dissolved during delignification in the cooking liquor [6, 281, 296]. A certain amount of base is needed for the neutralization of lignosulfonic acids and other acidic degradation products of the wood constituents. If in an acid sulfite cook, the base concentration is too low, the pH sharply decreases and the rate of competing lignin condensation reactions increases. This also accelerates the decomposition of cooking acid in the interior of the chips and results in dark, hard cores. These harmful reactions cause decreased delignification or may completely prevent it. Additionally, with respect to the combustion of spent liquors, the heat value of lignosulfonates is rather high, roughly one-half of that of oil, although the heat value of the soluble carbohydrate fraction is even much lower, roughly one-third of that of oil.

4.3.2 Reactions of the wood chemical constituents

Two types of reactions, sulfonation and hydrolysis, are responsible for delignification in sulfite pulping (Table 4.6) [6, 272, 274, 277, 279]. The former reaction generates hydrophilic sulfonic acid (-SO$_3$H) groups, while the latter reaction breaks aryl ether linkages between the phenylpropane units, which creates new free phenolic -OH groups and lowers the average molar mass. These both reactions increase the hydrophilicity of the feedstock lignin and facilitate its H$_2$O solubility. Compared to sulfonation, hydrolysis is fast under the conditions of acid sulfite pulping, although under these conditions, lignin is also sulfonated to a relatively high degree, thus promoting an extensive dissolution of lignin. In contrast, delignification proceeds slowly in neutral and alkaline sulfite pulping because the hydrolysis reactions are very slow compared to sulfonation and because the degree of sulfonation of lignin remains low.

The average degree of sulfonation of the undissolved softwood lignin as well as of most NSSC lignins, expressed as the molar ratio of -SO$_3$H groups to -OCH$_3$ groups, re-

Table 4.6: Main reactions and phenomena occurring in the feedstock chemical constituents during acid and bisulfite pulping (for details, see the text. Note that cellulose is practically stable under these conditions) [6].

LIGNIN
– Degradation (hydrolysis and M_w decreases) – Increase in hydrophilicity (sulfonation and liberation of phenolic groups)
→ H_2O/acid-soluble lignosulfonates

HEMICELLULOSES
– Cleavage of acetyl groups – Hydrolysis of glycoside bonds – Oxidation of aldoses to aldonic acids – Dehydration of monosaccharides
→ H_2O/acid-soluble mono-, oligo-, and polysaccharides, carboxylic acids, and furfural

EXTRACTIVES
– Hydrolysis of fatty acid esters (fats and waxes) – Dehydration – Sulfonation – Evaporation
→ H_2O/acid-soluble fragments and sulfite turpentine (*p*-cymene)

mains low (*i.e.*, -SO_3H/-OCH_3 about 0.3), whereas the corresponding value is about 0.5 or more for the dissolved lignosulfonates [6]. Additionally, this type of difference is typical for hardwood lignin, but the degree of sulfonation throughout is clearly lower than that of softwood lignin. However, in both cases, the major portion (80–90%) of the sulfur exists in the form of -SO_3H groups, although minor amounts of active sulfur chemical are also consumed in the formation of carbohydrate-based sulfonic acids.

Lignin

Most of the -SO_3H groups introduced into the lignin replace -OH or etherified substituents at the α-carbon atom of the side chain of phenylpropane units [6, 274, 277, 279, 281, 297–300]. Free phenolic structures (*i.e.*, nonetherified phenolic structures (aryl-OH)) are rapidly sulfonated at all pH values, whereas under acidic conditions, the lignin units are sulfonated, irrespective of whether they are free or etherified (aryl-O-R). During acid sulfite pulping, the α-hydroxyl and the α-ether groups are cleaved readily with the simultaneous formation of the intermediary carbonium ions (*i.e.*, benzylium ions) (Figure 4.14, reaction route a). Hence, the initial cleavage of open α-aryl ether bonds represents the only noteworthy fragmentation of lignin during acid sulfite pulping. However, although relatively few open α-aryl ether bonds are present in softwood lignin, their cleavage leads to a considerable fragmentation. The benzylium ions are then sulfonated by the attack of hydrated SO_2 or $HSO_3^{\ominus}$ present in the cooking liquor.

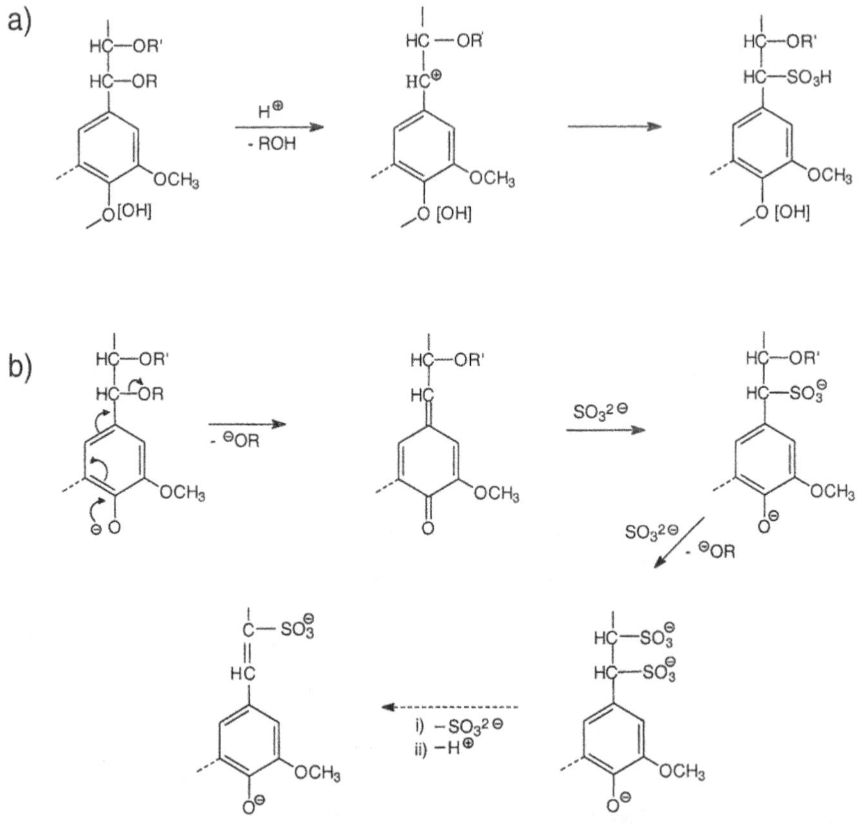

Figure 4.14.: Reactions of lignin in acid sulfite and bisulfite pulping (a) and neutral sulfite and alkali sulfite pulping (b) [6]. R is H or aryl group and R' is aryl group.

In neutral and alkaline sulfite pulping, the most important reactions of lignin are restricted to phenolic lignin units (Aryl-OH) and the principal initial reaction proceeds *via* the formation of a quinone-methide intermediate with the simultaneous cleavage of an α-hydroxyl group or an α-ether group (Figure 4.14, reaction route b) [6, 277, 282]. At least in noncyclic structures, the quinone methide is readily attacked by ions $SO_3^{2\ominus}$ and $HSO_3^{\ominus}$. The -SO_3H group formed at the α-carbon atom facilitates the nucleophilic displacement of the β-substituent in β-aryl ether structures by $SO_3^{2\ominus}$ or $HSO_3^{\ominus}$. The subsequent loss of the α-sulfonate group results in a styrene-β-sulfonic acid structure, especially at higher pH values. The cleavage of the α-aryl and β-aryl ether bonds generates new reactive phenolic units. The β-aryl ether bonds are apparently cleaved also in nonphenolic units during alkaline sulfite pulping. In this case, various condensation reactions are less important compared to those occurring during kraft pulping. Additionally, -OCH$_3$ groups that are completely stable under acidic conditions can be partly cleaved during neutral and alkali sulfite pulping, with the formation of methane sulfonic acid ions (CH$_3$SO$_3^\ominus$).

Under acidic conditions, various condensation reactions of carbonium ions compete with sulfonation, and their frequency is increased with increasing acidity [6, 277, 281, 282]. When the benzylium ions react with the weakly nucleophilic positions of other phenylpropane units, the C-C bonds are most commonly formed. These condensation reactions result in increased molar mass of the lignosulfonates, and the solubilization of lignin is retarded or inhibited. It is also possible that lignin can condense with reactive phenolic extractives. One example is the formation of harmful cross-links generated by pinosylvin and its monomethyl ethers (present in pine heartwood), which can act as nucleophilic agents. Additionally, similar cross-links between lignin entities can be generated by thiosulfate ions (S$_2$O$_3^{2\ominus}$) present in the cooking liquor [301]. This reaction leads to retarded delignification, and under exceptional circumstances, in complete inhibition ("black cook").

Although not investigated to the same extent as those formed during kraft pulping, a large number of low-molar-mass aromatic compounds are also apparently formed during alkaline sulfite delignification [6]. The typical monomeric compounds include 4-hydroxybenzoic acid, vanillin, vanillic acid, acetovanillone, dihydroconiferyl alcohol, syringol, syringaldehyde, syringic acid, and acetosyringone. Some dimeric compounds are also obtained.

Carbohydrates

Due to the sensitivity of glycosidic linkages toward acid hydrolysis, the most significant reaction of polysaccharides during acid sulfite and bisulfite pulping is the cleavage of glycosidic bonds between the monosaccharide moieties in the polysaccharide chains, giving rise to soluble monosaccharides as well as oligosaccharide fragments [6, 274, 277, 279, 281, 295, 302–304]. Like in kraft pulping, hemicelluloses are also, in this case, more readily attacked than cellulose because of their amorphous state and a relatively low degree of polymerization (DP). In practice, no cellulose is lost unless delignification is extended to very low lignin contents or under rather drastic conditions as those in the production of dissolving pulps. When hydrolysis has proceeded far enough, depolymerized hemicellulose fragments are dissolved in the cooking liquor and gradually tend to be hydrolyzed to lower-molar-compounds such as monosaccharides.

Besides the acetyl groups, the galactosidic bonds in the acetylated softwood galactoglucomannan are also completely hydrolyzed under normal acid sulfite pulping conditions [6]. Hence, the remaining component in the pulp is typically glucomannan, without any side groups. In the case of arabinoglucuronoxylan, the glucuronide bonds are exceptionally stable toward acid, although the glucuronic acid content of the remaining xylan fraction in the pulp is lower than that of the native xylan. This is probably due to the fact that the xylan fractions that are highly substituted with uronic acid groups are more easily dissolved in the cooking liquor. The acetylated hardwood glucuronoxylan is readily deacetylated during acid sulfite pulping and the fractions with low glucuronic acid contents are also, in this case, preferentially retained in the pulp. Other polysaccharides, such as pectins and starch, which are present in minor quantities in both softwoods and hardwoods, are dissolved already in the early stage of the cook. Additionally, a wide range of miscellaneous dehydration (*e.g.*, furans) and other degradation products are formed.

In addition to typical depolymerization *via* the cleavage of glycosidic bonds in polysaccharides, some other reactions take also place, including deacetylation (*i.e.*, the formation of CH_3CO_2H), oxidation of monosaccharides (15–20%) by $HSO_3^\ominus$ (*i.e.*, forming aldonic acids), and dehydration (*i.e.*, the formation of furan derivatives, such as furfural) [6, 277, 281, 305]. Hemicellulose yield losses are generally higher for hardwoods than for softwoods; the total yield in silver birch (*Betula pendula*) and Norway spruce (*Picea abies*) acid sulfite pulping is, respectively, 49% and 52% of the dry feedstock. However, after bisulfite and neutral sulfite cooking, a large portion of soluble carbohydrates remains as oligosaccharides and polysaccharides.

The formation of aldonic acids by $HSO_3^\ominus$ (Figure 4.15, reaction route a) is an important reaction, particularly because it gives rise to the formation of $S_2O_3^{2\ominus}$; these ions are harmful for delignification and also catalyze the formation of polythionate (*i.e.*, decomposition of the cooking acid) [6, 277, 281]. Besides the formation of aldonic acids, a minor fraction of the monosaccharides is converted to sugar sulfonic acids (Figure 4.15, reaction route b). These α-hydroxysulfonic acids form so-called "loosely combined SO_2" in sulfite spent liquors and are liberated slowly under titration or after certain treatments.

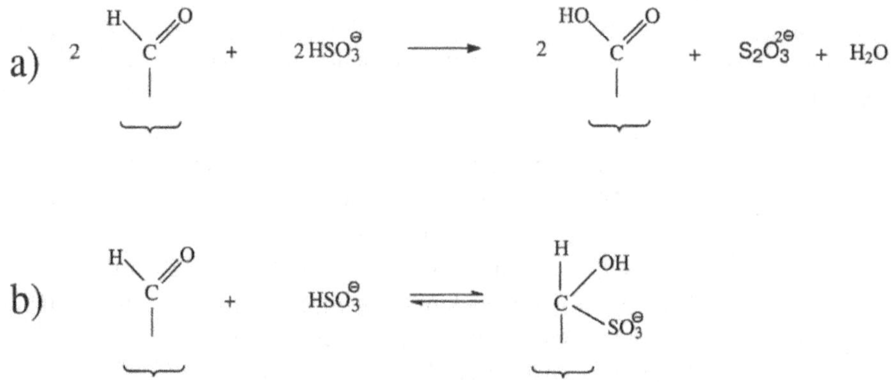

Figure 4.15: Oxidation of aldoses to aldonic acids (a) and formation of epimeric α-hydroxysulfonic acids (b) during acid sulfite and bisulfite pulping [6].

In general, more hemicellulose than lignin is dissolved in semichemical pulping [277]. Near-neutral pH is used in cooking to minimize carbohydrate losses, and the cooking liquor has a high buffering capacity ($HCO_3^\ominus$ – $CO_3^{2\ominus}$) to compensate for pH drops caused by the formation of free acids *via* the decomposition of hemicelluloses. Cellulose remains basically unchanged in semichemical pulping. During alkaline sulfite pulping with an excess of alkali present, degradation of polysaccharides by the peeling reaction also takes place [6, 281].

Extractives

Fatty acid esters are saponified to an extent, determined by the cooking conditions during acid sulfite and bisulfite pulping [281]. Some resin components can also become sulfonated, which leads to their increased hydrophilicity and better solubility. Additionally, dehydrogenation of certain extractives-derived compounds is possible. The formation of *p*-cymene (*i.e.*, sulfite turpentine) from α-pinene and quercetin from taxifolin are well-known reactions of this type. Due to their unsaturation, diterpenoids (*i.e.*, the resin acids) are probably polymerized to high-molar-mass products, which cause pitch problems in the subsequent paper production [257]. As indicated above, it is also possible that lignin can condense with reactive phenolic extractives; for example, pinosylvin and its monomethyl ethers in pine heartwood can act as nucleophilic agents, creating harmful cross-links. The reactions of extractives in alkaline sulfite pulping are probably rather like those generally occurring in alkaline pulping.

4.3.3 Spent liquor

A wide range of useful products can be obtained from acid sulfite spent liquors, but currently most of their organic materials are burned to generate energy and recover cooking chemicals [6, 274, 277, 306–311]. Since the dissolved organic solids represent a considerable fuel value, there are rather few industrial applications for the production of chemicals from spent liquor components. Typical sulfite spent liquors differ from alkaline ones in many respects (Table 4.7). Because of problems with the separation of extractives from spent liquors, the utilization of extractives has never received the same status as that concerning kraft pulping. Sulfite turpentine has traditionally been the sole extractives-derived by-product from acidic sulfite pulping; *p*-cymene can be separated from the digester gas relief condensates and purified by distillation. The crude product can be used within the mill as a resin-cleaning solvent, while the distilled product finds use in the paint and varnish industry.

High-molar-mass lignosulfonates in sulfite spent liquors can be separated in more or less pure form from low-molar-mass fragments by ultrafiltration [281, 312–314]. After separation and purification, the solution is concentrated by evaporation. Lignosulfonates are marketed in a powder form after spray-drying. They are useful for a great number

Table 4.7: Typical composition of Norway spruce (*Picea abies*) and silver birch (*Betula pendula*) spent acid sulfite liquors (kg/ton pulp) [6].

Component	Spruce	Birch
Lignosulfonates	510	435
Carbohydrates	270	380
Monosaccharides	215	305
Arabinose	10	5
Xylose	45	240
Galactose	30	5
Glucose	25	10
Mannose	105	45
Oligo- and polysaccharides	55	75
Carboxylic acids	70	130
Acetic acid	30	75
Aldonic acids	40	55
Extractives	40	40
Others	30	55

of applications, especially due to their adhesion and dispersion properties [306, 315–317]. Their primary utilization includes additives (in Portland cement concretes and oil well drilling muds), dispersing agents and binders (in textiles, products of printing industry, and mineral slurries), and in chemical purposes, such as the production of vanillin and phenolic resins. Minor uses in fertilizer applications, animal feeds, silages, insecticides, and herbicides exist as well.

Fermentation has played a dominating role in industrial processing of the carbohydrate fraction of sulfite spent liquors [6, 311]. These liquors have been used for producing ethanol (CH_3CH_2OH) [318, 319] and single-cell protein [320–324]. Production of protein by means of aerobic cultivation using either yeast (*Candida utilis*) or fungi (*Paecilomyces variotii*) has been realized in some sulfite mills. Besides hexose and pentose sugars, CH_3CO_2H, and aldonic acids are consumed by these microorganisms. Although fermentation processes were once an effective way of reducing the pollution load from the mill, this kind of by-product utilization is currently not economically attractive. The isolation of individual components from sulfite spent liquors is often possible only with tedious and complex separation methods. Therefore, the value of the final products must be high to compensate the separation cost. In contrast, the liquefaction of sulfite spent liquors, mainly to produce fuels, as an alternative way for their utilization, has been studied [325].

Different monosaccharides and their degradation products, such as furfural, can also be isolated from sulfite spent liquors and used in other applications [277, 281]. However, such processes have been of limited practical interest because of the complexity and expense of separating these substances. A characteristic of the spent liquors from neutral sulfite cooking of hardwood is the high proportion of CH_3CO_2H, in comparison with other organic compounds. A full-scale process has been developed for extracting CH_3CO_2H, with an organic solvent, from the spent liquor, after acidification [326]. The formic acid (HCO_2H) present in small quantities can be removed by azeotropic distillation if a pure product is needed.

4.4 Other delignification methods

4.4.1 Soda pulping

In the early 1800s, chemical pulp was first made in Europe by cooking straw in an aqueous solution with caustic soda (NaOH) in an open vessel [13, 295]. This precursor to the soda pulping process was applied to straw, since the chemical defibration of non-wood materials is generally easier than that of wood feedstocks. In the early 1850s, a more efficient method of using a closed vessel under pressure was introduced, which made possible the use of higher temperatures. According to a patent granted in 1854 to Charles Watt and Hugh Burgess, besides straw and other non-wood fiber feedstocks, wood raw materials could also be cooked in a still stronger aqueous solution of NaOH at 150 °C. The losses of the cooking chemical were replaced with soda (Na_2CO_3), whence the process was called "soda pulping". The first mill was started in 1866 in the United States. Their patent also included chlorine-based bleaching of the product. Furthermore, in 1865, a method for the recovery of the cooking liquors by incineration of the spent liquor (*i.e.*, soda black liquor) was patented. However, many of the early soda mills were gradually replaced by sulfite mills, and later by kraft mills, which produced, more effectively, pulps with increased strength properties.

If NaOH is the only cooking chemical, the delignification process, compared to the kraft process, is slower and leads to less and weaker quality pulp due to the alkaline degradation of carbohydrates [327]; strongly alkaline cooking conditions decompose carbohydrates by peeling reactions and alkaline hydrolysis (see Section 4.2.2 "Reactions of the wood chemical constituents"). In the search for alternative sulfur-free pulping methods, traditional soda processes, usually in the presence of some catalysts such as soda cooking, combined with a low addition of the AQ catalyst (*i.e.*, typically about 0.1% on dry feedstock) are still the most common alternatives [328–331]. AQ can be used as a catalyst, besides in soda process, also in kraft [332–336] and alkaline sulfite (see Section 4.3.1 "Process description") processes to increase delignification and decrease carbohydrate degradation. The general beneficial phenomena caused by AQ include increased delignification rates and improved selectivity, resulting in reduced alkali charges and improved pulp properties [337–339].

The chemical mechanism and effects behind alkaline AQ pulping are also well known [340–352]. In the typical reduction-oxidation cycle, AQ is first reduced to anthrahydroquinone (AHQ) by the polysaccharide end groups (*i.e.*, reducing end groups), which in turn, are oxidized to alkali-stable aldonic acid groups. AHQ then acts as an effective cleaving agent (*i.e.*, can partly replaces $HS^{\ominus}$ in kraft pulping) regarding the lignin β-aryl ether linkages in free phenolic phenylpropane units, and is simultaneously oxidized back to the initial AQ. The partly depolymerized lignin can be further degraded by NaOH at elevated temperatures. Hence, AQ also improves the conservation of carbohydrates during alkaline cooking, and the effect can be generally compared to that of polysulfide kraft pulping or modified kraft pulping by alkali concentration profiling [6, 332, 336, 353–357].

Soda-AQ pulping has been applied to softwood [358, 359] and hardwood [352,, 360, 362] feedstocks, and especially to non-wood raw materials [363–367]. Hence, the most common delignification methods for non-wood feedstocks are the soda and soda-AQ processes. Additionally, the characterization of black liquors from soda-AQ pulping of various wood [368–370] and non-wood [371–374] feedstocks has been studied. However, as indicated above, due to possible health risks, European Food Safety Authority (EFSA) and Confederation of European Paper Industries have recommended that the use of AQ should be decreased [292]. It is also possible that the delignification of non-woods (as well as of woods) with oxygen and alkali may offer a potential sulfur-free method [334]. Several investigations of one-stage delignification of wood by oxygen in the pH region of 7–9 have been made, especially in the 1950–1970s. The characteristic feature of these kinds of processes is extensive oxidation degradations of lignin and carbohydrates, leading to black liquors that have lower heating values than those from the kraft and soda-AQ pulping. Furthermore, the delignification rate in the oxygen-alkali pulping is usually slightly lower than that in the usual pulping methods. One sulfur-free approach is also to perform pre-hydrolysis with H_2O at elevated temperatures, with a subsequent soda cook and a single oxygen delignification step, for producing value-added cellulose [375].

Chemical recovery of non-wood soda pulping liquors is rather complex and has not been practiced until environmental pressures in the 1980s, when some mills started this practice with combustion of soda black liquor [376]. One of the main problems is that because large amounts of H_2O are required to wash pulps, much energy is needed to concentrate the dilute liquor from brown stock washers. Additionally, at the same yield level, black liquors from the soda pulping of non-wood materials typically contain more hemicellulose residues and less aliphatic carboxylic acids than kraft black liquors [377]. This practically means that soda black liquors are characteristically more viscous than kraft black liquors and their evaporation to a dry solids content of over 50% prior to combustion is difficult. Hence, liquor heat treatment after pulping is normally needed to reduce the viscosity of soda black liquor.

Sodium aluminum silicate scales are hard and glassy deposits [378]. They form, especially in alkaline delignification of non-wood feedstocks, a persistent layer that may grow very slowly on the heating surfaces, and even a thin layer reduces significantly the heat transfer. The scale consists primarily of alumunosilicate with a small amount of Na_2CO_3 and Na_2SO_4. The aluminum content of 0.02% dry solids is the limit above which one can expect problems with aluminum- and silicon-containing scaling.

The silicate mineral (silicon dioxide (SiO_2)) removal systems for black liquor operate in soda pulping mills. The major part of it can be removed from green liquor or black liquor, for example, by lowering of the pH of these liquors with CO_2-containing flue gases from the lime kiln or other sources [379–383].

4.4.2 Organosolv pulping

Pulping of non-wood feedstocks as well as wood raw materials with organic solvents, dating back to the beginning of the 1930s (Theodor N. Kleinert and Kurt von Tayenthal) [384], was not seriously considered for practical application until the 1980s [8, 10, 294, 385–396]. The basic idea was to use lignin solvents for this purpose, but the methods were systematically investigated later with a view to develop new industrial pulping processes. "Organosolv" (*i.e.*, solvent-based or solvolysis) pulping is a collective name for processes using an organic solvent, although many of them use an aqueous solution instead of inorganic species solvated in H_2O, as in kraft and sulfite pulping.

Both kraft and sulfite pulping have some serious drawbacks, such as environmental pollution and high investment costs [397]. Hence, it was considered that it might be possible with organosolv methods to avoid some of these problems, since they allow using both sulfur-free and chlorine-free conditions during pulping and bleaching. Especially in non-wood pulping, the main driving force for developing organosolv pulping has shifted from original energy-related considerations toward the possibility of less polluting and small-size (*i.e.*, economical) pulp mills with a simplified chemical recovery system as well as improved recovery and upgrading of by-product lignin and hemicelluloses [10, 396]. Many organosolv processes are acidic and can provide solutions to problems of conventional non-wood pulping, including those related to bleachability of pulp and chemical recovery of silicon-containing compounds; *i.e.*, these inorganic compounds dissolve in alkaline cooking liquors and are harmful in the recovery systems. Organosolv pulping can generally also be considered to represent one example of a biorefinery concept. For this reason, in the future, it may attract increased interest because its principal aim is not only effective production of chemical pulp but also the effective utilization of various by-products [8].

Organosolv pulping methods can be divided into six categories, according to their cooking chemistry [393]:

- methods involving thermal autohydrolysis that use the hydrolyzing effect of organic acids (mainly CH_3CO_2H) cleaved from the wood during cooking
- acid-catalyzed methods using acidic compounds to cause hydrolysis
- methods using phenols and acid catalysts (this topic could also be part of the previous category)
- alkaline organosolv cooking methods
- sulfite and sulfide cooking in organic solvents
- cooking, using oxidation of lignin in an organic solvent

The ambitious goals have given rise to many variations of organosolv pulping with organic solvents such as CH_3CH_2OH and CH_3OH as well as CH_3CO_2H and HCO_2H (Table 4.8). Other solvents include, for example, higher alcohols (propanol, butanol, glycol, and tetrahydrofurfuryl alcohol), CH_3CO_2H/ethyl acetate ($CH_3CO_2CH_2CH_3$), phenols (phenol and cresols), and many other systems [297, 394, 396]. Most of the organosolv processes have been intensively studied on a laboratory scale, and only few full-scale applications have emerged. Normally, the organic solvents have been used in the presence of H_2O and the introduction of suitable additives (*e.g.*, NaOH, H_2SO_4, HCl, $CaCl_2$, $MgCl_2$, $MgSO_4$, $AlCl_3$, and AQ) of the cooking liquor has enabled the production of pulps with a satisfactory delignification yield and strength. Additionally, some more complicated processes, such as ASAM (alkaline sulfite anthraquinone and methanol), have been developed.

Table 4.8: Examples of organosolv processes and their characteristic features.

Process[a]	Cooking liquor and temperature	Main feedstocks[b]	Reference examples
Alcell®	Aqueous CH_3CH_2OH (50 vol%), 185–200 °C, high pressure (3.4 MPa to 3.8 MPa)	NW and HW	Refs. [398–401]
APR	Aqueous CH_3CH_2OH (50 vol%), 190–200 °C	HW	Refs. [402, 403]
Organocell	Two stages: *i*) aqueous CH_3OH (up to 90 vol%) *ii*) aqueous CH_3OH (about 70 vol%) + NaOH (5–10%) and AQ (0.01–0.15%) 160–180 °C	SW and HW	Refs. [404–406]
ASAM	Aqueous CH_3OH (about 10 vol%) + NaOH, Na_2CO_3, and Na_2SO_3 + AQ (0.05–0.1%), 175 °C	SW, HW, and NW	Refs. [407–409]
NAEM	Aqueous CH_3OH (about 80 vol%) + small amounts of neutral alkali earth metal salts, 190–220 °C	SW, HW, and NW	Refs. [410, 411]
Acetocell	Aqueous CH_3CO_2H (about 85 vol%), 170–190 °C	SW, HW, and NW	Refs. [412–414]
Acetosolv	Aqueous CH_3CO_2H (about 90 vol%) + HCl (<1%), 60–160 °C	SW, HW, and NW	Refs. [415–419]
Formacell	$CH_3CO_2H/H_2O/HCO_2H$, typically in volume ratios of 75/15/10, 160–180 °C	HW and NW	Refs. [391, 420–423]

Table 4.8 (continued)

Process[a]	Cooking liquor and temperature	Main feedstocks[b]	Reference examples
MILOX	Three stages: *i)* aqueous HCO_2H (80–85 vol%) + H_2O_2 (1–2% of chips dry solids), 60 °C to 80 °C *ii)* HCO_2H, about 105 °C *iii)* HCO_3H + H_2O_2 (1–2% of the original chips dry solid), 60 °C	HW and NW	Refs. [424–428]

[a]Alcell® refers to the alcohol cellulose process, ARP to the alcohol pulping and recovery process, Organocell to the methanol-AQ-alkali process (or initially the MD process), ASAM to the alkaline sulfite anthraquinone and methanol process, NAEM to neutral alkali earth metal, and MILOX to the milieu pure oxidative process.
[b]NW refers to non-woods, HW to hardwoods, and SW to softwoods.

Although in organosolv pulping, many organic solvents have been used to achieve feedstock delignification, those most extensively studied have probably been alcohol-H_2O mixtures, primarily CH_3CH_2OH-H_2O [398–403, 429–432]. In the Alcell® process, the acid autohydrolysis of raw material takes place, since the cooking leads to the cleavage of acetyl groups (-$COCH_3$) to form free CH_3CO_2H and to cause delignification. Some hydrolysis of feedstock hemicelluloses, particularly xylan, also occurs during cooking. The Alcell® process is suitable for silicon-containing non-wood materials since silica does not dissolve in the acidic cooking liquor and remains in the fibers. This practically means no fouling of heat exchangers in the recovery unit. Cooking with an aqueous CH_3CH_2OH solution has also been performed to enhance delignification in the presence of acid [411, 433, 434] and alkali [430–432, 435–437], as well as alkali and AQ [430, 431, 438]. Additionally, the process of selective wood delignification by oxygen in aqueous CH_3CH_2OH has been investigated [439, 440].

Besides aqueous CH_3CH_2OH, aqueous CH_3OH has been used for delignification of various feedstock materials as such or in the presence of alkali [441–446]. The Organocell process [404–406] has some resemblance with the soda-AQ and kraft processes. As with soda and kraft cooking, the main delignifying species is the $HO^{\ominus}$ ion. However, unlike the soda process, the Organocell cooking is also suitable for softwood pulping due to the AQ addition. CH_3OH has a lower surface tension than H_2O. Hence, CH_3OH improves the ability of the cooking liquor to penetrate the wood chips and makes the lignin more soluble. The formation rate of aliphatic carboxylic acids due to alkaline conditions is also somewhat lower than that in kraft pulping [447, 448]. The main disadvantage of this process is that two parallel chemical recovery systems are needed; CH_3OH is recovered by evaporation and distillation and NaOH by causticization after combustion of the dissolved material. Additionally, the high pressure applied requires special cooking equipment and stringent safety measures.

The ASAM method [407–409], rather than being an actual organosolv process, is more a modified alkaline sulfite process, and hence, not sulfur-free. Due to these pulping conditions, it can produce both hardwood and softwood pulps, with a quality comparable with that of corresponding kraft pulps. It can also be used for pulping non-wood feedstocks. By varying the ratio of sulfite to sodium bases, it is possible to control the hemicellulose content of the pulp and affect the pulp yield and optical pulp properties. As in the case of the Organocell process, CH_3OH promotes impregnation of cooking liquor into the chips and increases the solubility of the AQ in the cooking liquor. CH_3OH also prevents lignin from participating in condensation reactions by methylation of reactive sites and enhances the solubility of the lignin decomposition products. The main advantage of the ASAM process is that, besides being suitable for all raw materials, the cooking yield is typically somewhat higher than in kraft cooking. Furthermore, the pulp is easy to bleach, even using only a short bleaching sequence, resulting in a strong pulp of high quality with good papermaking characteristics. The main disadvantages of the ASAM process are the sulfur-containing chemicals, the complex recovery system, and the high cooking pressure that requires special equipment and safety measures. For example, odor problems can also arise during the stripping and burning of H_2S in the recovery system.

The NAEM process [410, 411] is an organosolv-based process using a high concentration of alcohol, such as CH_3OH as the cooking medium, and neutral alkali earth metal as the catalyst. The process seems to have the advantage of digesting a wide variety of biomass and the pulp yield and pulp quality are comparable with those of the kraft process. Additionally, the dissolved wood components have been suggested to have great potential for conversion into value-added products. Some problems with the process include the high pressure, the handling of toxic CH_3OH, and the precipitation of lignin on a technical scale.

There have been several attempts to use only CH_3CO_2H or a mixture of CH_3CO_2H and HCO_2H, but some technical problems remain to be overcome to scale up the processes. The examples include the Acetocell process [412–414] with aqueous CH_3CO_2H and the Acetosolv process [415–419] with aqueous CH_3CO_2H, containing small amounts of HCl. One approach has also been ester pulping with the $CH_3CO_2H/CH_3CO_2CH_2CH_3/H_2O$ system, which provides an efficient chemical recovery system, with lignin recovered as a powder [449, 450]. In the Formacell process [394, 420–423], the chips are dried to about 20% moisture content and cooked in $CH_3CO_2H/H_2O/HCO_2H$. This process can be used to pulp hardwoods, softwoods, and grasses, but the pulp quality is inferior to that of kraft pulp, at least with softwoods. Since the acid hydrolysis is not selective, some feedstock polysaccharides also undergo hydrolysis and react further to form furfural. Under pulping conditions, CH_3CO_2H also reacts with the -OH groups of lignin and polysaccharides to produce the corresponding acetates. Additionally, the highly corrosive nature of the cooking acid requires the use of special steel, wherever the equipment is in contact with hot acid: *i.e.*, the digester, evaporation unit, distillation column, spray dryer, and some piping. The main advantages of the Formacell process are as follows: *i*) it involves no inorganic cooking or bleaching chemicals (*e.g.*, it is also possible to perform selective ozone delignification) and *ii*) it offers the possibility of obtaining sulfur-free by-products. The main disadvantages are the complex chemicals recovery, corrosion problems, and the need for safety equipment.

The MILOX process [424–428] occurs under sulfur-free and chlorine-free conditions, where HCO_2H reacts with the free aliphatic and phenolic -OH groups of lignin to produce formate esters. The polysaccharides also react with HCO_2H; the principal reaction being acid hydrolysis. Peroxyformic acid or performic acid (HCO_3H) is simple to prepare by an equilibrium reaction between HCO_2H and hydrogen peroxide (H_2O_2):

$$HCO_2H + H_2O_2 \leftrightarrow HCO_3H + H_2O \tag{4.6}$$

HCO_3H is a highly selective chemical that does not react with cellulose or other wood polysaccharides in the same way as HCO_2H. It oxidizes lignin and makes it more hydrophilic, and hence increases its solubility. Pulping of hardwood and softwood is possible, but in contrast to hardwood pulps, the production of softwood pulps is not profitable because of the high consumption of H_2O_2 and poor strength properties of the fibers. However, unlike in the pulping of wood, in the pulping of agricultural plants, two-stage MILOX cooking is more effective than three-stage cooking. The two-stage process uses cooking with HCO_2H alone, followed by treatment with HCO_2H and H_2O_2. However, HCO_2H is highly corrosive, especially at high temperatures, and the digester should be carbon steel, coated with zirconium. All other equipment, pipes, and valves in contact with hot HCO_2H are of duplex steel. The main advantages of the MILOX process include the low temperatures used and unpressurized reactors. The process is also suitable for pulping silicon-containing non-wood materials. Additionally, the bleaching needed can be a simple total chlorine-free sequence, and the chemical recovery system is relatively simple, together with sulfur-free by-products. The main disadvantages are, besides corrosion problems, the poor quality of softwood pulps and the drying needed for feedstocks.

The other organosolv processes investigated include, for example, the use of phenol [451, 452] and cresols [453, 454] as well as tetrahydrofurfuryl alcohol [455] as a solvent. Additionally, oxygen in acetone ($CH_3(CO)CH_3$)–H_2O media [456], aqueous *p*-toluenesulfonic acid treatment [457], and HCO_2H [458] have been used for different delignification purposes. The Lignol process is based on the Alcell biorefining technology using aqueous CH_3CH_2OH organosolv extraction [294, 396]. It was originally designed to produce fuel-grade CH_3CH_2OH and biochemicals, such as lignin, furfural, and CH_3CO_2H from biomass. The process uses aqueous CH_3CH_2OH (40–60 wt%) as a pulping liquor and H_2SO_4 as a catalyst. The Lignol process is capable of producing extracted softwood lignin and, due to its high purity, low molar mass, and abundance of reactive groups, it is suitable to produce lignin-based adhesives and other products.

The primary function of the organic solvent in organosolv cooking is to make the lignin more soluble in the cooking liquor [294, 390, 394, 397, 459]. However, in many cases, the solvent also participates in delignification reactions in one way or another.

The reactions taking place during alkaline organosolv methods are like those in the corresponding kraft pulping and alkaline sulfite pulping processes. Under alkaline conditions, the cleavage of β-ether linkages is more important than that of the α-ether linkage. At alkaline pH, α-aryl ether bonds break only in the phenyl propane units containing free phenolic -OH groups, whereas etherified phenol groups cannot undergo conversion to quinone methides and therefore, the α-bonds cannot be cleaved. In contrast, the β-aryl ether bonds can be broken, whether the phenyl -OH groups are free or etherified. Additionally, AQ promotes the cleavage of β-bonds. The easier delignification of hardwoods, compared to softwoods, is primarily a result of differences in β-ether reactivity, α-ether concentration, lignin content, and propensity for condensation reactions. In acidic organosolv pulping, the cleavage of the α-ether linkage in lignin is the most important delignification reaction but the cleavage of β-ether linkages also plays a role. The chemical bonds between lignin and carbohydrates are easy to break in acidic organosolv pulping.

In many cases, research with simple solvent systems as well as trials with more complicated cooking liquors have resulted in relatively high pulp yields but often with lower pulp strength properties than those of the kraft pulps. Additionally, in many cases, the chemical recovery (*i.e.*, the effective recovery of solvent and possible alkali added) of organosolv systems needs further development. However, some processes, for example, the Alcell® process using CH_3CH_2OH as a solvent and the Organocell process using CH_3OH and alkali, have reached pilot-scale and even full-scale production but no real breakthrough. Against this background, developing a new competitive process that would replace kraft pulping in a significant part or produce raw materials from biomass for the chemical industry still involves several major challenges. On the other hand, the present boom in the biorefinery business may boost research in the foreseeable future to develop new concepts of organosolv methods for fractionating effectively various lignocellulosic biomass feedstocks to utilize more effectively their chemical components.

4.5 Oxygen-alkali delignification and delignifying bleaching

4.5.1 Oxygen-alkali delignification

As a successive process stage of the kraft pulping process, oxygen-alkali delignification or simply oxygen delignification has acquired an important role in removing a substantial fraction of the residual lignin in unbleached pulp, prior to actual bleaching [460–469]. This process stage represents an effective and economically favorable way of reducing both the environmental load of bleach plant effluents and the use of expensive bleaching chemicals, such as ClO_2 or O_3. In general, oxygen is a beneficial delignification agent, since it is stable, nontoxic, relatively cheap, and readily available. The technology became possible on a full scale after the discovery of the inhibiting properties of magnesium salts against the severe degradation of polysaccharides taking place under oxygen-delignification conditions, although this degradation is still responsible for the main limitations of the use of oxygen delignification.

The industrial application of oxygen delignification has expanded very rapidly since the late 1960s and currently, it is a well-established technique [464, 469]. Most recent systems installed are based on the medium-consistency (MC, 10–14%) process. Delignification is usually conducted under pressure (*i.e.*, inlet pressure about 750 kPa and outlet pressure about 500 kPa) for 60 min and at relatively high temperatures (90–110 °C), compared to those of conventional bleaching (20–70 °C). High-consistency (HC, 25–30%) systems, forming most of the plants installed during the first 15 years of commercial applications, are still available. Magnesium salts (usually MgSO$_4$) are used, especially in the case of softwood kraft pulps, in low amounts (0.05–0.1% Mg$^{2\oplus}$ on oven-dried pulp).

All the main chemical pulp constituents, polysaccharides (*i.e.*, cellulose and hemicelluloses), lignin, and extractives, behave differently during oxygen delignification [461, 464]. As indicated above, one of the most important barriers to extending this technique is the intensive degradation of carbohydrates at higher degrees of delignification, resulting in inferior pulp strength properties and low pulp yields. Due to this non-selectivity, the delignification degree of kraft pulps from conventional cooking in oxygen delignification is typically in the range of 35–50%, if a suitable protector against carbohydrate degradation is added. However, in the modern systems with a two-stage oxygen stage, equipped with intermediate washing, a kappa reduction exceeding 60–70% can be achieved while still maintaining the pulp strength properties.

One of the most significant factors in the enhanced oxygen delignification as also in certain bleaching systems, is an effective mixing of chemicals [462], that is, mixing gaseous and liquid bleaching chemicals in the pulp matrix. Furthermore, effective washing of unbleached pulp is needed for increasing the effect of the oxygen stage and continue delignification to a lower kappa number. In practice, in the case of modified cooking, when the kappa number of hardwood and softwood pulp is, respectively, 15–20 and 20–25, the kappa number can be lowered by two-stage oxygen delignification to 8–10.

The principal environmental benefits of oxygen delignification are related to the fact that because sulfur-free chemicals are used, the recycling of oxygen-stage effluent to the recovery system *via* the brown stock washers is possible [464]. This practice reduces significantly the potential environmental impact of the bleach plant. The main advantages of oxygen delignification can be described as follows: *i*) notable decrease in biochemical oxygen demand (BOD), chemical oxygen demand (COD), adsorbable organic halogens (AOX) (*i.e.*, a decrease in the amount of chlorinated compounds), and color in the plant effluent, *ii*) a decrease in the consumption of oxidizing chemicals in the delignifying part of the bleaching sequence (*i.e.*, oxygen is less expensive than these bleaching chemicals), and *iii*) the possible use of oxidized white liquor instead of aqueous pure NaOH usually provides the necessary alkali for the oxygen stage at low cost.

To achieve the required chemical savings, the reaction products dissolved in the oxygen delignification stage must be washed from the pulp as thoroughly as possible. The most obvious disadvantages of oxygen delignification are as follows [462]: *i*) high capital costs due to the use of a reactor with two post-oxygen washing stages, *ii*) possibility of overloading the mill recovery system, and *iii*) non-selectivity at higher degrees of delignification. The increased load on the recovery boiler is typically about 4%, measured as total dry substance, and because this load contains highly oxidized substances, the corresponding increase in energy content is lower. If oxidized white liquor is used, the increased load on the causticizing department is about 5%.

Oxygen has become established as a weak oxidant in many industrial processes, including in conventional bleaching sequences and oxygen-alkali delignification, for several decades [460]. At normal temperatures and pressures, two atoms of oxygen (O) bind to form dioxygen (O_2), which is a colorless and odorless diatomic gas [470]. Dioxygen has a triplet electronic ground state (*i.e.*, a spin triplet state), with an electron configuration having two unpaired electrons; each of them has an opposite spin (Figure 4.16). This "triplet oxygen" is the most stable and common allotrope of oxygen. Hence, in its lowest energy state (*i.e.*, the ground state), molecular O_2 is a diradical, which has, as a free radical, a tendency to react with appropriate substrates at sites with high electron density [461, 462].

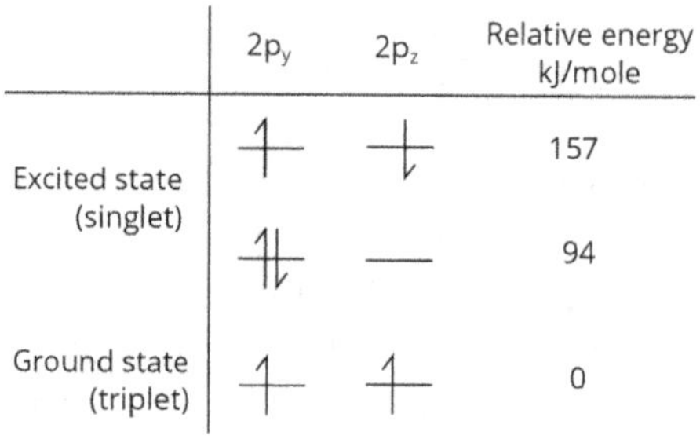

Figure 4.16: Energy states of oxygen molecule [461, 471].

The term "singlet oxygen" refers to several higher-energy species of molecular O_2, characterized by two paired or unpaired electrons with an antiparallel spin [470]. They can be obtained from triplet oxygen with chemicals, at higher temperatures, or photochemically, by irritation in presence of a suitable sensitizer. Singlet oxygen is much more reactive with common organic molecules than triplet oxygen and is responsible for many selective oxidation reactions occurring in Nature and in specific industrial processes. Singlet oxygen reacts electrophilically rather than in a free-radical fashion.

Because of its orbital structure, dioxygen (called simply "oxygen" below) can accommodate only one electron at a time when integrating with a substrate [464]. Although being less reactive than other free radicals, oxygen has a strong tendency to oxidize organic substances under alkaline conditions. However, due to several reactive intermediates formed in the reactions of oxygen with a substrate, the resulting reaction pattern is extremely complex and not fully understood. The complicated oxidation processes that take place during oxygen delignification include a multitude of radical chain reactions (Figure 4.17) involving a variety of organic species derived mainly from lignin and also from carbohydrates. Since molecular oxygen is not very reactive at room temperature, it requires activation of the substrate and higher temperatures to bring about the reaction. In oxygen delignification, free phenolic -OH groups in the residual lignin are ionized by providing alkaline conditions. Hence, the resulting sites with high electron density enhance the attack by oxygen.

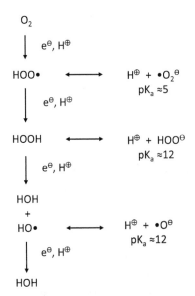

Figure 4.17: Formation of reactive intermediates during the stepwise reduction of oxygen to water [464].

When oxygen reacts with a substrate, such as lignin ($RO^{\ominus}$), it is reduced to H_2O by one-electron transfer in four successive stages involving intermediate parts as hydroperoxy radical ($HOO\bullet$), hydrogen peroxide ($HOOH$), and hydroxyl radical ($HO\bullet$) as well as their organic counterparts. When oxygen reacts with an organic radical ($R\bullet$), the corresponding organic intermediates, $ROO\bullet$, $ROOH$, and $RO\bullet (+H_2O)$, are formed. The pH conditions then determine the presence of the dissociated forms of these intermediates [462, 464, 467, 471]; superoxide radical anion ($\bullet OO^{\ominus}$), hydroperoxide anion ($HOO^{\ominus}$), and oxyl radical anion ($\bullet O^{\ominus}$). The experimental data available indicate that the undissociated species are clearly more reactive than the ionized ones and that the $HO\bullet$ radicals are strong oxidants. However, some of these intermediates are nonspecific oxidative agents to lignin, and in oxygen delignification as well as in H_2O_2 bleaching, it is necessary to control their formation for avoiding severe degradation of the polysaccharides [464].

Strongly alkaline conditions are needed to achieve appreciable delignification rates, which explains why free -OH of the residual lignin should be ionized [464, 472–479]. The main reaction has been suggested to proceed *via* a resonance-stabilized phenoxy radical, resulting in potential sites for the following step – conversion to a hydroperoxide. The hydroperoxide intermediates (*e.g.*, the corresponding anion) can subsequently undergo an intermolar nucleophilic reaction at an adjacent site, leading to structures that can be expected to enhance the solubility of the lignin in the alkaline medium. The first step of these reaction chains is the conversion of the ionized phenolic group to a phenoxy radical by the loss of a single electron to a suitable acceptor; in the case of molecular oxygen, the formation of $\cdot OO^{\ominus}$ takes place. Hence, the primary structures formed during oxygen delignification from phenols include oxidized structures, such as quinone- and muconic acid-type structures. Like in kraft pulping, some lignin monomers are also formed, although representing a minor fraction of the total lignin removed.

The volatile turpentine components are chemically stable during kraft pulping but the fatty acid esters of the native wood are hydrolyzed almost completely and some of the unsaturated fatty acids and resin acids are partly isomerized under alkaline pulping conditions [464, 480]. The amount of extractives that remain in the pulp after kraft pulping is relatively low; depending on the species, the typical values are 0.5–1.0% and 0.1–0.3% of the dry pulp for hardwood and softwood, respectively [467, 481–483]. The content of extractives has been found to be, respectively, 0.05–0.11% and 0.32–0.43% of the dry pulp, for softwood and hardwood, after oxygen delignification [482, 483].

Hence, only about half of the residual extractives in kraft pulp can be removed by oxygen delignification and this soluble part essentially represents several chromophore-containing extractives [484]. This phenomenon is important, especially in the case of hardwood pulps containing an enhanced amount of colorful neutral substances. However, due to their low concentration, it is obvious that extractives play a minor role in determining the outcome of oxygen delignification. It is also possible that the extractives react, to some extent, with oxygen under alkaline conditions [467, 485, 486]. Table 4.9 gives an example of the composition of the extractive-based fraction, formed during oxygen delignification. These fatty and resin acid components are also the prominent tall-oil constituents [488].

The term "selectivity" can be straightforwardly defined as the ratio of the removal of pulp lignin to that of pulp carbohydrates [467]. The structural pulp components, cellulose (65–75% of unbleached pulp) and hemicelluloses (20–30% of unbleached pulp) [464], contribute to pulp strength. Hence, it is desirable that these components do not react with oxygen or, in general, with other delignifying (*i.e.*, bleaching) chemicals [489].

In bleaching, since the expensive bleaching chemicals are very selective, only a moderate depolymerization of cellulose takes place, as indicated by the viscosity measurements [461, 464, 478, 489–491]. In contrast, in oxygen delignification, the residual lignin can be also removed from the pulp much more selectively than by kraft pulping. However, in this preliminary delignification stage after kraft pulping, a sharp viscosity

Table 4.9: Examples of the extractives determined after oxygen delignification of pine kraft pulp (% of total compounds) [487].

Compound	Content
Fatty acids	
Hexadecanoic acid	5
9,12-Octadecadienoic acid	5
9-Octadecenoic acid	20
Resin acids	
Abietic acid	5
Dehydroabietic acid	25
Neoabietic acid	10
Isopimaric acid	15
Sandraracopimaric acid	15

drop still takes place relatively fast, and on prolonged delignification, cellulose is gradually depolymerized, leading to loss in fiber strength and low pulp yield. Hence, during this delignification stage, besides the intensive depolymerization of cellulose, the losses in hemicelluloses are also considerable.

Due to the non-selectivity of oxygen delignification, practically, no more than about half of the residual lignin in kraft pulp entering this stage can be normally removed, if, especially in the case of softwood kraft pulps, a suitable protector against carbohydrate degradation is added (Figure 4.18). In the modern mills, with a two-stage oxygen stage equipped with intermediate washing, it is, however, possible to achieve a somewhat higher lignin reduction and still maintain the pulp strength properties.

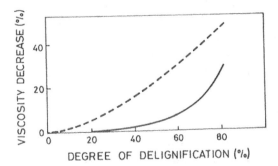

Figure 4.18: Depolymerization of cellulose (indicated by viscosity decrease) during oxygen delignification of pine kraft pulp [461]. Dotted line, no inhibitor has been added and full line, magnesium salt has been added as a protector.

The pulps contain trace quantities of transition metals, such as iron, copper, and manganese, which act as catalysts for peroxide decomposition during oxygen delignification [464, 492]. This decomposition generates harmful radical species (especially HO•), which are believed to be responsible, at higher degrees of delignification, for the major damage to carbohydrates. The transition metals can be removed, prior to oxygen delignification, by acid washing. However, another and more common approach is to add to the kraft pulp, different protectors that prevent the function of the transition metals, indirectly inhibiting carbohydrate degradation. In this respect, especially H_2O-soluble $MgSO_4$ or its heptahydrate (Epsom salt) have shown to be effective and economically favorable. $MgSO_4$ is normally used in low concentrations. 0.05-0.1% $Mg^{2\oplus}$, on oven-dried pulp. In the presence of alkali, $MgSO_4$ is readily precipitated as magnesium hydroxide ($Mg(OH)_2$). It is believed that this hydroxide either adsorbs the harmful transition metallic ions or forms inactive complexes with them and thus renders these metallic ions unavailable for the catalysis of peroxide decomposition.

The alkaline degradation reactions of cellulose under oxidative conditions, which take place in oxygen delignification, are analogous to those occurring in kraft pulping (see Section 4.2.2 "Reactions of the wood chemical constituents") [195, 461, 468]. However, with certain respects, the detailed mechanisms are somewhat different. Reactions that degrade polysaccharides can generally be divided into two categories [462, 464, 467, 468, 493, 494]: *i*) the cleavage of glycosidic bonds at any point along the polysaccharide chain and *ii*) the peeling reaction, by which end group units of the chain are successively removed (Figure 4.19). Although both types may take place during oxygen delignification, the former is more significant and promoted by traces of transition metals that randomly attack the polysaccharide chain. Due to cleavage of glycosidic bonds, this attack results in a decrease in pulp viscosity and ultimately to a decrease in pulp strength properties. In addition to the degradation reactions, oxidative stabilization of carbohydrates (*i.e.*, mainly cellulose) through formation of aldonic acid end groups occurs. To simplify the research systems, experiments have been carried out with model substances such as cellobiose (4-*O*-β-D-glucopyranosyl-D-glucopyranose or β–D–Glc$p$–$(1 \rightarrow 4)$–D–Glc$p$) [468, 495–497], birch xylan [498–501], cellobiitol [502, 503], and D-glucosone [504] as well as the formation of aldonic acid end groups in cellulose [505–508].

The initial step of the most important chain cleavage reaction, induced mainly by HO•, is the formation of a carbonyl group (>C=O) at the C_2-position of a monomeric sugar unit [461, 462, 464, 497]. This group formation (*i.e.*, the formation of oxycellulose [195]) facilitates a cleavage of the alkali-labile glycosidic bond at C_4 by β-alkoxy elimination, with the simultaneous formation of a new reducing end group (Figure 4.20, reaction route a). A comparable oxidation at C_3 and at C_6 can also lead to the same result. However, the initially formed carbonyl-containing unit does not necessarily have to lead to the cleavage of the polysaccharide chain. A competing reaction takes place when oxygen attacks its ionized keto form, forming a 2,3-diketo structure, which may be converted, without chain cleavage, to a furanosidic acid group or an open-chain structure containing two carboxylic acid groups (Figure 4.20, reaction route b).

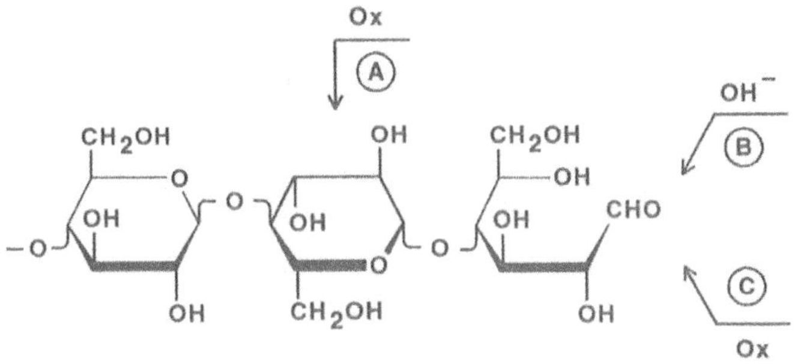

Figure 4.19: Different sites in cellulose chain to attack of agents under oxidative conditions [461]. Oxidation of any position (C_2, C_3 or C_6) within the chain moieties, resulting in alkali labile glycosidic linkages (A), peeling reaction starting from the reducing end group (B), and oxidation of the reducing end groups to carboxylic acid groups, preventing the peeling reaction (C).

Oxidative depolymerization of polysaccharides *via* random chain cleavage is not directly responsible for carbohydrate yield losses in oxygen delignification [461, 464, 467, 468, 471]. However, this cleavage generates new reducing end groups that are subjected to the alkali-catalyzed peeling reaction of the polysaccharide (hemicelluloses and cellulose) chains (see Section 4.2.2 "Reactions of the wood chemical constituents"). Furthermore, besides some other >C=O moieties, most of the terminal units in the polysaccharides of kraft pulps are alkali-stable metasaccharinic acid groups, and the peeling reaction cannot begin here. Additionally, the stabilization of carbohydrate chains *via* oxidation of aldehyde end groups to carboxylic acid end groups (see below) prevents the peeling reaction.

Although in different mass proportions, the peeling reaction leads to the same acids that are formed under non-oxidative conditions under kraft cooking [461, 464, 467, 468, 487, 492, 493, 499, 504, 509]. The most prominent aliphatic carboxylic acids formed in oxygen delignification from pine kraft pulp (*i.e.*, mainly from cellulose and glucomannan) are summarized in Table 4.10. Characteristic products from xylan in birch kraft pulps are glycolic, lactic, 3-hydroxypropanoic, 3-deoxy-tetronic, and xyloisosaccharinic acids [467, 499, 501, 510]. Some low-molar-mass acids are also partly formed from aliphatic side chains of lignin. The carbohydrate-derived yield loss in oxygen delignification is typically 2–3% of the kraft pulp. However, the situation is different for acid sulfite pulps because, in these cases, the terminal groups are reducing end units.

Oxidation of the reducing end groups to alkali-stable aldonic acid end groups also takes place readily (Figure 4.21) [461, 464, 504, 508]. The initial oxidative step in this case involves formation of an aldos-2-ulose (glycosone) end group, which, after a benzilic acid rearrangement, is transformed to an epimeric pair of aldonic acids; in the case of cellulose and glucomannan, mainly to mannonic acid. Additionally, the cleavage of C-C bonds occurs as well, leading predominantly to the formation of arabinonic and erythronic acid end groups.

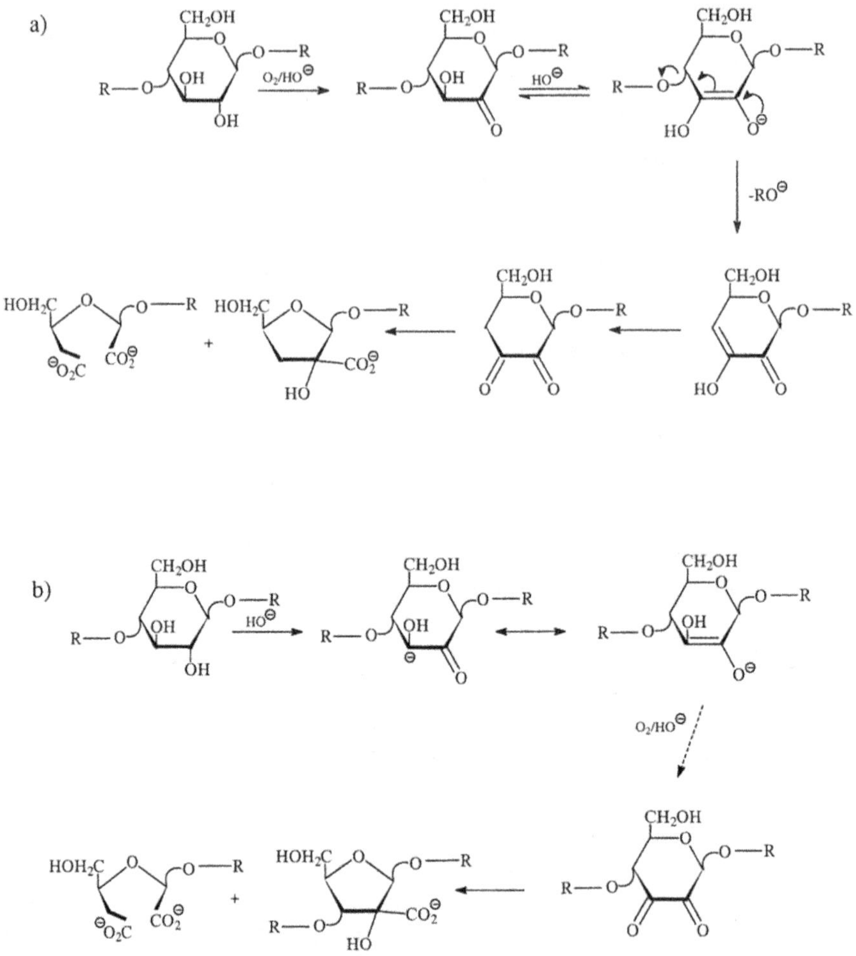

Figure 4.20: Cleavage of glycosidic bonds in polysaccharide chains (a) and oxidation of ketol structures along polysaccharide chains (b) during oxygen delignification [464]. R is part of the polysaccharide (cellulose) chain.

As indicated above, versatile degradation reactions of the pulp constituents taking place during oxygen delignification leads to a wide range of reaction products. The main soluble compounds are, besides degraded lignin fragments with varying molar masses, low-molar-mass aliphatic carboxylic acids (hydroxy monocarboxylic and hydroxy dicarboxylic acids, together with formic, acetic, and oxalic acids), carbohydrates (hemicellulose-derived fragments), CH_3OH, CO_2, and a minor number of extractives-derived compounds [464, 467, 487, 495, 511]. Furthermore, it has been suggested that, besides CO_2, various aliphatic monocarboxylic acids (e.g., formic, acetic, glycolic, and lactic acids) and especially, some dicarboxylic acids (e.g., oxalic, maleic, fumaric, succinic, malic, and malonic acids), are formed, apart from carbohydrates, also due to an oxidative breakdown of

Table 4.10: Formation of the main aliphatic carboxylic acids during the oxygen delignification[a] of pine kraft pulp [487].

Component	Content, % of acids
Monocarboxylic acids	80.1
Formic	31.4
Acetic	9.1
Glycolic	10.4
Lactic	1.1
3-Hydroxypropanoic	2.0
Glyceric	1.2
2-Deoxy-tetronic	6.0
3-Deoxy-tetronic	1.9
3,4-Dideoxy-pentonic	1.4
3-Deoxy-*erythro*-pentonic	2.1
3-Deoxy-*threo*-pentonic	6.5
Xyloisosaccharinic	0.8
α-Glucoisosaccharinic	1.1
β-Glucoisosaccharinic	3.6
Others	1.5
Dicarboxylic acids	12.7
Oxalic	5.4
Tartronic	1.6
Succinic	0.9
Glucoisosaccharinaric[b]	3.0
3-Deoxy-*threo*-pentaric	0.8
Others	1.0
Miscellaneous[c]	7.2
Total amount, % of the pulp	0.9

[a]The alkali charge is 3% of the pulp and the reaction time 60 min.
[b]Both *erythro* and *threo* isomers.
[c]Various mono- and dicarboxylic acids with low concentrations.

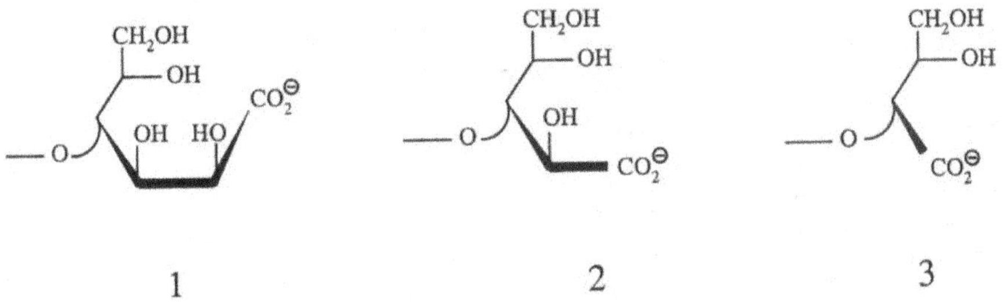

Figure 4.21: Aldonic acid end groups stabilizing cellulose chains during oxygen delignification [464]. Mannonic acid (1), arabinonic acid (2), and erythronic acid (3) end groups.

both aromatic ring and aliphatic side chain structures of lignin [467, 472, 473, 493, 511–516]. These findings have also been supported by the experimental data obtained from oxygen-alkali treatments of isolated lignin preparations. Besides oligomeric and polymeric lignin fragments, some lignin monomers are also formed, although they represent a minor fraction of the total lignin removed (0.2–0.4% of the pulp) [487]. In this case, the main source of CH_3OH are $-OCH_3$ groups of lignin [467, 475].

The prominent constitute in the effluent is CO_2 [511, 513, 517], which is originated mainly from lignin [513, 518] but also from certain carboxylic acids by decarboxylation [519, 520] as well as from carbohydrate degradation [491, 521]. Hence, it has been shown that the formation of CO_2, for example, correlates well with the amount of residual lignin in pulp as well as with the dissolved lignin [518, 522], whereas cellulose does not have such an effect on the amount of the CO_2 formed. However, it seems that if the degree of delignification proceeds over 65%, most of the CO_2 originates from carbohydrates [523].

The average composition of organic material in the effluent from oxygen delignification of the pine and birch kraft pulp is given in Table 4.11. Additionally, there is some inorganic material (*e.g.*, chemically bound sodium and sulfur, if oxidized white liquor is used for the oxygen stage) present. The overall composition of the spent liquors from delignification can be also roughly characterized by categorizing the soluble organic material in different groups. For comparison, Table 4.12 shows the typical average composition of the organic fraction from kraft pulping, oxygen-alkali delignification, and the TCF bleaching. These spent liquors differ from each other, clearly indicating the nature of each process. For example, when aiming at more closed pulp washing systems, TCF bleaching effluents could be ultimately burned together with black liquor and oxygen delignification effluent in the recovery furnace [464].

Table 4.11: Average composition of organic material in the effluent from oxygen delignification of the pine and birch kraft pulp (% of the total amount) [467].

Component	Pine	Birch
Lignin	23.0	16.9
Formic acid	3.6	4.3
Acetic acids	1.5	1.7
Hydroxy acids	6.3	9.2
Monocarboxylic acids	4.5	7.9
Dicarboxylic acids	1.8	1.3
Hemicelluloses	6.5	5.7
Methanol	2.6	2.5
Sodium	24.1	25.7
Others	32.4	34.0

Table 4.12: Typical composition of dissolved organics in different wood-derived spent liquors (% of the total amount) (modified from [464]).

Component	Kraft pulping	Oxygen delignification	TCF[a] bleaching
Lignin residues	45	45	10
Aliphatic carboxylic acids	45	25	40
Hemicellulose residues	<5	15	45
Other organics	5	15	<5

[a]TCF refers to totally chlorine-free.

4.5.2 Delignifying bleaching

Bleaching can basically be defined as a chemical process applied to chemical and mechanical pulps to increase their brightness [461, 465, 489, 491, 524–541]. Hence, to reach an acceptable brightness level, bleaching is performed either by removing the colored residual lignin of chemical pulps (*i.e.*, by delignifying or lignin-removing bleaching) or by converting and stabilizing chromophoric groups of mechanical pulps without loss of substance (*i.e.*, by lignin-preserving or lignin-retaining bleaching). However, in both cases, the evident prerequisite is that no significant losses in pulp strength take place.

When aiming at the desired brightness level of the pulp in bleaching, several factors have to be taken into account, for example, including the required pulp properties, production cost, and local environmental regulations as well as health and safety aspects [534]. Furthermore, in the case of specialty papers, where high purity is desired, the bleaching process not only removes lignin but also offers a means of improving the cleanliness of the pulp *via* removing extractives and other contaminants. The contaminants typically comprise inorganic impurities and insufficiently cooked particles, such as shives (*i.e.*, incompletely fiberized slivers of chips) and bark residues [6].

After kraft pulping and oxygen delignification, the residual lignin material of the pulp is typically 3–5% of the unbleached pulp by weight and is difficult to be removed [533, 541]. This is mainly due to many structural changes (*e.g.*, formation of new conjugated double bonds and condensation products) in the initial wood lignin occurring during delignification, and the presence of covalent lignin-carbohydrate linkages [542–544]. In contrast, since no such drastic chemical changes in the initial structure of lignin take place during delignification, the unbleached pulp from the acid sulfite process is easier to bleach [298]. The presence of -SO_3H groups in the residual lignin is also one reasons for the good bleaching response.

Many bleaching stages are practically needed to fully delignify and bleach a chemical pulp [533]. These stages form a bleaching sequence, which thus comprises a series of bleaching stages, usually with intermediate washing between them. Washing is generally also an important process step since the organic material dissolved in one bleaching stage can consume bleaching chemicals in the subsequent bleaching stages.

The main chemicals used in bleaching are *i*) oxidizing chemicals and *ii*) reducing bleach chemicals, together with *iii*) auxillary chemicals (Table 4.13). Table 4.14 shows the typical conditions applied during each bleaching stage. Regarding environmental aspects, bleaching chemicals are also classified into *i*) oxygen-based chemicals (*e.g.*, O_2, O_3, H_2O_2, and CH_3CO_3H) and *ii*) chlorine-containing chemicals (*e.g.*, Cl_2 and ClO_2). Hence, in the TCF bleaching, only the oxygen-based chemicals are used, although the use of elemental chlorine-free (ECF) bleaching (*i.e.*, the use of Cl_2 is replaced using ClO_2) is gradually increasing. Examples of the common bleaching sequences of kraft pulps are as follows: D_0-E_O-D_1-D_2 or $D_0E_OD_1D_2$ (ECF bleaching with vacuum washers), D_0-P_O-D_1-P_O or $D_0P_OD_1P_O$ (ECF bleaching with wash presses, can also be operated with the TCF sequence Q-P_O-Q-P_O), D_0-E_O-E_P-D_1-E_2-D_2 or $D_0E_OE_PD_1E_2D_2$ (TCF bleaching), and OP-Z-Q-PO or OPZQPO (an oxygen delignification stage prior to TCF bleaching). The term "bleach liquor" refers to a solution of a bleaching agent applied to the pulp, whereas the term "bleaching effluent" is generally used for a spent liquor obtained from the individual stage or the bleaching sequence.

Table 4.13: Description of the conventional chemicals used in delignifying bleaching (modified from [533]).

Agent	Formula and stage designation[a]		Form and main function
Oxidizing chemicals[b]			
Chlorine	Cl_2	C	Pressurized gas; chlorinates and oxidizes lignin
Ozone	O_3	Z	Pressurized gas mixture (10–12% O_3 in O_2); source is air or O_2; oxidizes lignin
Hydrogen peroxide Sodium peroxide	H_2O_2 Na_2O_2	P	2–5% solution in H_2O; oxidizes lignin and causes brightening effect
Chlorine dioxide	ClO_2	D	7–10 g/L solution in H_2O; oxidizes lignin, chlorination is usually carried out in the presence of ClO_2
Sodium hypochlorite	NaOCl	H	40–50 g/L in H_2O; causes brightening effect and oxidizes lignin
Reducing chemicals[c]			
Sodium dithionite (Zinc dithionite)	$Na_2S_2O_4$ $Zn_2S_2O_4$	Y	"Hydrosulfite solution" or made on-site from sodium borohydride (NaBH$_4$) solution, NaOH, and SO_2
Auxillary chemicals			
Sodium hydroxide	NaOH	E	5–10% solution; extraction of lignin and hydrolyses chlorinated lignin
Sulfur dioxide	SO_2	S	Solution; an acid treatment performed mainly after the final D, H, or P stages for destroying residual bleaching agents, removing of metal ions, and adjusting a pH favorably for brightness stability

Table 4.13 (continued)

Agent	Formula and stage designation[a]	Form and main function
Enzymes	Xylanases X	Catalyze xylan hydrolysis and facilitate lignin removal during the subsequent bleaching stages
Chelants	EDTA and DTPA[d] Q	Solid chemicals; remove metal ions, especially used for P stage[e]

[a]Other abbreviations generally used are, for example, as follows: A, acid (SO_2 and H_2SO_4) treatment; AZ, acid treatment reinforced with O_3; C_D (C/D or CD), CD treatment (mixture, principal agent Cl_2); (C+D), Cl_2 and ClO_2 are added simultaneously; DC, D followed by C (sequential with no wash in between); D_C (D/C or DC), DC treatment (mixture, principal agent ClO_2); E_O (EO), E reinforced with O_2; E_{OP} (EOP), E reinforced with O_2 and Na_2O_2; E_P (EP), E reinforced with Na_2O_2; N, neutralization; OO, double O stage (with no wash in between): Paa, peracetic acid (CH_3CO_3H) bleaching stage; P_O (PO), P reinforced with O_2; Px, peroxymonosulfuric acid; and W, water soak.
[b]For chemical pulps. P stage is also used for high-yield pulps. For the use of oxygen (O_2), see Section 4.5.1 "Oxygen-alkali delignification".
[c]For high-yield pulps, zinc dithionite is not commonly used anymore. The P stage is also used, besides for delignifying bleaching, for lignin-preserving bleaching.
[d]EDTA, ethylenediaminetetraacetic acid and DTPA, diethylenetriaminepentaacetic acid.
[e]Sodium silicate (Na_2SiO_3) and magnesium salts are also used for buffering and stabilizing, respectively, in H_2O_2 bleaching.

Table 4.14: Examples of process conditions applied in different bleaching stages [533].

Stage[a]	Consistency[b], %	Temperature, °C	Time, min	pH
C	LC	20–40	30–60	2–3
Z	HC or MC	20–60	<2	2–3
P	MC	50–100	90–180	9–11
D	MC	60–70	180–240	3–5
H	MC	30–40	90–120	9–11

[a]For abbreviations, see Table 4.13.
[b]LC, low consistency (up to 4%), MC, medium consistency (10–14%), and HC, high consistency (25–30%).

In contrast to kraft cooking chemicals, bleaching chemicals cannot be recovered and hence, they represent a significant part of the production costs of a bleached mill [533]. As shown in Table 4.13, bleaching chemicals are either used as solutions in H_2O (H_2O_2, ClO_2, and CH_3CO_3H) or as gas (O_2, Cl_2, and O_3). Some of them are bought by the mill and others, such as ClO_2 and O_3, must be prepared on site. The way these chemicals are mixed with pulp suspension, primarily depends on their nature (*i.e.*, liquid or gas), the operation conditions (*i.e.*, temperature and pressure), and the specific design of the bleaching towers and reactors within different bleaching sequences.

The chemistry belonging to this delignifying bleaching chapter outlines only generally this topic; the main reason being only its logical connection to chemical pulping and oxygen delignification. More detailed information and background data on bleaching can be found in the selected textbooks [526, 527, 534-536].

Each delignifying chemical has a specific impact on residual lignin and is selective. For example, one classification of the primary reactions of these chemicals with lignin indicates that Cl_2 and O_3 are the most powerful bleaching agents as they can react with almost all structure types of lignin, whereas O_2, ClO_2, and H_2O_2 are limited to certain lignin structures [473, 533, 534, 544–547]. In general, due to the complexity of the structure of residual lignin and the wide variety of reactive bleaching species present in different delignifying stages, the bleaching reactions that take place are highly complex. The reactions can be simply classified into those occurring in the carbon atoms of the propane side chain, at the aromatic nucleus, at the -OCH_3 group, and at the free or non-free phenolic -OH groups. In principle, the main delignifying bleaching chemicals basically react according to radical mechanisms (O_2 and ClO_2), or electrophiles (Cl_2, O_3, and CH_3CO_3H), or nucleophiles (Na_2O_2 and $NaOCl$).

As is typical for acidic-acid bleaching systems, Cl_2 reacts with aromatic lignin structures at their partially negatively charged sites, principally by electrophilic substitution under simultaneous formation of HCl by oxidative decomposition of aromatic rings to dicarboxylic acids (*i.e.*, muconic acid derivatives), by oxidative cleavages of aryl-alkyl ether linkages (*i.e.*, demethylation), by electrophilic side-chain displacement, and by oxidation of side-chain structures [533]. Additionally, some Cl_2 can be introduced by addition to >C=C< structures present in the side chain (*e.g.*, in stilbene and styrene structures). Although, in the C stage, the aromatic structure of the residual lignin is extensively destroyed, only a partial depolymerization takes place. Hence, a minor portion of lignin is degraded during the C stage (this stage is performed as the first stage in the conventional bleaching sequence) to low-molar-mass fragments and an effective dissolution of chlorinated lignin does not occur until in the subsequent alkaline E stage. During this E stage, a partial nucleophilic substitution of -Cl substituents by $HO^\ominus$ ions takes place and the residual lignin is converted to more H_2O-soluble degradation products.

Compared to Cl_2, the currently generally used ClO_2 is a much more selective oxidant that attacks predominantly lignin structures having a free phenolic -OH group or side-chain structures with electron-rich >C=C< bonds under industrial reaction conditions [533]. The reaction products are similar muconic acid structures as those obtained in the C stage but only slightly chlorinated. The C stage has been typically carried out in the presence of ClO_2 because ClO_2 acts as a radical scavenger (*e.g.*, radicals are harmful to carbohydrates). However, ClO_2 has been gradually replaced by Cl_2 in the first stage of the multistage bleaching sequence, and in the ECF bleaching, it is normally used in the final stage. Although, ClO_2 has long been known to be an excellent delignifying and bleaching agent that selectively oxidizes lignin as well as extractives, it is, due to toxicity and its high reactivity in the gaseous form, also difficult to handle in full-scale applications.

Due to the acidic conditions in the Z bleaching stage as well as in the D bleaching stage, the initial step of delignifying the bleaching reactions is an attack of the bleaching chemical on sites of high electron density, followed by nucleophilic reactions [533]. The strongly electrophilic character of O_3 promotes its reaction with aliphatic side-chain >C=C< bonds and the aromatic ring to form muconic acid derivatives and other compounds containing o-quinoidic and partially oxidized structures. Most free phenolic -OH groups are oxidized and many -CO_2H groups are introduced together with the formation of CH_3OH from the -OCH_3 groups.

Bleaching in alkaline medium comprises oxidation with Na_2O_2 and NaOCl. The relatively expensive H stage has traditionally been used in the later stages of delignifying bleaching sequences, leading to a markedly increased brightness of the fiber product [533]. In contrast to Cl_2 and ClO_2, the negatively charged hypochlorite anion ($ClO^\ominus$) is a strong nucleophile, mainly attacking the positively charged sites of the residual lignin. These reactive sites belong especially to quinoid and other enone structures formed during the preceding oxidation reactions. It cleaves C-C bonds in such structures, leading to many low-molar-mass products.

Bleaching chemicals may also influence pulp carbohydrates, which results in their partial depolymerization and oxidation [527, 533]. These reactions affect the strength properties of the pulp as well as its brightness stability. Among all common bleaching chemicals, ClO_2 is the most selective lignin oxidant, but reactions of Cl_2 with pulp carbohydrates are also of importance, although much slower than those with lignin. It seems that the reactions of Cl_2 with carbohydrates proceed mainly via radical mechanisms because these reactions are retarded by ClO_2 and other radical scavengers; the corresponding reactions with lignin, instead are presumably more ionic character, occurring via the positive end of the polarized Cl_2 molecule ($^{\delta\ominus}Cl\text{-}Cl^{\delta\oplus}$). Hence, as indicated above, the significant oxidation of polysaccharides in the C stage is often prevented by the addition of ClO_2 because it acts a radical scavenger. Additionally, of the generation of reactive radicals, it is difficult to avoid degradation of polysaccharides, especially during oxygen delignification but also in the Z stage. In the latter case, the most harmful radical intermediates are HO• and HOO• radicals, formed by the direct decomposition of O_3 in H_2O or indirectly by the reaction of O_3 with an organic substrate.

The main oxidizing attack of bleaching chemicals takes place within the polysaccharide chains, but can also be directed to the reducing end groups of the polysaccharide chains [533]. However, in kraft pulps, although new aldehydic (-CHO) end groups are formed via the cleavage of glycosidic linkages, the end groups are mainly of the -CO_2H type. Hence, the most harmful reaction in any bleaching systems is the oxidation of C_2-, C_3-, and C_6-positions of a monosaccharide moiety to a >C=O group because these structures create alkali-labile glycosidic bonds in the polysaccharide chains. The generation of >C=O groups during the acidic bleaching stages results in a pronounced degradation of polysaccharides during the subsequent alkaline E stage via the peeling reactions. The action of NaOCl is also rather nonspecific. Apart from the oxidation of -CHO end groups to $-CO_2H$ end groups, C-C linkages between C_2 and C_3 are cleaved, leading to a ring-opening and formation of dicarboxylic acids.

The small amounts of extractives remaining in the pulps after oxygen delignification and bleaching are detrimental to the pulp quality (*e.g.*, the brightness stability, especially in birch pulps) [533]. Most of these extractives are known to be sticky materials that are difficult to remove in washing stages, and can leave sticky deposits on process equipment [257]. Hence, it is important to perform bleaching under conditions that lead to the lowest possible content of extractives in the pulp. On the other hand, it is significant to avoid bleaching reactions, generating toxic or otherwise harmful products from extractives that are carried over to bleaching effluents. Regarding this aspect, much attention has been paid to the analysis of the lipophilic extractives-derived compounds because of their toxicity to fish and other organisms. Besides compounds from extractives, some harmful compounds are also degradation products of lignin and carbohydrates. For example, in the conventional Cl_2 bleaching of pulp, polychlorinated phenolic compounds were typically found in rather high concentrations in bleaching effluents, whereas the concentrations of these compounds in most cases are below the detection limit in effluents from the modern ECF bleaching.

In older types of bleaching sequences (*i.e.*, contain the C stage), wood extractives are chlorinated; Cl_2 typically participates in addition reactions with unsaturated constituents (*e.g.*, fatty acids), leading to compounds that are difficult to remove from the pulp in the subsequent bleaching stages [533]. In contrast to Cl_2, ClO_2 primarily reacts with extractives by oxidation and therefore, the formation of chlorinated compounds is normally very low in the ECF bleaching. In this case, introduction of $-CO_2H$ groups to extractives results in their increased hydrophilicity, thus improving the solubility of the reaction products. Hence, the D stage gives rise, for example, to an almost complete removal of unsaturated fatty acids in all the chemical pulps but has a minor effect on the removal of saturated fatty acids. Additionally, there are indications that after the P stage, for example, the content of unsaturated fatty acids in the pulps is usually only slightly lower than that in the unbleached pulps.

Accumulation of the most detrimental metals in bleaching processes may cause several problems in the processes, of which the most evident ones are inorganic scale formation *via* catalytic degradation of bleaching chemicals and metal salt precipitation [548]. The most detrimental metal ions for the bleaching processes are iron, copper, and manganese, because they catalyze decomposition of peroxy chemicals [549]. Efficient bleaching of pulps always also requires the effective control of these metal ions in the pulps and the wash waters. This can be achieved either by deactivating the harmful metal ions inside the process or by removing them from the process in a specific treatment and washing stage. The deactivation of metals in pulp bleaching without removing them is essential in the bleaching of mechanical pulps and, generally, with pulps having high lignin content. On the other hand, abundant scale formation requires frequent stops for cleaning either mechanically or chemically. A periodical chemical cleaning is typically done with a highly acidic or alkaline wash liquid containing a metal sequestrant or chelating agent circulated through the selected process section.

Magnesium, often applied in the form of $MgSO_4$, has been found to deactivate the catalytic effects of iron [550, 551], copper [552], and even manganese when the H_2O_2 solution contains pulp (see also Section 4.5.1 "Oxygen-alkali delignification") [553]. The role of magnesium in bleaching is radical scavenging and/or forming binuclear structures with other metals present. The binuclear structures [554] are of the type (Mg-O-M), where M is Si, Mg, or Zn. Additionally, it has been known that sodium silicate (Na_2SiO_3) (*i.e.*, water glass) has beneficial effects in the P stage [552, 553, 555]. Its primary role is to deactivate transition metal ions, thus lowering their catalytic activity toward H_2O_2 decomposition. The inactivation of manganese could happen through the formation of manganese silicate ($MnSiO_3$). In the presence of magnesium, silicate can act as a stabilizer [556]. Especially, when the process waters are recycled, the drawback of using silicates as H_2O_2 stabilizers is the potential formation of silicate precipitates. Therefore, organic H_2O-soluble polymers and monomers have been introduced as H_2O_2 stabilizers. Additionally, the closure of water systems in pulp and paper mills results in the accumulation of metals in the circulating process waters. When increasing the concentration in a process flow, the deposits of $CaCO_3$, calcium oxalate, and barium sulfate ($BaSO_4$) will start to form in areas determined by the local process conditions (*i.e.*, pH, temperature, and supersaturation level) [557]. The main source of calcium and barium is the wood itself and therefore, there will be variations in their intake amount, depending on the wood sourcing region.

The acidic D stages in chemical pulp bleaching provide an efficient opportunity for metal ion removal [558]. The acidic conditions release the bound metals, and the functional groups in the pulp are thereafter, for the most part, in their hydrogen form. However, at subsequent treatment stages, the pH is increased, and the pulp encounters dilution and wash waters that may contain significant amounts of metal ions. This may cause substantial metal sorption again onto the pulp fibers, based on their concentration and their order of sorption affinity. Hence, it is of the highest importance to ensure that, especially, iron and manganese levels entering a bleaching stage either with the pulp or with the dilution waters are well managed. This is typically achieved by solubilizing these metals as H_2O-soluble chelate-metal complexes from the pulp matrix prior to a washing stage where the metals can be separated from the pulp, prior to the actual bleaching stage. This practice is especially important if the bleaching sequence includes oxygen-based bleaching chemicals that are more susceptible to the detrimental effects of these metal ions.

In general, a chelating agent is a chemical compound that reacts with metal ions to form stable and H_2O-soluble metal complexes [548]. Chelating agents are used in several H_2O intensive industrial, agricultural, and domestic applications for metal management, scale control/removal, and corrosion control and most chelating agents lose their complexing power under strongly alkaline conditions. Additionally, in the P stage, they influence the decomposition reactions of H_2O_2 [558, 559]. The most common chelating agents in pulping and bleaching, DTPA and EDTA, have the best efficiency (Figure 4.22). However, due to their low biodegradability, the bioaccumulation of DTPA and EDTA has become an increasing concern. If the chelating agent does not rapidly degrade in the environment, there is a risk that it may bind heavy metals in sewage sludge or in river and lake sediments, and resuspend those metals into the H_2O phase; as a result, the exposure of aquatic species to these metals is increased [560].

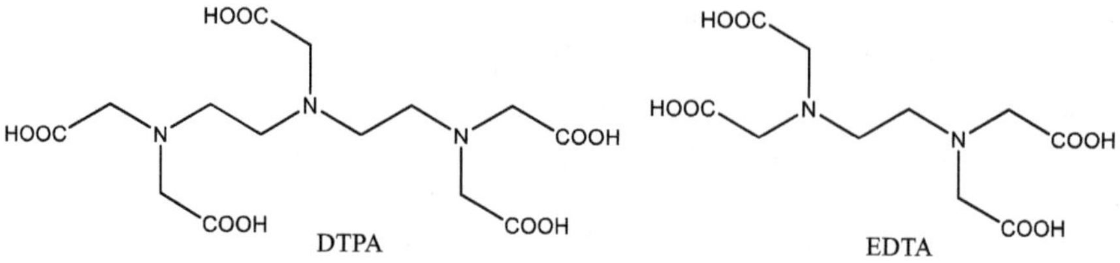

Figure 4.22: Chemical structures of diethylenetriaminepentaacetic acid (DTPA) and ethylenediaminetetraacetic acid (EDTA).

References

[1]   Rydholm, S.A. 1965. *Pulping Processes*. Interscience Publishers, New York, NY, USA. Pp. 576–649.

[2]   Fengel, D., and Wegener, G. 1989. *Wood – Chemistry, Ultrastructure, Reactions*. Walter de Gruyter, Berlin, Germany. Pp. 437–447.

[3]   Grace, T.M., Leopold, B., Malcolm, E.W., and Kocurek, M.J. (Eds.). 1989. *Pulp and Paper Manufacture, Volume 5, Alkaline Pulping*. 3rd edition. The Joint Textbook Committee of the Paper Industry, TAPPI&CPPA, USA and Canada. 637 p.

[4]   Sjöström, E. 1993. *Wood Chemistry – Fundamentals and Applications*. 2nd edition. Academic Press, San Diego, CA, USA. Pp. 140–161.

[5]   Biermann, C.J. 1996. *Handbook of Pulping and Papermaking*. 2nd edition. Academic Press, San Diego, CA, USA. Pp. 86–91.

[6]   Alén, R. 2000. Basic chemistry of wood delignification. In: Stenius, P. (Ed.). *Forest Products Chemistry*. Fapet, Helsinki, Finland. Pp. 58–104.

[7]   Sixta, H., Potthast, A., and Krotschek, A.W. 2006. Chemical pulping processes. In: Sixta, H. (Ed.). 2006. *Handbook of Pulp*. Wiley-VCH, Weinheim, Germany. Pp. 109–391.

[8]   Alén, R. 2011. Principles of biorefining. In: Alén, R. (Ed.). *Biorefining of Forest Resources*. Paper Engineers' Association, Helsinki, Finland. Pp. 55–114.

[9]   Fardim, P. (Ed.). 2011. *Chemical Pulping Part 1, Fibre Chemistry and Technology*. 2nd edition. In: Paper Engineers' Association, Helsinki, Finland. 748 p.

[10] Alén, R. 2015. Pulp mills and wood-based biorefineries. In: Pandey, A., Höfer, R., Taherzadeh, M., Nampoothiri, K.M., and Larroche, C. (Eds.). *Industrial Biorefineries & White Biotechnology*. Elsevier, Amsterdam, The Netherlands. Pp. 91–126.

[11] Smook, G. 2016. *Handbook for Pulp and Paper Technologists*. 4th edition. TAPPI Press Atlanta, GA. Pp. 65–83.

[12]   Särkkä, T., Gutiérrez-Poch, M., and Kuhlberg, M. (Eds.). 2018. *Technological Transformation in the Global Pulp and Paper Industry 1800–2018 – Comparative Perspectives*. Springer, Cham, Switzerland. 299 p.

[13]   Alén, R. 2018. Manufacturing cellulosic fibres for making paper: A historical perspective. In: Särkkä, T., Gutiérrez-Poch, M., and Kuhlberg, M. (Eds.). *Technological Transformation in the Global Pulp and Paper Industry 1800–2018 – Comparative Perspectives*. Springer, Cham, Switzerland. Pp. 13–34.

[14]   Statista. 2022. (https://www.statista.com/statistics/1178289/production-of-chemical-pulp-worldwide/).

[15]   Alén, R. 2018. *Carbohydrate Chemistry – Fundamentals and Applications*. World Scientific, Singapore. 586 p.

[16]   Clements, L.D., and Van Dyne, D.L. 2006. The lignocellulosic biorefinery – a strategy for returning to a sustainable source of fuels and industrial organic chemicals. In: Kamm, B., Gruber, P.R., and Kamm, M. (Eds.). *Biorefineries – Industrial Processes and Products, Volume 1*. Wiley-VCH, Weinheim, Germany. Pp. 115–128.

[17]   Kamm, B., Kamm, M., Gruber, P.R., and Kromus, S. 2006. Biorefinery systems – an overview. In: Kamm, B., Gruber, P.R., and Kamm, M. (Eds.). *Biorefineries – Industrial Processes and Products, Volume 1*. Wiley-VCH, Weinheim, Germany. Pp. 3–40.

[18]   Katzen, R., and Schell, D.J. 2006. Lignocellulosic feedstock biorefinery: History and plant development for biomass hydrolysis. In: Kamm, B., Gruber, P.R., and Kamm, M. (Eds.). *Biorefineries – Industrial Processes and Products, Volume 1*. Wiley-VCH, Weinheim, Germany. Pp. 129–138.

[19]   Koukoulas, A.A. 2007. Cellulosic biorefineries – charting a new course for wood use. *Pulp & Paper Canada* 108(6):17–19.

[20]   Clark, J.H., and Deswarte, E.I. 2008. The biorefinery concept – an integrated approach. In: Clark, J.H., and Deswarte, E.I. (Eds.). *Introduction to Chemicals from Biomass*. John Wiley & Sons, New York, NY, USA. Pp. 1–20.

[21]   Sixta, H., and Schild, G. 2009. A new generation kraft process. *Lenzinger Berichte* 87:26–37.

[22]   Cherubini, F. 2010. The biorefinery concept: Using biomass instead of oil for producing energy and chemicals. *Energy Convers Management* 51(7):1412–1421.

[23]   Hamaguchi, M., Cardoso, M., and Vakkilainen, E. 2012. Alternative technologies for biofuels production in kraft pulp mills – potential and prospects. *Energies* 5:2288–2309.

[24]   Perkins, J.K., and Cowan, B. 1989. In: Grace, T.M., Leopold, B., Malcolm, E.W., and Kocurek, M.J. (Eds.). *Pulp and Paper Manufacture, Volume 5, Alkaline Pulping*. 3rd edition. The Joint Textbook Committee of the Paper Industry, TAPPI&CPPA, USA and Canada. Pp. 241–278.

[25]   Mimms, A., Kocurek, M.J., Pyatte, J.A., and Wright, E.E. (Eds.). 1989. *Kraft Pulping – A Compilation of Notes*. TAPPI Press, Atlanta, GA, USA. Pp. 55–76.

[26]   Alén, R. 2021. Kraft pulping. In: Alén, R. (Ed.). *Pulping and Biorefining – ForestBioFacts, Digital Learning Environment*. Paperi ja Puu Ltd., Helsinki, Finland.

[27]   Croon, I., and Enström, B. 1962. The hemicelluloses in sulphate pulps from scots pine. *Svensk Papperstidning* 65:595–599.

[28]   Meier, H. 1962. On the behavior of wood hemicelluloses under different pulping conditions. Part 2. Spruce hemicellulose. *Svensk Papperstidning* 65:589–594.

[29]   Simonson, R. 1963. The hemicellulose in the sulfate pulping process. Part 1. The isolation of hemicellulose fractions from pine sulfate cooking liquors. *Svensk Papperstidning* 66:839–845.

[30]   Meller, A. 1965. The retake of xylan during alkaline pulping – a critical appraisal of the literature. *Holzforschung* 19(4):118–124.

[31]   Simonson, R. 1965. The hemicellulose in the sulfate pulping process. Part 3. The isolation of hemicellulose fractions from birch sulfate cooking liquors. *Svensk Papperstidning* 68(8):275–280.

[32]   Simonson, R. 1971. The hemicellulose in the sulfate pulping process. *Svensk Papperstidning* 74(21):691–700.

[33] Gierer, J. 1970. The reactions of lignin during pulping a description and comparison of conventional pulping processes. *Svensk Papperstidning* 73(18):571–596.

[34] Gierer, J. 1980. Chemical aspects of kraft pulping. *Wood Science and Technology* 14:241–266.

[35] Gierer, J. 1982. The chemistry of delignification. A general concept. Part I. *Holzforschung* 36:43–51.

[36] Gierer, J. 1982. The chemistry of delignification. A general concept. Part II. *Holzforschung* 36:55–64.

[37] Gellerstedt, G., and Lindfors, E.-L. 1984. Structural changes in lignin during kraft cooking. Part 4. Phenolic hydroxyl groups in wood and kraft pulps. *Svensk Papperstidning* 87(15):R 115–R 118.

[38] Gierer, J. 1985. Chemistry of delignification. Part 1: General concept and reactions during pulping. *Wood Science and Technology* 19(4):289–312.

[39] Iversen, T., and Westmark, U. 1985. Lignin-carbohydrate bonds in pine lignins dissolved during kraft pulping. *Cellulose Chemistry and Technology* 19:531–536.

[40] Gellersted, G., Gustafsson, K., and Northey, R.A. 1988. Structural changes in lignin during kraft cooking. Part 8. Birch lignins. *Nordic Pulp and Paper Research Journal* 3(2):87–94.

[41] Alén, R., and Vikkula, A. 1989. Formation of lignin monomers during kraft pulping of birch wood. *Cellulose Chemistry and Technology* 23:579–583.

[42] Alén, R., and Vikkula, A. 1989. Formation of lignin monomers during alkaline delignification of softwood. *Holzforschung* 43(6):397–400.

[43] Niemelä, K. 1990. *Low-Molecular Weight Organic Compounds in Birch Kraft Black Liquor*. Doctoral Thesis. *Annales Academiæ Scientiarum Fennicæ, Series A, II. Chemica*, Vol 229. Helsinki, Finland. 142 p.

[44] Tai, D.-s., Chen, C.-L., and Gratzl, J.S. 1990. Chemistry of delignification during kraft pulping of bamboos. *Journal of Wood Chemistry and Technology* 10(1):75–99.

[45] Niemelä, K., and Alén, R. 1999. Characterization of pulping liquors. In: Sjöström, E., and Alén, R. (Eds.). *Analytical Methods in Wood Chemistry, Pulping, and Papermaking*. Springer Verlag, Heidelberg, Germany. Pp. 193–231.

[46] Majtnerová, A., Gellerstedt, G., and Zhang, L. 2005. On the delignification mechanism in kraft pulping. In: *Proceedings of the 13th International Symposium on Wood, Fibre and Pulping Chemistry (ISWFPC)*, May 16–19, 2005, Auckland, New Zealand. Pp. 161–165.

[47] Axegård, P., and Wikén, J.-E. 1983. Delignification studies – factors affecting the amount of "residual lignin". *Svensk Papperstidning* 86(15):R178–R184.

[48] Iversen, T., and Wännström, S. 1986. Lignin-carbohydrate bonds in a residual lignin isolated from pine kraft pulp. *Holzforschung* 40(1):19–22.

[49] Hortling, B., Ranua, M., and Sundquist, J. 1990. Investigation of the residual lignin in chemical pulps. *Nordic Pulp and Paper Research Journal* 5(1):33–37.

[50] Jiang, Z.-H., and Argyropoulos, D.S. 1999. Isolation and characterization of residual lignins in kraft pulps. *Journal of Pulp and Paper Science* 25(1):25–29.

[51] Backa, S., Gustavsson, C., Lindström, M.E., and Ragnar, M. 2004. On the nature of residual lignin. *Cellulose Chemistry and Technology* 38(5–6):321–331.

[52] Koda, K., Gaspar, A., Yu, L., and Argyropoulos, D.S. 2005. Molecular weight – functional group relations in softwood residual kraft lignins. In: *Proceedings of the 13th International Symposium on Wood, Fibre and Pulping Chemistry (ISWFPC)*, May 16–19, 2005, Auckland, New Zealand. Pp. 473–480.

[53] Kleppe, P.J. 1970. Kraft pulping. *TAPPI* 53(1):35–47.

[54] Vroom, K. 1957. The "H" factor: A means of expressing cooking times and temperatures as a single variable. *Pulp and Paper Magazine of Canada* 58:228–231.

[55] Aurell, R., and Hartler, N. 1965. Kraft pulping of pine. Part 1. *Svensk Papperstidning* 68(3):59–68.

[56] Bhaskaran, T.A., and von Koeppen, A. 1970. The degradation of wood carbohydrates during sulphate pulping. *Holzforschung* 24(1):14–19.

[57] Malinen, R., and Sjöström, E. 1975. The formation of carboxylic acids from wood polysaccharides during kraft pulping. *Paperi ja Puu* 57(11):728–730,735–736.

[58] Löwendahl, L., Petersson, G., and Samuelson, O. 1976. Formation of carboxylic acids by degradation of carbohydrates during kraft cooking of pine. *TAPPI* 59(9):118–121.

[59] Niemelä, K., Alén, R., and Sjöström, E. 1985. The formation of carboxylic acids during kraft and kraft-anthraquinone pulping of birch wood. *Holzforschung* 39(3):167–172.

[60] Alén, R., Lahtela, M., Niemelä, K., and Sjöström, E. 1985. Formation of hydroxy carboxylic acids from softwood polysaccharides during alkaline pulping. *Holzforschung* 39(4):235–238.

[61] Alén, R., Hentunen, P., Sjöström, E., Paavilainen, L., and Sundström, O. 1988. A new control method for the kraft pulping process: a technique based on gas-liquid chromatography of the degradation products. In: *Proceedings of the 1990 TAPPI Pulping Conference, Book 3*, October 30–November 11, 1988. Pp. 535–540.

[62] Alén, R., Hentunen, P., Sjöström, E., Paavilainen, L., and Sundström, O. 1991. A new approach for process control of kraft pulping. *Journal of Pulp and Paper Science* 17(1):J6–J9.

[63] Alén, R., Käkölä, J., and Pakkanen, H. 2008. Monitoring of kraft pulping by a fast analysis of aliphatic carboxylic acids. *Appita Journal* 61(3):216–219.

[64] Alén, R. 1993. Analysis of low-molecular-mass compounds as an aid for monitoring oxygen delignification. *Cellulose Chemistry and Technology* 27:281–286.

[65] Alén, R. 1997. Analysis of degradation products. a new approach to characterizing the combustion properties of kraft black liquors. *Journal of Pulp and Paper Science* 23(4):J62–J66.

[66] Alén, R., Mäkelä, A., Sjöström, E., Hartus, T., Paavilainen, L., and Sundström, O. 1990. Analysis of lignin degradation products as a new tool for monitoring the kraft pulping process. In: *Proceedings of the 1990 TAPPI Pulping Conference, Book 1*, October 14–17, 1990. Pp. 55–60.

[67] Aho, W.O. 1983. Advances in chemical pulping processes. *Progress in Biomass Conversion* 4:149–181.

[68] Kleppe, P.J. 1986. Advances in chemical pulping. TAPPI *Journal* 69(3):50–52.

[69] Kerr, A.J. 1970. The kinetics of kraft pulping – process in the development of a mathematical model. *Appita Journal* 24(3):180–188.

[70] Kerr, A.J., and Uprichard, J.M. 1976. The kinetics of kraft pulping – refinement of a mathematical model. *Appita Journal* 30(1):48–54.

[71] Yan, J., and Johnson, D.C. 1981. Delignification and degelation: Analogy in chemical kinetics. *Journal of Applied Polymer Science* 26:1623–1635.

[72] Gustafson, R.G., Sleicher, C.A., McKean, W.T., and Finlayson, B.A. 1983. Theoretical model of the kraft pulping process. *Industrial & Engineering Chemistry Process Design and Development* 22(1):87–96.

[73] Paulonis, M.A., and Krishnagopalan, A. 1991. Adaptive inferential control of kraft batch digesters as based on pulping liquor analysis. *TAPPI Journal* 74(6):169–175.

[74] Vanchinathan, S., and Krishnagopalan, G.A. 1995. Kraft delignification kinetics based on liquor analysis. *TAPPI Journal* 78(3):127–132.

[75] Fan, Y., Gartshore, I.S., and Salcudean, M.E. 2006. A general kraft pulping reaction model. *Appita Journal* 59(3):237–241.

[76] Bethge, P.O., and Ehrenborg, L. 1967. Identification of volatile organic compounds in kraft mill emissions. *Svensk Papperstidning* 70(10):347–350.

[77] Wilson, D.F., and Hrutfiord, B.F. 1971. SEKOR IV. formation of volatile organic compounds in the kraft pulping process. *TAPPI* 54(7):1094–1098.

[78] Wilson, D.F., Johanson, L.N., and Hrutfiord, B.F. 1972. Methanol, ethanol, and acetone in kraft pulp mill condensate streams. *TAPPI* 55(8):1244–1246.

[79] Blackwell, B.R., MacKay, W.B., Murray, F.E., and Oldham, W.K. 1979. Review of kraft foul condensates. sources, quantities, chemical composition, and environmental effects. *TAPPI* 62(10):33–37.

[80] Chai, X.-S., Dhasmana, B., and Zhu, J.Y. 1998. Determination of volatile organic compound contents in kraft mill streams using headspace gas chromatography. *Journal of Pulp and Paper Science* 24(2):50–54.

[81] Zhu, J.Y., Yoon, S.-H., Liu, P.-H., and Chai, X.-S. 2000. Methanol formation during alkaline wood pulping. *TAPPI Journal* 83(7):69–81.

[82] Clayton, D.W. 1963. The alkaline degradation of some hardwood 4-O-methyl-D-glucuronoxylans. *Svensk Papperstidning* 66(4):115–123.

[83] Sarkanen, K.V., Chirkin, G., and Hrutfiord, B.F. 1963. Base-catalyzed hydrolysis of aromatic ether linkages in lignin: 1. the rate of hydrolysis of methoxyl groups by sodium hydroxide. *TAPPI* 46(6):375–379.

[84] Rönnholm, A.A.A. 1977. Removal of methanol from the sulphate process. *Paperi ja Puu* 59(11):721–732.

[85] Greis, O., and Pehu-Lehtonen, L. 2021. Methanol purification. In: Alén, R. (Ed.). *Pulping and Biorefining – ForestBioFacts, Digital Learning Environment*. Paperi ja Puu Ltd., Helsinki, Finland.

[86] Bryant, P.S., Robarge, K., and Edwards, L.L. 1993. Transition-metal profiles in open and closed kraft fiber lines. *TAPPI Journal* 76(10):148–159.

[87] Gu, Y., and Edwards, L. 2004. Prediction of metals distribution in mill processes, part 2: Fiber line metals profiles. *TAPPI Journal* 3(2):13–20.

[88] Gu, Y., and Edwards, L. 2004. Prediction of metals distribution in mill processes, part 3 of 3: NPE management in kraft chemical recovery. *TAPPI Journal* 3(3):9–15.

[89] Karhu, J. 2008. *Equilibria and Balances of Metal Ions in Kraft Pulping*. Doctoral Thesis. Åbo Akademi University, Laboratory of Analytical Chemistry, Åbo, Finland. 74 p.

[90] Lundqvist, F., Brelid, H., Saltberg, A., Gellerstedt, G., and Tomani, P. 2006. Removal of non-process elements from hardwood chips prior to kraft cooking. *Appita Journal* 59:493–499.

[91] Ferreira, L.M.G.A., Soares, M.A.R., Egas, A.P.V., and Castro, J.A.A.M. 2003. Selective removal of chloride and potassium in kraft pulp mills. *TAPPI Journal* 2(4):21–25.

[92] Hamaguchi, M., and Vakkilainen, E.K. 2011. Influence of chlorine and potassium on operation and design of chemical recovery equipment. *TAPPI Journal* 10(1):33–39.

[93] Keitaanniemi, O., and Virkola, N.-E. 1982. Undesirable elements in causticizing systems. *TAPPI* 65(7):89–92.

[94] Taylor, K., and McGuffie, B. 2007. Investigation of non-process element chemistry at Elk Falls mill – green liquor clarifier and lime cycle. *Pulp & Paper Canada* 108(2):27–31.

[95] Francey, S., Tran, H., and Berglin, N. 2011. Global survey on lime kiln operation, energy consumption, and alternative fuel usage. *TAPPI Journal* 10(8):19–26.

[96] Lachenal, D. 1996. Hydrogen peroxide as a delignifying agent. In: Dence, C.W., and Reeve, D.W. (Eds.). *Pulp Bleaching – Principles and Practice*. TAPPI Press, Atlanta, GA, USA. Pp. 347–361.

[97] Dyer, T.J., and Ragauskas, A.J. 2006. Deconvoluting chromophore formation and removal during kraft pulping: Influence of metal cations. *Appita Journal* 59(6):452–458.

[98] Saltberg, A., Brelid, H., and Theliander, H. 2006. Removal of metal ions from wood chips during acidic leaching 1: Comparison between Scandinavian softwood, birch and eucalyptus. *Nordic Pulp and Paper Research Journal* 21(4):507–512.

[99] Saltberg, A., Brelid, H., and Theliander, H. 2006. Removal of metal ions from wood chips during acidic leaching 2: Modeling leaching of calcium ions from softwood chips. *Nordic Pulp and Paper Research Journal* 21(4):513–519.

[100] Tran, H. (Ed.). 2020. *Kraft Recovery Boilers*. 3rd edition. TAPPI Press, Atlanta, GA, USA. 375 p.

[101] Hart, P.W. (Ed.). 2022. *Chemical Recovery in the Alkaline Pulping Processes*. 4th edition. TAPPI Press, Atlanta, GA, USA.

[102] Alén, R. 2015. Pulp mills and wood-based biorefineries. In: Pandey, A., Höfer, R., Taherzadeh, M., Nampoothiri, K.M., and Larroche, C. (Eds.). *Industrial Biorefineries & White Biotechnology*. Elsevier, Amsterdam, The Netherlands. Pp. 91–126.

[103] Vakkilainen, E. 1999. Chemical recovery. In: Gullichsen, J., and Fogelholm, C.-J. (Eds.). *Chemical Pulping*. Fapet, Helsinki, Finland. Pp. B95–B132.

[104] Hupa, M., Solin, P., and Hyöty, P. 1987. Combustion behavior of black liquor droplets. *Journal of Pulp and Paper Science* 13:J67–J72.

[105] Whitty, K., Backman, R., and Hupa, M. 1994. An empirical rate model for black liquor char gasification as a function of gas composition and pressure. *AIChE Symposium Series (American Institute of Chemical Engineers)* 90(302):73–84.

[106] Miller, P.T. 1986. *Swelling of Kraft Black Liquor – An Understanding of the Associated Phenomena during Pyrolysis*. Doctoral Thesis. The Institute of Paper Chemistry, Appleton, WI, USA. 126 p.

[107] Milanova, E. 1988. Variables affecting the swelling of kraft black liquor solids. *Journal of Pulp and Paper Science* 14(4):J95–J102.

[108] Miller, T., Clay, D.T., and Lonsky, W.F.W. 1989. The influence of composition on the swelling of kraft black liquor during pyrolysis. *Chemical Engineering Communications* 75:101–120.

[109] Frederick, W.J., Noopila, T., and Hupa, M. 1991. Swelling of spent pulping liquor droplets during combustion. *Journal of Pulp and Paper Science* 17(5):J164–J170.

[110] Alén, R. 1994. Swelling behaviour of kraft black liquors and its organic constituents. *Bioresource Technology* 49:99–103.

[111] Frederick, W.J., and Hupa, M. 1994. The effects of temperature and gas composition on swelling of black liquor droplets during devolatilization. *Journal of Pulp and Paper Science* 20(10):J274–J280.

[112] Chen, C., Alén, R., Lehtimäki, E., and Louhelainen, J. 2017. A salt-induced mechanism for the swelling of black liquor droplet during devolatilization. *Fuel* 202:338–344.

[113] Chen, C., Pakkanen, H., and Alén, R. 2017. Role of lignin and sodium carbonate on the swelling behavior of black liquor droplets during combustion. *Holzforschung* 73(3)179–185.

[114] Chen, C. 2017. *Combustion Behavior of Black Liquors – Droplet Swelling and Influence of Liquor Composition*. Doctoral Thesis. University of Jyväskylä, Laboratory of Applied Chemistry, Jyväskylä, Finland. 39 p.

[115] Kubes, G.J. 1984. The effect of wood species on kraft recovery furnace operation – an investigation using differential thermal analysis. *Journal of Pulp and Paper Science* 10(3):J63–J68.

[116] Söderhjelm, L., Hupa, M., and Noopila, T. 1989. Combustibility of black liquors with different rheological and chemical properties. *Journal of Pulp and Paper Science* 15(4):J117–J122.

[117] Frederick, W.J., Noopila, T., and Hupa, M. 1991. Combustion behavior of black liquor at high solids firing. *TAPPI Journal* 74(12):163–170.

[118] van Heiningen, A.R.P., Arpiainen, V.T., and Alén, R. 1994. Effect of liquor type and pyrolysis rate on the steam gasification reactivities of black liquor. *Pulp & Paper Canada* 95:T358–T363.

[119] Wintoko, J., Theliander, H., and Richards, T. 2007. Experimental investigation of black liquor pyrolysis using single droplet TGA. *TAPPI Journal* 6(5):9–15.

[120] Vakkilainen, E., Lampinen, P., and Nieminen, M. (Eds.). 2014. *Continuous Development of Recovery Boiler Technology – 50 Years of Cooperation in Finland*. Finnish Recovery Boiler Committee, Vantaa, Finland. 135 p.

[121] Noopila, T., Alén, R., and Hupa, M. 1991. Combustion properties of laboratory-made black liquors. *Journal of Pulp and Paper Science* 17(4):J105–J109.

[122] Alén, R., Hupa, M., and Noopila, T. 1992. Combustion properties of organic constituents of kraft black liquors. *Holzforschung* 46(4):337–342.

[123] Alén, R. 2004. Combustion behavior of black liquors from different delignification conditions. In: *Proceedings of International Recovery Boiler Conference, 40 Years Recovery Boiler Co-Operation in Finland*, May 12–14, 2004, Haikko Manor, Porvoo, Finland. Pp. 31–42.

[124] Empie, H.J., Lien, S.J., Yang, W., and Samuels, D.B. 1995. Effect of black liquor type on droplet formation from commercial spray nozzles. *Journal of Pulp and Paper Science* 21(2):J63–J67.

[125] Kankkunen, A. 2014. Understanding of black liquor sprays. In: Vakkilainen, E., Lampinen, P., and Nieminen, M. (Eds.). *Continuous Development of Recovery Boiler Technology – 50 Years of Cooperation in Finland*. Finnish Recovery Boiler Committee, Vantaa, Finland. Pp. 33–45.

[126] Kymäläinen, M., Forssén, M., DeMartini, N., and Hupa, M. 2001. The fate of nitrogen in the chemical recovery process in a kraft pulp mill. Part II: Ammonia formation in green liquor. *Journal of Pulp and Paper Science* 27(3):75–81.

[127] Tamminen, T., Forssén, M., and Hupa, M. 2002. Dust and flue gas chemistry during rapid changes in the operation of black liquor recovery boilers – part 3: Gaseous emissions. *TAPPI Journal* 1(7):25–29.

[128] DeMartini, N., Forssén, M., Niemelä, K., Samuelsson, Å., and Hupa, M. 2004. Release of nitrogen species from the recovery processes of three kraft pulp mills. *TAPPI Journal* 3(10):3–8.

[129] Brink, A., Engblom, M., and Hupa, M. 2008. Nitrogen oxide emissions formation in a black liquor boiler. *TAPPI Journal* 7(11):28–32.

[130] Vähä-Savo, N., DeMartini, N., and Hupa, M. 2012. Fate of biosludge nitrogen in black liquor evaporation and combustion. *TAPPI Journal* 11(9):53–59.

[131] Tamminen, T., Kiuru, J., Kiuru, R., Janka, K., and Hupa, M. 2002. Dust and flue gas chemistry during rapid changes in the operation of black liquor recovery boilers: Part 1 – dust formation. *TAPPI Journal* 1(5):27–31.

[132] Tamminen, T., Laurén, T., Janka, K., and Hupa, M. 2002. Dust and flue gas chemistry during rapid changes in the operation of black liquor recovery boilers: Part 2 – dust composition. *TAPPI Journal* 1(6):25–29.

[133] Vakkilainen, E., Backman, R., Forssén, M., and Hupa, M. 1999. Effects of liquor heat treatment on black liquor combustion properties. *Pulp & Paper Canada* 108(8):24–30.

[134] Louhelainen, J., Alén, R., and Mullen, T. 2003. Combustion properties of thermochemically treated softwood and hardwood kraft black liquors. *Nordic Pulp and Paper Research Journal* 18(2):150–157.

[135] Vähä-Savo, N., DeMartini, N., Ziesig, R., Tomani, P., Theliander, H., Välimäki, E., and Hupa, M. 2014. Combustion properties of reduced-lignin black liquors. *TAPPI Journal* 13(8):81–89.

[136] Hupa, M., Forssén, M., Backman, R., Stubbs, A., and Bolton, R. 2002. Fireside behavior of black liquors containing boron. *TAPPI Journal* 1(1):48–52.

[137] Ledung, L., Ulmgren, P., and Hupa, M. 1997. Combustion properties of black liquors containing additions of totally chlorine free bleach plant filtrates. *Nordic Pulp and Paper Research Journal* 12(3):145–149.

[138] Vakkilainen, E.K., Kankkonen, S., and Suutela, J. 2008. Advanced efficiency options: Increasing electricity generating potential from pulp mills. *Pulp & Paper Canada* 109(4):14–18.

[139] Lindberg H., and Ulmgren, P. 1986. The chemistry of the causticizing reaction – effects on the operation of the causticizing department in a kraft mill. *TAPPI* 69(3):126–130.

[140] Nohlgren, I. 2004. Non-conventional causticization technology: A review. *Nordic Pulp and Paper Research Journal* 19(4):470–480.

[141] Kiiskilä, E., and Virkola, N.-E. 1978. Recovery of sodium hydroxide from alkaline pulping liquors by autocausticising. *Paperi ja Puu* 61(3):129–132.

[142] Janson, J. 1979. Autocausticizing alkali and its use in pulping and bleaching. *Paperi ja Puu* 61(8):495–504.

[143] Janson, J. 1980. Pulping processes based on autocausticizable borate. *Svensk Papperstidning* 83(14):392–395.

[144] Kiiskilä, E. 1980. Recovery of sodium hydroxide from alkaline pulping liquors by causticizing molten sodium carbonate with amphoteric oxides. *Paperi ja Puu* 62(5):339–350.

[145] Tran, H., Mao, X., Cameron, J., and Bair, C.M. 1999. Autocausticizing of smelt with sodium borates. *Pulp & Paper Canada* 100(9):T283–T287.

[146] Bujanovic, B., Cameron, J., and Yilgor, N. 2004. Some properties of kraft and kraft-borate pulps of different wood species. *TAPPI Journal* 3(6):3–5.

[147] Björk, M., Sjögren, T., Lundin, T., Rickards, H., and Kochesfahani, S. 2005. Partial borate autocausticizing trial increases capacity at Swedish mill. *TAPPI Journal* 4(9):15–19.

[148] Sarkanen, K.V., and Ludwig, C.H. (Eds.). 1971. *Lignins – Occurrence, Formation, Structure and Reactions.* Wiley-Interscience, New York, NY, USA. 916 p.

[149] Bose, S.K., Omori, S., Kanungo, D., Francis, R.C., and Shin, N.-H. 2009. Mechanistic differences between kraft and soda/aq pulping. Part 1: Results from wood chips and pulps. *Journal of Wood Chemistry and Technology* 29:214–226.

[150] Kanungo, D., Francis, R.C., and Shin, N.-H. 2009. Mechanistic differences between kraft and soda/aq pulping. Part 2: Results from lignin model compounds. *Journal of Wood Chemistry and Technology* 29:227–240.

[151] Ljungren, S. 1980. The significance of aryl ether cleavage in kraft delignification of softwood. *Svensk Papperstidning* 83(13):363–369.

[152] Douglass, I.B., and Price, L. 1966. A study of methyl mercaptan and dimethyl sulfide formation in kraft pulping. *TAPPI* 49(8):335–342.

[153] McKean, W.T. Jr., Hrutfiord, B.F., Sarkanen, K.V., Price, L., and Douglass, I.B. 1967. Effect of kraft pulping conditions on the formation of methyl mercaptan and dimethyl sulfide. *TAPPI* 50(8):400–405.

[154] Sarkanen, K.V., Hrutfiord, B.F., Johanson, L., and Gardner, H.S. 1970. Kraft odor. *TAPPI* 53(5):766–783.

[155] Brakash, C.B., and Murray, F.E. 1976. Analysis of malodorous sulphur compounds in kraft mill aqueous solutions. *Svensk Papperstidning* 79(15):501–504.

[156] Silander, R. 1981. Reduction of the emission of odorous compounds in chemical pulping. *Paperi ja Puu* 63(2):65–68.

[157] Moilanen, P., Hynninen, P., and Rissanen, S. 1981. The odour elimination system of a new kraft pulp mill – design and construction and some start-up experiences. *Paperi ja Puu* 63(4a):259–268.

[158] Fullerton, T.J. 1987. The condensation reactions of lignin model compounds in alkaline pulping liquors. *Journal of Wood Chemistry and Technology* 7(4):441–462.

[159] Pakkanen, H., and Alén, R. 2012. Molecular mass distribution of lignin from alkaline pulping of hardwood, softwood, and wheat straw. *Journal of Wood Chemistry and Technology* 32:279–293.

[160] Laurichesse, S., and Avérous, L. 2014. Chemical modification of lignins towards biobased polymers. *Progress in Polymer Science* 39:1266–1290.

[161] Ebers, L.-S., Arya, A., Bowland, C.C., Glasser, W.G., Chmely, S.C., Naskar, A.K., and Laborie, M.-P. 2021. 3D printing of lignin: Challenges, opportunities and roads onward. *Biopolymers* 112(6):e23431.

[162] Clayton, D., Easty, D., Einspahr, D., Lonsky, W., Malcolm, E., McDonough, T., Schroeder, L., and Thompson, N. 1989. Chemistry of alkaline pulping. In: Grace, T.M., Leopold, B., Malcolm, E.W., and Kocurek, M.J. (Eds.). *Pulp and Paper Manufacture. Volume 5, Alkaline Pulping.* 3rd edition. The Joint Textbook Committee of the Paper Industry, TAPPI&CPPA, USA and Canada. Pp. 1–128.

[163] Fengel, D., and Wegener, G. 1989. *Wood – Chemistry, Ultrastructure, Reactions.* Walter de Gruyter, Berlin, Germany. Pp. 482–525.

[164] Sjöström, E. 1993. *Wood Chemistry – Fundamentals and Applications.* 2nd edition. Academic Press, San Diego, CA, USA. Pp. 204–224.

[165] Heinze, T.J., and Glasser, W.G. (Eds.). 1998. Cellulose derivatives: Modification, characterization, and nanostructures. In: *ACS Symposium Series 688*, American Chemical Society, Washington, DC, USA. 361 p.

[166] Kamide, K. 2005. *Cellulose and Cellulose Derivatives.* Elsevier Science, London, United Kingdom. 652 p.

[167] Alén, R. 2011. Cellulose derivatives. In: Alén, R. (Ed.). *Biorefining of Forest Resources.* Paper Engineers' Association, Helsinki, Finland. Pp. 305–354.

[168] Kiliani, H. 1885. Ueber Isosaccharin. *Chemische Berichte* 18:631–641.

[169] Ruff, O. 1902. Ueber den Abbau der Rhamnon- und Isosaccharin-Säure. *Chemische Berichte* 35:2360–2370.

[170] Kiliani, H., and Herold, F. 1905. Ueber Dioxy-Propenyltricarbonsäure und α,γ-Dioxy-Glutarsäure. *Chemische Berichte* 38:2671–2676.

[171] Kiliani, H. 1908. Über die Produkte aus Milchzucker und Calciumhydroxyd. *Chemische Berichte* 41:2650–2658.

[172] Nef, J.U., Hedenburg, O.F., and Glattfeld, J.W.E. 1917. The method of oxidation and the oxidation products of L-arabinose and of L-xylose in alkaline solutions with air and with cupric hydroxide. *Journal of the American Chemical Society* 39(8):1638–1652.

[173] Isbell, H.S. 1944. Interpretation of some reactions in the carbohydrate field in terms of consecutive electron displacement. *Journal of Research of the National Bureau of Standards* 32:45–59.

[174] Schmidt, O. 1935. The mechanism of some important organic reactions. *Chemical Reviews* 17(2):137–154.

[175] Evans, W.L. 1942. Some less familiar aspects of carbohydrate chemistry. *Chemical Reviews* 31(3):537–560.

[176] Warshowsky, B., and Sandstrom, W.M. 1952. The action of oxygen on glucose in the presence of potassium hydroxide. *Archives of Biochemistry and Biophysics* 37(1):46–55.

[177] Sjöström, E. 1977. The behavior of wood polysaccharides during alkaline pulping processes. *TAPPI* 60(9):151–154.

[178] Reintjes, M., and Cooper, G.K. 1984. Polysaccharide alkaline degradation products as a source of organic chemicals. *Industrial & Engineering Chemistry Process Design and Development* 23(1):70–73.

[179] Blažej, A., and Košík, M. 1985. Degradation reactions of cellulose and lignocellulose. In: Kennedy, J.F., Phillips, G.O., Wedlock, D.J., and Williams, P.A. (Eds.). *Cellulose and Its Derivatives*. Ellis Horwood, Chichester, England. Pp. 97–117.

[180] Krochta, J.M., Hudson, J.S., and Tillin, S.J. 1988. Kinetics of alkaline thermochemical degradation of polysaccharides to organic acids. In: Soltes, E., and Milne, T. (Eds.). *Pyrolysis Olis from Biomass: Producing, Analyzing, and Upgrading. ACS Symposium Series* 376:119–128.

[181] Sjöström, E. 1991. Carbohydrate degradation products from alkaline treatment of biomass. *Biomass and Bioenergy* 1(1):61–64.

[182] Johansson, D. 2008. *Carbohydrate Degradation and Dissolution During Kraft Cooking – Modelling of Kinetic Results*. Doctoral Thesis. Karlstad University, Chemical Engineering, Karlstad, Sweden. 48 p.

[183] Pakkanen, H., and Alén, R. 2013. Alkali consumption of aliphatic carboxylic acids during alkaline pulping of wood and nonwood feedstocks. *Holzforschung* 67(6):643–650.

[184] Alén, R. 2018. *Carbohydrate Chemistry – Fundamentals and Applications*. World Scientific, Singapore. Pp. 474–483.

[185] Franzon, O., and Samuelson, O. 1957. Degradation of cellulose by alkali cooking. *Svensk Papperstidning* 60(23):872–877.

[186] Meller, A. 1960. The chemistry of alkaline degradation of cellulose and oxidized celluloses I. *Holzforschung* 14(3):78–89.

[187] Alfredsson, B., Gedda, L., and Samuelson, O. 1961. A comparison between alkali cooking of cellulose and hot alkali treatment of hydrocellulose. *Svensk Papperstidning* 64(19):694–698.

[188] Haas, D.W., Hrutfiord, B.F., and Sarkanen, K.V. 1967. Kinetic study on the alkaline degradation of cotton hydrocellulose. *Journal of Applied Polymer Science* 11:587–600.

[189] Alfredsson, B., and Samuelson, O. 1968. Hydroxy acids formed by alkali treatment of hydrocellulose. *Svensk Papperstidning* 71(19):679–686.

[190] Johansson, M., and Samuelson, O. 1975. End-wise degradation of hydrocellulose during hot alkali treatment. *Journal of Applied Polymer Science* 19:3007–3013.

[191] Löwendahl, L., Petersson, G., and Samuelson, O. 1976. Formation of dicarboxylic acids during hot alkali treatment of hydrocellulose. *Cellulose Chemistry and Technology* 10:471–477.

[192] Johansson, M., and Samuelson, O. 1978. Endwise degradation of hydrocellulose in bicarbonate solution. *Journal of Applied Polymer Science* 22(3):615–623.

[193] Krochta, J.M., Hudson, J.S., and Drake, C.W. 1984. Alkaline thermochemical degradation of cellulose to organic acids. *Biotechnology & Bioengineering Symposium* 14:37–54.

[194] Niemelä, K. 1990. The formation of hydroxy monocarboxylic acids and dicarboxylic acids by alkaline thermochemical degradation of cellulose. *Journal of Chemical Technology and Biotechnology* 48:17–28.

[195] Knill, C.J., and Kennedy, J.F. 2003. Degradation of cellulose under alkaline conditions. *Carbohydrate Polymers* 51(3):281–300.

[196] Niemelä, K. 1986. The conversion of cellulose into carboxylic acids by a drastic alkali treatment. *Biomass* 11:215–221.

[197] Krochta, J.M., Tillin, S.J., and Hudson, J.S. 1988. Thermochemical conversion of polysaccharides in concentrated alkali to glycolic acid. *Applied Biochemistry and Biotechnology* 17:23–32.

[198] Niemelä, K. 1988. The conversion of cellulose into carboxylic acids by a drastic oxygen-alkali treatment. *Biomass* 15:223–231.

[199] Casebier, R.L., and Hamilton, J.K. 1965. Alkaline degradation of glucomannans and galactoglucomannans. *TAPPI* 48(11):664–669.

[200] Hansson, J.-Å., and Hartler, N. 1970. Alkaline degradation of pine glucomannan. *Holzforschung* 24(2):54–59.

[201] Young, R.A., and Liss, L. 1978. A kinetic study of the alkaline endwise degradation of gluco- and galactomannans. *Cellulose Chemistry and Technology* 12:399–411.

[202] Niemelä, K., and Sjöström, E. 1986. Alkaline degradation of mannan. *Holzforschung* 40(1):9–14.

[203] Whistler, R.L., and Corbett, W.M. 1956. Behavior of xylan in alkaline solution: The isolation of a new C5 saccharinic acid. *Journal of the American Chemical Society* 78(5):1003–1005.

[204] Hansson, J.-Å., and Hartler, N. 1968. Alkaline degradation of xylans from birch and pine. *Svensk Papperstidning* 71(9):358–365.

[205] Johansson, M., and Samuelson, O. 1977. Reducing end groups in birch xylan and their alkaline degradation. *Wood Science and Technology* 11:251–263.

[206] Niemelä, K. 1990. Conversion of xylan, starch, and chitin into carboxylic acids by treatment with alkali. *Carbohydrate Research* 204:37–49.

[207] Stén, M., and Mustola, T. 1973. The alkaline degradation of 4-*O*-methyl-*D*-glucose. *Cellulose Chemistry and Technology* 7(4):359–369.

[208] Blythe, D.A., and Schroeder, L.R. 1985. Degradation of a nonreducing cellulose model, 1,5-anhydro-4-*O*-β-D-glucopyranosyl-D-glucitol, under kraft pulping conditions. *Journal of Chemistry and Technology* 5(3):313–334.

[209] Davidson, G.F. 1934. The solution of chemically modified cotton cellulose in alkaline solutions, I: In solutions of sodium hydroxide particularly at temperatures below the normal. *Journal of the Textile Institute* 25:T174–T196.

[210] Malinen, R., and Sjöström, E. 1972. Studies on the reactions of carbohydrates during oxygen bleaching. *Paperi ja Puu* 54(8):451–468.

[211] Green, J.W., Thompson, N.S., Pearl, I.A., and Swanson, J.W. 1976. Study of the carbohydrate peeling and stopping reactions under the conditions of oxygen-alkali pulping. *Project Report*. Georgia Institute of Technology, Institute of Paper Science and Technology, Appleton, WI, USA.

[212] Dryselius, E., Lindberg, B., and Theander, O. 1958. Alkaline hydrolysis of glycosidic linkages. III. an investigation of some methyl α- and β-glycopyranosides. *Acta Chemica Scandinavica* 12(2):340–342.

[213] Lai, Y.-Z., and Sarkanen, K.V. 1967. Kinetics of alkaline hydrolysis of glycosidic bonds in cotton cellulose. *Cellulose Chemistry and Technology* 1:517–527.

[214] Sykes, P. 1970. *A Guidebook to Mechanism in Organic Chemistry*. 3rd edition. Longman, London, England. Pp. 201–202.

[215] Johansson, M., and Samuelson, O. 1974. The formation of end groups in cellulose during alkali cooking. *Carbohydrate Research* 34:33–43.

[216] Mitikka, M., Teeäär, R., Tenkanen, M., Laine, J., and Vuorinen, T. 1995. Sorption of xylans on cellulose fibers. In: *Proceedings of the 8th Symposium on Wood and Pulping Chemistry, Volume 3*, June 6–9, 1995, Helsinki, Finland. Pp. 231–236.

[217] Saake, B., Busse, T., and Puls, J. 2005. The effect of xylan adsorption on the properties of sulfite and kraft pulps. In: *Proceedings of the 13th International Symposium on Wood, Fibre and Pulping Chemistry (ISWFPC)*, May 16–19, 2005, Auckland, New Zealand. Pp. 141–146.

[218] Croon, I., and Enström, B. 1961. The 4-O-methyl-D-glucuronic acid groups of birch xylan during sulfate pulping. *TAPPI* 44(12):870–874.

[219] Aurell, R., and Karlsson, K. 1964. The 4-O-methyl-D-glucuronic acid content of xylan isolated from birch kraft pulps. *Svensk Papperstidning* 67(15):167–169.

[220] Sjöström, E., Juslin, S., and Seppälä, E. 1969. A gas chromatographic method for the identification of 4-O-methyl-D-glucuronic acid groups in wood xylan. *Acta Chemica Scandinavica* 23:3610–3611.

[221] Allison, R.W., Timonen, O., McGrouther, K.G., and Suckling, I.A. 1999. Hexenuronic acid in kraft pulps from radiata pine. *Appita Journal* 52(6):448–453.

[222] Daniel, A.I.D., Neto, C.P., Evtuguin, D.V., and Silvestre, A.J.D. 2003. Hexenuronic acid contents of *eucalyptus globulus* kraft pulps: Variation with pulping conditions and effect on ECF bleachability. *TAPPI Journal* 2(5):3–8.

[223] Simão, J.P.F., Egas, A.P.V., Baptista, C.M.S.G., and Carvalho, M.G. 2005. Heterogeneous kinetic model for the methylglucuronic and hexenuronic acids reactions during kraft pulping of *eugalyptus globulus*. Industrial and Engineering Chemistry Research 44:2997–3002.

[224] Li, J., and Gellerstedt, G. 1997. The contribution of kappa number from hexenuronic acid groups in pulp xylan. *Carbohydrate Research* 302:213–218.

[225] Rosenau, T., Potthast, A., Zwirchmayr, N.S., Hettegger, H., Plasser, F., Hosoya, T., Bacher, M., Krainz, K., and Dietz, T. 2017. Chromophores from hexeneuronic acids: Identification of hexa-derived chromophores. *Cellulose* 24:3671–3687.

[226] Zwirchmayr, N.S., Hosoya, T., Hettegger, H., Bacher, M., Krainz, K., Dietz, T., Henninges, U., Potthast, A., and Rosenau, T. 2017. Chromophores from hexeneuronic acids: Chemical behavior under peroxide bleaching conditions. *Cellulose* 24:3689–3702.

[227] Rosenau, T., Potthast, A., Zwirchmayr, N.S., Hosoya, T., Hettegger, H., Bacher, M., Krainz, K., Yoneda, Y., and Dietz, T. 2017. Chromophores from hexeneuronic acids (HexA): Synthesis of model compounds and primary degradation intermediates. *Cellulose* 24:3703–3723.

[228] Kawamura, A., Igarashi, H., Uchida, Y., Yaegashi, I., and Iwasaki, M. 2020. Relationship between cooking/bleaching conditions and hexenuronic acid content in kraft pulp. *International Journal of Scientific and Research Publications (IJSRP)* 10(8):767–787.

[229] Jiang, Z.-H., van Lierop, B., and Berry, R. 2000. Hexenuronic acid groups in pulping and bleaching chemistry. *TAPPI Journal* 83(1):167–175.

[230] Forrström, A., Gellerstedt, G., Jour, P., and Li, J. 2005. On selective removal of hexenuronic acid (HexA) by oxidative bleaching of eucalyptus O_2-delignified kraft pulp. In: *Proceedings of IPBC '05*, June 14–16, 2005, Stockholm, Sweden. Pp. 309–312.

[231] Kawamura, A., Igarashi, H., Uchida, Y., Yaegashi, I., and Iwasaki, M. 2020. Relationship between cooking/bleaching conditions and hexenuronic acid content in kraft pulp. *International Journal of Scientific and Research Publications (IJSRP)* 10(8):767–787.

[232] Teleman, A., Harjunpää, V., Tenkanen, M., Buchert, J., Hausalo, T., Drakenberg, T., and Vuorinen, T. 1995. Characterisation of 4-deoxy-β-L-*threo*-hex-4-enopyranosyluronic acid attached to xylan in pine kraft pulp and pulping liquor by ^{1}H and ^{13}C NMR spectroscopy. *Carbohydrate Research* 272:55–71.

[233] Gellerstedt, G., and Li, J. 1996. An HPLC method for the quantitative determination of hexeneuronic acid groups in chemical pulps. *Carbohydrate Research* 294:41–51.

[234] Teleman, A., Hausalo, T., Tenkanen, M., and Vuorinen, T. 1995. Identification of the acidic degradation products of hexenuronic acid and characterisation of hexenuronic acid-substituted xylooligosaccharides by NMR spectroscopy. *Carbohydrate Research* 280:197–208.

[235] Tenkanen, M., Gellerstedt, G., Vuorinen, T., Teleman, A., Perttula, M., Li, J., and Buchert, J. 1999. Determination of hexenuronic acid in softwood kraft pulps by three different methods. *Journal of Pulp and Paper Science* 25(9):306–311.

[236] Jiang, Z.-H., Audet, A., Sullivan, J., van Lierop, B., and Berry, R. 2001. A new method for quantifying hexenuronic acid groups in chemical pulps. *Journal of Pulp and Paper Science* 27(3):92–97.

[237] Saariaho, A.-M., Hortling, B., Jääskeläinen, A.-S., Tamminen, T., and Vuorinen, T. 2003. Simultaneous quantification of residual lignin and hexenuronic acid from chemical pulps with UV resonance Raman spectroscopy and multivariate calibration. *Journal of Pulp and Paper Science* 29(11):363–370.

[238] Jääskeläinen, A.-S., Saariaho, A.-M., and Vuorinen, T. 2005. Quantification of lignin and hexenuronic acid in bleached hardwood kraft pulps: A new calibration method for UVRR spectroscopy and evaluation of the conventional methods. *Journal of Wood Chemistry and Technology* 25:51–65.

[239] Brännström, H., and Halmemies, E. 2021. Tall oil-based products. In: Alén, R. (Ed.). *Pulping and Biorefining – ForestBioFacts, Digital Learning Environment*. Paperi ja Puu Ltd., Helsinki, Finland.

[240] Ekman, R. 2000. Resin during storage and in biological treatment. In: Back, E.L., and Allen, L.H. (Eds.). *Pitch Control, Wood Resin and Deresination*. TAPPI Press, Atlanta, GA, USA. Pp. 185–204.

[241] Halmemies, E.S., Brännström, H.E., Nurmi, J., Läspä, O., and Alén, R. 2021. Effect of seasonal storage on the single-stem bark extractives of Norway spruce (*Picea abies*). *Forest* 12:736–770.

[242] Halmemies, E.S., Alén, R., Hellström, J., Läspä, O., Nurmi, J., Hujala, M., and Brännström, H.E. 2021. Behaviour of extractives in Norway spruce (*Picea abies*) bark during pile storage. *Molecules* 27(4):article 1186.

[243] Höfer, R. 2015. The pine biorefinery platform chemicals value chain. In: Pandey, A., Höfer, R., Taherzadeh, M., Nampoothiri, K.M., and Larroche, C. (Eds.). *Industrial Biorefineries & White Biotechnology*. Elsevier, Amsterdam, The Netherlands. Pp. 127–155.

[244] Olm, L. 1984. Pitch problems and their control in kraft mills using hardwoods from temperate and tropical zones: A literature survey. *Appita Journal* 37(6):479–483.

[245] Hillis, W.E., and Sumimoto, M. 1989. Effect of extractives on pulping. In: Rowe, J.W. (Ed.). *Natural Products of Woody Plants, Vol. II*. Springer-Verlag, Heidelberg, Germany. Pp. 880–920.

[246] Ström, G., Stenius, P., Lindström, M., and Ödberg, L. 1990. Surface chemical aspects of the behavior of soaps in pulp washing. *Nordic Pulp and Paper Research Journal* 5(1):44–51.

[247] Allen, L.H. 2000. Pitch control in pulp mills. In: Back, E.L., and Allen, L.H. (Eds.). *Pitch Control, Wood Resin and Deresination*. TAPPI Press, Atlanta, GA, USA. Pp. 265–288.

[248] Back, E.L. 2000. Deresination in pulping and washing. In: Back, E.L., and Allen, L.H. (Eds.). *Pitch Control, Wood Resin and Deresination*. TAPPI Press, Atlanta, GA, USA. Pp. 205–230.

[249] Holmbom, B. 2000. Resin reactions and deresination in bleaching. In: Back, E.L., and Allen, L.H. (Eds.). *Pitch Control, Wood Resin and Deresination*. TAPPI Press, Atlanta, GA, USA. Pp. 231–244.

[250] Del Río, J.C., Romero, J., and Gutiérrez, A. 2000. Analysis of pitch deposits produced in kraft pulp mill using a totally chlorine free bleaching sequence. *Journal of Chromatography A* 874:235–245.

[251] Freire, C.S.R., Silvestre, A.J.D., and Pascoal Neto, C. 2005. Lipophilic extractives in *Eucalyptus globulus* kraft pulps. behavior during ECF bleaching. *Journal of Wood Chemistry and Technology* 25:67–80.

[252] Marques, G., Del Rio, J.C., and Gutiérrez, A. 2009. Fate of lipophilic extractives from several non-wood species during alkaline pulping and TCF/ECF bleaching. In: *Proceedings of the 15th International Symposium on Wood, Fibre and Pulping Chemistry (ISWFPC)*, June 15–18, 2009, Oslo, Norway. Pp. 130–133.

[253] Marques, G., Del Río, J.C., and Gutiérrez, A. 2010. Lipophilic extractives from several nonwoody lignocellulosic crops (flax, hemp, sisal, abaca) and their fate during alkaline pulping and TCF/ECF bleaching. *Bioresource Technology* 101(1):260–267.

[254] Jusner, P., Barbini, S., Schiehser, S., Bacher, M., Schwaiger, E., Potthast, A., and Rosenau, T. 2022. Impact of residual extractives on the thermal stability of softwood kraft pulp. *Cellulose* 29:8797–8810.

[255] Ekman, R., and Holmbom, B. 1990. Studies on the behavior of extractives in mechanical pulp suspensions. *Nordic Pulp and Paper Research Journal* 5(2):96–102.

[256] Allen, L.H. 2000. Pitch control in paper mills. In: Back, E.L., and Allen, L.H. (Eds.). *Pitch Control, Wood Resin and Deresination*. TAPPI Press, Atlanta, GA, USA. Pp. 307–328.

[257] Alén, R., and Selin, J. 2007. Deposit formation and control. In: Alén, R. (Ed.). *Papermaking Chemistry*. Paper Engineers' Association, Helsinki, Finland. Pp. 163–180.

[258] Baumassy, M. 2014. The tall oil industry: 100 years of innovation. In: *Proceedings of the 2014 PCA International Conference*, September 21–23, 2014, Seattle, WA, USA.

[259] Holmbom, B. 2011. Extraction and utilization of non-structural wood and bark components. In: Alén, R. (Ed.). *Biorefining of Forest Resources*. Paper Engineers' Association, Helsinki, Finland. Pp. 178–224.

[260] da Silva Rodrigues-Corrêa, K.C., de Lima, J.C., and Fett-Neto, A.G. 2013. Oleoresins from pine: Production and industrial uses. In: Ramawat, K.P., and Mérillon, J.-M. (Eds.). *Natural Products: Phytochemistry, Botany and Metabolism of Alkaloids, Phenolics and Terpenes*. Springer-Verlag, Heidelberg, Germany. Pp. 4037–4060.

[261] Phun, L., Snead, D., Hurd, P., and Jing, F. 2017. Industrial applications of pine-chemical-based materials. In: Tang, C., and Ryu, C.Y. (Eds.). *Sustainable Polymers from Biomass*. Wiley-VCH, Weinheim, Germany. Pp. 151–180.

[262] Silvestre, A.J., and Gandini, A. 2008. Terpenes: Major sources, properties and applications. In: Belgacem, M.N., and Gandini, A. (Eds.). *Monomers, Polymers and Composites from Renewable Resources*. Elsevier, Amsterdam, The Netherlands. Pp. 17–38.

[263] Brännström, H., and Halmemies, E. 2021. Tall oil-based products. In: Alén, R. (Ed.). *Pulping and Biorefining – ForestBioFacts, Digital Learning Environment*. Paperi ja Puu Ltd., Helsinki, Finland.

[264] Laine, J., and Stenius, P. 2007. Internal sizing of paper. In: Alén, R. (Ed.). *Papermaking Chemistry*. Paper Engineers' Association, Helsinki, Finland. Pp. 122–162.

[265] Höfer, R. (Ed.). 2022. *Renewable Resources for Surface Coatings, Inks and Adhesives*. Royal Society of Chemistry, Cambridge, United Kingdom. 854 p.

[266] Wang, B. 2018. Tall oil, its chemistries and applications. *Inform* 29(1):26–30.

[267] Krogerus, B. 2007. Papermaking additives. In: Alén, R. (Ed.). *Papermaking Chemistry*. Paper Engineers' Association, Helsinki, Finland. Pp. 54–121.

[268] Watkins, S.H. 1971. Rosin and rosin size: Preparation and properties. In: Swanson, J.W. (Ed.). *Internal Sizing of Paper and Paperboard. TAPPI Monographs 33*. TAPPI Press, Atlanta, GA, USA. p. 5.

[269] Gess, J.M. 1996. The sizing pf paper with rosin and alum at acid pHs. In: Roberts, J.C. (Ed.). *Paper Chemistry*. 2nd edition. Blackie Academic & Professional, London, United Kingdom. Pp. 121–139.

[270] Koebner, A. 1983. Separation of tall oil head fraction into fatty acids and unsaponifiables. *Journal of Wood Chemistry* 3(4):413–420.

[271] Peters, D., and Stojcheva, V. 2017. Crude tall oil low ILUC risk assessment – comparing global supply and demand. *Report SISNL17494*, ECOFYS Netherlands B.V., Utrecht, The Netherlands. 23 p.

[272] Rydholm, S. 1965. *Pulping Processes*. Interscience Publishers, New York, NY, USA. Pp. 439–576.

[273] Glennie, D.W. 1971. Reactions in sulfite pulping. In: Sarkanen, K.V., and Ludwig, C.H. (Eds.). *Lignin – Occurrence, Formation, Structure and Reactions*. Wiley-Interscience, New York, NY, USA. Pp. 597–637.

[274] Bryce, J.R.G. 1980. Sulfite pulping. In: Casey, J.P. (Ed.). *Pulp and Paper Chemistry and Chemical Technology, Volume 1*. 3rd edition. Wiley-Interscience, New York, NY, USA. Pp. 291–376.

[275] Ingruber, O.V., Kocurek, M.J., and Wong, A. (Eds.). 1985. *Sulfite Science & Technology, Volume 4*. 3rd edition. The Joint Textbook Committee of the Paper Industry, TAPPI&CPPA, USA and Canada. 352 p.

[276] Fengel, D., and Wegener, G. 1989. *Wood – Chemistry, Ultrastructure, Reactions*. Walter de Gruyter, Berlin, Germany. Pp. 452–462.

[277] Sjöström, E. 1993. *Wood Chemistry – Fundamentals and Applications*. 2nd edition. Academic Press, San Diego, CA, USA. Pp. 119–140.

[278] Biermann, C.J. 1996. *Handbook of Pulping and Papermaking*. 2nd edition. Academic Press, San Diego, CA, USA. Pp. 91–96.

[279] Sixta, H., Potthast, A., and Krotschek, A.W. 2006. Chemical pulping processes. In: Sixta, H. (Ed.). 2006. *Handbook of Pulp*. Wiley-VCH, Weinheim, Germany. Pp. 392–510.

[280] Smook, G. 2016. *Handbook for Pulp and Paper Technologists*. 4th edition. TAPPI Press Atlanta, GA, USA. Pp. 65–73.

[281] Alén, R. 2021. Acid sulphite pulping. In: Alén, R. (Ed.). *Pulping and Biorefining – ForestBioFacts, Digital Learning Environment*. Paperi ja Puu Ltd., Helsinki, Finland.

[282] Alén, R. 2021. Neutral and alkaline sulphite pulping. In: Alén, R. (Ed.). *Pulping and Biorefining – ForestBioFacts, Digital Learning Environment*. Paperi ja Puu Ltd., Helsinki, Finland.

[283] Kettunen, J., Virkola, N.-E., and Yrjälä, I. 1979. The effect of anthraquinone on neutral sulphite and alkaline sulphite cooking of pine. *Paperi ja Puu* 61:685–690,693–694,699–700.

[284] Virkola, N.-E., Pusa, R., and Kettunen, J. 1981. Neutral sulfite AQ pulping as an alternative to kraft pulping. *TAPPI* 64(5):103–107.

[285] Ingruber, O.V., Stradal, M., and Histed, J.A. 1982. Alkaline sulphite-anthraquinone pulping of eastern Canadian woods. *Pulp & Paper Canada* 83(12):79–88.

[286] Tikka, P., Tulppala, J., and Virkola, N.-E. 1982. Neutral sulfite AQ pulping and bleaching of the pulps. In: *Proceedings of the 1982 International Sulfite Pulping Conference*, October 20–22, 1982, Toronto, Ontario, Canada.

[287] Isotalo, I. 1983. Neutral sulphite anthraquinone (NS-AQ) pulp as raw material for kraftliner board. *Paperi ja Puu* 65(9):526–536.

[288] Fleming, B.I., Barbe, M.C., Miles, K., Page, D.H., and Seth, R.S. 1984. High-yield softwood pulps by neutral sulphite-anthraquinone pulping. *Journal of Pulp and Paper Science* 10(5):J113–J118.

[289] Tay, C.H., Fairchild, R.S., and Imada, S.E. 1985. A neutral-sulfite/SAQ chemimechanical pulp for newsprint. *TAPPI Journal* 68(8):98–103.

[290] Tay, C.H., and Imada, S.E. 1986. Chemimechanical pulp from jack pine by sulphite/quinone pulping. *Journal of Pulp and Paper Science* 12(3):J60–J66.

[291] Chen, H.-T., Ghazy, M., Funaoka, M., and Lai, Y.-Z. 1994. The influence of anthraquinone in sulfite delignification of Norway spruce. I. liquor pH and sulfonation. *Cellulose Chemistry and Technology* 28:47–54.

[292] European Food Safety Authority (EFSA). 2012. Reasoned opinion on the review of the existing maximum residue levels (MRLs) for anthraquinone according to article 12 of regulation (EC) No 396/2005. *EFSA Journal* 10:2761.

[293] Sjöström, E., Haglund, P., and Janson, J. 1962. Changes in cooking liquor composition during sulphite pulping. *Svensk Papperstidning* 65:855–869.

[294] Gustafsson, J., Alén, R., Engström, J., Korpinen, R., Kuusisto, P., Leavitt, A., Olsson, K., Piira, J., Samuelsson, A., and Sundquist, J. 2011. Pulping. In: Fardim, P. (Ed.). *Chemical Pulping Part 1, Fibre Chemistry and Technology*. 2nd edition. Paper Engineers' Association, Helsinki, Finland. Pp. 187–381.

[295] Alén, R. 2018. *Carbohydrate Chemistry – Fundamentals and Applications*. World Scientific, Singapore. Pp. 26–31.

[296] Forss, K.G., and Fremer, K.-E. 2003. *The Nature and Reactions of Lignin – A New Paradigm*. Oy Nord Print Ab, Helsinki, Finland. Pp. 227–315.

[297] Lindgren, B.O. 1952. The sulphonatable groups of lignin. *Svensk Papperstidning* 55:78–89.

[298] Gellerstedt, G., and Gierer, J. 1971. The reactions of lignin during acidic sulphite pulping. *Svensk Papperstidning* 74:117–127.

[299] Gellerstedt, G. 1976. The reactions of lignin during sulfite pulping. *Svensk Papperstidning* 79:537–543.

[300] Hanhikoski, S., Warsta, E., Varhimo, A., Niemelä, K., and Vuorinen, T. 2016. Sodium sulphite pulping of Scots pine under neutral and mildly alkaline conditions (NS pulping). *Holzforschung* 70(7):603–609.

[301] Goliath, M., and Lindgren, B. 1961. Reactions of thiosulphate during sulphite cooking. Part 2. Mechanism of thiosulphatesulphidation of vanillyl alcohol. *Svensk Papperstidning* 64:469–471.

[302] Annergren, G.E., and Rydholm, S.A. 1959. On the behavior of the hemicelluloses during sulfite pulping. *Svensk Papperstidning* 62:737–746.

[303] Janson, J., and Sjöström, E. 1964. Behaviour of xylan during sulphite cooking of birchwood. *Svensk Papperstidning* 67:764–771.

[304] Deshpande, R., Sundvall, L., Grundberg, H., and Germgård, O. 2016. The influence of different types of bisulfite cooking on pine wood components. *BioResources* 11(3):5961–5973.

[305] Pfister, K., and Sjöström, E. 1977. The formation of monosaccharides and aldonic and uronic acids during sulphite cooking. *Paperi ja Puu* 59:711–720.

[306] Collins, T.T. Jr., and Shick, P.E. 1971. Additives aid sulphite recovery: By-products depend on process. *Paper Trade Journal* 155(April 19):52–56.

[307] Pearl, I.A. 1982. Utilization of by-products of the pulp and paper industry. *TAPPI* 65(5):68–73.

[308] DeZylva Adhihetty, T.L. 1983. Utilization of spent sulphite liquor. *Cellulose Chemistry and Technology* 17:395–399.

[309] Lai, L.X., and Bura, R. 2012. The sulfite mill as a sugar-flexible future biorefinery. *TAPPI Journal* 11(8):27–35.

[310] Llano, T., García-Quevedo, N., Quijorna, N., Viguri, J.R., and Coz, A. 2015. Evolution of lignocellulosic macrocomponents in the wastewater streams of a sulfite pulp mill: A preliminary biorefining approach. *Hindawi Publishing Corporation, Journal of Chemistry*, Article ID 102534.

[311] Alén, R. 2021. Spent Liquor. In: Alén, R. (Ed.). *Pulping and Biorefining – ForestBioFacts, Digital Learning Environment*. Paperi ja Puu Ltd., Helsinki, Finland.

[312] Hurme, T. 1998. *Ultrafiltration of Acid Sulphite Spent Liquor to Produce Carbohydrates*. Doctoral Thesis. Åbo Akademi University, Department of Physical Chemistry, Åbo, Finland. 131 p.

[313] Collins, J.W., Boggs, L.A., Webb, A.A., and Wiley, A.A. 1973. Spent sulphite liquor reducing sugar purification with dynamic membranes. *TAPPI* 56(6):121–124.

[314] Bansal, I.K., and Wiley, A.J. 1975. Membrane processes for fractionation and concentration of spent sulphite liquors. *TAPPI* 58(1):125–130.

[315] Goheen, D.W. 1971. Low molecular weight chemicals. In: Sarkanen, K.V., and Ludwig, C.H. (Eds.). *Lignin – Occurrence, Formation, Structure and Reactions*. Wiley-Interscience, New York, NY, USA. Pp. 797–831.

[316] Hoyt, C.H., and Goheen, D.W. 1971. Polymeric products. In: Sarkanen, K.V., and Ludwig, C.H. (Eds.). *Lignin – Occurrence, Formation, Structure and Reactions*. Wiley-Interscience, New York, NY, USA. Pp. 833–865.

[317] Forss, K., and Fuhrman, A. 1976. KARATEX – the lignin-based adhesive for plywood, particle board and fibre board. *Paperi ja Puu* 58:817–824.

[318] Wayman, M., and Parekh, S.K. 1987. Fermentation to ethanol of pentose-containing spent sulphite liquor. *Biotechnology and Bioengineering* 29(9):1144–1150.

[319] Asadollahzadeh, M.T., Ghasemian, A., Saraeian, A.R., Resalati, H., Lennartsson, P.R., and Taherzadeh, M.J. 2017. Ethanol and biomass production from spent sulfite liquor by filamentous fungi. *International Journal of Biological, Biomolecular, Agricultural, Food and Biotechnological Bioengineering* 11(10):718–724.

[320] Andersen, R.F. 1979. Production of food yeast from spent sulphite liquor. *Pulp and Paper Magazine of Canada* 80(4):43–45.

[321] Forss, K., and Passinen, K. 1976. Utilization of the spent sulphite liquor components in the pekilo protein process and the influence of the process upon the environmental problems of a sulphite mill. *Paperi ja Puu* 58(9):608–618.

[322] Ingman, M. 1980. Pekilo-Prozeß, Protein aus Sulfitzellstoffablauge. *Wochenblatt für Papierfabrication* 6:193–196.

[323] Smith, M.T., Cameron, D.R., and Duff, S.J. 1997. Comparison of industrial yeast strains for fermentation of spent sulphite pulping liquor fortified with wood hydrolyzate. *Journal of Industrial Microbiology and Biotechnology* 18(1):18–21.

[324] Asadollahzadeh, M.T., Ghasemian, A., Saraeian, A.R., Resalati, H., and Taherzadeh, M.J. 2019. Production of fungal biomass protein by filamentous fungi cultivation on liquid waste streams from pulping process. *BioResources* 13(3):5013–5031.

[325] El-Saied, H., and Oelert, H.-H. 1980. Liquefaction of the lignohemicellulosic waste from sulphite spent liquor. *Cellulose Chemistry and Technology* 14:507–516.

[326] Biggs, W.A. Jr., Wise, J.T., Cook, W.R., Baxley, W.H., Robertson, J.D., and Copenhaver, J.E. 1961. The commercial production of acetic and formic acids from NSSC black liquor. *TAPPI* 44(6):385–392.

[327] Alén, R. 2021. Soda pulping. In: Alén, R. (Ed.). *Pulping and Biorefining – ForestBioFacts, Digital Learning Environment*. Paperi ja Puu Ltd., Helsinki, Finland.

[328] Kubes, G.J., Fleming, B.I., MacLeod, J.M., and Bolker, H.I. 1980. Alkaline pulping with additives. a review. *Wood Science and Technology* 14:207–228.

[329] Carlson, U., and Samuelson, O. 1981. Quinones as additives during soda cooking. *Journal of Wood Chemistry and Technology* 1(1):43–59.

[330] Wandelt, P., and Surewicz, W. 1983. Catalyzed alkaline sulfur-free pulping. selection of the catalyst and its dose. *Cellulose Chemistry and Technology* 17:543–552.

[331] Biermann, C.J. 1996. *Handbook of Pulping and Papermaking*. Academic Press, San Diego, CA, USA. Pp. 55–100.

[332] Pekkala, O. 1982. On the extended delignification using polysulphide or anthraquinone in kraft pulping. *Paperi ja Puu* 64(11):735–744.

[333] Gädda, L. 1982. Kraft pulping with anthraquinone, effect on delignification of the tracheid cell wall. *Paperi ja Puu* 64(12):793–797.

[334] Kotani, E.K., Hatton, J.V., and Hunt, K. 1983. Improved kraft pulping of white spruce bark/wood mixtures with anthraquinone. *Paperi ja Puu* 65(9):542–545.

[335] Malachowski, P.S., Poniatowski, S.E., and Walkinshaw, J.W. 1989. The effects of anthraquinone on the reaction rate of the kraft pulping of Northeastern white pine. *TAPPI Journal* 72(6):207–210.

[336] Li, Z., Ma, H., Kubes, G.J., and Li, J. 1998. Synergistic effect of kraft pulping with polysulphide and anthraquinone on pulp-yield improvement. *Journal of Pulp and Paper Science* 24(8):237–241.

[337] Eachus, S.W. 1983. Effect of soda-anthraquinone pulping conditions on holocellulose fibers. *TAPPI Journal* 66(2):85–88.

[338] Evans, R., Henderson, V.T., Nelson, P.F., and Vanderhoek, N. 1983. The soda-AQ semichemical pulping process. *Appita* 37(1):60–64.

[339] Löwendahl, L., and Samuelson, O. 1978. Carbohydrate stabilization during soda pulping with addition of anthraquinone. *TAPPI* 61(2):19–21.

[340] Abbot, J., and Bolker, H.I. 1984. The influence of a second additive on the catalytic action of anthraquinone during delignification. *Svensk Papperstidning* 87(12):R69–R73.

[341] Ruoho, K., and Sjöström, E. 1978. Improved stabilization of carbohydrates by the oxygen-quinone system. *TAPPI* 61(7):87–88.

[342] Algar, W.H., Farrington, A., Jessup, B., Nelson, P.F., and Vanderhoek, N. 1979. The mechanism of soda-quinone pulping. *Appita* 33(1):33–37.

[343] Gierer, J., Lindberg, O., and Norén, I. 1979. Alkaline delignification in the presence of anthraquinone/anthrahydroquinone. *Holzforschung* 33:213–214.

[344] Samuelson, O. 1980. Carbohydrate reactions during alkaline cooking with addition of quinones. *Pulp & Paper Canada* 81(8):T188–T190.

[345] Cassidy, R.F., Falk, L.E., and Dence, C.W. 1981. The reactions of radiolabelled anthraquinone with wood constituents in soda pulping. *Svensk Papperstidning* 84:R94–R99.

[346] Gierer, J., Kjellman, M., and Norén, I. 1983. Alkaline delignification in the presence of anthraquinone/anthrahydroquinone. Part 2: Experiments using extracted wood shavings from *Pinus sylvestris*. *Holzforschung* 37:17–22.

[347] Sjöström, E. 1993. *Wood Chemistry – Fundamentals and Applications*. 2nd edition. Academic Press, San Diego, CA, USA. Pp. 156–157.

[348] Vuorinen, T. 1993. The role of carbohydrates in alkali anthraquinone pulping. *Journal of Wood Chemistry and Technology* 13(1):97–125.

[349] de Groot, B., van Dam, J.E.G., and Van't Riet, K. 1995. Alkaline pulping of hemp woody core: Kinetic modelling of lignin, xylan and cellulose extraction and degradation. *Holzforschung* 49(4):332–342.

[350] Chai, X.-S., Samp, J., Hou, Q.X., Yoon, S.-H., and Zhu, J.Y. 2007. Possible mechanism for anthraquinone species diffusion in alkaline pulping. *Industrial & Engineering Chemistry Research* 46:5245–5249.

[351] Dimmel, D., and Gellerstedt, G. 2010. Chemistry of alkaline pulping. In: Heitner, C., Dimmel, D., and Schmidt, J. (Eds.). *Lignin and Lignans – Advances in Chemistry*. CRC Press, Boca Raton, FL, USA. Pp. 350–391.

[352] Prinsen, P., Rencoret, J., Gutiérrez, A., Liitiä, T., Tamminen, T., Colodette, J.L., Berbis, M.A., Jiménez-Barbero, J., Martínez, Á.T., and Del Río, J.C. 2013. Modification of the lignin structure during alkaline delignification of eucalyptus wood by kraft, soda-AQ, and soda-O_2 cooking. *Industrial & Engineering Chemistry Research* 52:15702–15712.

[353] Abenius, P.-H., Ishizu, A., Lindberg, B., and Theander, O. 1967. Reactions between D-glucose and polysulphide cooking liquor. *Svensk Papperstidning* 70(19):612–615.

[354] Germgård, U., Annergren, G., and Olsson, B. 2000. Fibrelines. In: Fardim, P. (Ed.). *Chemical Pulping Part 1, Fibre Chemistry and Technology*. 2nd edition. Paper Engineers' Association, Helsinki, Finland. Pp. 675–728.

[355] Yoon, S.-H., Cullinan, H., and Krishnagopalan, G.A. 2011. Polysulfide-borohydride modification of southern pine alkaline pulping integrated with hydrothermal pre-extraction of hemicelluloses. *TAPPI Journal* 10(7):9–16.

[356] Paananen, M., and Sixta, H. 2015. High-alkali low-temperature polysulfide pulping (HALT) of Scots pine. *Bioresource Technology* 193:97–102.

[357] Paananen, M., Rovio, S., Liitiä, T., and Sixta, H. 2015. Stabilization, degradation, and dissolution behavior of Scots pine polysaccharides during polysulfide (K-PS) and polysulfide anthraquinone (K-PSAQ) pulping. *Holzforschung* 69(9):1049–1058.

[358] Basta, J., and Samuelson, O. 1978. Sodium hydroxide cooking with addition of anthraquinone. *Svensk Papperstidning* 81(9):285–290.

[359] Eckert, R.C., Pfeiffer, G.O., Gupta, M.K., and Lower, C.R. 1984. Soda anthraquinone pulping of douglas fir. *TAPPI Journal* 67(11):104–108.

[360] MacLeod, J.M., and Cyr, N. 1983. Soda-AQ pulps from hardwoods – physical properties and bleachability. *Pulp & Paper Canada* 84(4):T81–T84.

[361] Venica, A.D., Chen, C.-L., and Gratzl, J.S. 2008. Soda-AQ delignification of poplar wood. Part 1: Reaction mechanism and pulp properties. *Holzforschung* 62:627–636.

[362] Venica, A.D., Chen, C.-L., and Gratzl, J.S. 2008. Soda-AQ delignification of poplar wood. Part 2: Further degradation of initially dissolved lignins. *Holzforschung* 62:637–644.

[363] Feng, Z., and Alén, R.J. 2001. Soda-AQ pulping of wheat straw. *Appita Journal* 54(2):217–220.

[364] Feng, Z., and Alén, R.J. 2001. Soda-AQ pulping of reed canary grass. *Industrial Crops and Products* 14:31–39.

[365] Atik, C. 2002. Soda-AQ pulping of okra stalk. *Cellulose Chemistry and Technology* 36(3–4):353–356.

[366] González-García, S., Moreira, M.T., Artal, G., Maldonado, L., and Feijoo, G. 2010. Environmental impact assessment of non-wood based pulp production by soda-anthraquinone pulping process. *Journal of Cleaner Production* 18:137–145.

[367] Dai, K., and Zhai, H. 2012. Effect of black liquor replacement in wheat straw soda-AQ cooking and lignin structure of pulps. *TAPPI Journal* 11(5):43–47.

[368] Samuelson, O., and Sjöberg, L.-A. 1978. Spent liquors from sodium hydroxide cooking with addition of anthraquinone. *Cellulose Chemistry and Technology* 12:463–472.

[369] Mašura, V. 1982. Alkaline degradation of spruce and beech wood. *Wood Science and Technology* 16:155–164.

[370] van der Klashorst, G.H., and Strauss, H.F. 1987. Properties and potential utilization of industrial eucalyptus soda/anthraquinone lignin. Part I. Isolation, identification and origin of the low molecular mass lignin fragments present in an industrial soda/anthraquinone spent pulping liquor. *Holzforschung* 41(2):123–131.

[371] Venter, J.S.M., and van der Klashorst, G.H. 1989. The recovery of by-products and pulping chemicals from industrial soda bagasse spent liquors. *TAPPI Journal* 72(3):127–132.

[372] Feng, Z., Alén, R., and Louhelainen, J. 2001. Characterisation of black liquors from soda-aq pulping of wheat straw. *Appita Journal* 54(2):234–238,244.

[373] Feng, Z., Alén, R., and Niemelä, K. 2002. Formation of aliphatic carboxylic acids during soda-aq pulping of kenaf bark. *Holzforschung* 56(4):388–394.

[374] Feng, Z., Alén, R., and Niemelä, K. 2002. Formation of aliphatic carboxylic acids during soda-aq pulping of wheat straw and reed canary grass. *Paperi ja Puu* 84(2):119–122.

[375] Karlström, K., Sjögren, B., Vorwerg, W., and Volkert, B. 2014. Sulphur-free cooking for value added cellulose. *Cellulose Chemistry and Technology* 48(9–10):781–786.

[376] Bajpai, P. 2018. *Biermann's Handbook of Pulp and Paper. Volume 1: Raw Material and Pulp Making*. Elsevier, Amsterdam, The Netherlands. Pp. 19–74.

[377] Louhelainen, J. 2003. *Changes in the Chemical Composition and Physical Properties of Wood and Non-Wood Black Liquors during Heating*. Doctoral Thesis. University of Jyväskylä, Laboratory of Applied Chemistry, Jyväskylä, Finland. 68 p.

[378] Holmlund, K., and Parviainen, K. 1999. Evaporation of black liquor. In: Gullichsen, J., and Fogelholm, C.-J. (Eds.). *Chemical Pulping*. Fapet, Helsinki, Finland. Pp. B37–B93.

[379] Kulkarni, A.G., Kolambe, S.L., Mathur, R.M., and Pant, R. 1984. Studies on desilication of bamboo kraft black liquor. *IPPTA* 21(1):37–44.

[380] Chen, J.-Z., Yu, J.-L., and Chen, J.-Z. 1992. A new method and its mechanisms of silica-removal from rice black liquor by microbial conversions. *Cellulose Chemistry and Technology* 26(3):345–350.

[381] Kulkarni, A.G., Mathur, R.M., and Dixit, A.K. 2005. Desilication of wheat straw black liquor. In: *Proceedings of the 13th International Symposium on Wood, Fibre and Pulping Chemistry (ISWFPC)*, May 16–19, 2005, Auckland, New Zealand. Pp. 615–621.

[382] Pekarovic, J., Pekarovicova, A., and Jouce, T.W. 2005. Desilication of agricultural residues – the first step prior pulping. *Appita Journal* 58(2):130–134.

[383] Xia, X., Du, M., and Geng, X. 2013. Removal of silicon from green liquor with carbon dioxide in the chemical recovery process of wheat straw soda pulping. *TAPPI Journal* 12(3):35–40.

[384] Kleinert, T.N., and von Tayenthal, K. 1931. Über Neuere Versuche zur Trennung von Cellulose und Inbrusten Verschiedener Hölzer. *Zeitschrift Für Angewandte Chemie* 44(39):788–791.

[385] Sarkanen, K.V. 1980. Acid-catalyzed delignification of lignocellulosics in organic solvents. In: Sarkanen, K.V., and Tillman, D.A. (Eds.). *Progress in Biomass Conversion, Volume 2*. Academic Press, New York, NY, USA. Pp. 127–144.

[386] Lora, J.H., and Aziz, S. 1985. Organosolv pulping: A versatile approach to wood refining. *TAPPI Journal* 68(8):94–97.

[387] Johansson, A., Aaltonen, O., and Ylinen, P. 1987. Organosolv pulping – methods and pulp properties. *Biomass* 13:45–65.

[388] Laxén, T. 1987. The characteristics of organosolv pulping discharges. *Paperi ja Puu* 69(5):417–421.

[389] Aziz, S., and Sarkanen, K. 1989. Organosolv pulping – a review. *TAPPI Journal* 72(3):169–175.

[390] Sarkanen K.V. 1990. Chemistry of solvent pulping. *TAPPI Journal* 73(10):215–219.

[391] Davis, J.L., and Young, R.A. 1991. Microwave-assisted solvent pulping. *Holzforschung* 45 (Supplement):71–77.

[392] Sierra, A.C., Salvador, A.R., and Soria, F.G.-O. 1991. Kinetics of wood extraction with solvents. *TAPPI Journal* 74(5):191–196.

[393] Hergert, H.L., and Pye, E.K. 1992. Recent history of organosolv pulping. In: *TAPPI Notes–1992 Solvent Pulping Symposium*. TAPPI Press, Atlanta, GA, USA. Pp. 9–26.

[394] Sundquist, J. 1999. Organosolv pulping. In: Gullichsen, J., and Fogelholm, C.-J. (Eds.). *Chemical Pulping*. Fapet, Helsinki, Finland. Pp. B410–B427.

[395] Laure, S., Leschinsky, M., Fröhling, M., Schultmann, F., and Unkelbach, G. 2014. Assessment of an organosolv lignocellulose biorefinery concept based on a material flow analysis of a pilot plant. *Cellulose Chemistry and Technology* 48(9–10):793–798.

[396] Alén, R. 2021. Organosolv pulping. In: Alén, R. (Ed.). *Pulping and Biorefining – ForestBioFacts, Digital Learning Environment*. Paperi ja Puu Ltd., Helsinki, Finland.

[397] Ligero, P., Villaverde, J.J., De Vega, A., and Bao, M. 2008. Delignification of *Eucalyptus globulus* saplings in two organosolv systems (formic and acetic acid). preliminary analysis of dissolved lignins. *Industrial Crops and Products* 27:110–117.

[398] Pye, E.K., and Lora, J.H. 1991. The Alcell™ process – a proven alternative to kraft pulping. *TAPPI Journal* 74(3):113–118.

[399] Winner, S.R., Goyal, G.C., Pye, E.K., and Lora, J.H. 1991. Pulping of agriculture residues by the alcell® process. In: *Proceedings of the TAPPI Pulping Conference*, November 3–7, 1991, Orlando, FL, USA. Pp. 435–439.

[400] Ni, Y., and Hu, Q. 1995. Alcell® lignin solubility in ethanol-water mixtures. *Journal of Applied Polymer Science* 57:1441–1446.

[401] Girard, R., and van Heiningen, A.R.P. 1997. Yield determination and mass balance closure of batch ethanol-based solvent cookings. *Pulp & Paper Canada* 98(8):65–68.

[402] Katzen, R., Frederickson, R., and Brush, B.F. 1980. The alcohol pulping & recovery process. *Chemical Engineering Progress* 76(2):62–67.

[403] Lora, J.H., and Aziz, S. 1985. Organosolv pulping: A versatile approach to wood refining. *TAPPI Journal* 68(8):94–97.

[404] Lindner, A., and Wegener, G. 1988. Characterization of lignins from organosolv pulping according to the organocell process. Part 1. Elemental analysis, nonlignin portions and functional groups. *Journal of Wood Chemistry and Technology* 8:323–340.

[405] Dahlmann, G., and Schroeter, M.C. 1990. The organocell process – pulping with the environment in mind. *TAPPI Journal* 73(4):237–240.

[406] Schroeter, M.C. 1991. Possible lignin reactions in the organocell pulping process. *TAPPI Journal* 74(10):197–200.

[407] Kordsachia, O., Reipschläger, B., and Patt, R. 1980. ASAM pulping of birch wood and chlorine free pulp bleaching. *Paperi ja Puu* 72(1):44–50.

[408] Zimmermann, M., Patt, R., Kordsachia, O., and Hunter, W.D. 1991. ASAM pulping of douglas-fir followed by a chlorine-free bleaching sequence. *TAPPI Journal* 74(11):129–134.

[409] Paananen, M., Rovio, S., Liitiä, T., and Sixta, H. 2015. Effect of hydroxide and sulfite ion concentration in alkaline sulfite anthraquinone (ASA) pulping – a comparative study. *Holzforschung* 69(6):661–666.

[410] Paszner, L., and Behera, N.C. 1989. Topochemistry of softwood delignification by alkali earth metal salt catalysed organosolv pulping. *Holzforschung* 43(3):159–168.

[411] Paszner, L., and Cho, H.J. 1989. Organosolv pulping: Acidic catalysis options and their effect on fiber quality and delignification. *TAPPI Journal* 72(2):135–142.

[412] Young, R.A., and Davis, J.L. 1986. Organic acid pulping of wood. Part II. Acetic acid pulping of aspen. *Holzforschung* 40(2):99–108.

[413] Neumann, N., and Balser, K. 1993. ACETOCELL – an innovative process for pulping, totally free from sulphur and chlorine. *Papier* 47(10): 16–24.

[414] Maldonado-Bustamante, S.R., Mondaca-Fernández, I., Gortares-Moroyoqui, P., Berg, A., Balderas-Cortés, J.J., Meza-Montenegro, M.M., Brown-Bojórquez, F., and Arvayo-Enríquez, H. 2022. The effectiveness of the organosolv process in wheat straw delignification optimizing temperature and time reaction. *Cellulose* 29:7151–7161.

[415] Parajó, J.C., Alonso, J.L., and Santos, V. 1995. Kinetics of *Eucalyptus* wood fraction in acetic acid-HCl-water media. *Bioresource Technology* 51(2–3):153–162.

[416] Vázquez, G., Antorrena, G., and González, J. 1995. Acetosolv pulping of *Eucalyptus globulus* wood. Part I. The effect of operational variables on pulp yield, pulp lignin content and pulp potential glucose content. *Holzforschung* 49:69–74.

[417] Vila, C., Santos, V., and Parajó, J.C. 2003. Recovery of lignin and furfural from acetic acid-water-HCl pulping liquors. *Bioresource Technology* 90(3):339–344.

[418] Nimz, H., Granzow, C., and Berg, A. 2007. Acetosolv pulping. *Holz als Roh- und Werkstoff* 44:362.

[419] Ferrer, A., Vega, A., Rodríquez, A., and Jiménez, L. 2013. Acetosolv pulping for the fractionation of empty fruit bunches from palm oil industry. *Bioresource Technology* 132:115–120.

[420] Saake, B., Lehnen, R., Schmekal, E., Neubauer, A., and Nimz, H.H. 1998. Bleaching of formacell pulp from aspen wood with ozone and peracetic acid in organic solvents. *Holzforschung* 52:643–650.

[421] Villaverde, J.J., Li, J., Ek, M., Ligero, P., and de Vega, A. 2009. Native lignin structure of *miscanthus x giganteus* and its changes during acetic and formic acid fractionation. *Journal of Agricultural and Food Chemistry* 57:6262–6270.

[422] Nayeem, J., Jahan, Md.S., and Rahman, Md.M. 2014. Formic acid/acetic acid/water pulping of agricultural wastes. *Cellulose Chemistry and Technology* 48(1–2):111–118.

[423] Hidayati, S., Suroso, E., Satyajaya, W., and Jryani, D.A. 2018. Chemistry and structure characterization of bamboo pulp with formacell pulping. In: *IOP Conference Series: Materials Science and Engineering* 532:article 012024.

[424] Sundquist, J. 1986. Bleached pulp without sulphur and chlorine chemicals by a peroxyacid/alkaline peroxide method – an overview. *Paperi ja Puu* 68(9):616–620.

[425] Sundquist, J., Laamanen, L., and Poppius, K. 1988. Problems of non-conventional pulping processes in the light of peroxyformic acid cooking experiments. *Paperi ja Puu* 70(2)143–148.

[426] Poppius-Levlin, K., Mustonen, R., Huovila, T., and Sundquist, J. 1991. MILOX pulping with acetic acid/peroxyacetic acid. *Paperi ja Puu* 73(2):154–158.

[427] Muurinen, E. 2000. *Organosolv Pulping, a Review and Distillation Study Related to Peroxyacid Pulping.* Doctoral Thesis. University of Oulu, Department of Process Engineering, Oulu, Finland. 314 p.

[428] Rousu, P. 2003. *Holistic Model of a Fibrous Production of Non-Wood Origin.* Doctoral Thesis. University of Oulu, Department of Process and Environmental Engineering, Oulu, Finland. 203 p.

[429] Pereira, H., Oliveira, M.F., and Miranda, I. 1986. Kinetics of ethanol-water pulping and pulp properties of *Eucalyptus globulus* Lab. *Appita* 39(6):455–458.

[430] Lönnberg, B., Laxén, T., and Sjöholm, R. 1987. Chemical pulping of softwood chips by alcohols. 1. Cooking. *Paperi ja Puu* 69(9):757–762.

[431] Lönnberg, B., Laxén, T., and Bäcklund, A. 1987. Chemical pulping of softwood chips by alcohols. 2. Bleaching and beating. *Paperi ja Puu* 69(10):826–830.

[432] Thykesson, M., Sjöberg, L.-A., and Ahlgren, P. 1997. Pulping of grass and straw. *Nordic Pulp and Paper Research Journal* 12(2):128–134.

[433] Paszner, L., and Chang, P.-C. 1983. Organosolv delignification and saccharification process for lignocellulosic plant materials. U.S. Patent 4,409,032.

[434] Billa, E., Kurek, B., Koutsoula, E., Monties, B., and Koukios, E. 2000. Effect of manganese peroxidase on the cell wall components of wheat straw and sorghum organosolv pulps. *Cellulose Chemistry and Technology* 34:131–137.

[435] Green, J., and Sanyer, N. 1982. Alkaline pulping in aqueous alcohols and amines. *TAPPI* 65(5):133–137.

[436] Marton, R., and Granzow, S. 1982. Ethanol-alkali pulping. *TAPPI* 65(6):103–106.

[437] Gautam, A., Kumar, A., and Dutt, D. 2016. Effects of ethanol addition and biological pretreatment on soda pulping of *Eulaliopsis binate*. *Journal of Biomaterials and Nanobiotechnology* 7:78–90.

[438] Valladares, J., and Rolz, C. 1984. Pulping of sugarcane bagasse with a mixture of ethanol-water solution in presence of sodium hydroxide and anthraquinone. In: *Proceedings of the 1984 TAPPI Pulping Conference, Book 1*, November 12–14, 1984, San Francisco, CA, USA. Pp. 83–88.

[439] Deinko, I.P., Makarova, O.V., and Zarubin, M.Ya. 1992. Delignification of wood by oxygen in low-molecular-weight alcoholic media. *TAPPI Journal* 75(9):136–140.

[440] Neto, C.P., Evtuguin, D., and Robert, A. 1994. Chemicals generated during oxygen-organosolv pulping of wood. *Journal of Wood Chemistry and Technology* 14(3):383–402.

[441] Nakano, J., Takatsuka, C.y., and Daima, H. 1976. Studies on alkali-methanol cooking (part 1). behavior of dissolution of lignin and carbohydrate. *Japan TAPPI* 30(12):650–655.

[442] Nakano, J., Daima, H., Hosoya, S., and Ishizu, A. 1981. Studies on alkali-methanol cooking. In: *Proceedings of the Ekman-Days 1981, Volume 2, June 9–12, 1981*, Stockholm, Sweden. Pp. 72–77.

[443] Tirtowidjojo, S., Sarkanen, K.V., Pla, F., and McCarthy, J.L. 1988. Kinetics of organosolv delignification in batch- and flow-through reactors. *Holzforschung* 42(3):177–183.

[444] Fengel, G., Wegener, G., and Greune, A. 1989. Studies on the delignification of spruce wood by organosolv pulping using SEM-EDXA and TEM. *Wood Science and Technology* 23:123–130.

[445] Bennani, A., Rigal, L., and Gaset, A. 1991. Refining of lignocellulose by organosolv processes. Part I: Isolation, characterisation and utilization of hemicellulose extracted from Norway spruce. *Biomass and Bioenergy* 1(5):289–296.

[446] Quinde, A., and Paszner, L. 1992. Behavior of the major resin- and fatty acids of slash pine (*pinus elliottii* engelm.) during organosolv pulping. *Holzforschung* 46(6):513–522.

[447] Alén, R. 1988. Formation of aliphatic carboxylic acids from hardwood polysaccharides during alkaline delignification in aqueous alcohols. *Cellulose Chemistry and Technology* 22(4):443–448.

[448] Alén, R. 1988. Formation of aliphatic carboxylic acids from softwood polysaccharides during alkaline delignification in aqueous alcohols. *Cellulose Chemistry and Technology* 22(5):507–512.

[449] Young, R.A., Fredman, T., Keith, T., and Nelson, J. 1981. Pulping of wood with organic acids and esters. In: *Proceedings of the 1987 International Wood and Pulping Chemistry Symposium*, Paris, France, April 21, 1987.

[450] Young, R.A. 1989. Ester pulping: A status report. *TAPPI Journal* 72(4):195–200.

[451] Chum, H.L. 1986. Key structural features of acetone-soluble phenol-pulping lignins by ^{1}H and ^{13}C N.M.R. spectroscopy. *Holzforschung* 40(Supplement):115–123.

[452] Funaoka, M., and Abe, I. 1989. Rapid separation of wood into carbohydrate and lignin with concentrated acid-phenol system. *TAPPI Journal* 72(8):145–149.

[453] Sakakibara, A., Edashige, Y., Sano, Y., and Takeyama, H. 1984. Solvolysis pulping with cresol-water system. *Holzforschung* 38(3):159–165.

[454] Sano, Y., Endo, M., and Sakashita, Y. 1989. Solvolysis pulping of softwoods. *Journal of the Japan Wood Research Society* 35(9):807–812.

[455] Johansson, A., Ebeling, K., Aaltonen, O., Aaltonen, P., and Ylinen, P. 1987. Wood pulping with aqueous tetrahydrofurfuryl alcohol. *Paperi ja Puu* 69(6):500–504.

[456] Zarubin, M.Ya., Dejneko, I.P., and Evtuguine, D.V. 1989. Delignification by oxygen in acetone-water media. *TAPPI Journal* 72(11):163–168.

[457] Li, P., Ji, H., Shan, L., Dong, Y., Long, Z., Zou, Z., and Pang, Z. 2020. Insights into delignification behavior using aqueous *p*-toluenesulfonic acid treatment: Comparison with different biomass species. *Cellulose* 27:10345–10358.

[458] Ferrer, A., Vega, A., Ligero, P., and Rodríquez, A. 2011. Pulping of empty fruit bunches (EFB) from the palm oil industry by formic acid. *BioResources* 6(4):4282–4301.

[459] McDonough, T.J. 1993. The chemistry of organosolv delignification. *TAPPI Journal* 76(8):186–193.

[460] van Lierop, B., Liebergott, N., Teodorescu, G., and Kubes, G.J. 1986. Oxygen in bleaching sequences – an overview. *Pulp & Paper Canada* 87(5):T193–T197.

[461] Sjöström, E. 1993. *Wood Chemistry – Fundamentals and Applications*. 2nd edition. Academic Press, San Diego, CA, USA. Pp. 166–198.

[462] McDonough, T.J. 1993. Oxygen delignification. In: Dence, C.W., and Reeve, D.W. (Eds.). *Pulp Bleaching – Principles and Practice*. TAPPI Press, Atlanta, GA, USA. Pp. 213–239.

[463] Crawford, R.J., Rovell-Rixx, D.C., Jett, S.W., Jain, A.K., and Dillard, D.S. 1995. Emissions of volatile organic compounds and hazardous air pollutants from oxygen delignification systems. *TAPPI Journal* 78(5):81–91.

[464] Alén, R. 2000. Oxygen-alkali delignification. In: Stenius, P. (Ed.). *Forest Products Chemistry*. Fapet, Helsinki, Finland. Pp. 86–93.

[465] Gellerstedt, G. 2001. Pulping chemistry. In: Hon, D.N.-S., and Shiraishi, N. (Eds.). *Wood and Cellulosic Chemistry*. 2nd edition. Marcel Dekker, New York, NY, USA. Pp. 859–905.

[466] Ala-Kaila, K., Li, J., Sevastyanova, O., and Gellerstedt, G. 2003. Apparent and actual delignification response in industrial oxygen-alkali delignification of birch kraft pulp. *TAPPI Journal* 2(10):23–27.

[467] Salmela, M. 2007. *Description of Oxygen-Alkali Delignification of Kraft Pulp Using Analysis of Dissolved Material*. Doctoral Thesis. University of Jyväskylä, Laboratory of Applied Chemistry, Jyväskylä, Finland. 71 p.

[468] Adibi, A.H.L. 2020. *Oxidative Reactions of Cellulose under Alkaline Conditions*. Doctoral Thesis. University of Jyväskylä, Laboratory of Applied Chemistry, Jyväskylä, Finland. 102 p.

[469] Alén, R. 2021. Oxygen-alkali delignification. In: Alén, R. (Ed.). *Pulping and Biorefining – ForestBioFacts, Digital Learning Environment*. Paperi ja Puu Ltd., Helsinki, Finland.

[470] Cotton, F.A., and Wilkinson, G. 1988. *Advanced Inorganic Chemistry*. 5th edition. John Wiley & Sons, New York, NY, USA. Pp. 449–452.

[471] Sjöström, E. 1981. The chemistry of oxygen delignification. *Paperi ja Puu* 63(6–7):438–442.

[472] Young, R.A., and Gierer, J. 1976. Degradation of native lignin under oxygen-alkali conditions. *Applied Polymer Symposia* 28:1213–1223.

[473] Gierer, J., and Imsgard, F. 1977. The reactions of lignin with oxygen and hydrogen peroxide in alkaline media. *Svensk Papperstidning* 80(16):510–518.

[474] Campion, S.H., and Suckling, I.D. 1998. Effect of lignin demethylation on oxygen delignification. *Appita Journal* 51(3):209–212.

[475] Pasco, M.F., and Suckling I.D. 1998. Chromophore changes during oxygen delignification of radiata pine kraft pulp. *Appita Journal* 51(2):138–146.

[476] Tong, G., Matsumoto, Y., and Meshitsuka, G. 2000. Analysis of progress of oxidation reaction during oxygen-alkali treatment of lignin 2: Significance of oxidation reaction of lignin during oxygen delignification. *Journal of Wood Science* 46:371–375.

[477] Tong, G., Yokoyama, T., and Matsumoto, Y. 2000. Analysis of progress of oxidation reaction during oxygen-alkali treatment of lignin I: Method and its application to lignin oxidation. *Journal of Wood Science* 46:32–39.

[478] Tao, L., Genco, J.M., Cole, B.J.W., and Fort, R.C. Jr. 2011. Selectivity of oxygen delignification for southern softwood kraft pulps with high lignin content. *TAPPI Journal* 10(8):29–39.

[479] Alén, R. 2021. Reactions of lignin – oxygen-alkali delifnification. In: Alén, R. (Ed.). *Pulping and Biorefining – ForestBioFacts, Digital Learning Environment*. Paperi ja Puu Ltd., Helsinki, Finland.

[480] Alén, R. 2021. Reactions of extractives – oxygen-alkali delifnification. In: Alén, R. (Ed.). *Pulping and Biorefining – ForestBioFacts, Digital Learning Environment*. Paperi ja Puu Ltd., Helsinki, Finland.

[481] Gellerstedt, G. 1996. Chemical structure of pulp components. In: Dence, C.W., and Reeve, D.W. (Eds.). *Pulp Bleaching – Principles and Practice*. TAPPI Press, Atlanta, GA, USA. Pp. 91–111.

[482] Shin, S.-J., Schroeder, L.R., and Lai, Y.-Z. 2004. Impact of residual extractives on lignin determination in kraft pulps. *Journal of Wood Chemistry and Technology* 24(2):139–151.

[483] Shin, S.-J., Schroeder, L.R., and Lai, Y.-Z. 2006. Understanding factors contributing to low oxygen delignification of hardwood kraft pulps. *Journal of Wood Chemistry and Technology* 26(1):5–20.

[484] Shin, S.-J., and Kim, C.-H. 2006. Residual extractives in unbleached aspen and pine kraft pulps and their fate on oxygen delignification. *Nordic Pulp and Paper Research Journal* 21(2):260–263.

[485] Dence, C.W. 1996. Chemistry of chemical pulp bleaching. In: Dence, C.W., and Reeve, D.W. (Eds.). *Pulp Bleaching – Principles and Practice*. TAPPI Press, Atlanta, GA, USA. Pp. 125–159.

[486] Holmbom, B. 2000. Resin reactions and deresination in bleaching. In: Back, E.L., and Allen, L.H. (Eds.). *Pitch Control, Wood Resin and Deresination*. TAPPI Press, Atlanta, GA, USA. Pp. 231–244.

[487] Alén, R., and Sjöström, E. 1991. Formation of low-molecular-mass compounds during the oxygen delignification of pine kraft pulp. *Holzforschung* 45(Suppl.):83–86.

[488] Holmbom, B. 2011. Extraction and utilization of non-structural wood and bark components. In: Alén, R. (Ed.). *Biorefining of Forest Resources*. Paper Engineers' Association, Helsinki, Finland. Pp. 176–224.

[489] Dence, C.W., and Reeve, D.W. (Eds.). 1993. *Pulp Bleaching — Principles and Practice*. TAPPI Press, Atlanta, GA, USA. 868 p.

[490] Bajpai, P. 2005. *Developments in Environmental Management – Environmentally Benign Approaches for Pulp Bleaching*. Elsevier, Amsterdam, The Netherlands. 278 p.

[491] Suess, H.U. 2010. *Pulp Bleaching Today*. Walter de Gruyter, Berlin, Germany. 310 p.

[492] Löwendahl, L., and Samuelson, O. 1974. Influence of iron and cobalt compounds upon oxygen-alkali treatment of cellulose. *Svensk Papperstidning* 77(16):593–602.

[493] Malinen, R. 1974. *Behavior of Wood Polysaccharides during Oxygen-Alkali Delignification*. Doctoral Thesis. Helsinki University of Technology, Laboratory of Wood Chemistry, Otaniemi, Finland. 9 p. Published in *Paperi ja Puu* 57(4a)(1975):193–196,199–201,203–204.

[494] Alén, R. 2021. Reactions of carbohydrates – oxygen-alkali delifnification. In: Alén, R. (Ed.). *Pulping and Biorefining – ForestBioFacts, Digital Learning Environment*. Paperi ja Puu Ltd., Helsinki, Finland.

[495] Samuelson, O., and Thede, L. 1968. Influence of oxygen upon glucose and cellobiose in strongly alkaline medium. *Acta Chemica Scandinavica* 22(6):1913–1923.

[496] Rowell, R.M., Somers, P.J., Barker, S.A., and Stacey, M. 1969. Oxidative alkaline degradation of cellobiose. *Carbohydrate Research* 11:17–25.

[497] Malinen, R., and Sjöström, E. 1972. Studies on the reactions of carbohydrates during oxygen bleaching. Part I. oxidative alkaline degradation of cellobiose. *Paperi ja Puu* 54(8):451–468.

[498] Hansson, J.-Å., and Hartler, N. 1968. Effects of reducing and oxidizing agents on the resistance of xylans from birch and pine towards alkaline degradation. *Svensk Papperstidning* 71(19):669–673.

[499] Kolmodin, H., and Samuelson, O. 1973. Oxygen-alkali treatment of hemicellulose. 2. Experiments with birch xylan. *Svensk Papperstidning* 76(2):71–77.

[500] Löwendahl, L., Petersson, G., and Samuelson, O. 1975. Dicarboxylic acids produced by oxygen-alkali treatment of birch xylan. *Acta Chemica Scandinavica B* 29(4):526–527.

[501] Malinen, R., and Sjöström, E. 1975. Studies on the reactions of carbohydrates during oxygen bleaching. Part V. Degradation of xylose, xylosone, xylo-oligosaccharides and birch xylan. *Paperi ja Puu* 57:101–102,107–109,111–114.

[502] Samuelson, O., and Stolpe, L. 1973. Degradation of cellobiitol and glucose by oxygen-alkali treatment. *Acta Chemica Scandinavica* 27(8):3061–3068.

[503] Samuelson, O., and Stolpe, L. 1974. Degradation of carbohydrates during oxygen bleaching. 3. metal compounds as inhibitors in model experiments with cellobiitol. *Svensk Papperstidning* 77(1):16–21.

[504] Ericsson, B., Lindgren, B.O., and Theander, O. 1973. Degradation of cellulose during oxygen bleaching. oxidation and alkaline treatment of d-glucosone. *Cellulose Chemistry and Technology* 7:581–591.

[505] Samuelson, O., and Stolpe, L. 1969. Aldonic acid end groups in cellulose after oxygen bleaching. I. model experiments with hydrocellulose. *TAPPI* 52(9):1709–1711.

[506] Rowell, R.M. 1971. Oxidative alkaline degradation III. Stopping reaction. *Pulp and Paper Magazine of Canada* 72(7):74–77.

[507] Heikkilä, H., and Sjöström, E. 1975. Introduction of aldonic acid end-groups into cellulose by various oxidants. *Cellulose Chemistry and Technology* 9:3–11.

[508] Yang, B.Y., and Montgomery, R. 1996. Alkaline degradation of glucose: Effect of initial concentration of reactants. *Carbohydrate Research* 280(1):27–45.

[509] Kolmodin, H., and Samuelson, O. 1972. Oxygen-alkali treatment of hydrocellulose. *Svensk Papperstidning* 75(9):369–372.

[510] Samuelson, O., and Sjöberg, L. 1974. Spent liquor from oxygen-alkali cooking of birch. *Cellulose Chemistry and Technology* 8(1):39–48.

[511] Pfister, K., and Sjöström, E. 1979. Characterization of spent bleaching liquors. Part 5. Composition of material dissolved during oxygen-alkali delignification. *Paperi ja Puu* 61(8):525–528.

[512] Renard, J.J., Mackie, D.M., and Bolker, H.I. 1975. Delignification of wood using pressurized oxygen. Part II. Kinetics of lignin oxidation. *Paperi ja Puu* 57(11):786–788,791–794,799–804.

[513] Soultanov, V., Krutov, S.M., and Zarubin, M.J. 2000. Chemical investigation of oxygen-alkaline delignification. In: *Proceedings of Sixth European Workshop on Lignocellulosics and Pulp*, September 3–6, 2000, Bordeaux, France. Pp. 595–598.

[514] Soultanov, V., Krutov, S.M., and Zarubin, M.J. 2003. Chemical investigation of oxygen-alkaline delignification. In: *Proceedings of 12th International Symposium on Wood and Pulping Chemistry, Volume 2*, June 9–12, 2003, Madison, WI, USA. Pp. 155–158.

[515] Grangaard, D.H. 1963. Decomposition of lignin. II. oxidation with gaseous oxygen under alkaline conditions. In: *Proceedings of 144th Meeting of American Chemical Society*, March 31–April 15, 1963, Los Angeles, CA, USA.

[516] Merriman, M.M., Choulet, H., and Brink, D.L. 1996. Oxidative degradation of wood. I. Analysis of products of oxygen oxidation by gas chromatography. *TAPPI Journal* 79(1):34–39.

[517] Scholander, E., Durst, W.B., Pearce, G., and Dence, C.W. 1974. Characterization of spent alkali/ oxygen bleaching liquor. *TAPPI* 57(3):142–145.

[518] Murphy, C., van de Ven, T.G.M., and Heitner, C. 2003. CO_2 evolution during H_2O_2 bleaching of lignin-containing pulp. In: *Proceedings of 12th International Symposium on Wood and Pulping Chemistry, Volume 1*, June 9–12, 2003, Madison, WI, USA. Pp. 101–104.

[519] Fessenden, R.J., and Fessenden, J.S. 1994. *Organic Chemistry.* 5th edition. Brooks/Cole Publishing Company, Pacific Grove, CA, USA.

[520] Huang, C.-L., Wu, C.-C., and Lien, M.-H. 1997. Ab initio studies of decarboxylations of the β-keto carboxylic acids $XCOCH_2COOH$ (X = H, OH, and CH_3). *Journal of Physical Chemistry A* 101(42):7867–7873.

[521] Guay, D.F., Cole, B.J.W., Fort, R.C. Jr., Genco, J.M., and Hausman, M.C. 2000. Mechanisms of oxidative degradation of carbohydrates during oxygen delignification. I. Reaction of methyl β-D-glucopyranoside with photochemically generated hydroxyl radicals. *Journal of Wood Chemistry and Technology* 20(4):375–394.

[522] Chai, X.-S., Zhu, J.Y., and Luo, Q. 2003. Minor sources of carbonate in kraft pulping and oxygen delignification processes. *Journal of Pulp and Paper Science* 29(2):59–63.

[523] Fu, S., Chai, X., Hou, Q., and Lucia, L.A. 2004. Chemical basis for a selectivity threshold to the oxygen delignification of kraft softwood fiber as supported by the use of chemical selectivity agents. *Industrial & Engineering Chemistry Research* 43(10):2291–2295.

[524] Rydholm, S.A. 1965. *Pulping Processes*. Interscience Publishers, New York, NY, USA. Pp. 837–1104.

[525] Barton, R., Tredway, C., Ellis, M., and Sullivan, E. 1987. Hydrosulfite bleaching. In: Kocurek, M.J. (Ed.). *Mechanical Pulping*. 3rd edition. TAPPI Press, Atlanta, GA, USA. Pp. 227–237.

[526] Patrick, K.L. (Ed.). 1991. *Bleaching Technology for Chemical and Mechanical Pulps*. Miller Freeman, San Francisco, CA, USA. 169 p.

[527] Singh, R.P. (Ed.). 1991. *The Bleaching of Pulp*. 3rd edition. TAPPI Press, Atlanta, GA, USA. 694 p.

[528] Sjöström, E. 1993. *Wood Chemistry – Fundamentals and Applications*. 2nd edition. Academic Press, San Diego, CA, USA. Pp. 198–203.

[529] Biermann, C.J. 1996. *Handbook of Pulping and Papermaking*. 2nd edition. Academic Press, San Diego, CA, USA. Pp. 123–136.

[530] Dence, C.W. 1996. Chemistry of mechanical pulp bleaching. In: Dence, C.W., and Reeve, D.W. (Eds.). *Pulp Bleaching – Principles and Practice*. TAPPI Press, Atlanta, GA, USA. Pp. 161–181.

[531] Ellis, M.E. 1996. Hydrosulfite (dithionite) bleaching. In: Dence, C.W., and Reeve, D.W. (Eds.). *Pulp Bleaching – Principles and Practice*. TAPPI Press, Atlanta, GA, USA. Pp. 491–512.

[532] Kappel, J. 1999. *Mechanical Pulps: From Wood to Bleached Pulp*. TAPPI Press, Atlanta, GA, USA. Pp. 263–326.

[533] Alén, R. 2000. Bleaching. In: Stenius, P. (Ed.). *Forest Products Chemistry*. Fapet, Helsinki, Finland. Pp. 93–102.

[534] Chirat, C., Hostachy, J.-C., Paloniemi, J., Pelin, K., Pohjanvesi, S., Nordén, S., Vesala, R., and Wennerström, M. 2011. Bleaching. In: Fardim, P. (Ed.). *Chemical Pulping Part 1, Fibre Chemistry and Technology*. 2nd edition. Paper Engineers' Association, Helsinki, Finland. Pp. 457–599.

[535] Bajpai, P. 2012. *Environmentally Benign Approaches for Pulp Bleaching*. 2nd edition. Elsevier, Amsterdam, The Netherlands. 416 p.

[536] Hart, P.W., and Rudie, A.W. 2012. *The Bleaching of Pulp*. 5th edition. TAPPI Press, Atlanta, GA, USA. 798 p.

[537] Bajpai, P. 2013. *Bleach Plant Effluents from the Pulp and Paper Industry*. Springer, Heidelberg, Germany. 88 p.

[538] Makarov, S.V., and Silaghi-Dumitrescu, R. 2013. Sodium dithionite and its relatives: Past and present. *Journal of Sulfur Chemistry* 34(4):444–449.

[539] Makarov, S.V., Horváth, A.K., Silaghi-Dumitrescu, R., and Gao, Q. 2017. *Sodium Dithionite, Rongalite and Thiourea Oxides – Chemistry and Application*. World Scientific, Singapore. 219 p.

[540] Isoaho, J.P. 2019. *Dithionite Bleaching of Thermomechanical Pulp – Chemistry and Optimal Conditions*. Doctoral Thesis. University of Jyväskylä, Laboratory of Applied Chemistry, Jyväskylä, Finland. 73 p.

[541] Alén, R. 2021. Delignifying or Lignin-removing Bleaching. In: Alén, R. (Ed.). *Pulping and Biorefining – ForestBioFacts, Digital Learning Environment*. Paperi ja Puu Ltd., Helsinki, Finland.

[542] Gierer, J. 1970. The reactions of lignin during pulping. *Svensk Papperstidning* 73:571–595.

[543] Karlsson, O., Pettersson, B., and Westermark, U. 2001. The use of cellulases and hemicellulases to study lignin-cellulose as well as lignin-hemicellulose bonds in kraft pulps. *Journal of Pulp and Paper Science* 27(6):196–201.

[544] Lachenal, D., Benattar, N., Allix, M., Marlin, N., and Chirat, C. 2005. Bleachability of alkaline pulps. Effect of quinones present in residual lignin. In: *Proceedings of 13th ISWFPC, Vol. 2*, Auckland, New Zealand, May 16–19, 2005. Pp. 23–27.

[545] Gierer, J. 1986. Chemistry of delignification. Part 2: Reactions of lignin during bleaching. *Wood Science and Technology* 20:1–33.

[546] Gierer, J. 1990. Basic principles of bleaching. Part 1: Cationic and radical processes. *Holzforschung* 44(5):387–394.

[547] Lachenal, D., and Muguet, M. 1992. Degradation of residual lignin in kraft pulp with ozone. application to bleaching. *Nordic Pulp and Paper Research Journal* 7(1):25–29.

[548] Aksela, R., and Konn, J. 2021. Metal management in pulping and bleaching. In: Alén, R. (Ed.). *Pulping and Biorefining – ForestBioFacts, Digital Learning Environment*. Paperi ja Puu Ltd., Helsinki, Finland.

[549] Wuorimaa A., Jokela, R., and Aksela, R. 2006. Recent developments in the stabilization of hydrogen peroxide bleaching of pulps: An overview. *Nordic Pulp and Paper Research Journal* 21(3):451–459.

[550] Colodette, J.L., Rothenberg, S., and Dence, C.W. 1988. Factors affecting hydrogen peroxide stability in the brightening of mechanical and chemimechanical pulps. Part I: Hydrogen peroxide stability in the absence of stabilizing systems. *Journal of Pulp and Paper Science* 14(6):J126–132.

[551] Brown, D.G., and Abbot, J. 1994. Magnesium as a stabilizer for peroxide bleaching of mechanical pulp. *Appita Journal* 47(3):211–216.

[552] Colodette, J.L., Rothenberg, S., and Dence, C.W. 1989. Factors affecting hydrogen peroxide stability in the brightening of mechanical and chemimechanical pulps. Part II: Hydrogen peroxide stability in the presence of sodium silicate. *Journal of Pulp and Paper Science* 15(1):J3–J10.

[553] Brown, D.G., and Abbot, J. 1995. Effects of metal ions and stabilisers on peroxide decomposition during bleaching. *Journal of Wood Chemistry and Technology* 15(1):85–111.

[554] Abbot, J. 1991. Catalytic decomposition of alkaline hydrogen peroxide in the presence of metal ions: Binuclear complex formation. *Journal of Pulp and Paper Science* 17(1):J10–J17.

[555] Burton, J.T. (1986). An investigation into the roles of sodium silicate and Epsom salt in hydrogen peroxide bleaching. *Journal of Pulp and Paper Science* 12(4):J95–J99.

[556] Liden, J., and Öhman L.-O. 1998. On the prevention of Fe- and Mn-catalyzed H_2O_2 decomposition under bleaching conditions. *Journal of Pulp and Paper Science* 24(9):269–276.

[557] Ulmgren, P. 1997. Non-process elements in a bleached kraft pulp mill with a high degree of system closure – state of the art. *Nordic Pulp and Paper Research Journal* 12(1):32–41.

[558] Salem, I.A., El-Maazawi, M., and Zaki, A.B. 2000. Kinetics and mechanisms of decomposition reaction of hydrogen peroxide in presence of metal complexes. *International Journal of Chemical Kinetics* 32:643–666.

[559] Kumar, G.S.V., and Mathew, B. 2004. Catalase-like activity of polystyrene-supported schiff base-metal complexes in hydrogen peroxide decomposition. *Journal of Applied Polymer Science* 92(2):1271–1278.

[560] Jessop, P.G., Ahmadpour, F., Buczynski, M.A., Burns, T.J., Green, N.B. II, Korwin, R., Long, D., Massad, S.K., Manley, J.B., and Omidbakhsh, N. 2015. Opportunities for greener alternatives in chemical formulations. *Green Chemistry* 17: 2664–2678.

5 Integrated forest biorefining

5.1 Introduction

As indicated in Chapter 1 "Introduction to biorefining" biorefineries are defined as facilities designed for processing lignocellulosic biomass feedstocks into versatile products showing potential for replacing fuels, chemicals, and other products currently manufactured from fossil resources [1–16]. Hence, the main principle of the biorefinery concept is to create a cost-effective and sustainable way to reduce the reliance on fossil fuels, reduce the greenhouse gas (GHG) emissions, and promote transitioning toward viable bioeconomy, relying on renewable feedstocks (*i.e.*, carbon dioxide (CO_2)-neutral materials). It can be simply concluded that modern biorefineries are analogous to conventional petroleum-based refineries (*i.e.*, petroleum refineries utilizing fossil resources) in producing a variety of products, but in this case from renewable biomass feedstocks, in an environmentally sustainable way. In practice, the general aim is to maximize the value of lignocellulosic biomass materials and, on the other hand, to minimize the formation of waste.

Pulp and paper production is globally an important branch of industry and chemical pulp mills use different technological innovations to fractionate and convert lignocellulosic biomass into a wide range of products, including cellulose (pulp), lignin-based materials, aliphatic carboxylic acids from hemicelluloses, and extractives-derived by-products [12, 17]. Against this background, it can be concluded that the first full-scale biorefineries (*i.e.*, chemical and thermochemical biorefineries) employing one type of feedstock were already operated in the pulp and paper industry about 160 years ago and were capable of green manufacturing several end products.

However, it can be also considered that modern pulp mills could serve as a promising platform for even more efficient use of wood and other lignocellulosic biomass materials. Hence, besides building new stand-alone biorefineries, modification of the existing lignocellulosic biomass processing plants, such as chemical pulp mills, into integrated biorefineries has gained considerable interest during the last decades [16, 18–21]. For example, the full-scale partial recovery of various organic degradation products (*i.e.*, lignin and aliphatic carboxylic acids) from cooking liquors has been an interesting alternative to using them as fuel, provided feasible methods of separation exist and the products can be marketed [17, 22]. Another general approach has been to apply the pre-extraction of hemicelluloses prior to delignification for obtaining suitable raw materials for different conversion methods [23–28]. One attraction of this integrated biorefinery concept is that the main product, cellulose fiber, suitable for versatile use in paper and paperboard, is already a well-established product. Additionally, besides the production of chemicals and other organic materials, the use of various organic fractions from cooking liquors and pulp mill effluents for producing biofuels is also possible [29–31].

Figure 5.1 shows a simplified representation of the conversion route "from forest to paper" in a case where delignification is based on the kraft process. The total con-

https://doi.org/10.1515/9783110608366-005

version can be divided into three distinct phases including *i)* timber procurement, *ii)* pulping, and *iii)* papermaking. However, with respect to novel integrated biorefining aspects, only the first two phases offer attractive possibilities. The main benefit of the process integrated to the forest industry and especially to operating pulp mills lies in the economic aspects, as integrating an additional biorefinery process (*e.g.*, pretreatment operations for recovering hemicelluloses) into an established process, where the biomass collection, storage, handling, and fractionation is already practiced, could lead to significant cost reductions for biorefineries [16]. Additionally, the integrated forest biorefining strategy offers potential ways for pulp and paper mills to diversify and increase their revenues by expanding their product portfolio and simultaneously face the challenges caused by declining demand of pulp and paper products, global competition, and high energy prices. This approach also allows more efficient utilization of whole lignocellulosic biomass raw material.

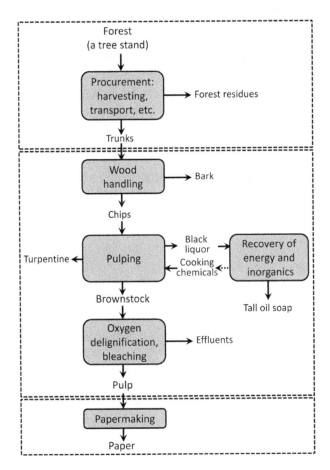

Figure 5.1: Schematic presentation of conversion of forest wood resources into paper products using kraft pulping for delignification [16].

Timber procurement covers, besides wood harvesting and logging, all the activities needed to produce timber from the forest for pulping mills [16, 32]. These activities include the cutting and off-road hauling of timber to roadside storage points or landing and transportation of timber from the landing to the mill yard. Procurement activities also require many other organizational facilities. It is obvious that forest industry companies use good forest management practices in harvesting trees to ensure a sustainable source of raw material. It can be estimated that especially, in the Nordic region 35–40% of the original dry tree biomass remains as harvesting residues (*i.e.*, branches, stumps, roots, needles, or foliage) [12]. Furthermore, it should be noted that the debarked dry wood in kraft pulping yields only about 50% pulp corresponding to about 25% of the original dry wood material.

Wood and harvesting residues, in addition to the material from debarking at a mill, can be processed in a large number of ways using different mechanical (*e.g.*, pulverization and pelletizing), chemical (*e.g.*, acid hydrolysis followed by fermentation) (see Chapter 3 "Chemical and biochemical conversion"), and thermochemical (*i.e.*, pyrolysis, gasification, hydrothermal carbonization, liquefaction, and combustion) (see Chapter 7 "Thermochemical conversion") methods, to produce energy and various chemical products [12, 17, 33–39]. However, in all cases, a prerequisite for planning economic processes for the manufacture of useful products is that the detailed chemical composition of feedstock materials as well as the process chemistry is known and all the constituents of feedstock material are taken into consideration.

5.2 Pretreatments of lignocellulosic feedstocks prior to pulping

5.2.1 General approach

Carbohydrates and their derivatives have a large potential as starting raw materials for manufacturing useful low-molar-mass and high-molar-mass products [17, 40]. The conventional approach of using chemically lignocellulosic biomass materials and other carbohydrate-containing biomasses comprises their straightforward degradation into fermentable water (H_2O)-soluble sugars; it can be accomplished by suitable pretreatment followed by acid or enzymatic hydrolysis (see Chapter 3 "Chemical and biochemical conversion") [33–35, 41, 42]. The hemicelluloses in lignocellulosic biomass materials are typically more readily hydrolysable by acids than cellulose, and their removal also enhances the reactivity of cellulose in the residual solids. In chemical pulping (see Chapter 4 "Chemical pulping-based methods"), majority of the hemicelluloses, along with lignin, are dissolved in varying forms into the cooking liquor during delignification [43, 44]. Hence, one potential approach would be to remove a prominent part of hemicelluloses prior to delignification instead of recovering their degradation products from the spent cooking liquors. However, this approach is principally not a recent idea, as industrial pre-hydrolysis of hemicelluloses, especially prior to hardwood kraft pulping, has been used for many years in the production of dissolving pulp [45, 46]. Additionally, in some cases, it may be an advantage to recover the hemicelluloses mostly as low-molar-mass sugars rather than trying to extract them in large quantities without considerable degradation.

The pulp and paper industry is facing major challenges and will need to find new products with a moderate added value in order to remain and increase competitive. The integration of various pretreatment stages prior to pulping may be one of the feasible possibilities. In developing potential concepts for the recovery of dissolved organic solids either in the pretreatment stage or from the black liquor in kraft pulping, certain limiting general factors, including both technical and economic factors, need to be considered [12, 17]. The main product is always cellulose fiber, and its strength properties must be maintained without interfering with the effective recovery of cooking chemicals and extractives. However, these prerequisites practically dictate that only partial recovery of the dissolved material is possible. Additionally, it would be advantageous, if there was the possibility of producing sulfur-free by-products and applying straightforward separation techniques. In many cases, one of the general aims would be to "maximize" the recovery of carbohydrate-derived material with low heating value, while "minimizing" the recovery of lignin-derived material with high heating value.

In case of the pretreatment approach the typical process is hot-H_2O extraction (HWE) catalyzed by acids and bases (Figure 5.2). Depending on the pH level used, it is possible to produce a wide range of products that are, especially in case of alkaline conditions, also present in the cooking liquors after delignification. In general, pretreatment processes of wood chips have been investigated under a variety of conditions and from several points of view.

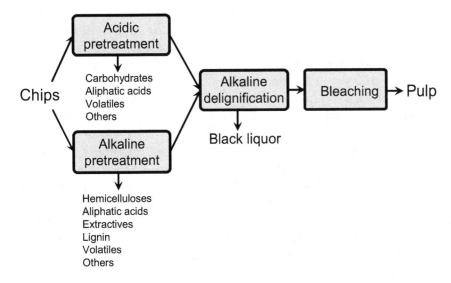

Figure 5.2: General principles of pretreatment stages applied in kraft pulping [17].

5.2.2 Acidic pretreatments

Autohydrolysis

H_2O is widely used in many industrial processes and everyday life and clear advantages of its use are that it is an abundant, environmentally friendly, and non-toxic solvent [47]. Furthermore, there are already established methods for its purification and recycling, hence complex new processes are not needed. In principle, there are two main ways to use HWE: *i*) hot H_2O can be used to dissolve various lignocellulosic biomass-containing compounds, which can then be found in the hot-H_2O extract and *ii*) the use of higher-temperature H_2O to dissolve and even break down lignocellulosic biomass materials. After these reactions a wide range of degradation compounds are released and dissolved in the hot-H_2O extract.

Hot-H_2O extraction has mainly been used in analytical purposes for determining compounds in different materials; the methods are typically called "pressurized hot-H_2O extraction" (PHWE) [47, 48] and "subcritical H_2O extraction" (or "superheated H_2O extraction") [49]. The principle behind these methods is to increase temperature in a way that desired compounds can be dissolved for detailed analysis. HWE-type methods have been applied, for example, to hardwoods [50–54], softwoods [55, 56], non-woods [57, 58], bark [59, 60], and certain specific materials [61–64]. Subcritical H_2O extraction has also been used for many different purposes [65–67].

When temperature increases over 100 °C, H_2O reacts with the feedstock material by degrading it partly *via* hydrolysis [12]. H_2O is mainly reacting with carbohydrates, which consist of various monosaccharide moieties chemically linked within carbohydrate chains, by breaking down the glycosidic bonds between monosaccharide units. Additionally, some organic acids (*i.e.*, typically acetic acid (CH_3CO_2H) from acetyl ($-COCH_3$) groups, deacetylation) are readily released when biomasses are heated in H_2O and thus, the hydrolysis is catalyzed by hydronium ions ($H_3O^\oplus$). The hydrolyzed H_2O-soluble fragments are then moved into the H_2O extract. The common terms for such simple treatments using only lignocellulosic biomass, H_2O, and heat, are "autohydrolysis", "pre-hydrolysis", or "hydrothermolysis".

In general, if HWE is performed at temperatures under and near H_2O boiling at 100 °C, the compounds that are inside lignocellulosic biomass cells are mostly extracted [67]. However, when temperature increases to 140–150 °C, the structural part of lignocellulosic materials starts to degrade and is released. An extreme example of PHWE is supercritical H_2O extraction where temperature is >374 °C and pressure >218 atm. Under these conditions, lignocellulosic biomass degrades almost completely during the treatment. Hence HWEs can be mild or harsh and the severity of these treatments primarily depends on temperature and treatment duration [68]. Milder treatments are used to extract hemicelluloses as oligosaccharides and polysaccharides from wood. In contrast, harsher treatments degrade hemicelluloses mainly to H_2O-soluble sugars, which can partly react further to furan derivatives [52]. Softwoods (*i.e.*, typically sawdust and bark that are mainly burned for energy in sawmills) can be extracted with hot H_2O for separating partially wood hemicelluloses, mainly gluco-mannan, whereas in case of hardwoods, the principal hemicellulose component that can be similarly extracted is xylan [47]. Both these hemicellulose constituents have characteristic properties and can be used, for example, in food applications, such as yoghurt and oat products [69] or in alkyd paints [70].

Spruce material is typically extracted with hot H_2O at 160–180 °C (the final pH 3.5–4.0) in a batch system; at 180 °C 25% of the ground spruce material can be removed, whereas at the same temperature the yield from the spruce chips is almost 10% lower [55]. In case of the ground material, the total yield is not much influenced after an extraction time of 20 min, whereas at 160 °C the yield increases from 10% to 20% of the initial material in the extraction time range from 20 to 100 min.

Continuous HWE of softwood has also been studied with a laboratory-scale flow-through system having a H_2O flow rate of 1 mL/min and an extraction time of about 40 min (Figure 5.3); at 120–140 °C the yield is low but as the temperature increases to >200 °C almost all the hemicelluloses are extracted [56]. This kind of extraction has also been scaled up from laboratory to a 300 L vessel [71]. In this case, HWE extraction has been carried out at 170 °C for 60 min and 80% of the original hemicelluloses of spruce sawdust can be extracted. Hence, smaller-scale extractions and larger-scale extractions have similar yields.

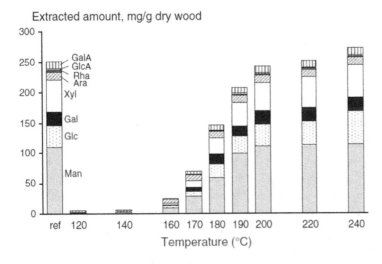

Figure 5.3: Amounts and composition of total hardwood carbohydrates in water extracts obtained from the continuous hot-water extraction [56]. For abbreviations and structures of monosaccharides, see Figure 2.6.

The main aim of the softwood bark extraction is to recover, besides its tannins, also its carbohydrates [47]. In this case, extraction temperatures are usually lower than those used for wood; for spruce bark typically <100 °C [59, 60]. The highest yields of tannins in laboratory experiments have normally been 6–8% of the extracted bark with an extraction time of 60–80 min. Extracted bark can be further processed by pyrolysis producing, for example, solid biochar.

Birch wood has been extracted in a laboratory-scale batch system, and at 180 °C most hemicelluloses can be removed within 40 min [52, 71]. This material has also been extracted in laboratory and pilot scales at 160 °C indicating that 50% of xylan can be separated (Figure 5.4). However, in all cases, xylan degrades to xylose-based sugars as also to furfural, to some extent.

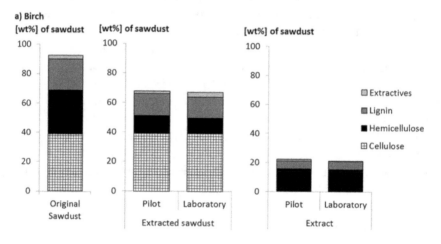

Figure 5.4: Data on pilot-scale and laboratory-scale hot-water extractions of birch sawdust [71].

Applications in chemical pulping

In integrated chemical pulping concepts, the acidic pretreatment processes have a profound effect on the chemical composition of wood feedstock and thus, on the subsequent delignification behavior of the pretreated wood chips [12, 17, 72]. This kind of acidic pretreatment stage has been applied under varying conditions to softwoods [51, 73], hardwoods [52, 74], and non-woods [57, 75], mainly prior to kraft pulping. Autohydrolysis is of special interest because, compared to pre-hydrolysis with dilute mineral acids, H_2O is the only reagent, making it an environmentally friendly and inexpensive process. However, the acidic pretreatment has been also studied prior to alkaline pulping using autohydrolysis with the addition of dilute acid [76–78]. Additionally, steam-explosion pretreatment of non-wood materials prior to alkaline pulping has been investigated [79–81].

Table 5.1 presents an example of the chemical composition of hardwood hydrolyzates and softwood hydrolyates. The acidic conditions in autohydrolysis are created through the cleavage of acetyl groups from hemicelluloses (i.e., the release of acetic acid) and the formation of some formic acid (HCO_2H) from carbohydrate-degradation products. Hence, the hydrolysates with pH 3–4 from conventional autohydrolysis assays contain a mixture of various carbohydrates together with a minor concentration of other organics (a few inorganics are also present), including the above mentioned aliphatic carboxylic acids as well as furans (5-(hydroxymethyl)furfural (HMF) and furfural) and some heterogeneous fractions of lignin-derived and extractives-derived materials.

Table 5.1: Chemical composition of hydrolyzates[a] from birch and pine acidic pretreatments (% of the initial wood dry matter) [17].

Component	Birch	Pine
Carbohydrates	13.1	10.2
Monosaccharides	1.4	1.8
Oligo- and polysaccharides	11.3	7.8
Uronic acids	0.4	0.6
Furans	0.1	0.1
Volatile acids	1.6	0.5
Acetic acid	1.5	0.4
Formic acid	0.1	0.1
Lignin and others	2.1	1.1
Total	16.9	11.9

[a]The chips are treated at 150 °C for 90 min, liquor-to-wood ratio 5 L/kg.

In principle, pentosans (*e.g.*, mainly present in hardwood autohydrolysates) and hexosans (*e.g.*, mainly present in softwood autohydrolysates) can be converted, after degradative hydrolysis, to monosaccharides (*i.e.*, pentose and hexose sugars, respectively), which then can be converted into many useful products, including alcohols (*e.g.*, ethanol and butanol) and carboxylic acids, among several other products [12, 41]. Some of these primary products are important building-block chemicals ("platform chemicals") that can be, due to the multiple functional groups present in most cases, subsequently converted further into a wide range of useful derivatives or used for other materials. It has been asserted [82, 83] that some noncarbohydrate materials in hydrolysates may be harmful, especially when considering the biochemical utilization of hydrolysate-based carbohydrates. Hence, besides detailed characterization of these hydrolysates, their fractionation and purification have been carried out [8, 84–90].

The effect of acidic pretreatment on delignification processes has been studied from several points of view; investigations typically deal with pulping performance [74, 91–96], fiber properties [91, 92, 97–101], and bleaching [96, 102]. For example, autohydrolysis and mild acidic pretreatments have generally been proposed to enhance the delignification rates of both hardwoods and softwoods when compared to those of the untreated feedstocks, resulting in savings in the costs of pulping and bleaching chemicals, reduced cooking times, and lower energy demand [72, 91, 93, 95, 96, 102]. The enhanced delignification of the pretreated wood chips has been explained by the improved penetration of cooking chemicals caused by the increased pore volume and permeability of the fiber cell wall together with the hydrolytic pre-hydrolysis-kraft cleavage of lignin structure (including lignin-carbohydrate complexes) as well as the removal of hemicelluloses and their degradation products. Hence, it has also been noted [96, 97] that the kappa number of the produced pulp can be lowered significantly by applying a pretreatment stage prior to pulping. The most noticeable negative effects caused by autohydrolysis and mild acidic pretreatments on pulping and pulp quality include the reduced yield of pulp, compared to the reference kraft cooks produced without pre-extraction, reduced refining response, and decreased strength properties (i.e., tensile and burst strengths) of the produced pulps [91, 98, 103].

The main aim of the traditional pre-hydrolysis-kraft process is to produce high-purity dissolving pulp suitable for manufacturing cellulose derivatives and regenerated cellulose industrially [12, 45, 77, 104]. This process is a modification of the typical kraft pulping process, with the addition of a prehydrolysis stage to extract hemicellulose fraction from wood chips prior to cooking [105–110]. The purpose is to remove the noncellulosic carbohydrates (i.e., hemicelluloses) as completely as possible and to produce cellulosic pulp, containing a low (3–4%) hemicellulose content. The subsequent purification processes, such as hot and cold caustic extraction, cause considerable increases in production costs due to high yield loss and high chemical charges but they are still needed, because even small amounts of noncellulosic polysaccharides may influence the processability and properties of the final product. It is known that 85–88% of the total dissolving pulp is manufactured from wood by pre-hydrolysis-kraft process and acid sulfite processes combined with subsequent purification stages [111–114]. However, besides to the production of high-grade dissolving pulp, the pre-hydrolysis-kraft process allows the recovery of various organic components, such as CH_3CO_2H and carbohydrates, from the pre-hydrolysis liquors [109, 110].

5.2.3 Alkaline pretreatments

Carbohydrate losses are high in the beginning of the alkaline cook, when delignification is still proceeding slowly [17, 43, 44, 115]. Besides deacetylation (as well as minor dissolution of hemicelluloses), the most important reactions responsible for the loss of carbohydrates and the reduction of their degree of polymerization (DP) can generally be divided into two categories: i) the peeling reaction, by which end group units of the chain are successively removed and ii) the cleavage of glycosidic bonds at any point along the polysaccharide chain (for details, see Chapter 4 "Chemical pulping-based methods"). Additionally, a stabilization reaction (i.e., the stopping reaction) of polysaccharides competes with peeling reaction and prevents further cleavage of the chains arising from the peeling process; the ratio between the peeling reaction and the stopping reaction rates is in the rough magnitude of 50–90:1 [116, 117]. All these degradation reactions lead to the formation of low-molar-mass aliphatic carboxylic acids, besides so-called "volatile acids" (i.e., HCO_2H and CH_3CO_2H), hydroxy monocarboxylic acids with a minor amount of hydroxy dicarboxylic acids [118]. Of the lignocellulosic biomass carbohydrates, hemicelluloses are clearly more susceptible toward alkaline degradation when compared to cellulose, which is due to the highly crystalline structure and high DP of cellulose. In practice, the rather low selectivity of softwood kraft pulping means that at the end of the heating-up period (somewhat after the lignin extraction phase [119]), the mass ratio of the aliphatic acid fraction to the lignin fraction is 1.1–1.2, whereas the corresponding ratio is 0.8–0.9 in the final black liquor [120].

Under alkaline conditions, especially various alkyl aryl linkages of lignin are readily cleaved, facilitating enhanced dissolution of lignin [43, 44]. The degradation reactions can generally be classified into three main categories: *i*) cleavage of α- and the β-aryl ether linkages in free phenolic structures, *ii*) cleavage of β-aryl ether linkages in nonphenolic structures, and *iii*) cleavage of methyl aryl ether bonds (demethylation) (for details, see Chapter 4 "Chemical pulping-based methods"). As the β-*O*-4 and α-*O*-4 ether links represent the most abundant inter-unitary linkages in native wood lignin (even up to 65%), it is not surprising that their cleavage is critical for efficient fractionation of lignocellulosic biomass materials. Besides the cleavage of different ether structures, also some other minor lignin reactions, such as the cleavage of C-C bonds as well as condensation and chromophore formation, may also take place depending on the conditions applied. All these reactions are responsible for altering the DP of lignin. The average molar mass of the kraft lignin dissolved during the initial phase (about 1,900 Da and 2,200 Da for softwood and birch lignin, respectively) is lower than that observed in the final black liquor (about 3,100 Da and 2,600 Da for softwood and birch lignin, respectively) [121]. Additionally, it is generally known that hydrogen sulfide ions ($HS^\ominus$) react primarily with lignin, whereas carbohydrate reactions are only affected by alkalinity (*i.e.*, by $HO^\ominus$) [43].

The general aim of alkaline pretreatment of lignocellulosic biomass materials is to remove lignin and facilitate the subsequent treatment process, which can be either chemical pulping or enzymatic hydrolysis [122–131]. Due to the material removal from the feedstock, various morphological phenomena, such as changes in specific surface area and porosity take place, thus improving the accessibility of the remaining feedstock. In case of alkaline pulping, the alkaline pretreatment can be considered the first phase of the total delignification process. Hence, it can be also concluded that in carrying out this early stage of delignification under alkaline conditions, a pronounced relative amount of aliphatic carboxylic acids in relation to lignin can be obtained [12]. However, one clear advantage is that the fractions of sulfur-free lignin as well as extractives with relatively high purities can be recovered from the hydrolysates formed. This kind of integrated biomass concept is also in accordance with the limiting factors listed in Section 5.2.1 "General approach".

Alkaline pretreatment of wood chips prior to kraft pulping represents one of the most attractive integrated biorefinery processes combined with chemical pulping [132–136]. This process can, for example, utilize the equipment and well-developed systems for chemical recovery, simultaneously lowering the capital cost of pretreatment and gaining more benefits for pulp mills. In a modern kraft process, when surplus energy generated from the burning of black liquor is available, partial recovery of dissolved organic solids becomes more attractive. Alkaline pretreatments can typically shorten the cooking times and lower the need for alkali charge of the subsequent cooking process. Green liquor (mainly $Na_2CO_3 + Na_2S$ in H_2O) [133, 137–141], and white liquor (mainly NaOH + Na_2S in H_2O) [142, 143] can be used for pretreatments as such, but also NaOH in H_2O has been used for various wood and non-wood feedstocks prior to kraft or soda-anthraquinone (AQ) pulping [12, 17, 74, 131, 132, 144–149], Alkaline pretreatments based on green liquor and white liquor utilize proven technologies for the sophisticated recovery of all the used cooking chemicals at a very high efficiency, and they can be rather straightforwardly integrated into existing pulp mill operations [138]. However, alkaline pretreatments with NaOH in H_2O are usually more effective on hardwoods, agricultural wastes, and herbaceous crops when compared to softwoods [131]. Although high NaOH charges may be needed to achieve sufficient lignin removal, especially from softwoods, the unconsumed alkali can be recovered and reused in alkali-based pulp mills with well-developed chemical recovery systems.

In general, alkaline pretreatments are carried out under milder conditions when compared to corresponding pretreatments performed with acidic chemicals; they are often conducted at lower temperatures and even under atmospheric pressures [122]. On the other hand, the treatment times are typically relatively long. When compared to HWE and acidic pretreatment methods, alkaline pretreatments have advantages, as they maintain the yield and quality of the produced pulp in the final delignification [144, 146, 150–152], Table 5.2 represents typical alkaline pretreatments with relatively high alkali charges carried out for silver birch under the same conditions as those shown in Table 5.1 but in this case, with alkali addition. It is obvious that compared to acidic pretreatments, alkaline pretreatments are generally more effective for solubilization of lignin and extractives.

Data in Table 5.2 indicate that in these alkaline hydrolysates, the mass ratio of aliphatic carboxylic acids to lignin is 2.6–3.0. However, besides the different carbohydrates-derived degradation products to lignin ratios, the types of hemicelluloses also have their characteristic effects on the alkaline treatments [130, 131]. It is worth emphasizing that in this case shown in Table 5.2, due to deacetylation of the acetyl groups of xylan [43], about 70% of the acids are volatile acids (mainly CH_3CO_2H). Softwood glucomannan is generally degraded by the alkaline peeling reactions rapidly under alkaline conditions. However, the deacetylated and solubilized oligomeric hardwood-rich xylan is more stable toward alkaline degradation when compared to softwood-rich glucomannan, due to the stabilizing effect of the 4-O-methylglucuronic acid side groups adjacent to the reducing end of the xylan chain [43, 44]. Additionally, one of the important mechanisms of the alkaline hydrolysis has been suggested to be the saponification of the intermolecular ester bonds cross-linking hemicelluloses and lignin [153].

Table 5.2: Chemical composition of soluble organic material during alkaline pretreatment[a] of birch (% of the initial wood dry matter) [17].

Component	6[b]	8[b]
Aliphatic carboxylic acids	7.4	9.1
Volatile acids[c]	5.3	6.0
Hydroxy carboxylic acids[d]	2.1	3.1
Lignin	2.5	3.5
Other organics[e]	2.4	3.5
Total	12.3	16.1

[a]The chips are treated at 150 °C for 90 min, liquor-to-wood ratio 5 L/kg.
[b]Alkali charge (% of NaOH) on oven-dried feedstock.
[c]Formic and acetic acids.
[d]Mainly glycolic, lactic, 2-hydroxybutanoic, 3,4-dideoxy-pentonic, 3-deoxy-pentonic, xyloisosaccharinic, and glucoisosaccharinic acids. For their chemical structures, see Section 5.3.1 "Chemical composition".
[e]Mainly extractives and hemicellulose residues.

As xylan is the dominant hemicellulose component in many commercially utilized hardwoods, alkaline pretreatment has been found to represent a more suitable extraction method for hardwoods than for softwoods. Besides the chemical nature of the hardwood hemicelluloses, more open vascular structure of hardwoods renders them more amenable to chemical pretreatments [131]. However, carbohydrate losses caused by peeling reactions can be avoided to some extent by stabilizing the reducing end groups (*i.e.*, aldehyde groups (-CHO) in form of hemiacetals) of carbohydrate chains by simple oxidation or reduction to carboxyl acid (-CO$_2$H) or alcohol (-CH$_2$OH) groups, respectively [154]. This can be made by utilizing simple reductants or oxidants, such as sodium borohydride (NaBH$_4$) or hydrogen peroxide (H$_2$O$_2$). Lignin fraction produced during the alkaline pretreatment with NaOH in H$_2$O is sulfur-free (unlike lignin produced during conventional kraft cooking) and it can be used for many applications including, for example, fuel additives, bio-based polymers (*i.e.*, resins and adhesives), high-purity carbon fiber, or low-molar-mass phenols [17, 155–158],

In practice, if about 15% of the dissolved feedstock material (Table 5.2) is withdrawn, the approximate amounts of volatile acids, hydroxy carboxylic acids, lignin, and extractives would be, respectively, 72,500, 35,000, 40,000, and 35,000 tons in the case of a kraft mill producing 500,000 oven-dried tons of unbleached pulp annually. The withdrawal of the pretreatment spent liquor may also slightly influence the combustion properties of black liquor in the recovery furnace as well as black liquor viscosity and pulp strength [12, 159]. However, it can be generally concluded that acidic pretreatments would have a more negative effect on pulp strength properties compared to alkaline pretreatments, which simultaneously also enhance the impregnation of cooking alkali in the next cooking stage. Furthermore, depending on the treatment conditions (*i.e.*, milder alkali charge and temperature) the main target of alkaline pretreatment can also be the dissolution of hemicelluloses [25, 133, 140, 143, 147, 160].

5.3 Black liquor

5.3.1 Chemical composition

The spent liquors from chemical pulping contain varying amounts of organic compounds derived from all the wood constituents (cellulose, hemicelluloses, lignin, and extractives) [12, 17, 43, 118]. The nature and amount of these compounds depend on the feedstock material used for the pulp production as well as on the pulping method applied. During kraft pulping, roughly half of the initial feedstock material degrades and dissolves in the cooking liquor ("black liquor") [44, 161–163], Hence, it is obvious that kraft black liquor as such forms the most important by-product of kraft pulping. Most kraft black liquors have relatively similar compositions and many of them (*i.e.*, from the common softwood- and hardwood-based conventional processes) have been traditionally characterized at least in terms of their main compound groups. However, during recent years, black liquors have also been analyzed in more detail for numerous reasons. The investigations have been focused mainly on the following areas: *i*) basic understanding of pulping chemistry, *ii*) on-line or off-line cooking control, *iii*) production of useful by-products, and *iv*) environmental impacts. Additionally, the determination of physical properties of various black liquors for the further processing of these liquors is of great importance.

As shown in Section 5.2.3 "Alkaline pretreatments", under alkaline conditions, a substantial amount of hemicelluloses is converted into aliphatic carboxylic acids, which together with lignin degradation fragments and extractives are removed from the initial chips and dissolved into the black liquor [12, 17, 43, 44, 161–164], After the recovery of most extractives-based compounds, the remaining black liquor mainly contains, besides inorganic substances, carbohydrate degradation products (aliphatic carboxylic acids and hemicellulose residues), lignin, and residual extractives (Table 5.3). Several different peeling reactions of carbohydrates produce, besides "volatile" HCO_2H and CH_3CO_2H, "nonvolatile" hydroxy monocarboxylic acids (*e.g.*, glycolic, lactic, 2-hydroxybutanoic, 3,4-dideoxy-pentonic, 3-deoxy-pentonic, xyloisosaccharinic, and glucoisosaccharinic acids) and a minor amount of hydroxy dicarboxylic acids (Figure 5.5). These acids are present in black liquor as sodium carboxylates.

The most common aliphatic carboxylic acids shown in Table 5.3 and Figure 5.5 are typically those determined in black liquors. For a specific need, other low-molar-mass compounds have also been analyzed [118, 165]. These compounds include *i*) uncommon carboxylic acids (belonging to volatile fatty acids, hydroxy monocarboxylic acids, dimeric hydroxy monocarboxylic acids, dicarboxylic acids, hydroxy dicarboxylic acids, and hydroxy tricarboxylic acids) and miscellaneous compounds (*e.g.*, cate-

Table 5.3: Typical composition of dry matter of Scots pine (*Pinus sylvestris*) and silver birch (*Betula pendula*) kraft black liquors (% of the total dry matter) [43].

Component	Pine	Birch
Lignin[a]	31	25
HMM (>500 Da) fraction	28	22
LMM (<500 Da) fraction	3	3
Aliphatic carboxylic acids	29	33
Formic acid	6	4
Acetic acid	4	8
Glycolic acid	2	2
Lactic acid	3	3
2-Hydroxybutanoic acid	1	5
3,4-Dideoxy-pentonic acid	2	1
3-Deoxy-pentonic acid	1	1
Xyloisosaccharinic acid	1	3
Glucoisosaccharinic acid	6	3
Others	3	3
Other organics	7	9
Extractives	4	3
Carbohydrates[b]	2	5
Miscellaneous compounds	1	1
Inorganics[c]	33	33
Sodium bound to organics	11	11
Inorganic compounds	22	22

[a]HMM and LMM refer to high- and low-molar-mass, respectively.
[b]Mainly hemicelluloses-derived fragments.
[c]Depending on the amount of "dead-load inorganics", this mass proportion may be higher.

chols, cyclopentenones, hydroxycyclopentenones, and thiophenes), all originated from carbohydrates, *ii*) a great amount of aromatic compounds from lignin, and *iii*) various extractives-based compounds.

The quantification of lignin's molar mass poses challenges and varying results can be obtained depending on the type of lignin and the methods employed [164]. As a general trend, it is known that the milled wood lignin (MWL) preparations have weight average molar mass (M_w) of between 15,000 Da and 20,000 Da, which is four to five times that of kraft lignin [166]. On the other hand, soluble hardwood kraft lignin has typically lower M_w than softwood kraft lignin [167]. Table 2.10 in Section 2.3.3 "Lignin" shows examples of M_w values; 1,900 Da to 4,100 Da and 2,200–3,400 Da for softwood and birch kraft lignin, respectively [121]. The reported maximum M_w of soluble lignin from the sulfur-free soda-AQ cook of birch is 3,300–4,400 Da [168].

The phenolic groups (Ø-OH) of lignin have pK_a values 9.5–10.5 (the pH of black liquor is about 13) and the pK_a values of aliphatic carboxylic acids (R-CO$_2$H) are 3–5 [17, 43]. Hence, these compound groups exist in the form of sodium salts in black liquor.

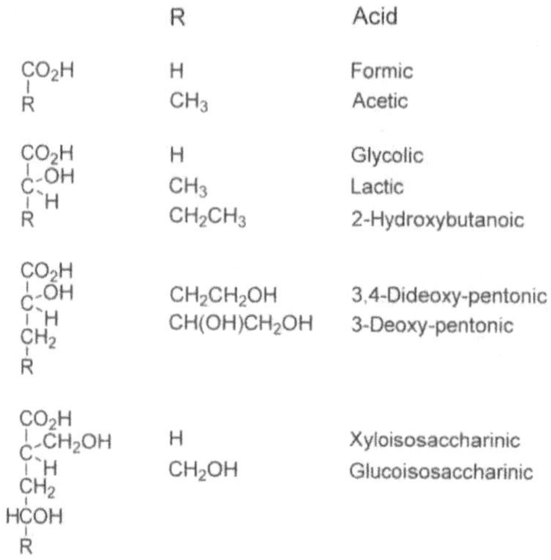

	R	Acid
CO₂H | R	H CH₃	Formic Acetic
CO₂H |—OH C ⁓ | ⁓H R	H CH₃ CH₂CH₃	Glycolic Lactic 2-Hydroxybutanoic
CO₂H |—OH C ⁓ | ⁓H CH₂ | R	CH₂CH₂OH CH(OH)CH₂OH	3.4-Dideoxy-pentonic 3-Deoxy-pentonic
CO₂H |—CH₂OH C ⁓ | ⁓H CH₂ | HCOH | R	H CH₂OH	Xyloisosaccharinic Glucoisosaccharinic

Figure 5.5: Abundant aliphatic carboxylic acids in black liquors (see Table 5.3).

Ordinary phenolic groups can also be expected to play a major role with respect to the properties of lignin in black liquor [164]. Additionally, with respect to many utilization applications of kraft lignin, it should be noted that kraft lignin contains some chemically bound sulfur (2–3%).

Besides the major organic components and sodium (as well as sulfur) chemically bound to them, a significant amount of other inorganic components is present in kraft black liquors [161]. Detailed quantitative data on the composition of the inorganic fraction are of importance in many respects, particularly, when predicting the combustion behavior of black liquor in the recovery furnace to recover the cooking chemicals [118]. In general, the substances in black liquor derive from two sources: *i*) wood feedstock and *ii*) white (cooking) liquor charged to the digester.

Besides the active cooking chemicals NaOH and Na₂S, the white liquor also contains appreciable amounts of other inorganic compounds and all these components (Table 5.4) are referred to collectively as "dead load" [161, 169]. These ineffective components include Na_2CO_3, sodium sulfate (Na_2SO_4), and sodium sulfite (Na_2SO_3), and sodium thiosulfate ($Na_2S_2O_3$). Of them, the most prominent ones, Na_2CO_3 and Na_2SO_4, mainly result from incomplete causticizing and incomplete reduction in the recovery furnace, respectively. Although the dead-load chemicals do not take part in the pulping reactions, their high concentration levels in the white liquor are undesirable. They can cause scaling in the digester and especially, in the evaporators, as well as increase the load of the recovery furnace generally.

The typical composition of inorganic fraction in kraft black liquor is shown in Table 5.5. However, this inorganic fraction is somewhat nebulous since some of the sodium is associated with inorganic anions and some of it with organic material (about 65%). There are also "others" (total amount about 10%), mainly including potassium

Table 5.4: Typical composition of white liquor [161].

Compound	Concentration range, g/L (as Na_2O)	Average amount, % of total
NaOH	81–120	53
Na_2S	30–40	21
Na_2CO_3	11–44	15
Na_2SO_4	4.4–18	5
Na_2SO_3	2.0–6.9	3
$Na_2S_2O_3$	4.0–8.9	3

salts and chlorides [118]. Additionally, small amounts of Si, Ca, Fe, Al, Mg, Mn, and P are also present, and in all cases, their comprehensive analysis is complicated and not suitable for routine purposes. Table 5.6 shows an example of the elemental composition of kraft black liquor. On the elemental basis, black liquors contain mainly carbon, oxygen, sodium, sulfur, and hydrogen with minor amounts of other elements [170, 171].

Table 5.5: Typical composition of inorganics in kraft black liquor [161].

Compound	Concentration range, g/L (as Na_2O)	Average amount, % of total
NaOH	1.0–4.5	7
Na_2S	1.6–5.6	19
Na_2CO_3	5.0–8.2	36
Na_2SO_4	0.5–6.0	13
Na_2SO_3	0.4–3.8	9
$Na_2S_2O_3$	1.8–5.1	16

The aqueous liquor that results when the inorganic smelt from the recovery furnace is dissolved in H_2O is called "green liquor" (Table 5.7). This term comes from the color given by small amounts of darkish iron sulfide (FeS) present in the solution. Green liquor contains a high level of Na_2CO_3 and a low level of NaOH. Hence, it must be converted into white liquor (Table 5.4) by a causticizing process before its use for the next cook [161].

The commonly used methods for analyzing white liquor, black liquor, and green liquor inorganics are mainly based on conventional wet chemistry (*i.e.*, standard methods), although a gradual transition from the old gravimetric and titrimetric procedures to the more sophisticated techniques, such as various chromatographic and different spectrometric determinations, is occurring [118]. In case of organic constituents of black liquor, for example, the aliphatic carboxylic acids can be determined by various chromatographic methods [172–174] and lignin by ultraviolet (UV) spectrophotometry [175–177].

The black liquor is separated from the pulp after cooking during washing as an aqueous solution with a dry solids content of 14–18%, depending on the raw material

Table 5.6: Elemental composition of typical kraft black liquor [170].

Element	Content, % of the dry solids
Carbon	34–39
Oxygen	33–38
Hydrogen	3–5
Sodium	17–25
Sulfur	3–7
Potassium	0.1–2
Chlorine	0.2–2
Nitrogen	0.04–0.2
Others	0.1–0.3

Table 5.7: Typical composition of green liquor [161].

Compound	Concentration range, g/L (as Na_2O)	Average amount, % of total
NaOH	10–18	8
Na_2S	35–40	20
Na_2CO_3	78–135	60
Na_2SO_4	7.4–24	6
Na_2SO_3	4.2–7.6	3
$Na_2S_2O_3$	4.3–6.5	3

and the efficiency of the washing plant [178]. However, this weak black liquor contains too much H_2O to be used direct as fuel in the recovery boiler. The main purpose of the evaporation plant is to increase the dry solids content of the black liquor by evaporating H_2O until the liquor reaches a concentration that allows it to be burnt in the recovery boiler; the dry solids concentration is normally 70–80% (minimum 60–65%). Many modern installations operate at a level above 80% dry solids (see Section 4.2.1 "Process description").

5.3.2 Physical properties

The physical properties of black liquors that are of general interest are those relating to their processing characteristics [179–181]. These properties primarily include density (specific gravity), viscosity, surface tension, boiling point rise (BPR), heating value, and thermal conductivity. They all depend on the temperature and the dry solids content of black liquor. Hence, several physical properties of black liquor must be considered, for example, in designing the recovery of cooking chemicals for optimum performance.

The density of black liquor is important, when calculating flow characteristics and static heads [170, 180]. At very low solids content, the density is close to that of

H$_2$O at the same temperature and the deviation in the specific gravity *vs.* concentration isotherms from a linear relationship becomes more and more pronounced as the liquor concentration increases. Compared to the organic material of black liquor, the inorganic components have a stronger effect on the density and hence, higher the inorganic content, steeper is the initial slope of these isotherms. The dependence of density on temperature may be taken to be linear in the range from 16 °C to the boiling point. The temperature effect increases slightly as concentration increases. However, the temperature effects are minor, compared to the effect of changes in the dry solids content. In general, the density characteristics of different black liquors are rather similar [180]. For example, a black liquor with a concentration of 16% dry solids has a density of 1,050 kg/m^3 and at 70% dry solids it has a density of 1,430 kg/m^3 [178]. Both values are based on a reference temperature of 90 °C.

The viscosity of black liquor is a function of concentration and temperature, specific for each liquor but there are other affecting factors as well [170, 178, 182–190], It depends, for example, on the wood species and the cooking conditions and may vary considerably from mill to mill. Viscosity has also a great influence on liquor spraying properties in the recovery boiler. It is normally defined as the ratio of the shear stress to the shear rate for a fluid subjected to a shearing force [170]. Black liquor is rheologically complex [171]. It is reasonably Newtonian (*i.e.*, viscosity is independent of shear rate and depends only on temperature and solids content) at solids level below about 50%, but becomes non-Newtonian (*i.e.*, can exhibit thixotropic, pseudoplastic, or viscoelastic behavior – or normally, shows shear-thinning behavior) at higher solids content.

When black liquor at a dry solids content of about 15% enters the evaporator set, it is quite thin with a viscosity two or three times that of H$_2$O at the same temperature [170]. However, viscosity increases fast with an increase in dry solids. In some cases, the increase is very steep after a certain point [178]. An increase in temperature generally lowers the viscosity. The practical limit for handling the liquor is the pumping limit of 300–500 cP (1 cP = 1 mPa·s). The viscosity should always be below this level and is usually much lower in the evaporation plant. If the liquor is stored at atmospheric pressure, the limit for the final concentration is 70–75% dry solids at the maximum temperature of 115 °C. If the final product liquor is stored in a pressurized storage tank, the final concentration can be 75–85% dry solids and even higher at a storage temperature of 125–150 °C. These extremely high concentrations may require using medium-pressure steam in the evaporation plant.

The liquor heat treatment (LHT) process and the oxidative heat treatment process can be used to reduce the viscosity of black liquor [178, 191–197], They basically treat the liquor at elevated temperatures for an extended time. During this kind of treatment high-molar-mass hemicellulose residues and lignin are split and viscosity is permanently reduced. In the LHT process, heating temperature is typically 175–180 °C with a residence time of 30 min or 190–195 °C with a residence time of 15 min. Typically, the heat-treated liquor can be fired at about 5% higher dry solids concentration than that of the untreated liquor at the same temperature.

There are several different options for designing the LHT process and its integration into the evaporation process [178]. Figure 5.6 shows the LHT process when performed in a separate reactor. The liquor feed to the reactor is preheated with direct flash vapor from the flash cooler. Steam consumption in the LHT reactor can be reduced by increasing the number of preheater/flash cooler units connected in series. Heat treatment also releases methanol (CH_3OH) and a considerable amount of organic sulfur compounds in gaseous form, corresponding to 2–5 kg organic sulfur per ton of the produced pulp. These gases are vented to the foul condensate stripper or to the evaporator and further from the vacuum system to the concentrated non-condensable gases (CNCG) collection system.

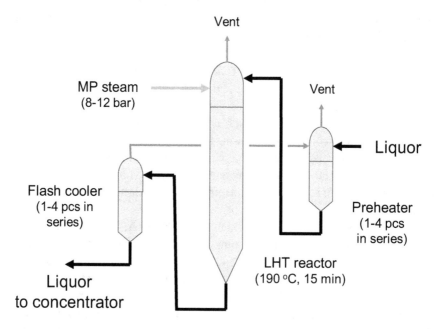

Figure 5.6: Liquor heat treatment (LHT) reactor. Modified from [178]. MP refers to medium-pressure.

The surface tension of a fluid is generally associated with the free energy of its surface [170]. The surface tension of solutions depends on temperature and the nature of the dissolved compounds. For example, the surface tension of H_2O is usually high: at 20 °C $72.8 \cdot 10^{-3}$ N/m and decreases linearly with temperature to a value of $58.9 \cdot 10^{-3}$ N/m at 100 °C. Normally, inorganic compounds slightly increase the surface tension when dissolved in H_2O. In contrast, some organic material, called "surface active agents", decrease the surface tension of H_2O. Even small amounts of these materials can decrease the surface tension by a factor of two or three. However, this effect is nonlinear and larger quantities of such organic materials in H_2O have only a very mild additional effect on surface tension.

The practical implication of surface tension in an evaporation plant is that a low surface tension will increase the tendency to foam [178]. Surface tension increases with an increase in dry solids concentration and decreases with an increase in temperature. Foaming is a problem primarily in the evaporator elements operating at low concentration. To overcome this phenomenon, the feed liquor concentration is often increased by recirculating a portion of the concentrated liquor (i.e., "sweetening" the feed liquor). However, crude tall oil and soap will decrease the surface tension. Therefore, they need to be separated from the liquor, if their contents are high, especially, when using softwood as raw material for pulping.

A liquid mixture that contains dissolved organic substances, inorganic substances, or both will boil at a higher temperature than H_2O at the same pressure [178]. This property is characteristically expressed as the BPR of black liquor, and it is specific for each type of black liquor and depends on the amount and composition of dissolved substances. The BPR increases with an increase in dry solids. For example, a liquor may have a BPR of less than 2 °C at the feed concentration of 20% dry solids, but 18 °C at the final product liquor concentration of 70% dry solids. BPR is normally measured at 1 atm pressure and then assumed to apply at all pressure levels encounter in an evaporator [180].

The net heating value (i.e., net calorific value) (NHV) accounts for the fact that H_2O in the combustion products is not condensed during combustion and steam generation [170, 198]. To obtain the NHV, the heat of vaporization of the H_2O is subtracted from the higher heating value (i.e., gross calorific value) (HHV), which is a measure of the amount of heat released. Black liquor releases heat when it burns in the recovery furnace. Typical values for the HHV of kraft black liquor lie between 13,400 kJ/kg of black liquor solids and 15,500 kJ/kg of black liquor solids. Both the organic components and the reduced sulfur (i.e., Na_2S and $Na_2S_2O_3$) contribute to the HHV. Other inorganic components act as diluents, lowering the heating value of black liquor. However, the NHV calculated from the measured HHV more nearly reflects the actual energy release. Table 5.8 illustrates the heating values of the main constituents of black liquor.

Table 5.8: Heating values of black liquor components (kJ/kg) [170].

Component	Heating value
Softwood lignin	26,900
Hardwood lignin	25,110
Carbohydrates-derived material	13,555
Extractives	37,710
Sodiun sulfide (Na_2S)	12,900
Sodium thiosulfate ($Na_2S_2O_3$)	5,790

Thermal conductivity is the property of any material that characterizes its ability to transfer heat [170, 188]. It is used in calculations of heat transfer rates during black liquor evaporation and in assessing the heat transfer within droplets as they dry and de-volatilize under high temperatures. Enthalpy data on black liquor are used for estimating the preheat requirements during evaporation [170], for estimating the heating times for droplet drying and pyrolysis during combustion, and for accurate energy balance around the recovery boiler. It can also be used with normal boiling point data to estimate the BPR of concentrated black liquor. The heat capacities of typical black liquor constituents are about half that for H_2O. As the dry solids content increases from zero, the heat capacity falls linearly from the value for pure H_2O to a value about half that of H_2O.

An essential part of sodium is chemically bound to the dissolved organic compounds (Table 5.3). The other inorganic material in kraft black liquors consists mainly of a variety of sodium salts with smaller amounts of potassium salts and minor quantities of other compounds (Table 5.5) [170]. At a concentration of below 45–60% dry solids content, the sodium salts are completely dissolved. When the total solids content is increased above this limit (*i.e.*, "solubility limit" or the critical concentration of the liquor) by evaporation of H_2O, these salts begin to precipitate. Hence, during concentration of the black liquor, a solubility limit point occurs, where Na_2CO_3, Na_2SO_4, and/or their double salt $Na_2SO_4{\cdot}2Na_2CO_3$ (dicarbonate) or $2Na_2SO_4{\cdot}Na_2CO_3$ (burkeite) begin to precipitate [178].

The crystallized salts may scale the heating surfaces very rapidly but are normally easily washable [178]. Normally, weak black liquor contains effective alkali (*i.e.*, NaOH + 0.5Na_2S) (see Chapter 4 "Chemical pulping-based methods") about 90 g NaOH/L giving the black liquor a pH value of 12–13. A clearly higher effective alkali than this value is not recommended, because the final concentrator may be exposed to rapid corrosion. If this residual alkali decreases below 10–20 g NaOH/L and the pH is below 11, also lignin compounds start to precipitate and foul the evaporator heating surfaces.

Nearly all the other inorganic species in kraft black liquor are insoluble as sulfates, carbonates, or hydroxides [170]. They are carried through the liquor recovery cycle as suspended solids and are mainly removed after the recovery furnace as green liquor dregs. It has been noticed that calcium ions form complexes with the dissolved organic matter in black liquor and the total dissolved concentration of calcium in kraft black liquor is higher than would be expected based on inorganic component solubility data. Calcium that reaches the recovery furnace precipitates as $CaCO_3$ in the dissolving tank and can lead to scaling problems both in this tank and in green liquor lines. Furthermore, aluminum, and silicon are problematic, since their solubilities in green liquor and white liquor increase with increasing pH. However, they can be removed quite effectively as green liquor dregs.

5.3.3 Fractionation

The possible utilization of organic solids in black liquor for nonfuel purposes is relatively complex and depends on several factors, including, for example, potential by-product markets, energy prices, and their fluctuations as well as a general inclination toward a more sustainable use of resources [12, 17, 164, 199–202], The present by-products are almost totally based on extractives, but the partial recovery of lignin and aliphatic carboxylic acids from black liquor appears to be more attractive as well. Especially in a modern kraft pulping process, when surplus energy generated from the burning of black liquor is available, the recovery of these dissolved materials becomes more realistic. Furthermore, it should be noted that black liquors typically contain a significant amount of aliphatic carboxylic acids, which have a relatively low heating value and represent a potentially interesting group of chemical compounds. Additionally, by applying the fractionation process to only a proportion of the total dissolved organic material normally formed during delignification, it is possible to increase the recovery capacity of an existing kraft mill. This fact is also of great importance to the kraft pulping industry because the recovery boiler is often a bottleneck in the mill.

As indicated in Section 5.3 "Black liquor", kraft black liquor is the most important by-product of kraft pulping, and the combustion of this heterogeneous biofuel is a significant source of energy [12, 17, 180]. Additionally, the recovery boilers producing bioenergy have developed significantly in the past 80 years, culminating with units that are among the largest biofuel-fired boilers in the world. In general, black liquor has several characteristic features, which make its combustion different from that of other fuels. For example, the high content of inorganic material (25–40%) and H_2O (15–35%) decreases the heating value (i.e., 12–15 MJ/kg dry solids: lignin 22–27 MJ/kg, aliphatic carboxylic acids 5–18 MJ/kg, and extractives about 35 MJ/kg) of the industrial black liquors to a clearly lower level than that of other common industrial fuels. However, due to the continuous development of the kraft pulping process, the efficiency has improved to the extent that the energy generated by the recovery boiler exceeds the consumption of the pulping process [178, 203, 204]. Therefore, most modern pulp mills use a condensing turbine to produce power from the excess steam. Another way of making use of the excess energy is to partially recover lignin, which is mostly responsible for the heating value of kraft black liquor and has a high energy density and low ash content.

Lignin
There are three main processes by which kraft lignin can be separated from black liquor [205]: i) acidification, ii) ultrafiltration, and iii) electrolysis. In kraft pulping the lignin is dissolved as various sodium salts due to the highly alkaline environment (pH about 13). There are two types of ionic groups in the lignin [17, 43, 164, 206]. The major type is the phenol group (Ø-OH), which has an acid dissociation constant (pK_a) of 9.5–10.5. The other type of ionic groups, the carboxylic acid (-CO_2H) group, releases its sodium at pH 3–5. The accurate pH values for the different functional groups vary with lignin type as well as the precipitation conditions, such as temperature and concentration. However, these figures indicate that all functional groups of lignin in the black liquor are in the ionized state. In general, the stability of colloidal kraft lignin is attributable to ionized hydrophilic groups, which impart an electrical charge to the colloidal particles. It is also

known that the colloidal particles lose their stability if the pH value of the black liquor is reduced to such an extent that a significant proportion of the ionized groups becomes transformed into unionized groups (i.e., Ø–ONa → Ø–OH).

It can be concluded that on acidification of black liquor, the sodium salts of lignin yield the "free lignin", which is insoluble in H_2O and hence, is precipitated. In contrast, on acidification the degradation products of carbohydrates, aliphatic carboxylic acids, in black liquor are H_2O soluble and they remain in solution, thus making possible to separate lignin-derived and carbohydrates-derived fractions [17, 207]. It is also obvious that acidification of black liquor with weak acids (i.e., typically with CO_2, pH to about 8) already results in a significant precipitation of lignin (about 80% of the initial lignin), whereas to achieve a more complete precipitation (about 90% of the initial lignin), it is necessary to reduce the pH value of black liquor below 3 (i.e., typically with H_2SO_4) [164, 208–210], However, this is economically not attractive, since sufficient acid must be added, not only to decompose the lignin salts, but also to neutralize the sodium salts of the aliphatic carboxylic acids. Hence, to get a sodium-free lignin it is necessary to re-duce the pH below that value and that is also the reason why the precipitated lignin at pH 8–10 is washed with H_2SO_4.

Lignin begins to separate in the form of a fine precipitate when the pH of the black liquor is reduced to about 11 by acidification [12, 17]. The presence of lignin-carbohydrate complexes (LCCs) may also play an important role with respect to the solubility proper-ties of kraft lignin since they can persist during cooking, and their hemicellulose compo-nents contain some H_2O-solubility-increasing -CO_2H groups [211–215]. Additionally, it is essential that the precipitated lignin can be readily separated from the acidified black liquor, and this is normally not possible by filtration in the ordinary manner. However, the filtration can be facilitated by heating the aqueous slurry of the precipitated lignin in a colloidal form to a temperature in the range of 80–100 °C (i.e., by coagulation to in-crease the average size of lignin particles) followed by cooling to about 60 °C or below prior to filtration [207, 208, 216, 217]. At pH of about 10, Na_2CO_3 in the black liquor occurs in equal proportions as sodium bicarbonate or sodium hydrogen carbonate ($NaHCO_3$) and Na_2CO_3 (pK_2 about 9.9), whereas about 99.9% of Na_2S is present as sodium hydrosul-fide (NaHS) and only 0.1% as hydrogen sulfide (H_2S) (pK_1 about 6.8) [218].

The precipitation of lignin applied to various alkaline spent cooking liquors to re-duce pH and substitute sodium ions with hydrogen ions has a long history. The general approach has been to use CO_2 from flue gases and from causticization with possible further addition of strong mineral acid (mainly H_2SO_4) for this purpose. The main topics have roughly dealt with general acidification of black liquors [164, 208, 219–229], coagu-lation and filtration of the precipitated lignin [208, 217, 230–233], and preliminary pro-cess considerations [208–210, 234–240]. Additionally, acidification has been applied specifically to eucalyptus kraft black liquors [207, 216, 241] and to other alkaline spent liquors, such as those from soda-AQ pulping [242–245]. Moreover, there have been vari-ous process concepts, for example, concerning non-wood soda pulping in which partial lignin precipitation and sodium recovery were accomplished with carbonation together with electrodialysis (*e.g.*, $Na_2CO_3 \rightarrow NaOH + CO_2$ and $R\text{-}CO_2Na \rightarrow RCO_2H + NaOH$) [246, 247].

There are a few companies traditionally producing kraft lignin by various separation processes but the most recent ones that have established their commercial full-scale applications are based on the LignoBoost^{TM} and the LignoForce System^{TM} technologies [204]. They both are quite similar and based on precipitation of lignin in black liquor with CO_2 and washing of the lignin with a weak H_2SO_4 solution.

In the LignoBoost^{TM} process [215, 229, 237, 238, 248–250], the kraft black liquor is first evaporated to a solids content of 30–45% and the tall soap skimmings are recovered (Figure 5.7). Lignin is then precipitated by CO_2 from black liquor and CO_2 injection is continued down to pH of about 10 where 60–70% of the original lignin is precipitated. The precipitated lignin is separated from the liquor in a press filter that produces a high-dry-solids cake of around 65% dry solids. The choice of liquor concentration in the precipitation stage is selected in a way that the precipitated amount of Na_2CO_3 and $NaHCO_3$ would not become too high and cause a high sodium carryover to the next stage of the process. The process also includes an effective purification phase, *i.e.*, the filter cake of precipitated lignin is redispersed and acidified (pH to about 3), and the resulting slurry is then filtered and washed by displacement washing with a weak H_2SO_4 solution. In the large-scale production, the lignin is finally dried in a ring drier. The typical elemental composition of the LignoBoost^{TM} lignin is shown in Table 5.9. The H_2S liberated is recovered by absorption in white liquor [251]. This kind of approach does not interfere much with the recovery of cooking chemicals. The consumption of CO_2 is 200–300 kg CO_2 per ton of isolated lignin and the consumption of H_2SO_4 is in the same range [204]. The first commercial LignoBoost^{TM} lignin separation plant with a capacity of 25,000 tons per year was implemented in 2013 in Domtar's Plymouth mill, NC, USA [204]. Stora Enso built the second commercial plant in 2015 in their Sunila mill in Finland with a capacity of 50,000 tons per year.

The process scheme of the LignoForce System^{TM} process is shown in Figure 5.8. In this case, in comparison to the LignoBoost^{TM} process, a filtration step and an oxidation step of the concentrated black liquor with a solids content of about 40% are introduced prior to lignin precipitation with CO_2 [251–253]. The oxidation performed using pure oxygen decreases (*i.e.*, a sulfide content to <0.5 g/L) the harmful totally reduced sulfur (TRS) compounds (*i.e.*, H_2S, methyl mercaptan (CH$_3$SH, MM), dimethyl sulfide (CH$_3$SCH$_3$, DMS), and dimethyl disulfide (CH$_3$SSCH$_3$, DMDS)) prior to further processing. The advantage of the oxidation is also the improved filtration properties of the lignin, compared to those obtained without oxidation. However, lignin and hemicellulose residues are also partly oxidized, which increases their H_2O solubility to some extent. The CO_2 demand for the acidification (*i.e.*, pH to about 9.5) corresponds to 300–400 kg CO_2 per ton of isolated lignin. Like in the LignoBoost^{TM} process, the lignin from the first filtration is acidified but due to the good filtration properties this can be done with the same filter, which is a simplification since both the re-slurry tank and the second filter are omitted [204]. The first commercial LignoForce System^{TM} lignin separation plant

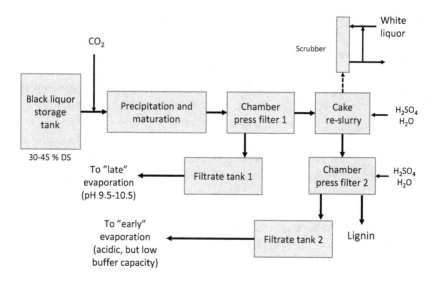

Figure 5.7: Principal scheme of the LignoBoost™ process. Modified from [250]. DS refers to dry solids.

Table 5.9: Elemental composition of LignoBoost™ lignin[a] (% of the dry solids) [250].

Element	Content
Carbon	64–66
Oxygen	26.0–27.5
Hydrogen	5.7–5.8
Nitrogen	0.1
Sodium	0.1–0.3
Sulfur	2.0–2.5

[a]High heating value 25.0–26.5 MJ/kg and lignin content 92–98%.

with an annual production of 10,500 tons was installed in 2016 in West Fraser's Hinton mill in Alberta, Canada [252].

The available kraft lignin flow in the world is 70–80 Mton/year and therefore, large efforts have been made to refine kraft lignin and to find applications for it [33, 164, 254–266], In principle, kraft lignin-type materials can be treated under a wide range of conditions, from thermochemical methods to versatile chemical processes. Hence, a great number of products can also be obtained, for example, solid and liquid fuels [262, 267, 268], carbon fiber and activated carbon [262, 269, 270], resins ("phenol mixtures") [251, 252, 271–273], and many other straightforward products without significant modification, including binders, surface, or dispersing agents, and emulsifiers [249]. Additionally, lignin can be used as a composite material [272, 274] for adsorption

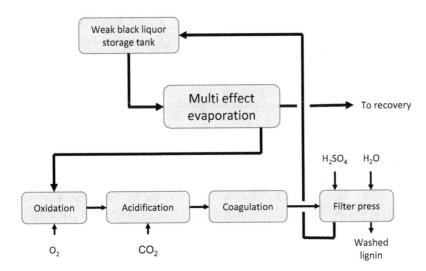

Figure 5.8: Principal scheme of the LignoForce System[TM] process. Modified from [252]. MEE refers to multi-effect evaporation.

of heavy metal ions [275] and for production of chemicals, such as vanillin [261, 262, 276]. However, the most obvious use of the continuous bulk production of lignin, such as the LignoBoost[TM] lignin, at a low cost is still as a biofuel in the form of powder, pellets, or mixed with other fuels. With respect to many fuel applications, it should be noted that kraft lignin contains some chemically bound sulfur (about 2%) and therefore, the combustion plant has to be equipped with a flue gas treatment system for handling the sulfur emissions.

In addition to precipitation, kraft lignin particles can be concentrated by ultrafiltration for different purposes [228, 234, 277–284], although the selection of membrane material seems to be of importance in avoiding problems with fouling and plugging [285–288]. In general, membrane-induced separation of kraft lignin from black liquor is influenced by many of the same factors as the acid precipitation-based separation, but the optimized chemical conditions tend to be very different; membrane separation methods typically work well under conditions favoring solubility, or at least colloidal stability of the lignin in the liquor [164]. However, although membrane-based separation processes can be cost-effective and efficient in many cases, there are two key limitations [164]: *i*) the processes get increasingly difficult to operate as the concentration of retained material increases [289] and *ii*) due to pore plugging and cake formation phenomena, the flux of permeate passing through a membrane tends to fall during continued usage [290–292]. It seems that ultrafiltration appears to be well suited for separation of aqueous solutions containing dissolved polymer matter [293] and hence, it could be used for concentrating black liquor, possibly as an alternative to evaporation [164].

Pore size is typically expressed as cut-off value (1–20 kDa), which is based on the molar mass of common protein molecules [293]. It is one of the key parameters in membrane separations, since there is often a strong relationship between the retained amount and the estimated pore size; *i.e.*, lower cut-off membranes lead to higher retention values [164]. However, ultrafiltration can be used to retain a major fraction of lignin in black liquor, although much of the lignin-related material passes through membranes, even when the molar mass cut-off value is at the low end for the membrane category. Different kinds of membrane materials are used, and the most significant distinction appears to be between polymeric and ceramic membranes [164, 270, 272, 294–297], Typically, membrane separations are performed at high pH values of black liquor and at such pH values, all phenolic groups are in their charged forms. Additionally, aliphatic carboxylic acids in black liquor exist in their dissociated, charged forms, thus contributing to the solubility; this can be expected to facilitate their permeation, as well as molecular size selection, through an ultrafiltration membrane.

Aliphatic carboxylic acids

As indicated in Section 5.3.1 "Chemical composition", besides lignin, large amounts of aliphatic carboxylic acids are formed from carbohydrates in the kraft pulp industry, and their partial recovery integrated in the kraft mill is also an interesting alternative to using them as fuel. The basic idea behind this integrated approach is simply the fact that about two-thirds of the total heat produced by the liquor organics originates from lignin, and only one-third stems from the remaining constituents (*i.e.*, mainly aliphatic acids) [12, 17, 180]. Of the numerous aliphatic acids, the most significant ones are shown in Figure 5.5 and their content in typical softwood and hardwood black liquors is given in Table 5.2.

Also in this approach, like in the case of lignin precipitation, one attraction is that the main product, cellulose fiber suitable for versatile utilization in paper or paperboard, is already a well-established product [160, 298, 299]. Hence, in developing processes for the potential recovery of aliphatic carboxylic acids from alkaline cooking liquors, certain limiting and economic factors should be considered. Examples of such influential factors are as follows [12]:

- Cellulosic fiber is the main product, and its strength properties must be maintained.
- An undistributed separation of extractives should remain.
- The process should not interfere much with the recovery of cooking chemicals; hence, only partial recovery of the aliphatic carboxylic acids (with respect to the normal operation of recovery furnace less than 25% of the total organics, *i.e.*, lignin and aliphatic carboxylic acids) is possible. However, this approach also increases the recovery capacity of an existing kraft mill, especially if the recovery boiler is often a bottleneck in the mill.
- There should be a possibility of producing sulfur-free by-products.
- Straightforward techniques (unit processes) can be applied in fractionation.

– Several alternative conversion methods to modify further the separated fragments into value-added products are readily possible and available.

In practice, if about 15% of the dissolved feedstock material is withdrawn from black liquor, the approximate removed amounts of aliphatic carboxylic acids, lignin, and extractives would be, respectively, 28,000, 25,000, and 22,000 tons in a kraft mill producing annually 800,000 o.d. tons of unbleached pulp.

Several peeling reactions of carbohydrates under alkaline conditions produce, besides HCO_2H and CH_3CO_2H, hydroxy monocarboxylic acids together with a minor amount of hydroxy dicarboxylic acids. These aliphatic carboxylic acids are present in black liquor as sodium carboxylates. The analysis of all the aliphatic acids in black liquors have been made by versatile chromatographic methods [118, 165, 172–174, 300–309], and the formation of these acids during the different stages of alkaline pulping has also been studied in a few investigations [305, 310–320].

The separation of aliphatic carboxylic acids and even individual components as a relatively pure fraction is possible in small volumes by various chromatographic methods, which, however, are practically suitable only for analytical and related purposes. In contrast, the realistic recovery of aliphatic carboxylic acids, either as a mixture or as main individual components, aiming at a full-scale application presents a complicated separation problem technically. Although the principles and main process phases of many possible alternatives have been clarified in detail on laboratory scale, any such processes have so far not been accomplished as a demonstration process. The key questions include the integration of the separation process to the normal pulp mill operations as well as the efficient purification of acids.

The laboratory-scale recovery of a crude fraction of aliphatic carboxylic acids has been reported by several investigations [208, 300, 302, 321–327], As an example, based primarily on the lignin precipitation approach, a simplified scheme shown in Figure 5.9 for the overall recovery of aliphatic carboxylic acids has been recently presented [12, 17, 328]. According to this process, after the recovery of tall soap skimmings during evaporation and lignin (i.e., by carbonation), the mother liquor is evaporated to crystallize out $NaHCO_3/Na_2CO_3$ suitable for causticization. However, it can be expected that only a minor proportion of aliphatic carboxylic acids in the residual liquor (pH about 8) can be utilized as such in the form of their sodium salts. Consequently, in the basic case [323, 327], the aliphatic acids are totally liberated (i.e., pH to about 2.5) from sodium by a strong mineral acid, H_2SO_4, with a simultaneous precipitation of about half of the remaining lignin and making it possible to recover most HCO_2H and CH_3CO_2H by evaporation. The Na_2SO_4 formed is then separately crystallized out almost completely by precipitation of the liquor under suitable conditions, for example, by using aqueous methanol (CH_3OH) [329]. It should be noted that hydroxy acids having hydroxyl groups at C_4 or C_5 (e.g., xyloisosaccharinic and glucoisosaccharinic acids) undergo lactonization under acidic conditions, whereas if CH_3OH is used, low-molar-mass hydroxy acids, such as glycolic, lactic, and 2-hydroxybuatanoic acids form methyl esters and can be readily

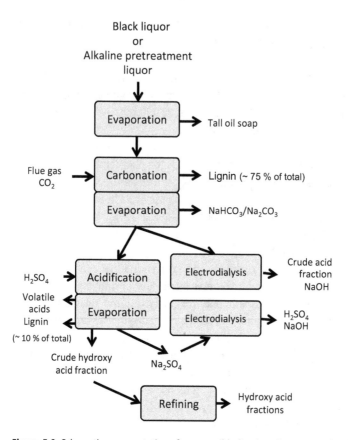

Figure 5.9: Schematic representation of one possible fractionation process for black liquor [17].

separated and fractionated by direct distillation, thus resulting in the formation of a crude residual fraction of higher-molar-mass hydroxy acids and their lactones [328, 330, 331]. Several modifications of this basic process scheme are also possible.

Volatile HCO_2H and CH_3CO_2H can be recovered almost completely after their liberation, and their mutual large-scale separation by azeotropic distillation with the aid of ethylene dichloride is possible [332]. In contrast, the purification of the fraction of hydroxy acids is somewhat problematic, and it seems realistic that only a rough fractionation of the main acids into two groups would be possible (if not making in the process scheme the methyl esters of glycolic, lactic, and 2-hydroxybutanoic acids, see above): *i*) low-molar -mass acids with 2–4 carbon atoms (*i.e.*, glycolic, lactic, and 2-hydroxybutanoic acids) and *ii*) high-molar-mass acids with 5 and 6 carbon atoms (*i.e.*, 3,4-dideoxy-pentonic, 3-deoxy-pentonic, xyloisosaccharinic, and glucoisosaccharinic acids). Their full-scale purification could be accomplished either by straightforward distillation under reduced pressure (0.067–0.173 kPa) [324, 325] or by ion-exclusion chromatographic techniques [333]. However, owing to difficulties, especially in the distillation of glucoisosaccharinic acid (the boiling point of α-glucoisosaccharino-1,4-lactone is 200 °C at 0.067 kPa) from other hydroxy acids and the instability of some of them (especially, glycolic and lactic acids) during heating [334], the overall distillation yields have been about 50% (pine) and about 75% (birch). This means that in this case, the overall recovery of aliphatic carboxylic acids (*i.e.*, volatile acids and nonvolatile acids) is about 70% and 85%, respectively.

In general, carbohydrate losses are high in the beginning of the alkaline cook, when delignification is still proceeding slowly [44]. This rather low selectivity of softwood kraft pulping means, for example, that at the end of the heating-up period (*i.e.*, somewhat after the lignin extraction phase) in the conventional cooking, the mass ratio of the aliphatic carboxylic acid fraction to the lignin fraction is higher than that in the final black liquor [120]. Hence, it can be concluded that one attractive possibility would be to withdraw black liquor from the early stage of pulping for the recovery of aliphatic acids. Another logical raw material for such recovery is the alkaline pretreatment liquor as shown in Figure 5.9.

In the concept (Figure 5.9), the difficulty is to find an appropriate way of handling the huge amount of Na_2SO_4 formed as a by-product [17]. Although Na_2SO_4 is basically a very cheap material, it has rather limited use. Additionally, it is chemically very stable and, for example, can be reduced to Na_2S only at high temperatures. However, electrochemical membrane process techniques (*e.g.*, electrodialysis: the reaction $Na_2SO_4 \rightarrow H_2SO_4 + NaOH$) [335–337] offer one interesting possibility of recovering and recycling at least part of H_2SO_4.

Electrolysis can be used for a great number of purposes even in the absence of membranes [164, 328, 338–340]. Electrodialysis is a unit operation where ions are transported through an ion-selective membrane with an electrical driving force. There are three types of membranes, which can be used separately or in combination: *i*) cation selective membranes (CSMs), *ii*) anion selective membranes (ASMs), and *iii*) bipolar membranes (BPMs). Both CSMs and ASMs hinder the passage of co-ions (*i.e.*, anions and cations, respectively), and the special BPMs are the combination of both the CSM and ASM. The BPMs have a distinct function compared to those of mono-polar membranes and can dissociate solvents; for example, H_2O is dissociated into $H^{\oplus}$ and $HO^{\ominus}$ [341]. Additionally, besides ion-selective membranes, nanofiltration (NF) membranes have been used in electrodialysis experiments. It is known that, due to fouling, the flux through a membrane at a constant transmembrane pressure tends to decrease during usage, which, however, greatly depends on the application [290, 292, 340, 342–344]. In case of electrochemical membrane separation processes, especially for black liquor, a pulsed electric field might be used for the cleaning of membranes, or the effective cleaning could be achieved with either NaOH solution or diluted black liquor [287, 288, 345].

Electrodialysis has been traditionally applied to remove ions in the desalination process of seawater and brackish water [346–348]. One of the growing trends toward the application of this technique has also been during the last three decades in the separation and purification of organic acids from their fermentation broths and related sources [349–351]. It is commonly used to separate, for example, HCO_2H [352], CH_3CO_2H

[353, 354], propanoic acid [355], lactic acid [356–358], citric acid [359–361], malic acid [362], tartaric acid [363], and glucuronic acid [364]. In the pulp industry, electrodialysis and electrolysis have been applied to the recovery of pulping chemicals from various alkaline cooking liquors [246, 247, 342, 343, 345, 365, 366], also including the recovery of lignin [287, 288, 367–370] and aliphatic carboxylic acids [328, 371–373]. Hence, the electrodialysis techniques can be also applied, for example, to the carbonated black liquor resulting in the formation of crude acid fraction and NaOH (*cf.*, the alternative process route shown in Figure 5.9).

Ion-exclusion chromatography is a technique suitable for the separation of relatively small organic and inorganic acids, especially those with a hydrophilic nature [328, 374–379]. This technique is based on the use of strong cation or anion exchange resins for the separation of ionic solutes from weakly ionized or neutral solutes and extensively applied in the sugar industries [380]. For example, free aliphatic carboxylic acids are usually separated on a column filled with a strongly acidic sulfonated polystyrene cation exchange resin in $H^{\oplus}$ form and cross-linked with divinylbenzene [333, 379]. In this technique, due to the Donnan effect, the ionic material is rejected by resin and passes quickly through, whereas the neutral substances are impeded and pass through more slowly. During the recent decades, there have been a lot of investigations reported to separate aliphatic carboxylic acids using ion-exclusion chromatography [333, 374, 375, 377, 381, 382]. The aliphatic carboxylic acids have also been recovered on laboratory scale using different adsorbents [383–386] or extracted with amine extractants [387–389]. Additionally, aliphatic acids and lignin have been separated on laboratory scale from alkaline cooking liquors without neutralization by subsequent methods, such as ultrafiltration, size-exclusion chromatography, ion-exchange, adsorption, and evaporation [390–393].

Aliphatic carboxylic acids can be used as single components, or as more or less purified mixtures, in many applications [394, 395]. Of this group HCO_2H, CH_3CO_2H, glycolic acid, and lactic acid are today commercially important and well-known chemicals with a great number of well-established uses and need not be discussed in more detail [41, 396–399], Today, no industrial utilization for 2-hydroxybutanoic, 3,4-dideoxy-pentonic, and 3-deoxy-pentonic acids has yet been established, although they seem to present suitable starting materials for the synthesis of many different organic chemicals and materials. Of the isosaccharinic acids, *i.e.*, xyloisosaccharinic acid and glucoisosaccharinic acid, only the use of the latter has been studied more effectively. Xyloisosaccharinic acid is formed by alkaline treatment of xylan and xylooligosaccharides [400–402] and apart from the analytical uses, other reactions, such as reduction (*i.e.*, production of polyalcohols) [403] and oxidation (*i.e.*, production of polycarboxylic acids) [404] have received little attention. On the contrary, glucoisosaccharinic acid (its α-form) can readily be obtained either as calcium salt or 1,4-lactone from calcium hydroxide treatment of lactose [402, 405–409]. Its different reduction [410, 411], oxidation [412], dehydration [413], and heat treatment [414] reactions have been investigated. The strong complexing properties are typical for α-glucoisosaccharinic acid [415–424] and some drugs and other bioactive substances can also be produced from it [425–429]. However, the β-form of glucoisosaccharinic acid is difficult to separate as a pure crystalline form (as a salt or lactone) and its utilization has not been studied. Additionally, a great number of potential specific applications have been described for hydroxy acid mixtures separated from black liquors, including, for example, surfactants [430]. One interesting approach has been obtained by direct esterification of α-glucoisosaccharinic acid with long-chain fatty acids (*e.g.*, with a conventional fatty acid fraction from crude tall oil) [431, 432].

References

[1]   Kamm, B., Gruber, P.R., and Kamm, M. 2005. *Biorefineries – Industrial Processes and Products: Status Quo and Future Directions*. Wiley-VCH, Weinheim, Germany. 497 p.

[2]   Pervaiz, M., and Sain, M. 2006. Biorefinery: Opportunities and barriers for petrochemical industries. *Pulp and Paper Canada* 107(6):31–33.

[3]   Mabee, W.E., and Saddler, W.E. 2006. The potential of bioconversion to produce fuels and chemicals. *Pulp and Paper Canada* 107(6):34–37.

[4]   Clements, L.D., and Van Dyne, D.L. 2006. The lignocellulosic biorefinery – a strategy for returning to a sustainable source of fuels and industrial organic chemicals. In: Kamm, B., Gruber, P.R., and Kamm, M. (Eds.). *Biorefineries – Industrial Processes and Products. Volume 1*. Wiley-VCH, Weinheim, Germany. Pp. 115–128.

[5]   Koukoulas, A.A. 2007. Cellulosic biorefineries – charting a new course for wood use. *Pulp and Paper Canada* 108(6):17–19.

[6]   Chambost, V., Eamer, R., and Stuart, P.R. 2007. Systematic methodology for identifying promising biorefinery products. *Pulp and Paper Canada* 108(6):31–35.

[7]   Clark, J.H., and Deswarte, E.I. 2008. *The Biorefinery Concept – An integrated Approach, in Introduction to Chemicals from Biomass*. John Wiley & Sons, New York, NY, USA. Pp. 1–20.

[8]   Amidon, T.E., and Liu, S. 2009. Water-based woody biorefinery. *Biotechnology Advances* 27:542–550.

[9]   Cherubini, F. 2010. The biorefinery concept: Using biomass instead of oil for producing energy and chemicals. *Energy Conversion and Management* 51(7):1412–1421.

[10] Thorp, B.A., and Akhtar, M. 2010. Is the biorefinery for real? *Paper360°* 5(4):8–12.

[11] Janssen, M., and Stuart, P. 2010. Drivers and barriers for implementation of the biorefinery. *Pulp and Paper Canada* 111(3):13–17.

[12] Alén, R. 2011. Principles of biorefining. In: Alén, R. (Ed.). *Biorefining of Forest Resources*. Paper Engineers' Association, Helsinki, Finland. Pp. 56–114.

[13] Christopher, L. (Ed.). 2012. *Integrated Forest Biorefineries: Challenges and Opportunities*. Royal Society of Chemistry, Cambridge, United Kingdom. 323 p.

[14] Liu, S., Lu, H., Hu, R., Shupe, A., Lin, L., and Liang, B. 2012. A sustainable woody biomass biorefinery. *Biotechnology Advances* 30(4):785–810.

[15] Liu, S., Abrahamson, L.P., and Scott, G.M. 2012. Biorefinery: Ensuring biomass as a sustainable renewable source of chemicals, materials and energy. *Biomass and Bioenergy* 39(4):1–4.

[16] Lehto, J. 2021. Integrated biorefineries. In: Alén, R. (Ed.). *Pulping and Biorefining – ForestBioFacts, Digital Learning Environment*. Paperi ja Puu Ltd., Helsinki, Finland.

[17] Alén, R. 2015. Pulp mills and wood-based biorefineries. In: Pandey, A., Höfer, R., Taherzadeh, M., Nampoothiri, K.M., and Larroche, C. (Eds.). *Industrial Biorefineries & White Biotechnology*. Elsevier, Amsterdam, The Netherlands. Pp. 91–126.

[18] Amidon, T. 2006. Forest Biorefinery: A new business model. *Pulp and Paper Canada* 107(3):19.

[19] Moshkelani, M., Marinova, M., Perrier, M., and Paris, J. 2013. The forest biorefinery and its implementation in the pulp and paper industry: Energy overview. *Applied Thermal Engineering* 50(2):1427–1436.

[20] Sanglard, M., Chirat, C., Jarman, B., and Lachenal, D. 2013. Biorefinery in a pulp mill: Simultaneous production of cellulosic fibers from *Eucalyptus globulus* by soda-anthraquinone cooking and surface-active agents. *Holzforschung* 67(5):481–488.

[21] Gomes, F.J.B., Santos, F.A., Colodette, J.L., Demuner, I.F., and Batalha, L.A.R. 2014. Literature review on biorefinery processes integrated to the pulp industry. *Natural Resources* 5:419–432.

[22] Benali, M., Périn-Levasseur, Z., Savulescu, L., Kouisni, L., Jemaa, N., Kudra, T., and Paleologou, M. 2014. Implementation of lignin-based biorefinery into a Canadian softwood kraft pulp mill: Optimal resources integration and economic viability assessment. *Biomass and Bioenergy* 67(8):473–482.

[23] Mendes, C.V.T., Carvalho, M.G.V.S., Baptista, C.M.S.G., Rocha, J.M.S., Soares, B.I.G., and Sousa, G.D.A. 2009. Valorisation of hardwood hemicelluloses in the kraft pulping process by using an integrated biorefinery concept. *Food and Bioproducts Processing* 87(3):197–207.

[24] Huang, H.-J., Ramaswamy, S., Al-Dajani, W.W., and Tschirner, U. 2010. Process modeling and analysis of pulp mill-based integrated biorefinery with hemicellulose pre-extraction for ethanol production: A comparative study. *Bioresource Technology* 101(2):624–631.

[25] Lyytikäinen, K., Saukkonen, E., Kajanto, I., and Käyhkö, J. 2011. The effect of hemicellulose extraction on fiber charge properties and retention behavior of kraft pulp fibers. *BioResources* 6(1):219–231.

[26] Huang, F., and Ragauskas, A. 2013. Integration of hemicellulose pre-extraction in the bleach-grade pulp production process. *TAPPI Journal* 12(10):55–61.

[27] Martin-Sampedro, R., Eugenio, M.E., Moreno, J.A., Revilla, E., and Villar, J.C. 2014. Integration of a kraft pulping mill into a forest biorefinery: Pre-extraction of hemicellulose by steam explosion versus steam treatment. *Bioresource Technology* 153(2):236–244.

[28] Ajao, O., Marinova, M., Savadogo, O., and Paris, J. 2018. Hemicellulose based integrated forest biorefineries: Implementation strategies. *Industrial Crops and Products* 126(12):250–260.

[29] Ragauskas, A.J., Nagy, M., and Kim, D.H. 2006. From wood to fuels – integrating biofuels and pulp production. *Industrial Biotechnology* 2(1):55–65.

[30] Thakur, I.S., and Nakagoshi, N. 2011. Production of biofuels from lignocellulosic biomass in pulp and paper effluents for low carbon society. *Journal of International Development and Cooperation* 18(1):1–12.

[31] Alén, R. 2013. Integrated possibilities of producing biofuels in chemical pulping. In: Ragauskas, A.J. (Ed.). *Materials for Biofuels*. World Scientific, Singapore. Pp. 317–338.

[32] Harstela, P. 1998. Timber procurement. In: Kellomäki, S. (Ed.). *Forest Resources and Sustainable Management*. Fapet Oy, Helsinki, Finland. Pp. 311–362.

[33] Herrick, F.W., and Hergert, H.L. 1977. Utilization of chemicals from wood: Retrospect and prospect. In: Loewus, F.A., and Runecles, V.C. (Eds.). *The Structure, Biosynthesis, and Degradation of Wood, Recent Advances in Phytochemistry*. Vol. *11*. Plenium Press, New York, NY, USA. Pp. 443–515.

[34] Goldstein, I.S. (Ed.). 1981. *Organic Chemicals from Biomass*. 2nd printing. CRC Press, Boca Raton, FL, USA. 310 p.

[35] Wise, D.L. (Ed.). 1983. *Organic Chemicals from Biomass*. The Benjamin/Cummins Publishing Company, London, United Kingdom. 465 p.

[36] Hakkila, P. 1989. *Utilization of Residual Forest Biomass*. Springer-Verlag, Heidelberg, Germany. 568 p.

[37] Zhu, J.Y. 2011. Forest biorefinery: The next century of innovation. *TAPPI Journal* 10(5):5.

[38] Hamaguchi, M., Cardoso, M., and Vakkilainen, E. 2012. Alternative technologies for biofuels production in kraft pulp mills – potential and prospects. *Energies* 5:2288–2309.

[39] Acharys, B., Sule, I., and Dutta, A. 2012. A review on advances of torrefaction technologies for biomass processing. *Biomass Conversion and Biorefinery* 2:349–369.

[40]   Alén, R. 2018. *Carbohydrate Chemistry – Fundamentals and Applications*. World Scientific, Singapore. 586 p.

[41]   Viikari, L., and Alén, R. 2011. Biochemical and chemical conversion of forest biomass. In: Alén, R. (Ed.). *Biorefining of Forest Resources*. Paper Engineers'. Association, Helsinki, Finland. Pp. 225–261.

[42]   Wang, Y., and Liu, S. 2012. Pretreatment technologies for biological and chemical conversion of woody biomass. *TAPPI Journal* 11(1):9–16.

[43]   Alén, R. 2000. Basic chemistry of wood delignification. In: Stenius, P. (Ed.). *Forest Products Chemistry*. Fapet, Helsinki, Finland. Pp. 58–104.

[44]   Sjöström, E. 1993. *Wood Chemistry – Fundamentals and Applications*. 2nd edition. Academic Press, San Diego, CA, USA. Pp. 114–164.

[45]   Alén, R. 2011. Cellulose derivatives. In: Alén, R. (Ed.). *Biorefining of Forest Resources*. Paper Engineers'. Association, Helsinki, Finland. Pp. 305–354.

[46]   Mateos-Espejel, E., Radiotis, T., and Jemaa, N. 2013. Implications of converting a kraft pulp mill to a dissolving pulp operation with a hemicellulose extraction stage. *TAPPI Journal* 12(2):29–38.

[47]   Kilpeläinen, P. 2021. Hot-water extraction. In: Alén, R. (Ed.). *Pulping and Biorefining – ForestBioFacts, Digital Learning Environment*. Paperi ja Puu Ltd., Helsinki, Finland.

[48]   Teo, C.C., Tan, S.N., Yong, J.W.H., Hew, C.S., and Ong, E.S. 2010. Pressurized hot water extraction (PHWE). *Journal of Chromatography A* 1217:2484–2494.

[49]   Smith, R.M. 2002. Extractions with superheated water. *Journal of Chromatography A* 975:31–46.

[50]   Garrote, G., Domínquenz, H., and Parajó, J.C. 1999. Mild autohydrolysis: An environmentally friendly technology for xylooligosaccharide production from wood. *Journal of Chemical Technology and Biotechnology* 74:1101–1109.

[51]   Tunc, M.S., and van Heiningen, R.P. 2008. Hemicellulose extraction of mixed southern hardwood with water at 150 °C: Effect of time. *Industrial & Engineering Chemistry Research* 47:7031–7037.

[52]   Borrega, M., Nieminen, K., and Sixta, H. 2011. Degradation kinetics of the main carbohydrates in birch wood during hot water extraction in a batch reactor at elevated temperatures. *Bioresource Technology* 102:10724–10732.

[53]   Kilpeläinen, P., Kitunen, V., Pranovich, A., Ilvesniemi, H., and Willför, S. 2013. Pressurized hot water flow-through extraction of birch sawdust with acetate pH buffer. *BioResources* 8(4):5202–5218.

[54]   Penttilä, P.A., Kilpeläinen, P., Tolonen, L., Suuronen, J.-P., Sixta, H., Willför, S., and Serimaa, R. 2013. Effects of pressurized hot water extraction on the nanoscale structure of birch sawdust. *Cellulose* 20(5):2335–2347.

[55]   Song, T., Pranovich, A., Sumerskiy, I., and Holmbom, B. 2008. Extraction of galactoglucomannan from spruce wood with pressurised hot water. *Holzforschung* 62:659–666.

[56]   Leppänen, K., Spetz, P., Pranovich, A., Hartonen, K., Kitunen, V., and Ilvesniemi, H. 2011. Pressurized hot water extraction of Norway spruce hemicelluloses using a flow-through system. *Wood Science and Technology* 45:223–236.

[57]   Kubikova, J., Lu, P., Zemann, A., Krkoška, P., and Bobleter, O. 2000. Aquasolv pretreatment of plant materials for the production of cellulose and paper. *Cellulose Chemistry and Technology* 34:151–162.

[58]   Mosier, N., Hendrickson, R., Ho, N., Sedlak, M., and Ladisch, M.R. 2005. Optimization of pH controlled liquid hot water pretreatment of corn stover. *Bioresource Technology* 96:1986–1993.

[59]   Kemppainen, K., Siika-aho, M., Pattathil, S., Giovando, S., and Kruus, K. 2014. Spruce bark as an industrial source of condensed tannins and non-cellulosic sugars. *Industrial Crops and Products* 52:158–168.

[60]   Rasi, S., Kilpeläinen, P., Rasa, K., Korpinen, R., Raitanen, J.-E., Vainio, M., Kitunen, V., Pulkkinen, H., and Jyske, T. 2019. Cascade processing of softwood bark with hot water extraction, pyrolysis and anaerobic digestion. *Bioresource Technology* 292:[121893].

[61]   Aubrey, A.D., Chalmers, J.H., Bada, J.L., Grunthaner, F.J., Amashukeli, X., Willis, P., Skelley, A.M., Mathies, R.A., Quin, R.C., Zent, A.P., Ehrenfreund, P., Amundson, R., Gavin, D.P., Botta, O., Barron, L., Blaney, D.L., Clark, B.C., Coleman, M., Hofmann, B.A., Josset, J.-L., Rettberg, P., Ride, S., Robert, F., Sephton, M.A., and Yen, A. 2008. The Urey instrument: An advanced *in situ* organic and oxidant detector for Mars exploration. *Astrobiology* 8:583–595.

[62]   Borrega, M., and Sixta, H. 2013. Purification of cellulosic pulp by hot water extraction. *Cellulose* 20:2803–2812.

[63]   Plaza, M., and Turner, C. 2015. Trends in analytical chemistry pressurized hot water extraction of bioactives. *Trends in Analytical Chemistry* 71:39–54.

[64]   Andersson, T. 2007. *Parameters Affecting the Extraction of Polycyclic Aromatic Hydrocarbons with Pressurised Hot Water*. Doctoral Thesis. University of Helsinki, Laboratory of Analytical Chemistry, Helsinki, Finland. 94 p.

[65]   Hartonen, K., Inkala, K., Kangas, M., and Riekkola, M.-L. 1997. Extraction of polychlorinated biphenyls with water under subcritical conditions. *Journal of Chromatography A* 785:219–226.

[66]   Drews, M.J., González-Pereyra, N.G., Mardikian, P., and De Viviés, P. 2013. The application of subcritical fluids for the stabilization of marine archaeological iron. *Studies in Conservation* 58:314–325.

[67]   Cocero, M.J., Cabeza, Á., Abad, N., Adamovic, T., Vaquerizo, L., Martínes, C.M., and Pazo-Cepeda, M.V. 2018. Understanding biomass fractionation in subcritical & supercritical water. *Journal of Supercritical Fluids* 133:550–565.

[68]   Rissanen, J.V., Grénman, H., Willför, S., Murzin, D.Y., and Salmi, T. 2014. Spruce hemicellulose for chemicals using aqueous extraction: Kinetics, mass transfer, and modeling. *Industrial & Engineering Chemistry Research* 53:6341–6350.

[69]   Valoppi, F., Maina, N., Allen, M., Miglioli, R., Kilpeläinen, P., and Mikkonen, K. 2019. Spruce galactoglucomannan-stabilized emulsions as essential fatty acid delivery systems for functionalized drinkable yogurt and oat-based beverage. *European Food Research and Technology* 245:1387–1398.

[70]   Mikkonen, K.S., Kirjoranta, S., Xu, C., Hemming, J., Pranovich, A., Bhattarai, M., Peltonen, L., Kilpeläinen, P., Maina, N., Tenkanen, M., Lehtonen, M., and Willför, S. 2019. Environmentally-compatible alkyd paints stabilized by wood hemicelluloses. *Industrial Crops and Products* 133:212–220.

[71]   Kilpeläinen, P., Hautala, S., Byman, O., Tanner, L., Korpinen, R., Lillandt, K.-J., Pranovich, A., Kitunen, V., Willför, S., and Ilvesniemi, H. 2014. Pressurized hot water flow-through extraction system scale up from the laboratory to the pilot scale. *Green Chemistry* 16:3186–3194.

[72]   Lehto, J. 2021. Acidic pretreatments prior to delignification. In: Alén, R. (Ed.). *Pulping and Biorefining – ForestBioFacts, Digital Learning Environment*. Paperi ja Puu Ltd., Helsinki, Finland.

[73]   Yoon, S.-H., MacEwan, K., and van Heiningen, A. 2008. Hot-water pre-extraction from loblolly pine (*Pinus taeda*) in an integrated forest products biorefinery. *TAPPI Journal* 7(6):27–32.

[74]   Huang, F., and Ragauskas, A. 2013. Integration of hemicellulose pre-extraction in the bleach-grade pulp production process. *TAPPI Journal* 12(10):55–61.

[75]   Vena, P.F., Brienzo, M., García-Aparicio, M.P., Görgens, J.F., and Rypstra, T. 2013. Hemicelluloses extraction from giant bamboo (*Bambusa balcooa* Roxburgh) prior to kraft or soda-AQ pulping and its effect on pulp physical properties. *Holzforschung* 67(8):863–870.

[76]   Conner, A.H., and Lorenz, L.F. 1986. Kinetic modelling of hardwood prehydrolysis. Part III. Water and dilute acetic acid prehydrolysis of southern red oak. *Wood and Fiber Science* 18(2):248–263.

[77]   Sixta, H., Potthast, A., and Krotschek, A.W. 2006. Chemical pulping processes. In: Sixta, H. (Ed.). *Handbook of Pulp*. Wiley-VCH Verlag, Weinheim, Germany. Pp. 325–366.

[78]   Wafa Al-Dajani, W., Tschirner, U.W., and Jensen, T. 2009. Pre-extraction of hemicelluloses and subsequent kraft pulping. Part II: Acid- and autohydrolysis. *TAPPI Journal* 8(9):30–37.

[79]   Montane, D., Farriol, X., Salvadó, J., Jollez, P., and Chornet, E. 1998. Application of steam explosion to the fraction and rapid vapor-phase alkaline pulping of wheat straw. *Biomass and Bioenergy* 14(3):261–276.

[80]   Montane, D., Farriol, X., Salvadó, J., Jollez, P., and Chornet, E. 1998. Fractionation of wheat straw by steam-explosion pretreatment and alkali delignification. Cellulose pulp and byproducts from hemicellulose and lignin. *Journal of Wood Chemistry and Technology* 18(2):171–191.

[81]   Martin-Sampedro, R., Eugenio, M.E., Moreno, J.A., Revilla, E., and Villar, J.C. 2014. Integration of a kraft pulping mill into a forest biorefinery: Pre-extraction of hemicellulose by steam explosion versus steam treatment. *Bioresource Technology* 153:236–244.

[82]   Olsson, L., and Hahn-Hägerdal, B. 1996. Fermentation of lignocellulosic hydrolysates for ethanol production. *Enzyme and Microbial Technology* 18:312–331.

[83]   Palmqvist, E., and Hahn-Hägerdal, B. 2000. Fermentation of lignocellulosic hydrolysates. I: Inhibition and detoxification. *Bioresource Technology* 74(1):17–24.

[84]   Sears, K.D., Beélik, A., Casebier, R.L., Engen, R.J., and Hergert, H.L. 1971. Southern pine prehydrolyzates: Characterization of polysaccharides and lignin fragments. *Journal of Polymer Science: Part C* 36:425–443.

[85]   Willför, S., and Holmbom, B. 2004. Isolation and characterization of water soluble polysaccharides from Norway spruce and Scots pine. *Wood Science and Technology* 38:173–179.

[86]   Carvalheiro, F., Duarte, L.C., and Gírio, F.M. 2008. Hemicellulose biorefineries: A review on biomass pretreatments. *Journal of Scientific and Industrial Research* 67:849–864.

[87]   Frederick, W.J. Jr., Lien, S.J., Courchene, C.E., DeMartini, N.A., Ragauskas, A.J., and Iisa, K. 2008. Co-production of ethanol and cellulose fiber from southern pine: A technical and economical assessment. *Biomass Bioenergy* 32:1293–1302.

[88]   Tunc, M.S., and van Heiningen, A.R.P. 2011. Characterization and molecular weight distribution of carbohydrates isolated from the autohydrolysis extract of mixed southern hardwoods. *Carbohydrate Polymers* 83(1):8–13.

[89]   Lehto, J., and Alén, R. 2012. Purification of hardwood-derived autohydrolysates. *BioResources* 7(2):1813–1823.

[90]   Bujanovic, B.M., Goundalkar, M.J., and Amidon, T.E. 2012. Increasing the value of a biorefinery based on hot-water extraction: Lignin products. *TAPPI Journal* 11(1):19–26.

[91]   Yoon, S.-H., and van Heiningen, A. 2008. Kraft pulping and papermaking properties of hot-water pre-extracted loblolly pine in an integrated forest products biorefinery. *TAPPI Journal* 7(7):22–27.

[92]   Kautto, J., Saukkonen, E., and Henricson, K. 2010. Digestibility and paper-making properties of prehydrolyzed softwood chops. *BioResources* 5(4).

[93]   Lu, H., Hu, R., Ward, A., Amidon, T.E., Liang, B., and Liu, S. 2012. Hot-water extraction and its effects on soda pulping of aspen woodchips. *Biomass and Bioenergy* 39(4):5–13.

[94]   Coelho dos Santos Muguet, M., Ruuttunen, K., Jääskeläinen, A.-S., Colodette, J.L., and Vuorinen, T. 2013. Defibration mechanisms of autohydrolyzed *Eucalyptus* wood chips. *Cellulose* 20(5):2647–2654.

[95]   Hamaguchi, M., Kautto, J., and Vakkilainen, E. 2013. Effects of hemicellulose extraction on the kraft pulp mill operation and energy use: Review and case study with lignin removal. *Chemical Engineering Research and Design* 91(2013):1284–1291.

[96]   Runge, T., and Zhang, C. 2013. Hemicellulose extraction and its effect on pulping and bleaching. *TAPPI Journal* 12(10):45–52.

[97]   Reguant, J., Martínez, J.M., Montané, D., Salvadó, J., and Farriol, X. 1997. Cellulose from softwood via prehydrolysis and soda/anthraquinone pulping. *Journal of Wood Chemistry and Technology* 17(1997):91–110.

[98] Duarte, G.V., Ramarao, B.V., Amidon, T.E., and Ferreira, P.T. 2011. Effect of hot water extraction on hardwood kraft pulp fibers (*Acer saccharum*, sugar maple). *Industrial & Engineering Chemistry Research* 50:9949–9959.

[99] Vila, C., Romero, J., Fransisco, J.L., Garrote, G., and Parajó, J.C. 2011. Extracting value from *Eucalyptus* wood before kraft pulping: Effects of hemicelluloses solubilization on pulp properties. *Bioresource Technology* 102(8):5251–5254.

[100] Duarte, G.V., Gamelas, J.A.F., Ramarao, B.V., Amidon, T.E., and Ferreira, P.T. 2012. Properties of extracted *Eucalyptus globulus* kraft pulps. *TAPPI Journal* 11(4):47–55.

[101] Saukkonen, E., Kautto, J., Rauvanto, I., and Backfolk, K. 2012. Characteristics of prehydrolysis-kraft pulp fibers from Scots pine. *Holzforschung* 66(7):801–808.

[102] Chirat, C., Boiron, L., and Lachenal, D. 2013. Bleaching ability of pre-hydrolyzed pulps in the context of a biorefinery mill. *TAPPI Journal* 12(11):49–53.

[103] Vena, P.F., Brienzo, M., García-Aparicio, M.P., Görgens, J.F., and Rypstra, T. 2013. Hemicelluloses extraction from giant bamboo (*Bambusa balcooa* Roxburgh) prior to kraft or soda-AQ pulping and its effect on pulp physical properties. *Holzforschung* 67(8):863–870.

[104] Biermann, C.J. 1996. *Handbook of Pulping and Papermaking*. Academic Press, San Diego, CA, USA. p. 55–100.

[105] Kenealy, W.R., Houtman, C.J., Laplaza, J., Jeffries, T.W., and Horn, E.G. 2007. Pretreatments for converting wood into paper and chemicals. In: Argyropoulos, D.S. (Ed.). *Materials, Chemicals, and Energy from Forest Biomass, ACS Symposium Series. Vol. 954.* American Chemical Society, Washington, DC, USA. Pp. 392–408.

[106] Li, H., Saeed, A., Jahan, M.S., Ni, Y., and van Heiningen, A. 2010. Hemicellulose removal from hardwood chips in the pre-hydrolysis step of the kraft-based dissolving pulp production process. *Journal of Wood Chemistry and Technology* 30:48–60.

[107] Liu, Z., Ni, Y., Fatehi, P., and Saeed, A. 2011. Isolation and cationization of hemicelluloses from pre-hydrolysis liquor of kraft-based dissolving pulp production process. *Biomass and Bioenergy* 35(5):1789–1796.

[108] Saeed, A., Jahan, M.S., Li, H., Liu, Z., Ni, Y., and van Heiningen, A. 2012. Mass balances of components dissolved in the pre-hydrolysis liquor of kraft-based dissolving pulp production process from Canadian hardwoods. *Biomass and Bioenergy* 39(4):14–19.

[109] Mateos-Espejel, E., Radiotis, T., and Jemaa, N. 2013. Implication of converting a kraft pulp mill to a dissolving pulp operation with a hemicellulose extraction stage. *TAPPI Journal* 12(2):29–38.

[110] Ahsan, L., Jahan, M.S., and Ni, Y. 2014. Recovering/concentrating of hemicellulosic sugars and acetic acid by nanofiltration and reverse osmosis from prehydrolysis liquor of kraft based hardwood dissolving pulp process. *Bioresource Technology* 155(3):111–115.

[111] Sixta, H., and Schild, G. 2009. A new generation kraft process. *Lenzinger Berichte* 87:26–37.

[112] Borrega, M., Tolonen, L.K., Bardot, F., Testova, L., and Sixta, H. 2013. Potential of hot water extraction of birch wood to produce high-purity dissolving pulp after alkaline pulping. *Bioresource Technology* 135(5):665–671.

[113] Radiotis, T., Zhang, X., Paice, M., and Byrne, V. 2014. Optimizing production of xylose and xylooligomers from wood chips. *Proceedings of Nordic Wood and Biorefinery Conference (NWBC)*, Stockholm, Sweden, 22.–24.3.2014. Pp. 92–99.

[114] Wang, H., Pang, B., Wu, K., Kong, F., Li, B., and Mu, X. 2014. Two stages of treatments for upgrading bleached softwood paper grade pulp to dissolving pulp for viscose production. *Biochemical Engineering Journal* 82(1):183–187.

[115] Aurell, R., and Hartler, N. 1965. Kraft pulping on pine. I. Changes in the composition of wood residue during the cooking process. *Svensk Papperstidning* 68(3):59–68.

[116] Johansson, M.H., and Samuelson, O. 1975. End-wise degradation of hydrocellulose during hot alkali treatment. *Journal of Applied Polymer Science* 19(11):3007–3013.

[117] Yang, B.Y., and Montgomery, R. 1996. Alkaline degradation of glucose: Effect of initial concentration of reactants. *Carbohydrate Research* 280(1):27–45.

[118] Niemelä, K., and Alén, R. 1999. Characterization of pulping liquors. In: Sjöström, E., and Alén, R. (Eds.). *Analytical Methods in Wood Chemistry, Pulping, and Papermaking*. Springer, Heidelberg, Germany. Pp. 193–231.

[119] Kleppe, P.J. 1970. Kraft pulping. *TAPPI* 53(1):35–47.

[120] Alén, R., Moilanen, V.-P., and Sjöström, E. 1986. Potential recovery of hydroxy acids from kraft pulping liquors. *TAPPI Journal* 69(2):76–78.

[121] Pakkanen, H., and Alén, R. 2012. Molecular mass distribution of lignin from the alkaline pulping of hardwood, softwood, and wheat straw. *Journal of Wood Chemistry and Technology* 32(4):279–293.

[122] Hu, G., Heitmann, J.A., and Rojas, O.J. 2008. Feedstock pretreatment strategies for producing ethanol from wood, bark, and forest residues. *BioResources* 3(1):270–294.

[123] McIntosh, S., and Vancov, T. 2010. Enhanced enzyme saccharification of *Sorghum bicolor* straw using dilute alkali pretreatment. *Bioresource Technology* 101(17):6718–6727.

[124] Ju, Y.-H., Huynh, L.-H., Kasim, N.S., Guo, T.-J., and Wang, J.H. 2011. Analysis of soluble and insoluble fractions of alkali and subcritical water treated sugarcane bagasse. *Carbohydrate Polymers* 83(2):591–599.

[125] Kallioinen, A., Hakola, M., Riekkola, T., Repo, T., Leskelä, M., Von Weymarn, N., and Siika-aho, M. 2013. A novel alkaline oxidation pretreatment for spruce, birch and sugar cane bagasse. *Bioresource Technology* 140(7):414–420.

[126] Bali, G., Meng, X., Deneff, J.I., Sun, Q., and Ragauskas, A.J. 2015. The effect of alkaline pretreatment methods on cellulose structure and accessibility. *ChemSusChem* 8(2):275–279.

[127] Karp, E.M., Resch, M.G., Donohoe, B.S., Ciesielski, P.N., O'Brien, M.H., Nill, J.E., Mittal, A., Biddy, M.J., and Beckham, G.T. 2015. Alkaline pretreatment of switchgrass. *ACS Sustainable Chemistry and Engineering* 3(7):1479–1491.

[128] Lehto, J., Pakkanen, H., and Alén, R. 2015. Characterization of lignin dissolved during alkaline pre-treatment of softwood and hardwood. *Journal of Wood Chemistry and Technology* 35(5):337–347.

[129] Xu, H., Li, B., and Mu, X. 2016. Review of alkali-based pretreatment to enhance enzymatic saccharification for lignocellulosic biomass conversion. *Industrial Engineering and Chemistry Research* 55(32):8691–8705.

[130] Baruah, J., Nath, B.K., Sharma, R., Kumar, S., Deka, R.C., Barua, D.C., and Kalita, E. 2018. Recent trends in the pretreatment of lignocellulosic biomass for value-added products. *Frontiers in Energy Research* 6(12):1–19.

[131] Lehto, J. 2021. Alkaline pretreatments prior to delignification. In: Alén, R. (Ed.). *Pulping and Biorefining – ForestBioFacts, Digital Learning Environment*. Paperi ja Puu Ltd., Helsinki, Finland.

[132] Yoon, S.-H., Tunc, M.S., and van Heiningen, A. 2011. Near-neutral pre-extraction of hemicelluloses and subsequent kraft pulping of southern mixed hardwoods. *TAPPI Journal* 10(1):7–15.

[133] Luo, J., Genco, J.M., and Zou, H. 2012. Extraction of hardwood biomass using dilute alkali. *TAPPI Journal* 11(6):19–27.

[134] Von Schenck, A., Berlin, N., and Uusitalo, J. 2013. Ethanol from Nordic wood raw material by simplified alkaline soda cooking pre-treatment. *Applied Energy* 102(2):229–240.

[135] Singh, R., Shukla, A., Tiwari, S., and Srivastava, M. 2014. A review on delignification of lignocellulosic biomass for enhancement of ethanol production potential. *Renewable and Sustainable Energy Reviews* 32(4):713–728.

[136] Kim, J.S., Lee, Y.Y., and Kim, T.H. 2016. A review on alkaline pretreatment technology for bioconversion of lignocellulosic biomass. *Bioresource Technology* 199(8):42–48.

[137] Um, B.-H., and Van Walsum, G.P. 2009. Acid hydrolysis of hemicellulose in green liquor pre-pulping extract of mixed northern hardwoods. *Applied Biochemistry and Biotechnology* 153:127–138.

[138] Jin, Y., Jameel, H., Chang, H., and Phillips, R. 2010. Green liquor pretreatment of mixed hardwood for ethanol production in a repurposed kraft pulp mill. *Journal of Wood Chemistry and Technology* 30(1):86–104.

[139] Walton, S.L., Hutto, D., Genco, J.M., van Walsum, G.P., and van Heiningen, A.R.P. 2010. Pre-extraction of hemicelluloses from hardwood chips using an alkaline wood pulping solution followed by kraft pulping of the extracted wood chips. *Industrial Engineering and Chemistry Research* 49(24):12638–12645.

[140] Yoon, S.-H., and van Heiningen, A. 2010. Green liquor extraction of hemicelluloses from southern pine in an integrated forest biorefinery. *Journal of Industrial and Engineering Chemistry* 16(1):74–80.

[141] Lundberg, V., Axelsson, E., Mahmoudkhani, M., and Berntsson, T. 2012. Process integration of near-neutral hemicellulose extraction in a Scandinavian kraft pulp mill – consequences for the steam and Na/S balances. *Applied Thermal Engineering* 43:42–50.

[142] Helmerius, J., von Walter, J.V., Rova, U., Berglund, K.A., and Hodge, D.B. 2010. Impact of hemicellulose pre-extraction for bioconversion on birch kraft pulp properties. *Bioresource Technology* 101(8):5996–6005.

[143] De Lopez, S., Tissot, M., and Delmas, M. 1996. Integrated cereal straw valorization by an alkaline pre-extraction of hemicellulose prior to soda-anthraquinone pulping. Case study of barley straw. *Biomass and Bioenergy* 10(4):201–211.

[144] Jun, A., Tschirner, U.W., and Tauer, Z. 2012. Hemicellulose extraction from aspen chips prior to kraft pulping utilizing kraft white liquor. *Biomass and Bioenergy* 37(2):229–236.

[145] Mosier, N., Wyman, C., Dale, B., Elander, R., Lee, Y.Y., Holtzapple, M., and Ladisch, M. 2005. Features of promising technologies for pretreatment of lignocellulosic biomass. *Bioresource Technology* 96(6):673–686.

[146] Al-Dajani, W.W., and Tschirner, U.W. 2008. Pre-extraction of hemicelluloses and subsequent kraft pulping. Part I: Alkaline extraction. *TAPPI Journal* 7(6):3–8.

[147] Park, Y.C., and Kim, J.S. 2012. Comparison of various alkaline pretreatment methods of lignocellulosic biomass. *Energy* 47(1):31–35.

[148] Longue Júnior, D., Ayoub, A., Venditti, R.A., Jameel, H., Colodette, J.L., and Chang, H.-M. 2013. Ethanol precipitation of hetero-polysaccharide material from hardwood by alkaline extraction prior to the kraft cooking process. *BioResources* 8(4):5319–5332.

[149] Johakimu, J.K., Jerome, A., Sithole, B.B., and Prabashni, L. 2016. Fractionation of organic substances from the South African *Eucalyptus grandis* biomass by a combination of hot water and mild alkaline treatments. *Wood Science and Technology* 50(2):365–384.

[150] Al-Dajani, W.W., and Tschirner, U.W. 2010. Pre-extraction of hemicelluloses and subsequent ASA and ASAM pulping: Comparison of autohydrolysis and alkaline extraction. *Holzforschung* 64(4):411–416.

[151] Chen, B.-Y., Chen, S.-W., and Wang, H.T. 2012. Use of different alkaline pretreatments and enzyme models to improve low-cost cellulosic biomass conversion. *Biomass and Bioenergy* 39(4):182–191.

[152] Sim, K., Youn, H.J., Cho, H., Shin, H., and Lee, H.L. 2012. Improvements in pulp properties by alkali pre-extraction and subsequent kraft pulping with controlling H-factor and alkali charge. *BioResources* 7(4):5864–5878.

[153] Sun, Y., and Cheng, J. 2002. Hydrolysis of lignocellulosic materials for ethanol production: A review. *Bioresource Technology* 83(1):1–12.

[154] Lehto, J., van Heiningen, A., Patil, R., Louhelainen, J., and Alén, R. 2021. Effect of sodium borohydride and hydrogen peroxide pretreatments on soda pulping of sugar maple (*Acer saccharum*). *Journal of Wood Chemistry and Technology* 41(2–3):128–136.

[155] El Mansouri, N.-E., Yuan, Q., and Huang, F. 2011. Characterization of alkaline lignins for use in phenol-formaldehyde and epoxy resins. *BioResources* 6(3):2647–2662.

[156] Bujanovic, B.M., Goundalkar, M.J., and Amidon, T.E. 2012. Increasing the value of a biorefinery based on hot-water extraction: Lignin products. *TAPPI Journal* 11(1):19–26.

[157] Mao, J.Z., Zhang, L.M., and Xu, F. 2012. Fractional and structural characterization of alkaline lignins from *Carex meyeriana* Kunth. *Cellulose Chemistry and Technology* 46(3–4):193–205.

[158] Xiao, L.-P., Shi, Z.-J., Xu, F., and Sun, R.-C. 2012. Characterization of lignins isolated with alkaline ethanol from the hydrothermal pretreated *Tamarix ramosissima*. *Bioenergy Research* 6(2):519–532.

[159] Veeramani, H., and Baba, M.A. 1983. Effect of alkali pretreatment on eucalypt chips on kraft black liquor viscosity and pulp strength. *Journal of Wood Chemistry and Technology* 3(1):17–34.

[160] van Heiningen, A. 2006. Converting a kraft pulp mill into an integrated forest biorefinery. *Pulp and Paper Canada* 107(6):38–43.

[161] Grace, T.M., Leopold, B., Malcolm, E.W., and Kocurek, M.J. (Eds). 1989. *Pulp and Paper Manufacture, Volume 5, Alkaline Pulping*. 3rd edition. The Joint Textbook Committee of the Paper Industry, TAPPI&CPPA, USA and Canada. p. 23–44.

[162] Biermann, C.J. 1996. *Handbook of Pulping and Papermaking*. 2nd edition. Academic Press, San Diego, CA, USA. Pp. 86–91.

[163] Smook, G. 2016. *Handbook for Pulp and Paper Technologists*. 4th edition. TAPPI Press, Atlanta, GA, USA. Pp. 74–83.

[164] Hubbe, M.A., Alén, R., Paleologou, M., Kannangara, M., and Kihlman, J. 2019. Lignin recovery from spent alkaline pulping liquors using acidification, membrane separation, and related processing steps: A review. *BioResources* 14(1):2300–2351.

[165] Niemelä, K. 1990. *Low-Molecular-Weight Organic Compounds in Birch Kraft Black Liquor*. Doctoral Thesis. *Annales Academiæ Scientiarum Fennicæ, Series A, II. Chemica*, Vol. 229. Helsinki, Finland. 142 p.

[166] Glasser, W.G., Barnett, C.A., and Sano, Y. 1983. Classification of lignins with different genetic and industrial origins. *Journal of Applied Polymer Science* 37:441–460.

[167] Goring, D.A.I. 1971. Polymer properties of lignin and lignin derivatives. In: Sarkanen, K.V., and Ludwig, C.H. (Eds.). *Lignins. Occurrence, Formation, Structure and Reactions*. Wiley-Interscience, New York, NY, USA. Pp. 695–768.

[168] Lehto, J., Pakkanen, H., and Alén, R. 2015. Molecular mass distribution of sulfur-free lignin from alkaline pulping preceded by hot-water-extraction. *Appita Journal* 68(2):149–157.

[169] Mimms, A., Kocurek, M.J., Pyatte, J.A., and Wright, E.E. (Eds.). 1989. *Kraft Pulping – A Compilation of Notes*. TAPPI Press, Atlanta, GA, USA. p. 55–76.

[170] Frederick, W.J. 1997. Black liquor properties. In: Adams, T.N. (Ed.). *Kraft Recovery Boilers*. TAPPI Press, Atlanta, GA, USA. Pp. 59–99.

[171] Tran, H. (Ed.). 2020. *Kraft Recovery Boilers*. 3rd edition. TAPPI Press, Atlanta, GA, USA. 375 p.

[172] Alén, R., Niemelä, K., and Sjöström, E. 1984. Gas-liquid chromatographic separation of hydroxy monocarboxylic acids and dicarboxylic acids on a fused-silica capillary column. *Journal of Chromatography* 301(1):273–276.

[173] Alén, R., Jännäri, P., and Sjöström, E. 1985. Gas-liquid chromatographic determination of lactic acid and C_1–C_6 volatile fatty acids as their benzyl esters on a fused-silica capillary column. *Finnish Chemical Letters* 12(5):190–192.

[174] Käkölä, J. 2009. *Fast Chromatographic Methods for Determining Aliphatic Carboxylic Acids in Black Liquors*. Doctoral Thesis. University of Jyväskylä, Laboratory of Applied Chemistry, Jyväskylä, Finland. 92 p.

[175] Wegener, G., Przyklenk, M., and Fengel, D. 1983. Hexafluoropropanol as valuable solvent for lignin in UV and IR spectroscopy. *Holzforschung* 37(6):303–307.

[176] Alén, R., and Hartus, T. 1988. UV spectrophotometric determination of lignin from alkaline pulping liquors. *Cellulose Chemistry and Technology* 22:613–618.

[177] Lin, S.Y. 1992. Ultraviolet spectrophotometry. In: Lin, S.Y., and Dence, C.W. (Eds.). *Methods in Lignin Chemistry*. Springer-Verlag, Heidelberg, Germany. Pp. 217–232.

[178] Parviainen, K., Jaakkola, H., and Nurminen, K. 2008. Evaporation of black liquor. In: Tikka, P. (Ed.). *Chemical Pulping Part 2, Recovery of Chemicals and Energy*. 2nd edition. Paper Engineers'. Association, Helsinki, Finland. Pp. 36–84.

[179] Söderhjelm, L., and Hausalo, T. 1996. Extensive analysis of strong black liquor. *Appita Journal* 49(4):263–268.

[180] Grace, T.M. 1989. Black liquor evaporation. In: Grace, T.M., Leopold, B., Malcolm, E.W., and Kocurek, M.J. (Eds.). *Pulp and Paper Manufacture*. Vol. *5, Alkaline Pulping*. 3rd edition. The Joint Textbook Committee of the Paper Industry, TAPPI&CPPA, USA and Canada. Pp. 477–530.

[181] Cardoso, M., De oliveira, É.D., and Passos, M.L. 2009. Chemical composition and physical properties of black liquors and their effects on liquor recovery operation in Brazilian pulp mills. *Fuel* 88:756–763.

[182] Oye, R., Langfors, N.G., Phillips, F.H., and Higgins, H.G. 1977. The properties of kraft black liquors from various eucalypts and mixed tropical hardwoods. *Appita Journal* 31(1):33–40.

[183] Söderhjelm, L. 1988. Viscosity of strong black liquor from birch pulping. *Paperi ja Puu* 70(4):348–351,353–355.

[184] Salin, J.-G. 1988. Spent liquor rheology and viscosity. *Paperi ja Puu* 70(8):721–727.

[185] Söderhjelm, L. 1988. Factors affecting the viscosity of strong black liquor. *Appita Journal* 41(5):389–392.

[186] Milanova, E., and Dorris, G.M. 1990. Effects of residual alkali content on the viscosity of kraft black liquors. *Journal of Pulp and Paper Science* 16(3):J94–J101.

[187] Söderhjelm, L., Sågfors, P.-E., and Janson, J. 1992. Black liquor viscosity. *Paperi ja Puu* 74(1):56–58.

[188] Ramamurthy, P., van Heiningen, A.R.P., and Kubes, G.J. 1993. Viscosity and thermal conductivity of black liquor. *TAPPI Journal* 76(11):175–179.

[189] Söderhjelm, L., and Sågfors, P.-E. 1994. Factors influencing the viscosity of kraft black liquor. *Journal of Pulp and Paper Science* 20(4):J106–J110.

[190] Zaman, A.A., and Fricke, A.L. 1995. Viscosity of softwood kraft black liquors at low solids concentrations: Effects of solids content, degree of delignification and liquor composition. *Journal of Pulp and Paper Science* 21(4):J119–J126.

[191] Kiiskilä, E., and Virkola, N.-E. 1987. Method of decreasing black liquor viscosity. Patent WO1987003315A1 04.06.

[192] Porter, J., Magnotta, V., Mullen, T., and Zielinski, J. 1998. Oxidative heat treatment for increasing recovery capacity: Concept & initial in-mill pilot evaluation. *TAPPI Proceedings of 1998 International Chemical Recovery Conference*. Pp. 193–211.

[193] Söderhjelm, L., Kiiskilä, E., and Sågfors, P.-E. 1999. Factors influencing heat treatment of black liquor. *Journal of Pulp and Paper Science* 25(10):367–371.

[194] Louhelainen, J., Alén, R., and Zielinski, J. 2002. Influence of the oxidative thermochemical treatment on the chemical composition of hardwood kraft black liquor. *TAPPI Journal* 1(10):9–13.

[195] Louhelainen, J., Alén, R., Zielinski, J., and Sågfprs, P.-E. 2002. Effects of oxidative and non-oxidative thermal treatments on the viscosity and chemical composition of softwood kraft black liquor. *Journal of Pulp and Paper Science* 28(9):285–291.

[196] Louhelainen, J.H., Alén, R.J., Feng, Z., and Pakkanen, H.K. 2003. Changes in the chemical composition of non-wood black liquors from alkaline pulping during heat treatment. *Appita Journal* 56(6):460–467,475.

[197] Louhelainen, J. 2003. *Changes in the Chemical Composition and Physical Properties of Wood and Nonwood Black Liquors during Heating*. Doctoral Thesis. University of Jyväskylä, Laboratory of Applied Chemistry, Jyväskylä, Finland. 68 p.

[198] Zaman, A.A., and Fricke, A.L. 1995. Effects of pulping conditions and black liquor composition on the heat of combustion of slash pine black liquor. *Advances in Pulp and Papermaking, AIChE Symposium Series* 91(307):154–161.

[199] Findley, M.E. 1960. Production of lignin and organic acids from kraft pulping black liquor. *TAPPI* 43(8):183A–188A.

[200] Wenzl, H., and Ingruber, O.V. 1967. Recovery of by-products of the kraft pulping process. *Paper Trade Journal* 13:53–57.

[201] Chambost, V., McNutt, J., and Stuart, P.R. 2009. Partnerships for successful enterprise transformation of forest industry companies implementing the forest biorefinery. *Pulp and Paper Canada* 110(5/6):19–24.

[202] Ghezzaz, H., Pelletier, L., and Stuart, P.R. 2012. Biorefinery implementation for recovery debottlenecking at existing pulp mills – Part I: Potential for debottlenecking. *TAPPI Journal* 11(7):17–25.

[203] Fardim, P. (Ed.). 2011. *Chemical Pulping Part 1, Fibre Chemistry and Technology*. Paper Engineers' Association, Helsinki, Finland. p. 748.

[204] Wimby, M. 2021. Fractionation – lignin. In: Alén, R. (Ed.). *Pulping and Biorefining – ForestBioFacts, Digital Learning Environment*. Paperi ja Puu Ltd., Helsinki, Finland.

[205] Kihlman, J. 2016. The sequential liquid-lignin recovery and purification process: Analysis of integration aspects for a kraft pulp mill. *Nordic Pulp and Paper Journal* 31(4):573–582.

[206] Sjöström, E. 1989. The origin of charge on cellulosic fibers. *Nordic Pulp and Paper Research Journal* 4(2):90–93.

[207] Merewether, J.W.T. 1962. Lignin XIV – the precipitation of lignin from kraft black liquor. *Holzforschung* 15(6):169–177.

[208] Alén, R., Patja, P., and Sjöström, E. 1979. Carbon dioxide precipitation of lignin from pine kraft black liquor. *TAPPI* 62(11):108–110.

[209] Wallmo, H. 2008. *Lignin Extraction from Black Liquor – Precipitation, Filtration and Washing*. Doctoral Thesis. Chalmers University of Technology, Forest Products and Chemical Engineering, Gothenburg, Sweden. p. 73.

[210] Zhu, W. 2015. *Precipitation of Kraft Lignin – Yield and Equilibrium*. Doctoral Thesis. Chalmers University of Technology, Department of Chemistry and Chemical Engineering. Gothenburg, Sweden. p. 63.

[211] Lawoko, M., Henriksson, G., and Gellerstedt, G. 2005. Structural differences between the lignin-carbohydrate complexes present in wood and in chemical pulps. *Biomacromolecules* 6(6):3467–3473.

[212] Lawoko, M., Henriksson, G., and Gellerstedt, G. 2006. Characterisation of lignin-carbohydrate complexes (LCCs) of spruce wood (*Picea abies* L.) isolated with two methods. *Holzforschung* 60(2):156–161.

[213] Gellerstedt, G., Tomani, P., Axegård, P., and Backlund, B. 2013. Lignin recovery and lignin-based products. In: Christopher, L. (Ed.). *Integrated Forest Biorefineries – Challenges and Opportunities*. Royal Society of Chemistry, Cambridge, United Kingdom. Pp. 180–210.

[214] Tarasov, D., Leitch, M., and Fatehi, P. 2018. Lignin-carbohydrate complexes: Properties, applications, analysis, and methods of extraction: A review. *Biotechnology for Biofuels* 11(1):1–28.

[215] Wallmo, H., Theliander, H., Jönssön, A.S., Wallberg, O., and Lindgren, K. 2009. The influence of hemicelluloses during the precipitation of lignin in kraft black liquor. *Nordic Pulp and Paper Research Journal* 24(2):165–171.

[216] Merewether, J.W.T. 1962. The precipitation of lignin from eucalyptus kraft black liquors. *TAPPI* 45(2):159–163.

[217] Ball, F.J., and Vardell, W.G. 1962. Continuous acidulation and coagulation of lignin in black liquor. U.S. Patent 3,048,576, August 7, 1962.

[218] Rydholm, S.A. 1965. *Pulping Processes*. Interscience Publishers, New York, NY, USA. Pp. 583–589.

[219] Pollak, A., Keilen, J.J. Jr., and Drum, L.F. 1949. Method of producing lignin from black liquor. U.S. Patent 2,464,828, Marc 22, 1949.

[220] Gray, K., Crosby, H.L., and Steinberg, J.C. 1956. Recovery of chemicals in wood pulp preparation. U.S. Patent 2,772,965, December 4, 1956.

[221] Ball, F.J., and Vardell, W.G. 1958. Decantation of lignin. U.S. Patent 2,9997,466, November 4, 1958.

[222] Giesen, J. 1958. Process for the recovery of lignin from black liquors. U.S. Patent 2,828,297, March 25, 1958.

[223] Nikitin, V.M., Obolenskaya, A.V., Skachkov, V.M., and Ivanenko, A.D. 1963. Precipitation of alkali lignin with carbon dioxide under pressure. *Bumazhnaya Promyshlennost'* 38(11):14–15.

[224] Fischer, F., and Wienhaus, O. 1982. Die Abtrennung des Lignins aus Sulfatschwarzlaugen als Vorstufe für die stoffwirtschatliche Nutzung dieses Polyphenols. Teil 1: Grundlagen. *Zellstoff Und Papier* 31(4):149–152.

[225] Fischer, F., and Wienhaus, O. 1982. Die Abtrennung des Lignins aus Sulfatschwarzlaugen als Vorstufe für die stoffwirtschatliche Nutzung dieses Polyphenols. Teil 2: Untersuchungen zur Gewinnung von Sulfatablaugenlignin aus Kiefern-Schwarzlauge. *Zellstoff und Papier* 31(5):198–204.

[226] Alén, R., Sjöström, E., and Vaskikari, P. 1985. Carbon dioxide precipitation of lignin from alkaline pulping liquors. *Cellulose Chemistry and Technology* 19:537–541.

[227] Kim, H., Hill, M.K., and Fricke, A.L. 1987. Precipitation of kraft lignin from black liquor. *TAPPI Journal* 70(12):112–116.

[228] Uloth, V., and Wearing, J. 1989. Kraft lignin recovery: Acid precipitation versus ultrafiltration. Part I: Laboratory test results. *Pulp and Paper Canada* 90(9):67–71.

[229] Gellerstedt, G., Tomani, P., Axegård, P., and Backlund, B. 2013. Lignin recovery and lignin-based products. In: Christopher, L. (Ed.). *Integrated Forest Biorefineries – Challenges and Opportunities*. Royal Society of Chemistry, Cambridge, UK. Pp. 180–210.

[230] Keilen, J.J. Jr., Ball, F.J., and Gressang, R.W. 1952. Method of coagulating colloidal lignates in aqueous dispersions. U.S. Patent 2,623,040, December 23, 1952.

[231] Whalen, D.M., and Tokoli, E.G. 1968. Lignin precipitation from black liquor in the presence of chloro, bromo or nitro containing hydrocarbons. U.S. Patent 3,546,200, June 18, 1968.

[232] Whalen, D.M. 1975. Simple method for precipitating easily filterable acid lignin from kraft black liquor. *TAPPI* 58(5):110–112.

[233] Brežny, R., Micko, M.M., and Paszner, L. 1988. Mild coagulation of aqueous suspensions of kraft lignin and its derivatives. *Holzforschung* 42(5):335–336.

[234] Uloth, V., and Wearing, J. 1989. Kraft lignin recovery: Acid precipitation versus ultrafiltration. Part II: Technology and economics. *Pulp and Paper Canada* 90(10):34–37.

[235] Loutfi, H., Blackwell, B., and Uloth, V. 1991. Lignin recovery from kraft black liquor: Preliminary process design. *TAPPI* 74(1):203–210.

[236] Moosavifar, A. 2008. *Lignin Extraction from Black Liquor – Properties of the Liquors and Sulphur Content in the Lignin*. Doctoral Thesis. Chalmers University of Technology, Forest Products and Chemical Engineering, Gothenburg, Sweden. 79 p.

[237] Wallmo, H., Richards, T., and Theliander, H. 2009. An investigation of process parameters during lignin precipitation from kraft black liquors: A step towards an optimised precipitation operation. *Nordic Pulp and Paper Research Journal* 24(2):158–164.

[238] Wallmo, H. 2008. *Lignin Extraction from Black Liquor – Precipitation, Filtration and Washing*. Doctoral Thesis. Chalmers University of Technology, Forest Products and Chemical Engineering, Gothenburg, Sweden. 73 p.

[239] Zhu, W. 2015. *Precipitation of Kraft Lignin – Yield and Equilibrium*. Doctoral Thesis. Chalmers University of Technology, Department of Chemistry and Chemical Engineering. Gothenburg, Sweden. 63 p.

[240] Sewring, T. 2019. *Precipitation of Kraft Lignin from Aqueous Solutions*. Doctoral Thesis. Chalmers University of Technology, Department of Chemistry and Chemical Engineering. Gothenburg, Sweden. 80 p.

[241] Merewether, J.W.T. 1962. Lignin XV – the coagulation of lignin salt from acidified kraft black liquor. *Holzforschung* 16(1):26–29.

[242] Dhingra, D.R., Bhatnagar, M.S., and Nigam, P.C. 1952. Lignin recovery from soda black liquor. *Indian Pulp and Paper* 7:311–315.

[243] Basu, S. 1971. Studies on carbonation of black liquor for lignin precipitation and its subsequent separation. *Indian Pulp & Paper Technology* (IPPTA) 8:207–214.

[244] Schulze, P., Seidel-Morgenstern, A., Lorenz, H., Leschinsky, M., and Unkelbach, G. 2016. Advanced process for precipitation of lignin from ethanol organosolv spent liquors. *Bioresource Technology* 199:128–134.

[245] Kumar, H., Alén, R., and Sahoo, G. 2016. Characterization of hardwood soda-AQ lignins precipitated from black liquor through selective acidification. *BioResources* 11(4):9869–9879.

[246] Radhamohan, K., and Basu, S. 1980. Electrodialysis in the regeneration of paper mill spent liquor. *Desalination* 33:185–200.

[247] Arulanantham, M.E.L.N., and Shanthini, R. 1997. Recovery of sodium hydroxide from *Embilipitiya* black liquor by electrodialysis. *Indian Pulp & Paper Technology* (IPPTA) 9(3):1–8.

[248] Tomani, P. 2010. The LignoBoost process. *Cellulose Chemistry and Technology* 44:53–58.

[249] Bajpai, P. 2013. *Biorefinery in the Pulp and Paper Industry*. Elsevier, Amsterdam, The Netherlands. p. 103.

[250] Björk, M., Rinne, J., Kotilainen, A., Korhonen, V., Wallmo, H., and Karlsson, H. 2015. Successful start-up of lignin extraction at Stora Enso Sunila mill. *Proceedings of Nordic Wood Biorefinery Confrence 2015*, Helsinki, Finland.

[251] Kouisni, L., Fang, Y.L., Paleologou, M., Ahvazi, B., Hawari, J., Zhang, Y.L., and Wang, X.-M. 2011. Kraft lignin recovery and its use in the preparation of lignin-based phenol formaldehyde resins for plywood. *Cellulose Chemistry and Technology* 45(7,8): 515–520.

[252] Kouisni, L., Holt-Hindle, P., Maki, K., and Paleologou, M. 2012. The LignoForce system TM: A new process for the production of high-quality lignin from black liquor. *Journal of Science and Technology for Forest Products* 2(4):6–10.

[253] Kouisni, L., Gagné, A., Maki, K., Holt-Hindle, P., and Paleologou, M. 2016. LignoForce System for the recovery of lignin from black liquor. Feedstock options, odor profile, and product characterization. *ACS Sustainable Chemistry & Engineering* 4(10):5152–5159.

[254] Lundquist, K., and Kirk, K.T. 1980. Fractionation-purification of an industrial kraft lignin. *TAPPI* 63(1):80–82.

[255] Coheen, D.W. 1983. Chemicals from lignin. In: Goldstein, I.S. (Ed.). *Organic Chemicals from Biomass*. 2nd printing. CRC Press, Boca Raton, FL, USA. Pp. 143–161.

[256] Falkehag, I. 1989. A systemic view of lignin uses. *Proceedings of the 1989 International Symposium on Wood and Pulping Chemistry*, May 22–25, 1989, Raleigh, NC, USA. Pp. 107–112.

[257] Glasser, W.G., and Sarkanen, S. 1989. *Lignin: Properties and Materials*. American Chemical Society, Washington, DC, USA. p. 545.

[258] Vasudevan, N., and Mahadevan, A. 1990. Degradation of black liquor lignin by microorganisms. *Holzforschung* 44(5):325–330.

[259] Pye, E.K. 2006. Industrial lignin production and applications. In: Kamm, B., Gruber, P.R., and Kamm, M. (Eds.). *Biorefineries – Industrial processes and Products, Volume 2*. Wiley-VCH, Weinheim, Germany. Pp. 165–200.

[260] Öhman, F., and Theliander, H. 2006. Washing lignin precipitated from kraft black liquor. *Paperi ja Puu* 88(5):287–292.

[261] Bozell, J., Holladay, J., Johnson, D., and White, J. 2007. *Top Value Added Chemicals from Biomass – Volume II: Results of Screening for Potential Candidates from Biorefinery Lignin*. U.S. Department of Energy, Oak Ridge, TN, USA. 79 p.

[262] Henriksson, G., Li, J., Zhang, L., and Lindström, M.E. 2010. Lignin utilization. In: Crocker, M. (Ed.). 2010. *Thermochemical Conversion of Biomass to Liquid Fuels and Chemicals*. RSC Publishing, Cambridge, United Kingdom. Pp. 222–262.

[263] Nagy, M., Kosa, M., Theliander, H., and Ragauskas, A.J. 2010. Characterization of CO_2 precipitated Kraft lignin to promote its utilization. *Green Chemistry* 12:31–34.

[264] Vishtal, A., and Kraslawski, A. 2011. Challenges in industrial applications of technical lignins. *BioResources* 6(3):3547–3568.

[265] Gellerstedt, G. 2015. Softwood kraft lignin: Raw material for the future. *Industrial Crops and Products* 77:845–854.

[266] Chio, C., Sain, M., and Qin, W. 2019. Lignin utilization; a review of lignin depolymerization from various aspects. *Renewable and Sustainable Energy Reviews* 107:232–249.

[267] Richardson, B., and Uloth, V.C. 1990. Kraft lignin: A potential fuel for lime kilns. *TAPPI Journal* 73(10):191–194.

[268] Richardson, B., Watkinson, A.P., and Barr, P.V. 1990. Combustion of lignin in a pilot lime kiln. *Proceedings of the 1990 Pulping Chemistry, Book 1*, October 14–17, 1990, Toronto, ON, Canada. Pp. 457–465.

[269] Li, Q., Ragauskas, A.J., and Yuan, J.S. 2017. Lignin carbon fiber: The path for quality. *TAPPI Journal* 16(3):107–108.

[270] Ayoub, A., Treasure, T., Hansen, L., Nypelö, T., Jameel, H., Khan, S., Chang, H.-M., Hubbe, M.A., and Venditti, R.A. 2021. Effect of plasticizers and polymer blends for processing softwood kraft lignin as carbon fiber precursors. *Cellulose* 28:1039–1053.

[271] McDonald, C. 2012. A proving ground for lignin. *Pulp and Paper Canada* 113(3):14–16.

[272] Kruus, K., and Hakala, T. 2016. *The Making Bioeconomy Transformation*. VTT Technical Research Centre of Finland Ltd., Espoo, Finland. Pp. 36–37.

[273] Glasser, W.G., Loos, R., Cox, B., and Cao, N. 2017. Melt-blown compostable polyester films with lignin. *TAPPI Journal* 16(3):111–121.

[274] Gericke, M., Bergrath, J., Schulze, M., and Heinze, T. 2022. Composite nanoparticles derived by self-assembling of hydrophobic polysaccharide derivatives and lignin. *Cellulose* 29:3613–3620.

[275] Zhang, Y., Ni, S., Wang, X., Zhang, W., Lagerquist, L., Qin, M., Willför, S., Xu, C., and Fatehi, P. 2019. Ultrafast adsorption of heavy metal ions onto functionalized lignin-based hybrid magnetic nanoparticles. *Chemical Engineering Journal* 372:82–91.

[276] Araújo, J.D.P. 2008. *Production of Vanillin from Lignin Present in the Kraft Black Liquor of the Pulp and Paper Industry*. Doctoral Thesis. University of Porto, Department of Chemical Engineering, Porto, Portugal. 318 p.

[277] Hill, M., and Fricke, A.L. 1984. Ultrafiltration studies on a kraft black liquor. *TAPPI* 67(6):100–103.

[278] Alén, R., Sjöström, E., and Vaskikari, P. 1986. Ultrafiltration studies on alkaline pulping liquors. *Cellulose Chemistry and Technology* 20:417–420.

[279] Lin, S.Y. 1992. Ultrafiltration. In: Stephen, Y.L., and Dence, C.D. (Eds.). *Methods in Lignin Chemistry*. Springer-Verlag, Heidelberg, Germany. Pp. 518–523.

[280] Wallberg, O., Jönsson, A.-S., and Wimmerstedt, R. 2003. Fractionation and concentration of kraft black liquor lignin with ultrafiltration. *Desalination* 154(2):187–199.

[281] Wallberg, O., Holmqvist, A., and Jönsson, A.-S. 2005. Ultrafiltration of kraft cooking liquors from a continuous cooking process. *Desalination* 180(1–3):109–118.

[282] Wallberg, O., and Jönsson, A.-S. 2006. Separation of lignin in kraft cooking liquor from a continuous digester by ultrafiltration at temperatures above 100 °C. *Desalination* 195:187–200.

[283] Sevastyanova, O., Helander, M., Chowdhury, S., Lange, H., Wedin, H., Zhang, L., Ek, M., Kadla, J.F., Crestini, C., and Lindström, M.E. 2014. Tailoring the molecular and thermo-mechanical properties of kraft lignin by ultrafiltration. *Journal of Applied Polymer Science* 131(18):40799/1–40799/11.

[284] Zhu, W.Z., Westman, G., and Theliander, H. 2016. Lignin separation from kraft black liquor by combined ultrafiltration and precipitation: A study of solubility of lignin with different molecular properties. *Nordic Pulp and Paper Research Journal* 31(2):270–278.

[285] Jin, W., Tolba, R., Wen, J.L., Li, K.C., and Chen, A.C. 2013. Efficient extraction of lignin from black liquor via a novel membrane-assisted electrochemical approach. *Electrochimica Acta* 107:611–618.

[286] Mattson, T., Lewis, W.J.T., Chew, Y.M.J., and Bird, M.R. 2015. In situ investigation of soft cake fouling layers using fluid dynamic gauging. *Food and Bioproducts Processing* 93:205–210.

[287] Haddad, M., Bazinet, L., Savadogo, O., and Paris, J. 2017. A feasibility study of a novel electro-membrane based process to acidify kraft black liquor extract lignin. *Process Safety and Environmental Protection* 106:68–75.

[288] Haddad, M., Mikhaylin, S., Bazinet, L., Savadogo, O., and Paris, J. 2017. Electrochemical acidification of kraft black liquor: Effect of fouling and chemical cleaning on ion exchange membrane integrity. *ACS Sustainable Chemistry & Engineering* 5(1):168–178.

[289] Humbert, D., Ebrahimi, M., and Czermak, P. 2016. Membrane technology for the recovery of lignin: A review. *Membranes* 6(3):article no. 42.

[290] Fane, A.G., and Fell, C.J.D. 1987. A review of fouling and fouling control in ultrafiltration. *Desalination* 62:117–136.

[291] Hubbe, M.A., Chen, H., and Heitmann, J.A. 2009. Permeability reduction phenomena in packed beds, fiber mats, and wet webs of paper exposed to flow of liquids and suspensions: A review. *BioResources* 4(1)405–451.

[292] Shi, X.F., Tal, G., Hankins, N.P., and Gitis, V. 2014. Fouling and cleaning of ultrafiltration membranes: A review. *Journal of Water Process Engineering* 1:121–138.

[293] Zeman, L.J., and Zydney, A.L. 2020. *Microfiltration and Ultrafiltration: Principles and Applications*. Bio-Green Books, New Delhi, India.

[294] Keyoumu, A., Sjödahl, R., Henriksson, G., Ek, M., Gellerstedt, G., and Lindström, M.E. 2004. Continuous nano- and ultra-filtration of kraft pulping black liquor with ceramic filters – a method for lowering the load on the recovery boiler while generating valuable side-products. *Industrial Crops and Products* 20(2):143–150.

[295] Holmqvist, A., Wallberg, O., and Jönsson, A.-S. 2005. Ultrafiltration of kraft black liquor from two Swedish pulp mills. *Chemical Engineering Research and Design* 83(8A):994–999.

[296] Toledano, A., Serrano, L., Garcia, A., Mondragon, I., and Labidi, J. 2010. Comparative study of lignin fractionation by ultrafiltration and selective precipitation. *Chemical Engineering Journal* 157(1)93–99.

[297] Arkell, A., Olsson, J., and Wallberg, O. 2014. Process performance in lignin separation from softwood black liquor by membrane filtration. *Chemical Engineering Research and Design* 92(9):1792–1800.

[298] Axegård, P. 2006. Utilization of black liquor and forestry residues in a pulp mill biorefinery. *Proceedings of Forest Based Sector Technology Platform Conference*, November 22–23, 2006, Lahti, Finland.

[299] Mateos-Espejel, E., Marinova, M., Schneider, S., and Paris, J. 2010. Simulation of a kraft pulp mill for the integration of biorefinery technologies and energy analysis. *Pulp and Paper Canada* 111(3):19–23.

[300] Green, J.W. 1956. Alkaline pulping. I. Saccharinic acids in kraft black liquor. *TAPPI* 39(7):472–477.

[301] Ackman, R.G. 1964. Fundamental groups in the response of flame ionization detectors to oxygenated aliphatic hydrocarbons. *Journal of Gas Chromatography* 3(6):173–179.

[302] Radej, Z., and Kristofova, Z. 1964. The fraction of hydroxy acids and lactones in black liquors. *Papir a Celulóza* 19(6):152–153.

[303] Verhaar, L.A.T., and de Wild, H.G.J. 1969. The gas chromatographic determination of polyhydroxy monocarbonic acids obtained by oxygenation of hexoses in aqueous alkaline solutions. *Journal of Chromatography A* 41(2):168–179.

[304] Pettersson, G. 1974. Gas-chromatographic analysis of sugars and related hydroxy acids as acyclic oxime and ester trimethylsilyl derivatives. *Carbohydrate Research* 33:47–61.

[305] Malinen, R., and Sjöström, E. 1975. The formation of carboxylic acids from wood polysaccharides during kraft pulping. *Paperi ja Puu* 57(11):728–730,735–736.

[306] Löwendahl, L., Petersson, G., and Samuelson, O. 1976. Formation of carboxylic acids by degradation of carbohydrates during kraft cooking of pine. *TAPPI* 59(9):118–121.

[307] Pettersson, G., 1977. Retention data in GLC analysis; carbohydrate-related hydroxy carboxylic acids and dicarboxylic acids as trimethylsilyl derivatives. *Journal of Chromatographic Science* 15(7):245–255.

[308] Hyppänen, T., Sjöström, E., and Vuorinen, T. 1983. Gas-liquid chromatographic determination of hydroxy carboxylic acids on fused-silica capillary column. *Journal of Chromatography* 261:320–323.

[309] Niemelä, K., and Sjöström, E. 1986. Simultaneous identification of aromatic and aliphatic low molecular weight compounds from alkaline pulping liquors by capillary gas-liquid chromatography – mass spectrometry. *Holzforschung* 40:361–368.

[310] Alén, R., Niemelä, K., and Sjöström, E., 1984. Modification of alkaline pulping to facilitate isolation of aliphatic acids. Part 1. Sodium hydroxide pretreatment of pine wood. *Journal of Wood Chemistry and Technology* 4(4):405–419.

[311] Alén, R., Niemelä, K., and Sjöström, E., 1985. Modification of alkaline pulping to facilitate isolation of aliphatic acids. Part 2. Sodium hydroxide pretreatment of birch wood. *Journal of Wood Chemistry and Technology* 5(3):335–345.

[312] Niemelä, K., Alén, R., and Sjöström, E. 1985. The formation of carboxylic acids during kraft and kraft-anthraquinone pulping of birch wood. *Holzforschung* 39(3):167–172.

[313] Alén, R., Lahtela, M., Niemelä, K., and Sjöström, E. 1985. Formation of hydroxy carboxylic acids from softwood polysaccharides during alkaline pulping. *Holzforschung* 39(4):235–238.

[314] Alén, R. 1988. Formation of aliphatic carboxylic acids from hardwood polysaccharides during alkaline delignification in aqueous alcohols. *Cellulose Chemistry and Technology* 22(4):443–448.

[315] Alén, R. 1988. Formation of aliphatic carboxylic acids from softwood polysaccharides during alkaline delignification in aqueous alcohols. *Cellulose Chemistry and Technology* 22(5):507–512.

[316] Alén, R., Hentunen, P., Sjöström, E., Paavilainen, L., and Sundström, O. 1991. A new approach for process control of kraft pulping. *Journal of Pulp and Paper Science* 17(1):J6–J9.

[317] Alén, R. 1997. Analysis of degradation products. a new approach to characterizing the combustion properties of kraft black liquors. *Journal of Pulp and Paper Science* 23(4):J62–J66.

[318] Feng, Z. 2001. *Alkaline Pulping of Non-Wood Feedstocks and Characterization of Black Liquors*. Doctoral Thesis. University of Jyväskylä, Laboratory of Applied Chemistry, Jyväskylä, Finland. 54 p.

[319] Alén, R., Käkölä, J., and Pakkanen, H. 2008. Monitoring of kraft pulping by a fast analysis of aliphatic carboxylic acids. *Appita Journal* 61(3):216–219.

[320] Pakkanen, H. 2012. *Characterization of Organic Material Dissolved during Alkaline Pulping of Wood and Non-Wood Feedstocks*. Doctoral Thesis. University of Jyväskylä, Laboratory of Applied Chemistry, Jyväskylä, Finland. 76 p.

[321] Reed, R.W. 1939. *The Use of Saccharinic Acids from Black Liquor as a Plasticizer for Glassine Paper*. M.Sc. Thesis. Institute of Paper Chemistry, Appleton, WI, USA. 107 p.

[322] Pilyugina, L.G., Komshilov, N.F., Bachurina, G.V., and Dyukkiev, E.F. 1964. Separation of lignin and organic acids from black liquor. *Khimicheskaya Pererabotka Drevesiny, Nauchno-Tekhnologisheskie Sbornik* 14:9–12.

[323] Alén, R., and Sjöström, E. 1980. Isolation of hydroxy acids from pine kraft black liquors, Part 1. Preparation of crude fraction. *Paperi ja Puu* 62(5):328–330.

[324] Alén, R., and Sjöström, E. 1980. Isolation of hydroxy acids from pine kraft black liquor, Part 2. Purification by distillation. *Paperi ja Puu* 62(8):469–471.

[325] Alén, R., and Sjöström, E. 1981. Isolation of hydroxy acids from alkaline birch black liquors. *Paperi ja Puu* 63(1):5–6,16.

[326] Sjöström, E. 1983. Alternatives for balanced production of fibers, chemicals, and energy from wood. *Journal of Applied Polymer Science: Applied Polymer Symposium* 37:577–592.

[327] Alén, R., Moilanen, V.-P., and Sjöström, E. 1986. Potential recovery of hydroxy acids from kraft pulping liquors. *TAPPI Journal* 69(2):76–78.

[328] Kumar, H. 2016. *Novel Concepts on the Recovery of By-Products from Alkaline Pulping*. Doctoral Thesis. University of Jyväskylä, Laboratory of Applied Chemistry, Jyväskylä, Finland. 61 p.

[329] Kumar, H., and Alén, R. 2014. Partial recovery of aliphatic carboxylic acids and sodium hydroxide from hardwood black liquor by electrodialysis. *Industrial & Engineering Chemistry Research* 53:9464–9470.

[330] Alén, R. 1981. *Isolation of Aliphatic Acids from Pine and Birch Black Liquors*. Doctoral Thesis. Helsinki University of Technology, Laboratory of Wood Chemistry, Otaniemi, Finland. 22 p.

[331] Kumar, H., and Alén, R. 2015. Recovery of aliphatic low-molecular-mass carboxylic acids from hardwood kraft black liquor. *Separation and Purification Technology* 142:293–298.

[332] Biggs, W.A. Jr., Wise, J.T., Cook, W.R., Baxley, W.H., Robertson, J.D., and Copenhaver, J.E. 1961. Commercial production of acetic and formic acids from NSSC black liquor. *TAPPI* 44:385–392.

[333] Alén, R., Sjöström, E., and Suominen, S. 1990. Application of ion-exclusion chromatography to alkaline pulping liquors, separation of hydroxy carboxylic acids from inorganic solids. *Journal of Chemical Technology & Biotechnology* 51:225–233.

[334] Alén, R., and Sjöström, E. 1980. Condensation of glycolic, lactic, and 2-hydroxybutanoic acids during heating and identification of the condensation products by GLC-MS. *Acta Chemica Scandinavica B* 34(9):633–636.

[335] Raucq, D., Pourcelly, G., and Gavach, G. 1993. Production of sulfur acid and caustic soda from sodium sulfate by electromembrane process. Comparison between electro-electrodialysis and electrodialysis on bipolar membrane. *Desalination* 91(12):163–175.

[336] Genders, D. 1995. Splitting sodium sulfate to caustic and sulfuric acid. *Watts New – Quarterly Newsletter* 1(1):1–6.

[337] Paleologou, M., Thibault, A., Wong, P.-Y., Thompson, R., and Berry, R.M. 1997. Enhancement of the current efficiency for sodium hydroxide production from sodium sulphate in a two-compartment bipolar membrane electrodialysis system. *Separation and Purification Technology* 11:159–171.

[338] Wilhelm, F.G. 2001. *Bipolar Membrane Electrodialysis – Membrane Development and Transport Characteristics*. Doctoral Thesis. University of Twente, Twente, The Netherlands. 235 p.

[339] Tanaka, Y. 2013. *Ion Exchange Membrane Electrodialysis: Fundamentals, Desalination, Separation – Water Resource Planning, Development and Management*. Nova Science Publishers, New York, NY, USA. 308 p.

[340] Yang, S.-T., and Lu, C. 2013. Extraction-fermentation hybrid (extractive fermentation): In: Ramaswamy, S., Huang, H.-J., and Ramarao, B.V. (Eds.). *Separation and Purification Technologies in Biorefineries*. John Wiley & Sons, West Sussex, United Kingdom. Pp. 409–437.

[341] Mani, K.N. 1991. Electrodialysis water splitting technology. *Journal of Membrane Science* 58:117–138.

[342] Mishra, A.K., and Bhattacharya, P.K. 1984. Alkaline black liquor treatment by batch electrodialysis. *The Canadian Journal of Chemical Engineering* 62(5):723–727.

[343] Mishra, A.K., and Bhattacharya, P.K. 1987. Alkaline black liquor treatment by continuous electrodialysis. *Journal of Membrane Science* 33(1):83–95.

[344] Xie, M., Shon, H.K., Gray, S.R., and Elimelech, M. 2016. Membrane-based processes for wastewater nutrient recovery: Technology, challenges, and future direction. *Water Research* 89:210–221.

[345] Haddad, M., Labrecque, R., Bazinet, L., Savadogo, O., and Paris, J. 2016. Effect of process variables on the performance of electrochemical acidification of kraft black liquor by electrodialysis with bipolar membrane. *Chemical Engineering Journal* 304:977–985.

[346] Gladkii, A.A., Vishnyakova, E.V., Tskhai, A.A., Miliganov, A.P., Ivanova, L.M., and Tananina, I.N. 1990. The use of electrodialysis to desalinate waste water in low-waste viscose manufacturing technology. *Fibre Chemistry* 2281:46–49.

[347] Greiter, M., Novalin, S., Wendland, M., Kulbe, K.-D., and Fischer, J. 2002. Desalination of whey by electrodialysis and ion exchange resins: Analysis of both processes with regard to sustainability by calculating their cumulative energy demand. *Journal of Membrane Science* 210:91–102.

[348] Tado, K., Sakai, F., Sano, Y., and Nakayama, A. 2016. An analysis on ion transport process in electrodialysis desalination. *Desalination* 378:60–66.

[349] Bailly, M., Roux-de balmann, H., Aimar, P., Lutin, F., and Cheryan, M. 2001. Production processes of fermented organic acids targeted around membrane operations: Design of the concentration step by conventional electrodialysis. *Journal of Membrane Science* 191:129–142.

[350] Huang, H.-J., and Ramaswamy, S. 2013. Overview of biomass conversion processes and separation and purification technologies in biorefineries. In: Ramaswamy, S., Huang, H.-J., and Ramarao, B.V. (Eds.). *Separation and Purification Technologies in Biorefineries.* John Wiley & Sons, West Sussex, United Kingdom. Pp. 1–36.

[351] Lopez, A.M., and Hestekin, J.A. 2013. Separation of organic acids from water using ionic liquid assisted electrodialysis. *Separation and Purification Technology* 116:162–169.

[352] Luo, G., Pan, S., and Liu, J. 2002. Use of the electodialysis process to concentrate a formic acid solution. *Desalination* 150(3):227234.

[353] Nomura, Y., Iwahara, M., and Hongo, M. 1988. Acetic acid production by an electrodialysis fermentation method with a computerized control system. *Applied and Environmental Microbiology* 54(1):137–142.

[354] Jones, R.J., Massanet-Nicolau, J., Guwy, A., Premier, G.C., Dinsdale, R.M., and Reilly, M. 2015. Removal and recovery of inhibitory volatile fatty acids from mixed acid fermentations by conventional electrodialysis. *Bioresource Technology* 189:279–284.

[355] Boyaval, P., Seta, J., and Gavach, C. 1993. Concentrated propionic acid production by electrodialysis. *Enzyme and Microbial Technology* 15(8):683–686.

[356] Boyaval, P., Corre, C., and Terre, S. 1987. Continuous lactic acid fermentation with concentrated product recovery by ultrafiltration and electrodialysis. *Biotechnology Letters* 9(3):207–212.

[357] Åkerberg, C., and Zacchi, G. 2000. An economic evaluation of the fermentative production of lactic acid from wheat flour. *Bioresource Technology* 75:119–126.

[358] Kim, Y.H., and Moon, S.-H. 2001. Lactic acid recovery from fermentation broth using one-stage electrodialysis. *Journal of Chemical Technology & Biotechnology* 76(2):169–178.

[359] Novalic, S., Okwor, J., and Kulbe, K.D. 1996. The characteristics of citric acid separation using electrodialysis with bipolar membranes. *Desalination* 105(3):277–282.

[360] Pinacci, P., and Radaelli, M. 2002. Recovery of citric acid from fermentation broths by electrodialysis with bipolar membranes. *Desalination* 148(1–3):177–179.

[361] Tongwen, X., and Weihua, Y. 2002. Citric acid production by electrodialysis with bipolar membranes. *Chemical Engineering and Processing* 41(6):519–524.

[362] Sridhar, S. 1988. Application of electrodialysis in the production of malic acid. *Journal of Membrane Science* 36:489–495.

[363] Andres, L.J., Riera, F.A., and Alvarez, R. 1997. Recovery and concentration by electrodialysis of tartaric acid from fruit juice industries waste waters. *Journal of Chemical Technology & Biotechnology* 70(3):247–252.

[364] Molnár, E., Nemestóthy, N., and Bélafi-Bakó, K. 2010. Utilisation of bipolar electrodialysis for recovery of galacturonic acid. *Desalination* 250:1128–1131.

[365] Potapenko, A.P., Arestova, G.A., and Gerasimchuk, L.A. 1973. Study of the principal parameters of electrodialysis of black liquor with the aid of ion-exchange membranes. *Zhurnal Prikladnoi Khimii* 46(3):535–539. (In Russian).

[366] Paleologou, M., Cloutier, J.-N., Ramamurthy, P., Berry, R.M., Azarniouch, M.K., and Dorica, J. 1994. Membrane technologies for pulp and paper applications. *Pulp and Paper Canada* 95(10):36–40.

[367] Prabhu, A.M., and Basu, S. 1980. Studies of electrodialytic recovery and decationization of lignin from kraft black liquor. *Proceedings of the 7th International Symposium on Fresh Water from the Sea, Vol. 2*, September 23–26, Amsterdam, The Netherlands. Pp. 425–431.

[368] Ghatak, H.R. 2009. Reduction of organic pollutants with recovery of value-added products from soda black liquor of agricultural residues by electrolysis. *TAPPI Journal* 8(7):4–10.

[369] Ghatak, H.R. 2009. Economic potential of black liquor electrolysis as a treatment option for small agro-based mills. *TAPPI Journal* 8(11):4–11.

[370] Jin, W., Tolba, R., Wen, J.L., Li, K.C., and Chen, A.C. 2013. Efficient extraction of lignin from black liquor via a novel membrane-assisted electrochemical approach. *Electrochimica Acta* 107:611–618.

[371] Rowe, J.W., and Gregor, H.H. 1986. Membrane processes for separation of organic acids from kraft black liquors. U.S. Patent 4,584,057, April 22, 1986.

[372] Kumar, H., and Alén, R. 2014. Partial recovery of aliphatic carboxylic acids and sodium hydroxide from hardwood black liquor by electrodialysis. *Industrial & Engineering Chemistry Research* 53(22):9464–9470.

[373] Patil, R., Genco, J., Pendse, H., and van Heiningen, A. 2016. Treating kraft mill extract using bipolar membrane electrodialysis for the production of acetic acid. *TAPPI Journal* 15(3):215–226.

[374] Tanaka, K., and Fritz, J.S. 1986. Separation of aliphatic carboxylic acids by ion-exclusion chromatography using a weak-acid eluent. *Journal of Chromatography A* 361:151–160.

[375] Fritz, J.S. 1991. Principles and applications of ion-exclusion chromatography. *Journal of Chromatography A* 546:111–118.

[376] Glód, B.K. 1997. Ion exclusion chromatography: Parameters influencing retention. *Neurochemical Research* 22(10):1237–1248.

[377] Haddad, P.R., and Jackson, P.E. 2003. *Ion Chromatography: Principles and Applications*. Elsevier, Amsterdam, The Netherlands. p. 776.

[378] Weiss, J. 2004. *Handbook of Ion Chromatography*. 3rd edition. Wiley-VCH, Weinheim, Germany. 894 p.

[379] Fritz, J.S., and Gjerde, D.T. 2009. *Ion Chromatography*. 4th edition. Wiley-VCH, Weinheim, Germany. 385 p.

[380] Springfield, R.M., and Hester, R.D. 1999. Continuous ion-exclusion chromatography system for acid/sugar separation. *Separation Science and Technology* 34(6,7):1217–1241.

[381] Ohta, K., Tanaka, K., and Haddad, P.R. 1996. Ion-exclusion chromatography of aliphatic carboxylic acids on an unmodified silica gel column. *Journal of Chromatography A* 739(1,2):359–365.

[382] Tanaka, K., Ding, M.-Y., Helaleh, M.I.H., Taoda, H., Takahashi, H., Hu, W., Hasebe, K., Haddad, P.R., Fritz, J.S., and Sarzanini, C. 2002. Vacancy ion-exclusion chromatography of carboxylic acids on a weakly acidic cation-exchange resin. *Journal of Chromatography A* 956(1,2):209–214.

[383] Davison, B.H., Nghiem, N.P., and Richardson, G.L. 2004. Succinic acid adsorption from fermentation broth and regeneration. *Applied Biochemistry and Biotechnology* 113–116:653–668.

[384] Aşçi, Y.S., and İnci, İ. 2009. Extraction of glycolic acid from aqueous solutions by Amberlite LA-2 in different solvents. *Journal of Chemical & Engineering Data* 54(10):2791–2794.

[385] Aşçi, Y.S., and İnci, İ. 2010. Extraction equilibria of succinic acid from aqueous solutions by Amberlite LA-2 in various diluents. *Journal of Chemical & Engineering Data* 55(2):847–851.

[386] Uslu, H., İnci, İ., and Bayazit, S.S. 2010. Adsorption equilibrium data for acetic acid and glycolic acid onto Amberlite IRA-67. *Journal of Chemical & Engineering Data* 55(3):1295–1299.

[387] Tamada, J.A., Kertes, A.S., and King, C.J. 1990. Extraction of carboxylic acids with amine extractants. 1. Equilibria and law of mass action modeling. *Industrial & Engineering Chemistry Research* 29:1319–1326.

[388] Tamada, J.A., and King, C.J. 1990. Extraction of carboxylic acids with amine extractants. 2. Chemical interactions and interpretation data. *Industrial & Engineering Chemistry Research* 29:1327–1333.

[389] Tamada, J.A., and King, C.J. 1990. Extraction of carboxylic acids with amine extractants. 3. Effect of temperature, water coextraction, and process considerations. *Industrial & Engineering Chemistry Research* 29:1333–1338.

[390] Niemi, H., Lahti, J., Hatakka, H., Kärki, S., Rovio, S., Kallioinen, M., Mänttäri, M., and Louhi-Kultanen, M. 2011. Fractionation of organic and inorganic compounds from black liquor by combining membrane separation and crystallization *Chemical Engineering & Technology* 34(4):593–598.

[391] Hellstén, S., Lahti, J., Heinonen, J., Kallioinen, M., Mänttäri, M., and Sainio, T. 2013. Purification process for recovering hydroxy acids from soda black liquor. *Chemical Engineering Research and Design* 91(12):2765–2774.

[392] Sainio, T., Kallioinen, M., Niemelä, K., Hellstén, S., Lahti, J., Heinonen, J., and Mänttäri. 2014. Recovery of hydroxy acids from soda black liquor. *Proceedings of the 5th Nordic Wood Biorefinery Conference*, March 22–24, 2014, Stockholm, Sweden. Pp. 174–179.

[393] Mänttäri, M., Lahti, J., Hatakka, H., Louhi-Kultanen, M., and Kallioinen, M. 2015. Separation phenomena in UF and NF in the recovery of organic acids from kraft black liquor. *Journal of Membrane Science* 490:84–91.

[394] Sjöström, E. 1991. Carbohydrate degradation products from alkaline treatment of biomass. *Biomass and Bioenergy* 1:61–64.

[395] Alén, R. 1998. Utilisation of the aliphatic carboxylic acids formed as by-products in kraft pulping. *Kemia-Kemi* 15:565–569. (In Finnish).

[396] Alén, R., and Sjöström, E. 1980. Condensation of glycolic, lactic and 2-hydroxybutanoic acids during heating and identification the condensation products by GLC-MS. *Acta Chemica Scandinavica* B34:633–636.

[397] Alén, R., and Sjöström, E. 1985. Degradative conversion of cellulose-containing materials into useful products. In: Nevell, T.P., and Zeronian, S.H. (Eds.). *Cellulose Chemistry and its Applications*. Ellis Horwood, Chichester, England. Pp. 531–542.

[398] Datta, R., and Henry, M. 2006. Lactic acid: Recent advances in products, processes and technologies – a review. *Journal of Chemical Technology & Biotechnology* 81(7):1119–1129.

[399] Alén, R. 2009. *Collection of Organic Compounds – Properties and Uses*. Consalen Consulting, Helsinki, Finland. 1370 p. (In Finnish).

[400] Aspinall, G.O., Carter, M.E., and Los, M. 1956. The degradation of xylobiose and xylotriose by alkali. *Journal of Chemical Society* 4807–4810.

[401] Whistler, R.L., and Corbett, W.M. 1956. Behavior of xylan in alkaline solution: The isolation of a new C_5-saccharinic acid. *Journal of American Chemical Society* 78:1003–1005.

[402] Alén, R., and Valkonen, J. 1995. The crystal and molecular structures od five-carbon and six-carbon isosaccharino-1,4-lactones from alkaline pulping. *Acta Chemica Scandinavica* 49:536–539.

[403] Alén, R., and Sjöström, E. 1980. Separation and identification of the silylated reduction products from xyloisosaccharinic acid (3-deoxy-2-C-hydroxymethyltetronic acid) by GLC-MS. *Acta Chemica Scandinavica B* 34:387–388.

[404] Alén, R. 1986. Oxidation of xyloisosaccharinic acid by nitric acid. *Carbohydrate Research* 154:301–304.

[405] Sowden, J.C. 1957. The saccharinic acids. *Advances in Carbohydrate Chemistry* 12:35–79.

[406] Whistler, R.L., and BeMiller, J.H. 1963. α-D-Isosaccharino-1,4-lactone, action of lime water on lactone. *Methods in Carbohydrate Chemistry* 2:477–479.

[407] Norrestam, R., Werner, P.-E., and von Glehn, M. 1968. The crystal structure of calcium-α-D-glucoisosaccharinate and some extended Hückel calculations on α-D-glucoisosaccharinic acid. *Acta Chemica Scandinavica* 22:1395–1403.

[408] Werner, P.-E., Norrestam, R., and Rönnquist, O. 1969. The crystal structure of strontium 3-deoxy-2-C-hydroxymethyl-*erythro*-pentoate. *Acta Crystallographica B* 25:714–719.

[409] Cho, C., Rai, D., Hess, N.J., Xia, Y., and Rao, L. 2003. Acidity and structure of isosaccharinate in aqueous solution: A nuclear magnetic resonance study. *Journal of Solution Chemistry* 32:691–702.

[410] Feast, A.A.J., Lindberg, B., and Theander, O. 1965. Studies on the D-glucoisosaccharinic acids. *Acta Chemica Scandinavica* 19:1127–1136.

[411] Alén, R., and Sjöström, E. 1979. Separation of the silylated reduction products from α-D-glucoisosaccharinic acid by GLC-MS. *Acta Chemica Scandinavica B* 33:693–694.

[412] Alén, R. 1986. Oxidation of α-D-glucoisosaccharinic acid by nitric acid. *Carbohydrate Research* 161:156–160.

[413] Alén, R. 1985. Formation of acid-catalysed dehydration products from α-D-glucoisosaccharinic acid. *Carbohydrate Research* 144:163–168.

[414] Alén, R., and Oasmaa, A. 1988. Conversion of glucoisosaccharinic acid by heating under pressure. *Acta Chemica Scandinavica B* 42:563–566.

[415] Holgersson, S., Albinsson, Y., Allard, B., Boren, H., Pavasars, I., and Engvist, I. 1998. Effects of glucoisosaccharinate on Cs, Ni, Pm and Th sorption onto, and diffusion into cement. *Radiochimica Acta* 82:393–398.

[416] Rai, D., Rao, L., and Xia, Y. 1998. Solubility of crystalline calcium isosaccharinate. *Journal of Solution Chemistry* 27(12):1109–1122.

[417] Vercammen, K. 2000. *Complexation of Calcium, Thorium and Europium by α-Isosaccharinic Acid under Alkaline Conditions*. Doctoral Thesis. Swiss Federal Institute of Technology, Zürich, Switzerland. 169 p.

[418] Wieland, E., Tits, J., Dobler, J.P., and Spieler, P. 2002. The effect of α-isosaccharinic acid on the stability of and Th(IV) uptake by hardened cement paste. *Radiochimica Acta* 90:683–688.

[419] Evans, N.D.M. 2003. *Studies on Metal α-Isosaccharinic Acid Complexes*. Doctoral Thesis. University of Loughborough, Loughborough, England. 284 p.

[420] Allard, S. 2005. *Investigations of α-D-Isosaccharinate: Fundamental Properties and Complexation*. Doctoral Thesis. Chalmers University of Technology, Gothenburg, Sweden. 67 p.

[421] Allard, S., and Ekberg, C. 2006. Complexing properties of α-isosaccharinate: Stability constants, enthalpies and entropies of Th-complexation with uncertainty analysis. *Journal of Solution Chemistry* 25:1173–1186.

[422] Warvick, P., Evans, N., and Vines, S. 2006. Studies on some divalent metal α-isosaccharinic acid complexes. *Radiochimica Acta* 94(6,7):363–368.

[423] Svensson, M., Berg, M., Ifwer, K., Sjöblom, R., and Ecke, H. 2007. The effect of isosaccharinic acid (ISA) on the mobilization of metals in municipal solid waste incineration (MSWI) dry scrubber residue. *Journal of Hazardous Materials* 144:477–484.

[424] Gaona, X., Montoya, V., Colàs, M., Grivé, M., and Duro, L. 2008. Review of the complexation of tetravalent actinides by ISA and gluconate under alkaline to hyperalkaline conditions. *Journal of Contaminant Hydrology* 102:217–227.

[425] Florent, J.-C., Ughetto-Monfrin, J., and Monneret, C. 1987. Anthracyclinones. 2. Isosaccharinic acid as chiral template for the synthesis of (+)-4-demethoxy-9-deacetyl-9-(hydroxymethyl)daunomycinone and (-)-4-deoxy-γ-rhodomycinone. *Journal of Organic Chemistry* 52:1051–1056.

[426] Deguin, B., Florent, J.-C., and Monneret, C. 1991. Anthracyclinones. 5. Glucoisosaccharino-1,4-lactone as chiral template for the synthesis of new anthracyclinones. *Journal of Organic Chemistry* 56:405–411.

[427] Bennis, K., Gelas, J., and Thomassigny, C. 1995. Synthesis from lactose of a new enantiomerically pure polyhydroxylated pyrrolidines with branched structures. *Carbohydrate Research* 279:307–314.

[428] Thomassigny, C., Bennis, K., and Gelas, J. 1997. Synthesis of enantiomerically pure heterocycles: Access to hydroxylated piperidines from a sugar lactone. *Synthesis* (2):191–194.

[429] Bertounesque, E., Millal, F., Meresse, P., and Monneret, C. 1998. Synthesis of a highly functionalized γ-lactone as a precursor of 9-pentyl anthracyclines. *Tetrahedron: Asymmetry* 9:2999–3009.

[430] Reintjes, M., and Cooper, G.K. 1984. Polysaccharide alkaline degradation products as a source of organic chemicals. *Industrial and Engineering Chemistry Product Research* 23(1):70–73.

[431] Alén, R., Hartus, T., Korpela, A., and Sjöström, E. 1989. By-products from alkaline pulping liquors. Part 2. Preparation of surfactants from glucoisosaccharinic acid. *Proceedings of the 5th International Symposium on Wood and Pulping Chemistry*, May 22–25, 1989, Raleigh, NC, USA. Pp. 165–167 (posters).

[432] Kumar, H., and Alén, R. 2016. Microwave-assisted catalytic esterification of α-glucoisosaccharino-1,4-lactone with tall oil fatty acids. *Sustainable Chemical Processes* 4(4):1–5.

6 Cellulose derivatives

6.1 Approach to the concept of making cellulose derivatives

6.1.1 General background

As described in Section 2.3.1 "Cellulose", cellulose is the world's most abundant and important biopolymer (polysaccharide), which, as the major organic component in lignocellulosic biomass materials, has significant mechanical load-bearing functions for these materials [1–10]. Hence, it is also easy to understand that cellulose is one of the oldest known natural polymers, serving mankind for centuries, and has gradually established itself with a wide range of technical methods for its manufacturing in a more or less purified form as well as has a long record of industrial utilization applications. Cellulose consists of repeating β-D-glucopyranose (β-D-Glc$p$) moieties in a $^4C_1$ conformation, linked together by (1→4)-glycosidic bonds. The degree of polymerization (DP) of native wood cellulose is of the order of about 10,000, and is lower than that of cotton cellulose (about 15,000). The polydispersity (M_w/M_n) of cellulose is rather low (<2), thus indicating that the weight average molecular mass (M_w) and the average molecular mass (M_n) number do not deviate much from each other. Because of the strong tendency for intramolecular and intermolecular hydrogen bonding, bundles of native cellulose molecules aggregate to microfibrils, and further to fibrils, which contain highly ordered crystalline (60–75% of the total cellulose and 50–150 nm in length) or less ordered amorphous regions (25–50 nm in length).

The content of cellulose in plant materials and technical products varies, depending on the origin (Figure 6.1). Cellulose has conventionally been harvested as commercial fibers from the seed hairs of the cotton plant (linters contain up to 95% cellulose) or as bast fibers (consist of 70–75% cellulose), for example, from flax (linen), jute, hemp, sisal, and ramie [11]. The lignocellulosic biomass materials also contain (as a percentage of the dry solids) extractives (3–10%) and, especially, fiber plants and agricultural residues (*e.g.*, maize husks and stalks) 5–20% proteins, and inorganic compounds (primarily, silicon dioxide or silica, SiO_2). The moisture content of a living tree typically varies in the range of 30–60% of the total wood mass [12]. Additionally, cellulose is also utilized after the chemical separation of lignin, extractives, and most hemicelluloses, from wood, by alkaline kraft or acidic sulfite and bisulfite processes, predominantly as chemical pulp (production is annually almost 150 million tons, over 90% of it is based on kraft pulping) [13] in the form of paper and board. The corresponding annual production of high-yield pulps (*i.e.*, mechanical and semichemical pulps) as well as other pulps (mainly dissolving pulp) is about 30 million tons and about 10 million tons, respectively.

The cellulose content of dissolving pulp, based on acidic sulfite, multistage sulfite, and pre-hydrolysis kraft methods is about 90% [1, 5, 11, 14, 15]. This kind of high-

https://doi.org/10.1515/9783110608366-006

Table 6.1: Typical distribution of the main structural components of wood and non-wood feedstocks (% of the dry solids) [10].

Material	Cellulose	Hemicelluloses	Lignin
Softwoods	40–45	25–30	25–30
Hardwoods	40–45	30–35	20–25
Fiber plants (*e.g.*, cotton, flax, hemp, and sisal)	70–95	5–25	<5
Natural-growing reeds and agricultural residues	25–45	25–50	10–30
Chemical pulps	65–80	20–30	<5

cellulose-containing chemical pulp is normally used after bleaching for the production of cellulose derivatives (*e.g.*, cellulose esters and ethers) and regenerated celluloses (*e.g.*, rayon fibers), for example, for the manufacture of threads (textile fibers) and foils. The regenerated cellulose has almost the same DP and M_w/M_n as the initial cellulose in dissolving pulps, but its morphology is changed to some extent, and its microfibrils are fused into a relatively homogeneous macrostructure. Regenerated cellulose fibers have long been produced according to different processes, resulting in fibers with a wide range of mechanical properties. These fiber products as well as cellulose derivatives are known under many common trade names. Purified cellulose cotton linters or chemical cottons (*i.e.*, they are washed mechanically and chemically to remove proteins, waxes, and other polysaccharides) are generally of higher purity and have higher DP than purified cellulose from the dissolving grades of wood pulp.

The present use of cellulose for high-value products is still marginal, compared to the volume of paper and board products [9]. About a third of the worldwide production of purified cellulose is used as the base material for a great number of cellulose derivatives with pre-signed and wide-ranging properties. The characteristics of these chemical derivatives are based on the linear cellulose chain macromolecular structure as well as the groups introduced and their degree of substitution (DS). Cellulose derivatives, produced by chemical modification of native cellulose, typically have properties different from those of initial cellulose (*e.g.*, they can often be water (H_2O)-soluble). All these products have been created to meet numerous areas of applications. In contrast, in cases of regenerated celluloses, the derivatives are only intermediates for transforming the state of cellulose.

The product group of cellulose derivatives can be classified in different ways, for example, according to the type of the chemical treatment applied (see Sections 6.2 "Cellulose reactions with organic acids and carbon disulfide (CS_2)", 6.3 "Cellulose esters of inorganic acids", and 6.4 "Cellulose ethers") [9, 10, 16–26]. Another way of classification is based on the technical utilization, and the products mainly belong to fibers, films, and plastics. Additionally, typical more specific applications mainly include binders and additives for special mortars and paints, special chemicals for building and construction, paper additives, adhesives, and thickeners as well as synthetic resins. It is also worthy to mention that besides the traditional bulk utilization, certain more fine-chemical approaches have been developed. These applications comprise, for example, for preparing the stationary phases in chromatographic systems as well as anion and cation exchange materials or other inert matrices for binding various medical test indicators (see Section 6.5 "Specific cellulose products"). A relatively new application area includes the use of micro-sized and nano-sized fibrillated cellulose preparations as well as cellulose-based liquid crystals for different purposes. There are also microbial extracellular carbohydrates, such as the so-called "bacterial cellulose" [27–30]. It is of high purity and mainly synthesized by an acetic acid-producing bacterium, the common vinegar bacterium, *Gluconacetobacter xylinus* (formerly classified as *Acetobacter xylinum*), on nutrient media, containing glucose. The process has potential for large-scale production from alternative substrates and the bacterial cellulose has also high potential in many applications [31–35].

It can be claimed that the present production of nontoxic cellulose derivatives from renewable biomass feedstocks exhibits great chemical variability and potential possibilities [9]. Additionally, it has been generally stated that the establishment of petroleum chemistry and the ease of processing mineral oil into useful products have also traditionally retarded this kind of cellulose research. However, in the 1970s, with the growing shortage of raw materials as well as the uncertainties about oil prices and supplies, together with environmental concerns, active cellulose research has been revived by the chemical industry. On the other hand, the full potential of cellulose has not yet been exploited. This unfortunate fact is mainly due to the lack of environmental-friendly methods to extract cellulose from native resources, the limited number of common and efficient solvents, and the certain difficulty in modifying cellulose properties.

During the recent decades, due to their relatively high price, compared to petrochemical polymer replacements, many current cellulose polymers have still comparatively low volumes or niche applications [9, 36]. Although the manufacturing of all the types of possible bio-based polymers, such as polyamides (nylon), poly(butylene succinate) (PBS), poly(butylene terephthalate) (PBT), polyhydroxyalkanoates (PHAs), poly(lactic acid) (PLA), poly(trimethylene terephthalate) (PTT), polyurethane (PUR), and starch polymers, is an emerging field, characterized by new synergies and a broad variety of several technologies, the production of cellulosic plastics and fibers has stagnated. In spite of this trend, there are also a great number of cellulose derivatives that cannot be replaced by petrochemical-based polymers.

This chapter provides only a brief overview of the basic chemistry and the general aspects, rather than describing any detailed data on technical processes for producing typical cellulose derivatives and other related cellulose-based products. The emphasis is on the present applications, although even in the near future, many other similar products may become more important and attractive as well. Furthermore, the scope of the current research is not only to improve existing products but also to create new cellulose-based products, with a controlled chemical and supramolecular structure for new applications. In practice, this means that one aim of the forthcoming research within cellulose chemistry is also to design new cellulose derivatives with well-defined properties suitable, for example, for pharmaceutical, food, fine chemical, building, and other high-value applications. The key questions comprise, besides the discovery of alternative and economic synthesis paths for manufacturing novel products, the versatile elucidation of chemical structures, and properties dealing with these derivatives. A further potential aid to reach this goal is the effective utilization of molecular modelling to simulate molecular behavior on a computer to predict the parallel experimental results. Additionally, it is also obvious that this kind of approach concerning cellulose can be analogously applied to hemicelluloses, the next prominent naturally occurring carbohydrates-based polymers.

6.1.2 Reactivity and accessibility of cellulose

The properties of cellulose derivatives are influenced by the amount and distribution of the substituents in the cellulose molecule [9]. All β-D-glucopyranose units within the cellulose chain have principally three different reactive hydroxyl (-OH) groups: one primary group belonging to the hydroxymethyl (-CH$_2$OH) group at C$_5$, and two secondary groups attached to the carbon atoms C$_2$ and C$_3$. The rate and degree of conversion ("reactivity") strongly depend on the availability of these -OH groups, and this parameter is called as "accessibility" [1, 5, 22, 37]. In practice, accessibility means the relative ease of the reactants reaching the -OH groups. Due to its significance, there are various methods available for the determination of accessibility of cellulose to different reagents under varying conditions. On the other hand, because of the insolubility of cellulose, the chemical reactions of -OH groups proceed, at least in the initial phase of the reaction, in a heterogeneous reaction medium. In the heterogeneous reactions of cellulose, chemical conversion preferentially proceeds at the surface and in the low-ordered parts of the cellulose structure. However, it is possible that during several common reactions, the partially or/and totally substituted products formed are gradually dissolved and further reaction steps may then occur homogeneously.

It is evident that several factors affect the accessibility and reactivity of cellulose [9]. For example, one prerequisite for etherification (see Section 6.4 "Cellulose ethers") is the ionization of -OH groups and thus, the formation of reactive alkoxides (C-O$^{\ominus}$). Owing to the inductive effects of neighboring substituents, the tendency toward this kind of dissociation (*i.e.*, depends primarily on acidity) decreases according to the series C$_2$-OH > C$_3$-OH > C$_6$-OH. Hence, C$_2$-OH is more readily etherified than the other -OH groups. However, after substitution of C$_2$-OH, the acidity of C$_3$-OH usually increases, resulting in its higher reactivity. On the other hand, in esterification (see Sections 6.2 "Cellulose reactions with organic acids and carbon disulfide (CS$_2$)" and 6.3 "Cellulose esters of inorganic acids"), the primary -OH group in C$_6$-OH shows the highest reactivity. Additionally, being the least sterically hindered, C$_6$-OH groups show higher reactivity toward bulky substituents than do the other -OH groups.

The topochemistry as well as the morphology (*i.e.*, the ratio of crystalline regions to amorphous regions) are important factors in the chemical reactivity and accessibility of cellulose [9]. The -OH groups located in the amorphous regions are accessible and react readily, while those in crystalline regions with close packing and strong internal hydrogen bonding can be completely inaccessible. Hence, a pretreatment (preswelling) of the cellulose is needed in both etherification (by alkalis) and esterification (by acids). It may also be possible to increase accessibility by biochemical pretreatments. All these pretreatments result in the creation of new accessible -OH groups. The degree of crystallinity largely depends on the origin of the cellulose preparation, and relatively high variations are obtained by alternative determination techniques, even for the same sample.

Differences in the degree and the rate of conversion between regions of high and low degree of order are most distinct in reactions accompanied only by a small or medium degree of swelling [9]. In the strongly swollen or soluble state of cellulose, all the -OH groups are accessible to the reactant. In practice, the number of -OH groups available for reactions can be only about 10% in highly crystalline cellulose, but as much as about 90% in decrystallized cellulose. However, due to the random nature of the reaction, a homogeneous product is obtained by a complete substitution of -OH groups. In this case, the maximum DS 3 is reached. At any DS lower than this maximum value, the reaction results in random sequences of units, consisting of the unreacted, monosubstituted, disubstituted, and fully substituted moieties. Experiments have indicated that moderate differences in reactivity do not significantly influence the overall distribution of these units at different DS values. For example, when DS is 2, the distribution of unreacted, monosubstituted, disubstituted, and fully substituted moieties is, respectively, 5%, 20%, 45%, and 30% [21].

Different solvents cause interfibrillar swelling or intrafibrillar swelling or dissolution of cellulose [3, 5, 9, 17, 21, 38, 39]. The extent of swelling depends on the solvent and the nature of the cellulose used; for example, the degree of the chemical and mechanical treatments performed and the content of hemicelluloses and lignin. A prerequisite for swelling is breaking of internal hydrogen bonds between cellulose chains, mostly in the amorphous regions (*i.e.*, intercrystalline swelling) or in more effective case, both in the amorphous and crystalline regions (*i.e.*, intracrystalline swelling). When the penetrating agent causes intracrystalline swelling, cellulose-swelling agent complex is formed. Apart from some salts, it can be also accomplished by concentrated solutions of strong bases or acids. In the case of limited swelling, the swelling agents combine only with the ordered cellulose but do not fully destroy the interfibrillar hydrogen bonding. In contrast, in the case of unlimited swelling, the swelling agents are bulky and form complexes with cellulose, leading to breakage of the adjacent hydrogen bonds and separation of the cellulose chains with gradual dissolution.

Cellulose swells in electrolyte solutions due to the penetration of hydrated ions, which require more space than the H_2O molecules [9, 40]. The H_2O retention of cellulose fibers at a given relative humidity (*i.e.*, sorption of H_2O from the vapor state) generally varies, depending on whether the equilibrium has occurred by sorption or desorption (*i.e.*, hysteresis). Additionally, the H_2O uptake continuously decreases after repeating the moistening and drying of the fibers. When dry cellulose fibers are exposed to humidity, they adsorb H_2O and the cross section of fibers increases to some extent; at a humidity of 100%, 80%, 60%, 40%, and 20%, the H_2O contents (sorption) are roughly 23%, 10%, 6%, 4%, and 3% of cellulose, respectively, and in the case of desorption, the values are 24%, 12%, 8%, 5%, and 4% of cellulose, respectively [5]. At varying humidity values, in contrast to change in the fiber diameter, the dimensional change in the longitudinal direction is very small.

The most important swelling complex of cellulose (*i.e.*, alkali cellulose) is that formed with sodium hydroxide (NaOH) as well as with other hydroxides, such as CsOH, KOH, LiOH, and RbOH [10]. This kind of alkaline pretreatment is extremely important in making intermediates, which are, due to an increased penetration of the reagents into the swollen cellulose structure, more reactive than the original cellulose. For this reason, the activation of cellulose by NaOH increases its accessibility, particularly when the reaction is accompanied by strong swelling of the cellulose structure (*i.e.*, the subsequent reactions proceed in an alkaline medium). As a strong swelling agent, NaOH also causes changes in the cellulose crystalline structure (*i.e.*, changes in the polymorphous lattices [41]).

The traditional process for making alkali celluloses (*i.e.*, the treatment with an aqueous 16–18% NaOH solution at room temperature for varying times) was developed in 1844 and named "mercerization" after its inventor John Mercer [10, 42, 43]. It should be noted that since the solutions are alkaline, cellulose can be, to some extent, depolymerized in the presence of oxygen. It is also known that besides making different "cellulosates", extensive swelling can also be achieved in solutions of various acids and salts, although in these cases, formation of definite complexes is not straightforward [9]. However, when using strong acids, such as sulfuric acid (H_2SO_4) or hydrochloric acid (HCl), some hydrolysis of cellulose may take place, depending on the pretreatment conditions applied. In contrast, phosphoric acid (H_3PO_4) and nitric acid (HNO_3) do not hydrolyze cellulose, but HNO_3 reacts to some extent with it (*cf.*, the formation of nitrocellulose). Additionally, there are also other activation pretreatments, including solvent exchange, inclusion of structure-loosening additives, or even mechanical action, which can be used to increase the reactivity.

6.1.3 Conventional solvents for cellulose

Cellulose is typically soluble only in a few solvents, although some of them can be consequently used for improving cellulose reactivity and accessibility [9]. A new area of cellulose chemistry was started almost 50 years ago when novel nonaqueous solvent systems for cellulose, including numerous aprotic ones, were discovered. Traditionally, solvents for cellulose are central in the preparation of cellulose derivatives but they are also needed in laboratory work. However, those conventional solvents cannot be recovered and reused. Hence, since the 1970s, novel solvents for cellulose have been systematically sought for industrial purposes, especially for rayon and cellophane industries (*i.e.*, for preparing regenerated cellulose fibers). The dissolution of cellulose with proper solvents allows a comprehensive modification of cellulose in a homogeneous system. Additionally, some advanced derivatives of celluloses, which are also synthesized by new methods, cannot even be prepared under heterogeneous conditions.

The solvents for cellulose fall into two main categories [3, 5, 9]: *i*) non-derivatizing solvents and *ii*) derivatizing solvents. These both groups of solvent systems can be established under aqueous or nonaqueous media. The term "non-derivatizing solvent" denotes systems (*i.e.*, mainly containing conventional aqueous inorganic complex-forming compounds) dissolving cellulose only by intermolecular interactions, whereas the "derivatizing solvents" comprise all the systems where the dissolution of cellulose takes place in combination with the covalent derivatization into an unstable ester, ether, or acetal. The cellulose derivative formed in the latter case is readily decomposed to regenerated cellulose by changing the medium or the pH value of the system.

The group of the non-derivatizing solvents (Table 6.2) contains conventional aqueous inorganic complex-forming compounds (*i.e.*, solubility *via* the complex formation) [17, 22]. Other similar compounds are ("en" is ethyleneamine, "dien" diethylenetriamine, "pp" 1,3-propylenediamine, and "tren" tris(2-aminoethyl)amine) $[Pd(en)](OH)_2$, $[Zn(dien)](OH)_2$, $[Cu(pp)_2](OH)_2$, $[Cd(tren)](OH)_2$, and $[Ni(tren)](OH)_2$ [9]. Molten salt hydrates [44], such as lithium thiocyanate ($LiSCN\cdot2H_2O$) and lithium perchlorate ($LiClO_4\cdot3H_2O$), as well as aqueous solutions, such as saturated ammonium thiocyanate (NH_3/NH_4SCN) [45, 46] or calcium thiocyanate ($Ca(SCN)_2$) [47], can dissolve limited amounts of cellulose, and have been used earlier for this purpose. Additionally, for example, the organic liquid/salt-solvent systems, dimethylamine/lithium chloride (DMA/LiCl) [5, 48], *N,N*-dimethylacetamide/lithium chloride (DMAc/LiCl) [49–52], dimethyl sulfoxide/tetrabutylammonium fluoride ($\cdot3H_2O$) (DMSO/TBAF) [53–55], ethylenediamine/potassium thiocyanate (EDA/KSCN), 1,3-dimethyl-2-imidazolidone/lithium chloride (DMI/LiCl) [57], *N*-methyl-2-pyrrolidone/lithium chloride (NMP/LiCl) [58], and aqueous poly(ethylene glycol)/sodium hydroxide (PEG/NaOH) [59], belong to the first group of the solvents (Figure 6.1). An important recent example of non-derivatizing and nonaqueous solvent for cellulose is *N*-methylmorpholine *N*-oxide (NMMO) (*cf.*, the commercial Lyocell process) [5, 60–63]. A commercial process called "solvent spinning" has also been introduced to utilize an aqueous NMMO solution in the production of wood-based regenerated cellulose fibers (*i.e.*, Lyocell fibers). Additionally, the dissolution of cellulose in the NaOH/urea ($H_2N-CO-NH_2$) aqueous system by a two-step process [64, 65] or the effects of additives (*i.e.*, ZnO and $NaAlO_2$) on dissolution of cellobiose (*i.e.*, a simple model compound for cellulose) in the $NaOH/H_2N-CO-NH_2$ aqueous system [66] have been investigated.

Examples of aqueous and nonaqueous derivatizing and solubilizing solvents systems for cellulose include trimethylchlorosilane (($CH_3)_3SiCl$)/pyridine (cellulose derivative Cell-O-Si($CH_3)_3$), trifluoroacetic acid/trifluoroacetic acid anhydride ($CF_3CO_2H/CF_3(CO)_2O$, Cell-O-(O)CCF$_3$), H_3PO_4 (>85%)/H_2O (Cell-O-PO$_3H_2$), formic acid/zinc chloride ($HCO_2H/ZnCl_2$, Cell-O-(O)CH), nitrogen tetroxide/*N,N*-dimethylformamide (N_2O_4/DMF, Cell-O-N=O), di-

Table 6.2: Examples of aqueous complex-forming inorganic solvents for cellulose [9, 10].

Abbreviation	Structure[b]	Name	Color
Cuoxam or Schweizer's solution	$[Cu(NH_3)_4](OH)_2$	Cuprammonium hydroxide	Violet
Nioxam	$[Ni(NH_3)_6](OH)_2$	Nickel-ammonium hydroxide	Dark blue
Cuen or CED	$[Cu(en)_2](OH)_2$	Cupriethylenediamine hydroxide	Violet
Cadoxen	$[Cd(en)_3](OH)_2$	Cadmium-ethylenediamine hydroxide	Transparent
Cooxen	$[Co(en)_3](OH)_2$	Cobalt-ethylenediamine hydroxide	Dark red
Nioxen	$[Ni(en)_3](OH)_2$	Nickel-ethylenediamine hydroxide	Violet
Zincoxen	$[Zn(en)_3](OH)_2$	Zinc-ethylenediamine hydroxide	Transparent
EWNN[a]	$[FeT_3]Na_6$	Iron sodium tartrate	Greenish

[a]EWNN refers to Eisen-Weinsäure-Natrium-Komplex.
[b]"en" refers to ethylenediamine (-HNCH$_2$CH$_2$NH-) and "T" to tartrate ($C_4H_3O_6^{2\ominus}$).

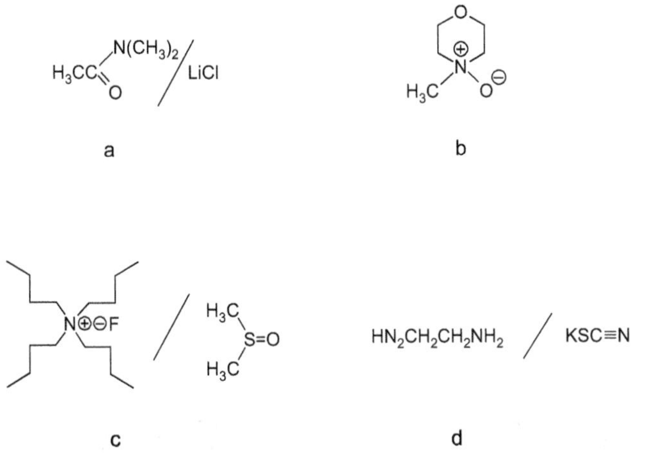

Figure 6.1: Examples of non-derivatizing solvent systems for cellulose [10]. *N,N*-Dimethylacetamide/lithium chloride (DMAC/LiCl) (a), *N*-methylmorpholine *N*-oxide (NMMO) (b), dimethyl sulfoxide/tetrabutylammonium fluoride (DMSO/TBAF) (c), and ethylenediamine/potassium thiocyanate (EDA/KSCN) (d).

methyl sulfoxide/paraformaldehyde (DMSO/PF, Cell-O-CH$_2$OH), trichloroacetaldehyde/dimethyl sulfoxide/tetraethylammonium chloride (CCl$_3$CHO/DMSO/TEAC, Cell-O-CH(OH)CCl$_3$), and CS$_2$/NaOH/H$_2$O (Cell-O-C(S)SNa) [5].

6.1.4 Ionic liquids and deep eutectic solvents for cellulose

Ionic liquids (ILs) (Figure 6.2) are nonaqueous solvents that are organic salts (*i.e.*, typically consist of an organic cation and an inorganic anion), existing as liquids at relatively low temperatures <100 °C and mostly non-volatile with moderate thermal stability [64, 67–81]. Room temperature ILs (RTILs) or those utilizing slight heating in a microwave oven are considered suitable to replace the conventional organic solvents in a wide range of applications due to their many advantageous properties. ILs also represent new solvents for cellulose in the preparation of regenerated cellulose and the subsequent synthesis of cellulose derivatives as well as the fractionation of lignocellulosic materials. In general, they are non-derivatizing solvents with strong intramolecular and intermolecular interactions.

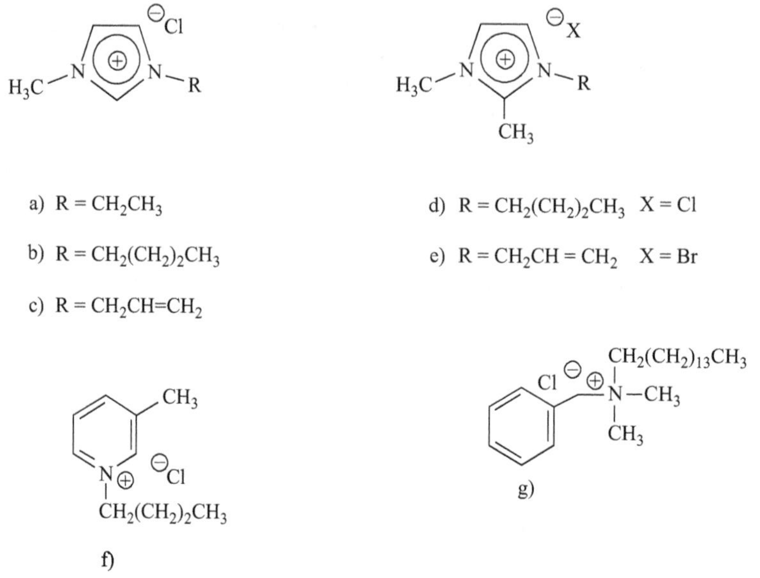

a) R = CH₂CH₃

b) R = CH₂(CH₂)₂CH₃

c) R = CH₂CH=CH₂

d) R = CH₂(CH₂)₂CH₃ X = Cl

e) R = CH₂CH = CH₂   X = Br

f)

g)

Figure 6.2: Typical ionic liquids (ILs) [10]. 1-Ethyl-3-methylimidazolium chloride (a), 1-butyl-3-methylimidazolium chloride (b), 1-allyl-3-methylimidazolium chloride (c), 1-butyl-2,3-dimethylimidazolium chloride (d), 1-allyl-2,3-dimethylimidazolium bromide (e), 1-*N*-butyl-3-methylpyridinium chloride (f), and benzyldimethyl(tetradecyl)ammonium chloride (g).

ILs typically dissolve cellulose in a more efficient and environmentally acceptable way than conventional methods in aqueous solution [9, 74, 81]. One of the primary driving forces behind investigation of novel solvents (especially ILs) for cellulose include environmental concerns (*i.e.*, a "green chemistry" approach) and, on the other hand, the ability to create liquid crystals in the new solvent. However, the "green character" of ILs can be partly questioned since most of them are synthesized from fossil resources and hence, their synthesis cannot be considered as being fully "green" [82]. The earliest described ILs are the protic ILs ethanolammonium nitrate in 1888 [83] and ethylammoniumnitrate in 1914 [84], whereas as early as 1934, Charles Graenacher suggested that molten *N*-ethylpyridinium chloride in the presence of nitrogen-containing bases could be used to dissolve cellulose [9]. However, this molten salt system, due to its relatively high melting point (118 °C), was considered at that time somewhat esoteric as well as a novelty of small practical value.

The most effective ILs are those containing anions that are strong hydrogen bond acceptors. The most important ILs are 1-ethyl-3-methylimidazolium [C_2mim]$^\oplus$ chloride (EMIMCl), 1-butyl-3-methylimidazolium [C_4mim]$^\oplus$ chloride (BMIMCl) (or tetrafluoroborate BMIM-BF$_4$), 1-butyl-2,3-dimethylimidazolium [C_4dmim]$^\oplus$ chloride (BDMIMCl), 1-allyl-3-methylimidazolium chloride (AMIMCl), 1-allyl-2,3-dimethylimidazolium [Admim]$^\oplus$ bromide, 1,3-dimethylimidazolium dimethyl phosphate (DMIMDMP), 1-ethyl-3-methylimidazolium dicyanamide, 1-N-butyl-3-methylpyridinium chloride, 1-N-ethyl-3-methylpyridinium ethyl sulfate, benzyldimethyl(tetradecyl)ammonium chloride, and 1-N-ethyl-3-hydroxymethylpyridinium ethyl sulfate.

Due to their diversity and exceptional properties, ILs are currently of versatile scientific interest for enhancing, for example, the reaction efficiency when making polysaccharide derivatives [85–88] or the pretreatment of polysaccharides, prior to their hydrolysis [89–91]. However, they may still significantly extend rapidly their range of applications within the carbohydrate chemistry. Additionally, cellulose-based fiber spinning processes using ILs have been studied [92–94]. Furthermore, it has been shown that cellulose can be easily regenerated from its IL solutions, for example, by the addition of H_2O, ethanol (CH_3CH_2OH), or acetone (CH_3COCH_3), although many other methods have been applied as well. Furthermore, it is obvious that novel research dealing with the economical production of ILs and the physical behavior (viscous or elastic) of different cellulose IL solutions, together with their toxicity as well as the potential utilization possibilities of various cellulosic fiber resources and IL recyclability (*cf.*, the Ioncell$^\circledR$ technology with [DBHN][OAc, 1,5-diazabicyclo[4.3.0]non-5-enum acetate] [95]) is needed. All these factors are of importance when further promoting the industrial application of ILs.

Lignocellulosic biomass materials can also be dissolved and fractionated by deep eutectic solvents (DESs), which can be considered a new class of IL analogues, because they share many characteristics and properties with ILs [96–102]. DESs consist of a eutectic mixture of Lewis or Brønsted acids and bases, which can contain a variety of anionic and/or cationic species; they contain large, non-symmetric ions that are capable of associating with each other through hydrogen bond interactions to form a eutectic mixture. The original DESs are typically prepared by the complexation of a quaternary ammonium salt (hydrogen bond acceptor, (HBA)) with a metal salt or a hydrogen bond donor (HBD), such as an amine or a carboxylic acid. In general, a eutectic system is a homogeneous mixture of two solid-phase chemicals and is characterized by a melting point lower than that of each individual component [103, 104]. One of the illustrative examples is a mixture of choline-chloride (ChCl, Ch refers to $HOCH_2CH_2N^\oplus(CH_3)_3$) and H_2N-CO-NH$_2$ in a molar ratio of 1:2 (*i.e.*, at a particular molar ratio called the "eutectic composition") [105]. It has a melting point of 12 °C, which is far less than the melting point of ChCl (302 °C) and H_2N-CO-NH$_2$ (133 °C). The DESs are called "deep" because the melting point curve of a eutectic mixture has a particularly deep crevice at the eutectic point. At the eutectic composition of the mixtures, the molar ratio of the components gives the lowest melting point.

DESs can be described by the general formula $Cat^{\oplus} X^{\ominus} zY$, where $Cat^{\oplus}$ is, in principle, any ammonium, phosphonium, or sulfonium cation, and X is a Lewis base, typically a halide anion [98]. Complex anionic species are formed between $X^{\ominus}$ and either a Lewis or Brønsted acid Y, and z refers to the number of Y molecules that interact with the anion. The most used and studied mixtures is $ChCl/H_2N\text{-}CO\text{-}NH_2$ and similar mixtures, and this kind of example can be simplified, interpreting the DESs as mixtures of HBD and HBA molecules. DESs are classified into four classes (Table 6.3), based on their starting materials, especially, depending on the nature of the complexing agent used. Type I DESs are prepared from non-hydrated metal halides such as $FeCl_2$, $AgCl$, $CuCl$, $LiCl$, $CdCl_2$, $CuCl_2$, $SnCl_2$, $ZnCl_2$, $LaCl_3$, YCl_3, and $SnCl_4$, and quaternary ammonium salts such as imidazolium salts, whereas type II DESs are prepared from hydrated metal halides and quaternary ammonium salts [102]. The benefit for utilizing hydrated metal halides lies in their relatively low cost and their insensitivity toward moisture, making them interesting alternatives for large-scale utilization. Type III DESs are characteristically composed of ChCl, combined with different HBDs, including different amides, alcohols, and carboxylic acids. Finally, type IV DESs are typically composed of $ZnCl_2$, combined with organic molecules such as $H_2N\text{-}CO\text{-}NH_2$, acetamide ($CH_3\text{-}CO\text{-}NH_2$), ethylene glycol ($HOCH_2CH_2OH$), and 1,6-hexanediol (($HOCH_2(CH_2)_4CH_2OH$).

Table 6.3: General formula for the classification of DESs [97, 98, 100].

Type	Components[a]	General formula[b]	Terms	Example[c]
I	Metal salt + organic salt	$Cat^{\oplus} X^{\ominus} zMCl_x$	M = Zn, Sn, Fe, Al, Ga, and In	$ZnCl_2 + ChCl$
II	Metal salt hydrate + organic salt	$Cat^{\oplus} X^{\ominus} zMCl_x \cdot yH_2O$	M = Cr, Co, Cu, Ni, and Fe	$CoCl_2 \cdot 6H_2O + ChCl$
III	HBD + organic salt	$Cat^{\oplus} X^{\ominus} zRZ$	Z = $CONH_2$, CO_2H, and OH	Urea + ChCl
IV	Zinc/aluminium chloride + HBD	$MCl_x + RZ = MCl^{\oplus}_{x-1} \cdot RZ + MCl^{\ominus}_{x+1}$	M = Al and Zn and Z = $CONH_2$ and OH	$ZnCl_2 + $ urea

[a]HBD refers to a hydrogen bond donor.
[b]$Cat^{\oplus}$ refers to a cation, any ammonium, phosphonium, or sulfonium cation; X to a Lewis base, generally a halide anion; R to an organic radical; z to the number of molecules that interact with the anion; and y to the number of H_2O molecules.
[c]Ch refers to choline ($HOCH_2CH_2N^{\oplus}(CH_3)_3$).

One particularly promising type of DESs that has recently gained large-scale interest comprises DESs completely manufactured from precursors (*e.g.*, organic acids, amino acids, and sugars) originating from natural sources such as primary metabolites found from plants [102, 106–113]. A variety of such natural deep eutectic solvents (NADESs) are considered the solvent in living cells, which explains their high potential for solubilizing natural products. As also in the case of DESs, the various compositions and wide possibilities for using different components in making NADESs lead to a broad range of physical properties. In the manufacture of NADESs, alcohols, amines, aldehydes, ketones, and carboxylic groups can act both as HBDs and HBAs. NADESs can be classified into four main classes [114]: *i*) derivatives from organic acids, *ii*) derivatives from ChCl, *iii*) mixtures of sugars, and *iv*) other combinations. Due to the numerous structural alternatives and the possibilities for adjusting their physicochemical properties, NADESs can be considered as promising designer solvents, finding their applications, for example, when selectively separating trace natural compounds from complex matrices. NADESs can be prepared from readily available and biodegradable components, and their safety, reusability, and low cost are the major driving forces that are pushing forward their utilization in analytical applications. Furthermore, negligible volatility, adjustable viscosity, polarity, and a high solubilization capacity favor their utilization in various green chemistry applications.

Compared to ILs, DESs are rather easily prepared from relatively inexpensive and toxicologically well-characterized components, allowing their easy transportation for large-scale processing [96, 98, 99, 101, 102, 104]. However, they are generally less chemically inert than ILs. DESs have low volatility and they are typically non-flammable, nontoxic, and biodegradable. Additionally, DESs normally do not inactivate enzymes, which make them valuable in biofuel processing, and they are not particularly sensitive to H_2O. The latter property also enables their use with wet biomass, offering a clear advantage, as the drying of biomass normally constitutes a substantial portion of the total processing costs.

The production of DESs simply involves the mixing of two components and requires only a gentle mixing of the components in the proper molar ratios at temperatures, which are usually 130 °C or less [98, 99, 101, 102, 115]. This practically maintains a comparatively low production cost with respect to conventional ILs and permits large-scale applications. Some examples of the inexpensive components used to make conventional DESs are formic acid (HCO_2H), lactic acid ($CH_3C(OH)CO_2H$), and acetic acid (CH_3CO_2H) as well as ChCl and betaine. HCO_2H, $CH_3C(OH)CO_2H$, and CH_3CO_2H are all common food additives, and can be sustainably produced from biomass, whereas ChCl is used in large quantities as an important additive in feed, especially for chickens, and betaine can be obtained from sugar beets.

The main research of DES research has concentrated on hydrophilic DESs, which limits their practical applications mainly to polar compounds [102, 116]. However, besides the conventional hydrophilic DESs, hydrophobic DESs and their applications have also been reported and they have emerged as an alternative extractive media capable of extracting nonpolar and hydrophobic compounds from different feedstocks [117–119]. The advantage of hydrophobic DESs is that they can be used for extractions conventionally performed with toxic organic solvents. As hydrophilic DESs, hydrophobic DESs can also be synthesized from relatively inexpensive materials [102]. ChCl, betaine, or menthol can be used as an HBA, whereas a variety of organic acids (*e.g.*, CH_3CO_2H, $CH_3C(OH)CO_2H$, octanoic acid ($CH_3CH_2(CH_2)_4CO_2H$), decanoic acid ($CH_3CH_2(CH_2)_6CH_2CO_2H$), and lauric acid ($CH_3CH_2(CH_2)_8CH_2CO_2H$) or alcohols (*e.g.*, 1-propanol ($CH_3CH_2CH_2OH$), glycerol ($HOCH_2C(OH)CH_2OH$), and 1-butanol ($CH_3(CH_2)_2CH_2OH$)) can be used as HBDs. The synthesis of hydrophobic DESs is similar to that of their hydrophilic counterparts; the constituents are mixed at ambient temperature or with moderate heating. The extraction efficiency of the hydrophobic DESs can be tailored by modifying the molar ratios of the constituents. Hydrophobic DESs have been used for extracting different biomass-based analytes such as polyprenyl acetates, pyrethroids, and polycyclic aromatic hydrocarbons [120, 121].

The physical properties of DESs can be designed by properly combining various quaternary ammonium salts (*e.g.*, ChCl) with different HBDs and thus, can be tailored for specific applications [96, 101–103, 121–123]. DESs are low melting mixtures and in most cases, they exist as liquid at or below 100 °C because the melting point is drastically reduced after mixing two components, as compared to the melting points of the original two components. Most of DESs are denser (*i.e.*, with typical densities of 1.10–1.40 g/cm^3) than H_2O and this notable difference of density might be attributed to a different molecular organization or packing of DESs. Viscosity generally depends on the chemical nature of DES components (and their molar ratio), temperature, and H_2O content. DESs are usually highly viscous at room temperature (>100 cP). This property is often attributed to the presence of an extensive hydrogen bond network between each component as well as van der Waals and electrostatic interactions, which together lead to a lower mobility of free spaces within the DES as suggested by the so-called "hole theory". Additionally, viscosity of most eutectic mixtures changes significantly as a function of temperature: as the temperature increases, the viscosity decreases. In general, when keeping in mind their potential applications as green solvents, DESs with low viscosities are highly desirable. However, H_2O can be part of DES and it can play an important role in overcoming the difficulties encountered with high viscosity DESs. Hence, the dilution of DES with H_2O allows the quantitative adjustment of physicochemical properties such as their conductivity, polarity, viscosity, and density, which facilitates their applications as solvents. The optimal H_2O content in DESs depends on the DES composition and on the polarity of the compounds.

DESs can be utilized in many application areas (*e.g.*, extraction of various compounds, organic synthesis, preparation of inorganic materials, electrochemistry, and processing of biodiesel and other fuels), although their use in biomass processing is still in the preliminary stage compared to the development of DESs in other areas, such as the electroplating industry [96, 99, 103]. DESs have been used, for example, for fractionating natural compounds and lignocellulosic biomass materials into their components [100, 124–127]. They have also been used for dissolving pulp and various cellulose fibers as well as microcrystalline cellulose and cellulose nanocrystals [128–131]. Besides their solvating effects, DESs have been successfully used as a reaction media and as catalysts for different chemical conversions [132]. Especially, the use of DESs as catalysts in various organic reactions has been reported, and in the future, this field of application may appear to attain more importance on industrial scale [125, 133, 134]. Additionally, recent advances in making these solvents recyclable and reusable favor their utilization in industrial applications [129, 130, 135–137].

6.2 Cellulose reactions with organic acids and carbon disulfide (CS$_2$)

Since cellulose can be regarded as a polyalcoholic compound, its functionalization is principally based on the reactions taking place at the three reactive -OH groups of its anhydroglucose units (AGUs) [9, 17, 19, 21–23, 138–142]. Hence, for example, the formation of esters with organic and inorganic acids as well as ethers is possible. Although cellulose can theoretically form a great number of organic esters, their industrial possibilities and markets are still rather limited, and the most common organic esters are typically those based on the reaction with a few low-molar-mass organic acids. On the other hand, because of some advantageous properties of certain cellulose esters, a clear recent growth rate in their global market can be observed. In general, cellulose can be esterified by reacting with acids, acid anhydrides, acid chlorides, or unsaturated agents, such as CS$_2$, H$_2$N-CO-NH$_2$, and phenyl isocyanate (C$_6$H$_5$-N=C=O) [139, 143]. In this subchapter, besides the most important examples of the end product cellulose esters based on organic acids, the production of viscose fiber *via* the esterification stage of cellulose are also briefly discussed. In each case, the chemistry of their formation, together with some aspects on their utilization is also discussed.

The most important organic ester, cellulose acetate Cell-O-(CO)CH$_3$, is formed by reacting cellulose, CH$_3$CO$_2$H, and acetic anhydride (H$_3$C(OC)-O-(CO)CH$_3$) with an acid catalyst, and was commercially available in the 1930s [9, 144]. Cell-O-(CO)CH$_3$ was the first ester derivative of cellulose; it was produced by Paul Schützenberger in 1865 by heating cotton and H$_3$C(OC)-O-(CO)CH$_3$ at 180 °C. Cellulose chemistry, as an individual branch of polymer chemistry, can be traced back to fundamental experiments in the 1920s and 1930s on acetylation and deacetylation of cellulose, resulting in the concept of polymer-analogous reactions.

The main production of Cell-O-(CO)CH$_3$ is still based on a solution process (*i.e.*, "solution acetylation") that includes a reaction between cellulose and H$_3$C(OC)-O-(CO)CH$_3$, dissolved in CH$_3$CO$_2$H, with H$_2$SO$_4$ as a catalyst [9, 17]:

$$(6.1)$$

In this topochemical heterogeneous reaction, successive layers of cellulose fibers react, and are solubilized into the medium, revealing new unreacted surface areas (Figure 6.3). The reaction rate is controlled by the diffusion of the reagents into the fiber matrix [9, 21]. Cellulose is first pretreated with CH_3CO_2H (i.e., the activation stage) in the presence of H_2SO_4 to swell the fibers, achieve a uniform reaction, and to adjust the DP to a suitable lower level (normally between 350 and 500). Acetylation is then performed for several hours at 50 °C with a mixture of $H_3C(OC)$-O-$(CO)CH_3$/CH_3CO_2H. A fully acetylated product with a DS of 3 (i.e., cellulose triacetate, CTA) is necessary to secure its complete solubility. The use of H_2SO_4 as a catalyst enables the production at lower temperatures, up to 50 °C. Typical specifications for acetylation-grade dissolving pulps include a low ash (<0.08%), metal (iron <10 ppm), extractives (<0.15%, diethyl ether-(CH_3CH_2-O-CH_2CH_3)-extractable compounds), and pentosans (<2.1%) content as well as a high alpha-cellulose content (>95.6%) [19].

The end product CTA ("primary acetate"), with an acetate content of 44.8% and a melting point of about 300 °C, is usually partially deacetylated without isolation in an aqueous CH_3CO_2H solution for some hours at 40–80 °C to an CH_3COCH_3-soluble product ("secondary acetate"), with a DS of about 2.5 ("cellulose 2.5-acetate"), an acetate content of about 40%, and a melting point of about 230 °C, or most commonly about 2 ("diacetate" or simply "acetate"), with an acetate content of about 35% [9, 21]. In this hydrolysis under controlled conditions, sulfate ions ($SO_4^{2\ominus}$, i.e., degradation products of sulfate half-ester groups) and a sufficient number of acetyl groups are removed, and a homogeneous distribution of the residual acetyl groups can be obtained. The product is then precipitated and washed with H_2O to remove acid impurities, and the H_2O is removed by centrifugation or pressing, and finally by drying. The product is readily soluble in CH_3COCH_3 (i.e., it is suitable for extrusion) and it can be converted by the so-called "dry spinning process" into filaments or films.

Both CTA and "acetate" are white, odorless, and nontoxic substances that tolerate a wide range of solvents [9, 21]. Cell-O-$(CO)CH_3$ is characteristically used in textiles (cf., acetate fiber or acetate rayon) and as a lining, composite fabrics (e.g., satins, knit fabrics, and taffetas), films, plastics, cigarette filters, lacquers, insulating foils, and reverse osmosis membranes. As an acetate fiber, it has many beneficial properties; for example, special dyes have been developed for CTA, since it does not accept dyes used for cotton and rayon. Fortisan[TM] is a trade name for high-tenacity cellulose fiber,

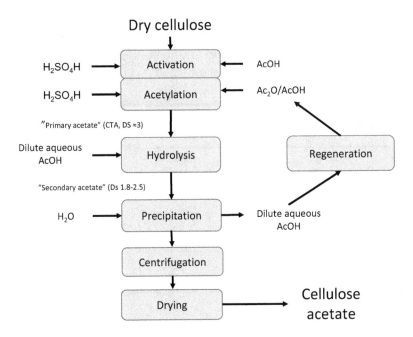

Figure 6.3: Simplified flow sheet of the conventional cellulose acetate process [9]. AcOH is acetic acid, Ac$_2$O acetic anhydride, CTA cellulose triacetate, and DP degree of substitution.

manufactured by partial saponification of stretched (*i.e.*, handled in steam under pressure to improve orientation) Cell-O-(CO)CH$_3$.

In addition to Cell-O-(CO)CH$_3$, cellulose propionate (Cell-O(CO)CH$_2$CH$_3$) and cellulose butyrate (Cell-O(CO)CH$_2$CH$_2$CH$_3$) (cellulose formate (Cell-O(CO)H) is unstable) are produced to some extent [23, 138, 141, 142, 145, 146]. They have some favorable properties compared to those of Cell-O-(CO)CH$_3$. Additionally, some mixed thermoplastic ester derivatives such as cellulose acetate propionate or cellulose acetopropionate (CAPr), cellulose acetate butyrate or cellulose acetobutyrate (CAB), and cellulose acetate phthalate or cellulose acetophthalate (CAP) are prepared commercially from cellulose with H$_3$C(OC)-O-(CO)CH$_3$ and the corresponding acid in the presence of H$_2$SO$_4$. These mixed esters have desirable properties not exhibited by "acetate" or CTA, and find applications mainly in lacquers, sheetings, molding plastics, film products, and hot-melt coatings.

Due to Cell-O-(CO)CH$_3$'s sensitivity to moisture, its limited compatibility with other synthetic resins and its relatively high processing temperature, a great number of cellulose esters of higher molar mass have been prepared (Table 6.4), but only a few have attained commercial use [142]. They can be easily prepared with procedures similar to those used for Cell-O-(CO)CH$_3$. In a series with increasing acyl chain length from C$_2$ to C$_6$, the melting point, density, and tensile strength decrease, while resistance to moisture and solubility in nonpolar solvents increase. Additionally, some mixed esters (*e.g.*, cellulose propionate-isobutyrate and cellulose propionate-valerate) have found use in plastic composites when good grease-repelling and H_2O-repelling properties are required [21]. Due to their ordered arrangement in solution, some cellulose esters dissolved in appropriate solvents show liquid crystalline characteristics like those of other rigid-chain polymers.

Table 6.4: Some properties of cellulose triesters [145].

Cellulose triester	Melting point, °C	Density, g/cm^3	Tensile strength, MPa	Moisture regain, % (95% rh)
Acetate C$_2$	306	1.28	71.6	7.8
Propionate C$_3$	234	1.23	48.0	2.4
Butyrate C$_4$	183	1.17	30.4	1.0
Valerate C$_5$	122	1.13	18.6	0.6
Caproate C$_6$	94	1.10	13.7	0.4
Heptylate C$_7$	88	1.07	10.8	0.4
Laurate C$_{12}$	91	1.00	5.9	0.3

In contrast, the preparation of cellulose esters of aromatic acids has so far acquired only little commercial interest and has mainly been limited to special cases, such as cellulose cinnamate, cellulose salicylate, cellulose phthalate, and cellulose terephthalate [9]. However, some mixed cellulose esters, containing a dicarboxylic moiety (especially, cellulose acetate phthalate), have technically useful properties, including solubility in alkalis and excellent film-forming characteristics. Additionally, various cellulose esters with organic acids, carrying sulfonic acid ($-SO_3H$) or phosphonic acid ($-PO_3H_2$) groups, have been prepared. In all these cases, homogeneous esterification in different solutions (*e.g.*, in DMA/LiCl solution) is possible (Figure 6.4).

During the 1890s, the development was stimulated not only by scientific curiosity, but also by the more practical intention to obtain the H_2O-insoluble cellulose material into the dissolved state by a suitable chemical transformation for preparing an endless cellulose thread, aiming at the ultimate preparation of an artificial silk [9]. By using, instead of a normal spinning jet (*i.e.*, spinneret), a slit and by adding $HOCH_2$-$(OH)CH_2OH$, which prevented the harmful formation of excess crystallinity, to the solution, plastic (film) cellophane could be produced.

The largest part of the cellulose-based artificial fibers, viscose rayon, is the first semi-synthetic fiber product and is manufactured by the so-called "viscose process", (xanthation), originally invented in 1892 by Charles Frederick Cross, Edward John Bevan, and Clayton Beadle [9, 142]. This product was commercialized in 1894 for textile purposes. Rayon is used today for many clothing articles and has been the most useful synthetic fiber or filament to human beings. The process, "viscose route to artificial silk", is simply based on the formation of cellulose xanthate Cell-O-CS$_2$H (or cellulose xanthogenate), which is then decomposed by spinning under acidic conditions. Cell-O-CS$_2$H

Cell—OH +

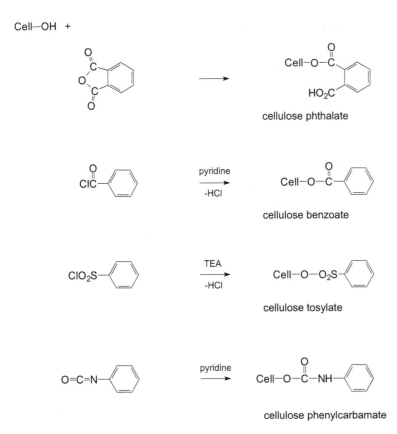

Figure 6.4: Typical examples of aromatic cellulose derivatives from homogeneous esterification [9]. TEA refers to triethylamine.

is a H$_2$O-soluble, unstable anionic ester, obtained by reacting cellulose with CS$_2$ ("sulfo-carbide") in an aqueous solution of NaOH.

High-quality regenerated cellulose, viscose fiber, can be obtained by the viscose process (Figure 6.5) in which the cellulose feedstock material, with a cellulose content of 91–96% and an ash content of <0.1%, is first mercerized to convert it to alkali cellulose [9, 17, 142, 147]. After this stage, the cellulose mass is pressed between rollers to remove excess liquid to obtain a cellulose followed by shredding or crumbling (the product is called "white crumb") and oxidative depolymerization ("aging" or "pre-ripening"), to obtain an appropriate DP level of 200–400 for the production of cellulose xanthate Cell-O-CS$_2$H with suitable properties. In the alkaline oxidation stage, the cellulose is exposed to ambient air or pressurized oxygen at 20–50 °C. Hence, the treatment of alkali cellulose with CS$_2$ leads finally to the formation of cellulose xanthate:

Formation of cellulose xanthate

$$\text{Cell–O}^{\ominus} + \underset{\text{S}}{\overset{\text{S}}{\text{C}}} \rightleftharpoons \text{Cell–O–C}\underset{\text{S}}{\overset{\text{S}^{\ominus}}{}}$$

Regeneration of cellulose from viscose

$$\text{Cell–O–C}\underset{\text{S}}{\overset{\text{SNa}}{}} + \text{H}^{\oplus} \rightleftharpoons \text{Cell–O–C}\underset{\text{S}}{\overset{\text{SH}}{}} + \text{Na}^{\oplus}$$

$$\downarrow$$

$$\text{Cell–OH} + \text{CS}_2$$

(6.2)

$$2\text{Cell–O–C}\underset{\text{S}}{\overset{\text{SNa}}{}} + \text{Zn}^{2\oplus} \rightleftharpoons \begin{array}{c} \text{Cell–O–C}\underset{\text{S}}{\overset{\text{S}}{}} \\ | \\ \text{Zn} \\ | \\ \text{Cell–O–C}\underset{\text{S}}{\overset{\text{S}}{}} \end{array} + \text{Zn} + 2\text{Na}^{\oplus}$$

$$\overset{\text{H}^{\oplus}}{\nearrow}$$

$$2\text{Cell–O–C}\underset{\text{S}}{\overset{\text{SH}}{}} + \text{Zn}^{2\oplus}$$

$$\searrow$$

$$2\text{Cell–OH} + 2\text{CS}_2$$

The xanthation at 25–30 °C for about three hours results in an orange-yellow product ("yellow crumb" with a DS of about 0.5) that is not completely soluble but is dissolved in dilute aqueous NaOH, usually under high-intensity mechanical action, leading to the formation of a viscous yellow solution (Figure 6.5) [9, 17, 142, 147]. This viscose solution contains, besides Cell-O-CS$_2$$^{\ominus}$ (about 8%) and free NaOH (6–7%), trithiocarbonate ($CS_3^{2\ominus}$) and carbonate ($CO_3^{2\ominus}$) at the 1% level, sulfide ($S^{2\ominus}$) and perthiocarbonate ($CS_4^{2\ominus}$) at the 0.1% level, and small amounts of thiosulphate ($S_2O_3^{2\ominus}$), dithiocarbonate ($COS_2^{2\ominus}$), and monothiocarbonate ($CO_2S^{2\ominus}$). The purpose of the viscose aging ("ripening") at room temperature approximately for 1–3 days is to adjust the viscosity of the solution to the level desired for the spinning process. Cellulose chains must be short enough to give manageable viscosities in the spinning solution but still long enough to impart good physical properties to the fiber product. The DS value is also reduced to some extent during this ripening, but the xanthate groups are redistributed more uniformly along the cellulose chains. The primary xanthate groups at C$_6$ are more stable than those at C$_2$ and C$_3$ positions, and the latter groups can be hydrolyzed 15–20 times faster under certain conditions [139, 147]. Hence, the extent of hydrolysis and the redistribution of xanthate

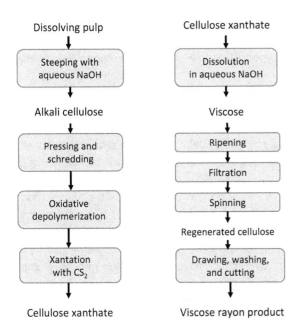

Figure 6.5: Simplified scheme of the viscose process [9]. For explanations, see the text.

groups vary during ripening. In the conventional spinning process, the viscose solution is pressed at about 40 °C through corrosion-resistant spinnerets having, depending on the product, 100–1000 holes (hole diameter of 50–100 μm) into a bath of aqueous H$_2$SO$_4$ and Na$_2$SO$_4$. Finally, the fine rayon filaments are mechanically treated to modify their physical properties suitable for textile fibers and washed to remove salts and other H$_2$O-soluble impurities.

The chemistry of the viscous process is rather complicated, although the mechanism behind the xanthation process is simply the introduction of a sufficient number of anionic COS$_2^\ominus$ groups to the cellulose chain to convert it to a homogeneous soluble derivative [9, 17, 142, 147]. However, only about 70% of the CS$_2$ input is consumed in the reaction of cellulose, while the rest is consumed under alkaline conditions in various side reactions. Furthermore, the conversion of the alkaline cellulose solution to filaments of cellulose by spinning in the presence of transition metal cations (*e.g.*, Zn$^{2\oplus}$, Ag$^\oplus$, or Hg$^{2\oplus}$). in a combination with special additives (*e.g.*, amines and/or polyethylene oxides), is a complex chemical process that is strongly affected by the previous steps of xanthation as well as the dissolution of cellulose xanthate. The regeneration reaction ("solidifying") during which the viscose is predominantly coagulated by metal ions and decomposed by the acid, occurs in two basic steps with different velocity constants in an acid bath: *i*) the formation of cellulose xanthogenic acid (a fast reaction) and then *ii*) the decomposition of this acid (a slow reaction) with the simultaneous regeneration of almost pure cellulose. In general, when the viscose is passed through a spinneret, the product is artificial fibers (*i.e.*, rayon), whereas in the case of a slit, cellophane can be obtained.

Nowadays, several commercial rayon fibers are named according to the process used to convert the cellulose into a soluble polymer and then regenerated [142]. The dominating step is still the viscose process. Despite certain shortcomings, such as the "excess consumption" of CS_2 as well as the need to handle sulfur-containing discharges, together with some slow unit operation stages [148], the main reasons for the clear dominance of this process are its versatility and adaptability to end-use requirements, combined with a long engineering experience and development that has led to many recent improvements.

It can be concluded that this conventional process converts the short wood-derived fibers of cellulose into endless filaments, or stable fibers, or films utilized primarily in food packaging [142]. Rayon fibers of different grades are used in a great variety of applications for textiles in apparel and home furnishing, but they also have significant uses, for example, as nonwoven materials and many other industrial products. They are generally easy to dye, soft, resistant to wrinkles (*i.e.*, they drape well), and highly absorbent. Depending on the process, it is also possible to adjust the product properties within a wide range. For example, a drawing stage when applied, leads to products that have about twice the strength and two-thirds of the stretch of regular rayon.

Cellulose reacts with H_2N-CO-NH_2 and with substituted ureas at relatively high temperatures to yield carbamates [140, 149–151]. In recent years, cellulose carbamate (Cell-O-$CONH_2$) has received increasing attention as a potential alkali-soluble intermediate, with a DS of 0.2–0.3, for the manufacture of regenerated cellulose products. This cellulose derivative is formed at elevated temperatures (135–180 °C) from cellulose and H_2N-CO-NH_2 *via* isocyanic acid (H-N=C=O) as an active intermediate [152, 153]:

$$H_2N\text{-CO-}NH_2 \rightarrow \text{H-N=C=O} + NH_3 \tag{6.3}$$

$$\text{Cell-OH} + \text{H-N=C=O} \rightarrow \text{Cell-O-}CONH_2 \tag{6.4}$$

Compared to the viscose process, the cellulose carbamate process has been claimed to be advantageous by avoiding the use of hazardous sulfur-containing compounds for derivatization, although higher NaOH consumption is typically needed [154]. When heating cellulose with thiourea (H_2N-CS-NH_2) at 180 °C, the reaction leads in an analogical way to cellulose thiocarbamate (Cell-O-CSNH$_2$) [140].

The most typical raw material for cellulose carbamates is dissolving pulp, but normal paper pulp and recycled cellulose fibrous feedstock, together with pretreated recycled textile-based materials, are suitable as well [151]. In the production, high accessibility is an essential prerequisite for the homogeneous substitution of cellulose material. Cellulose structure can be affected by means of various mechanical activations, but the activation pretreatment of cellulose can be also performed, for example, with NaOH, ammonia (NH_3), hot H_2O under pressure, and enzymes. Additionally, DESs are utilized in the carbamate process of certain feedstock materials [155]. However, the further optimization of the process for specific applications is still needed. In general, cellulose carbamate is especially suited for textile fibers, but also for films, foams, membranes, and many specific applications [151, 156–161].

Cellulose can be also dissolved by several other methods and solidified back as regenerated cellulose with its main properties (*i.e.*, DP and crystalline structure) different from those of the initial cellulose [17, 21, 22, 147]. For example, in 1857, Matthias Eduard Schweizer observed that cellulose is soluble in a solution of copper(II) hydroxide ($Cu(OH)_2$) in ammonium hydroxide (NH_4OH) and in the 1890s, the so-called "cuprammonium method" (the cupron process) was developed. In this method, regenerated cellulose filaments are obtained by spinning an alkaline cellulose solution containing cupritetramine hydroxide $[Cu(NH_3)_4](OH)_2$ in an excess of NH_4OH into an aqueous hardening bath that removes the copper and ammonia and neutralizes NaOH.

6.3 Cellulose esters of inorganic acids

Cellulose esters of different inorganic acids are cellulose derivatives that result from the esterification of the -OH groups of cellulose with simply inorganic acids [9, 10, 17, 18, 21, 22]. A prerequisite for the esterification is that the acids used can bring about a strong swelling, penetrating throughout the cellulose structure. Additionally, this reaction can be considered as a typical equilibrium reaction in which all steps can run in the reverse direction and an alcohol and acid react to form ester and H_2O.

Cellulose was used as a starting material for technically important derivatives, even before its polymeric nature was well understood [10]. Henri Braconnot was the first to prepare the oldest cellulose derivative of commercial importance, cellulose nitrate (Cell-O-NO_2), as early as in 1832. The early history of this inorganic ester is strongly associated with the militaries of most European nations during the second half of the nineteenth century. In practice, Christian Friedrich Schönbein developed in 1846, the preferred method of highly nitrated cellulose (*i.e.*, "cotton powder" or "gun cotton" with a nitrogen content of 12.0–13.6%) using HNO_3-H_2SO_4 mixtures, and Frederick Abel developed in 1863 a method of safely handling Cell-O-NO_2, thus making possible its use as an explosive.

Later in 1868, the trial-and-error discovery of the first semisynthetic thermoplastic "celluloid" was made by John Wesley Hyatt by combining Cell-O-NO_2, having a nitrogen content of 9–11% ("collodion"), with camphor (about 20% of the product) and a minor concentration of other plasticizers (*e.g.*, dibutyl sulfate) [10]. This significant discovery was originally made in a public competition with the objective of finding substitute materials for ivory in the production of billiard balls. Hence, Cell-O-NO_2 gradually became the progenitor of the industries of explosives, plastics, lacquers, photographic films, protective coatings, and cements.

The synthesis of Cell-O-NO_2 also formed the basis for the first industrial process of manufacturing "artificial silk" by spinning a Cell-O-NO_2 in CH_3CO_2H into a precipitation bath of cold CH_3CH_2OH [10, 151]. The full-scale regeneration of cellulose from Cell-O-NO$_2$ through denitration, dissolution, spinning, and regeneration was realized by Hilaire de Chardonnet in 1899, although he developed the method principles already in 1885. However, this invention can be traced even back to 1850 when Joseph Wilson Swan made electric lamp filaments by extrusion and to 1855, when Georges Audemars developed the preparation of Cell-O-NO$_2$-based artificial silk by a method impractical for commercial use; due to the high flammability of the nitrogen-containing artificial silk, it was quickly taken off the market. So-called "Chardonnet's artificial silk" was exhibited at the Paris Exhibition of 1899, and it gained the Grand Prix.

The current industrial production of Cell-O-NO$_2$ is still based on the typical heterogeneous equilibrium reaction between cellulose and HNO_3 in the presence of H_2O and H_2SO_4 (*i.e.*, "nitrating acid") [9, 10, 17, 18, 21, 22]. The reaction is retarded by H_2O; the removal of H_2O with H_2SO_4 forces the reaction to completion. The first reaction step involves the generation of the nitronium ion ($NO_2^{\oplus}$):

$$HNO_3 + 2H_2SO_4 \leftrightarrow NO_2^{\oplus} + 2HSO_4^{\ominus} + H_3O^{\oplus} \tag{6.5}$$

Variations of the nitrating acid as well as the temperature and reaction time determine product qualities of the resulting Cell-O-NO$_2$ [9]. Industrial-scale nitration of high-quality dissolving pulp or cotton linters can be performed either in a continuous reactor or in a batch process. In the classical latter case, a rather small quantity of cellulose, compared to the liquid phase of HNO_3-H_2O-H_2SO_4 (1:20–1:50), is vigorously agitated for 20–30 min. A proper ratio of HNO_3 to H_2SO_4 is chosen to yield the desired degree of nitration and to help maintain a fibrous product structure. At a DS of 1.9, 2.4, and 2.7, the relative percentage composition of the system HNO_3:H_2O:H_2SO_4 is 25.0:19.3:55.7, 25.0:15.2:59.8, and 25.0:8.5:66.5, respectively.

Most of the excess acid is separated after nitration by centrifuging and recycled [9, 17]. "Pre-stabilization" of the Cell-O-NO$_2$ consists of a series of washing and cooking with H_2O to remove last traces of adhering acid. With low-medium and medium-nitrated products, subsequent digestion under pressure at 130–150 °C reduces the chain length of the molecules (*i.e.*, the adjustment of viscosity) and equalizes the distribution of NO_2-subtituents (*i.e.*, "stabilization" or "post-stabilization"). However, highly nitrated products for explosives require careful stabilization to avoid uncontrolled decomposition. Finally, for further processing, H_2O is displaced by alcohol in the case of celluloid nitrates or lacquers; the former products are available in the form of fibers or flakes.

Products can be gelatinized by using softeners, such as phthalic acid esters [10]. It is also possible to make aqueous dispersions of the softened products for their further use in coatings. At a DS of 1.8–2.7, the product mass is generally 125–150% of the initial mass of cellulose. The continuous process, originally from the 1960s, has clear advantages, such as shorter processing time, higher product uniformity, and higher safety, compared to the conventional batch process. The DS range of commercial products is between 1.8 and 2.8, with a content of nitrogen between 10.5% and 13.6%, respectively.

Table 6.5 shows examples of commercial grades of Cell-O-NO$_2$. The products are white, odorless, tasteless, and rather hydrophobic substances, whose physical and chemical properties depend on DS. For example, their density typically varies from 1.5 g/cm^3 to 1.7 g/cm^3. Commercial products of Cell-O-NO$_2$ can be plasticized with a variety of conventional softeners; in this property, they are compatible with many synthetic polymers.

Table 6.5: Typical commercial grades of cellulose nitrate (solvents suitable for each derivative are indicated in parentheses) [17, 21].

Degree of substitution	Nitrogen content, %	Applications
1.8–2.0	10.5–11.0	Plastics and lacquers (ethanol)
1.9–2.1	10.9–11.3	Lacquers (ethanol/diethyl ether)
2.2–2.3	11.8–12.2	Lacquers, films, and coatings (esters)
2.2–2.8	11.8–13.6	Explosives (acetone)

Besides the only important commercial product Cell-O-NO$_2$, some other inorganic esters of cellulose, including cellulose sulfate (Cell-O-SO$_3$H), cellulose phosphate (Cell-O-PO$_3$H$_2$), cellulose borate (Cell-O$_3$B), and cellulose nitrite (Cell-O-NO) have also been systematically investigated [9, 17, 18, 22], but they have not achieved any commercial significance (Table 6.6). Cell-O-SO$_3$H can be prepared by a variety of reagent combinations; the straightforward treatment of cellulose with aqueous H$_2$SO$_4$ leads only to a very low conversion (with a maximum DS of 1.5), together with the formation of a large bulk of different degradation products. The stable sodium salt of Cell-O-SO$_3$H is a white, odorless, and tasteless powder that is completely H$_2$O-soluble at a DS of above 0.2–0.3 and exhibits good thermal stability up to 100 °C. Cell-O-SO$_3$H has found limited specialty use as thickeners for lacquers, cosmetics, pharmaceuticals, and food, as printing inks, coatings of photographic films, detergents, and oil drilling fluids. It reacts with hydrogen halides to yield deoxyhalogenated products with excellent flame-retardant properties.

Table 6.6: Examples of the inorganic esters of cellulose and their main routes of production [10].

Cellulose derivative	Reagent	Product
Cellulose nitrate	HNO$_3$/H$_2$SO$_4$	Cell-O-NO$_2$
Cellulose sulfate	SO$_3$ or H$_2$SO$_4$/SO$_3$ or ClSO$_3$/SO$_3$	Cell-O-SO$_3$H
Cellulose phosphate	H$_3$PO$_4$	Cell-O-PO$_3$H$_2$
Cellulose borate	H$_3$BO$_3$ or B(OR)$_3$	Cell-O$_3$B
Cellulose nitrite	N$_2$O$_4$	Cell-O-NO

Cell-O-PO$_3$H$_2$ is traditionally prepared as an ammonium salt by the reaction of cellulose with H$_3$PO$_4$ in molten H$_2$N-CO-NH$_2$ [9, 18, 22]. The other typical reagents are the derivatives of pentavalent phosphorous (as also in case of H$_3$PO$_4$) diphosphorous pentaoxide (P$_2$O$_5$) and phosphoryl chloride (POCl$_3$). Besides phosphate groups $\left(\text{Cell-O-P(O)(OH)}_2\right)$, phosphite groups $\left(\text{Cell-O-P(OH)}_2\right)$, and phosphinate groups $\left(\text{Cell-O-P(O)(OH)}\right)$ are formed. Many phosphorus-containing cellulose esters have been of increasing interest because of their flame-retarding properties and potential use in textiles as well as their ion-exchange capability (*i.e.*, as weak cation exchangers). Boron-containing cellulose esters have also been investigated to improve, for example, the flame retardancy or heat stability of cellulose. The main production methods of Cell-O$_3$B involve a direct esterification of cellulose with boric acid (H$_3$BO$_3$) (or (HBO$_2$)$_n$) or a transesterification of cellulose with boronic acid esters of lower aliphatic alcohols (boron alkoxides, B(OR)$_3$). Additionally, a strong cross-linking takes place during the "borylation". In general, there are also several organic flame retardants or coatings added to materials, such as cotton and viscous fibers, to inhibit, suppress or delay the production of flames [161]. However, often, the needed amounts may be high, typically 10–30% of fire-safe consumer products. Cell-O-NO has also gained some importance, mainly as an interesting intermediate in cellulose chemistry. A highly substituted product can be obtained by nitrogen tetroxide (N$_2$O$_4$, *i.e.*, NO$^{\oplus}$ + NO$_3^{\ominus}$) and it is a yellow, hygroscopic mass that decomposes rapidly in the presence of moisture with the release of nitrogen oxides (NO$_x$).

6.4 Cellulose ethers

Cellulose ethers comprise a class of cellulose derivatives that contains several commercially important members [9, 10, 21, 22, 27, 162, 163]. The most prominent ethers are H$_2$O soluble. Hence, they are widely used in many H$_2$O-based formulations as thickeners for adjusting the rheology of solutions (*i.e.*, the control of thickening, viscosity, and flow behavior), mainly in food applications, pharmaceuticals, cosmetics, drilling muds, latex paints, and building materials. The most significant properties of cellulose ethers are their solubility, together with their H$_2$O-binding ability (*i.e.*, absorbency and retention), chemical stability, and nontoxicity.

The introduction of ether groups into a cellulose molecule at very low DS levels leads to the swelling ability or solubility of the product, even in cold H$_2$O [9]. The type of constituents, DS, and the uniformity of substitution control the H$_2$O solubility and/or organic solvent solubility of cellulose ethers. For example, for methylcellulose (Cell-O-CH$_3$) and ethylcellulose (Cell-O-CH$_2$CH$_3$), the DS range typically needed for solubility in H$_2$O is 1.5–2.0 and 0.7–1.7, and in organic solvents, >2.5 and >2.2, respectively [17, 22].

Cellulose ethers can be prepared by treating alkali cellulose with a variety of different reagents. Among the synthesis routes under heterogeneous conditions, only three reaction ways are of commercial importance (Table 6.7) [9]: *i*) the reaction of -OH groups with an alkyl or aryl chloride in the presence of NaOH ("the alkali-consuming process",

1 mol NaOH/1 mol reagent) and often also, an inert diluent, according to the Williamson reaction, *ii*) the reaction of -OH groups with an alkylene oxide ("the ring-opening reaction" without alkali consumption), and *iii*) the reaction of -OH groups with α,β-unsaturated compounds, activated by electron-attracting groups (the Michael addition reaction).

Table 6.7: Examples of typical cellulose ethers obtained by different methods.

Reaction with alkyl or aryl chlorides		
Carboxymethylcellulose	Cell-O-CH$_2$CO$_2$H	CMC
Methylcellulose	Cell-O-CH$_3$	MC
Ethylcellulose	Cell-O-CH$_2$CH$_3$	EC
Propylcellulose	Cell-O-CH$_2$CH$_2$CH$_3$	PC
Benzylcellulose	Cell-O-CH$_2$C$_6$H$_5$	BC
Reaction with an alkylene oxide		
Hydroxyethylcellulose	Cell-(OCH$_2$CH$_2$)$_n$OH	HEC
Hydroxypropylcellulose	Cell-(OCH$_2$CH(CH$_3$))$_n$OH	HPC
Hydroxybutylcellulose	Cell-(OCH$_2$CH(CH$_2$CH$_3$))$_n$OH	HBC
Reaction with an α,β-unsaturated compounds		
Cyanoethylcellulose	Cell-O-CH$_2$CH$_2$C ≡ N	
Carbamoylethylcellulose	Cell-O-CH$_2$CH$_2$CONH$_2$	
Carboxyethylcellulose	Cell-O-CH$_2$CH$_2$CO$_2$H	CEC

In the reactions of alkali cellulose with alkyl or aryl chlorides, the purposes of the inert diluent are to suspend/disperse the cellulose raw material, moderate reaction kinetics, provide heat transfer, and facilitate recovery of the products [9]. Reactions are typically conducted at elevated temperatures (50–140 °C) and under nitrogen atmosphere to avoid oxidative degradation reactions of cellulose. However, in all cases, a great variety of side reactions takes place. After etherification, crude grades are simply dried, ground, and packed out, but purified grades require the removal of several by-products in a separate stage, prior to drying. There are also some commercially important cellulose ethers, which contain several types of functional groups ("mixed cellulose ethers").

The most common and simplest representatives of alkylcelluloses or cellulose alkyl ethers are Cell-O-CH$_3$ and Cell-O-CH$_2$CH$_3$. [9, 17, 21, 22], They both are mainly produced by a reaction ("alkylation") of alkali cellulose with the corresponding alkyl chlorides:

$$\text{Cell-O}^{\ominus} + \text{R-Cl} \rightleftharpoons \text{Cell-O-R} + \text{Cl}^{\ominus} \tag{6.6}$$

$$R = CH_3 \text{ or } CH_2CH_3$$

This Williamson reaction proceeds according to the S_N2 mechanism (bimolecular nucleophilic substitution) [9]. Under these alkaline conditions, CH_3OH or CH_3CH_2OH is also obtained as a by-product that can react further with alkyl chloride (R-Cl) to form, respectively, ethers CH_3-O-CH_3 and CH_3CH_2-O-CH_2CH_3. This by-product formation accounts for 20–30% of the R-Cl consumption, thus decreasing the reagent efficiency. Alkylcelluloses are white-to-yellowish and nontoxic solids that exhibit, depending on the substituent and DS, varying solubility in various media. The DS of Cell-O-CH_3 is typically in the range of 1.5–2.0 and it is H_2O soluble.

Cell-O-CH_2CO_2H (CMC) (or sodium carboxymethylcellulose Cell-O-CH_2CO_2Na (NaCMC)) is commercially the most important cellulose ether [9, 10, 17, 21, 164]. The basic chemistry of carboxymethylation has been well known for a long time. It was clarified in 1918 and later, efforts have mainly been directed toward optimization and rationalization of the process. In general, Cell-O-CH_2CO_2H is prepared from alkali cellulose with sodium monochloroacetate (ClCH$_2$CO$_2$Na) as a reagent; in the "semi-dry process", the solvent is either CH_3CH_2OH or CH_3OH and in the "slurry process", it is isopropanol or 2-propanol ($CH_3CH(OH)CH_3$):

$$\text{Cell-O}^{\ominus} + \underset{\text{Cl}}{CH_2CO_2^{\ominus}} \longrightarrow \text{Cell-O-CH}_2CO_2^{\ominus} + \text{Cl}^{\ominus} \tag{6.7}$$

The most important side reaction is the formation of sodium glycolate ($HOCH_2CO_2Na$) and sodium chloride (NaCl). The product is the sodium salt NaCMC, but usually designated simply as CMC. Another commercial form of CMC is the product with calcium as the counter ion (CaCMC). However, this product is not H_2O soluble; it swells substantially in aqueous media and hence, is useful as a tablet disintegrant.

In the carboxymethylation process, monochloroacetic acid (ClCH$_2$CO$_2$H) is added to the reaction slurry containing NaOH in sufficiently excess levels to neutralize it (i.e., the formation of ClCH$_2$CO$_2$Na) and to promote its reaction [9]. The reaction requires at least 0.8 mol of NaOH per mol of an AGU in the cellulose chain to remain the proper alkalinity throughout the etherification (e.g., at 60–80 °C for 90 min). The products consist of purified CMC powder (>95% CMC, about one-third of the total production), together with various unpurified, crude technical grade powders (55–75% CMC). Commercial CMCs are normally soluble in H_2O (DP 200–1,000), with DS values ranging from 0.4 to 1.4 (typically 0.5–0.8). At a lower DS of 0.05–0.25, the product is soluble only in aqueous NaOH (4–10%), whereas at a DS of 0.4–1.4, a good H_2O solubility is attained and at a DS of >2.2, the products are soluble in polar organic solvents.

Since its commercial production, CMC has found use in an ever-increasing number of applications [9]. It can be used in a variety of products such as in food products (mainly as a thickener but also as a stabilizer and H_2O-binder as well as a whipping agent in ice-creams and jellies), detergents (*e.g.*, as a soil antiredeposition aid), oil drilling muds (as a dispersion-stabilizing aid and viscosity-adjusting aid), textiles (*e.g.*, as a warp size), paper products (as a papermaking chemical), paper coatings, pharmaceuticals (as a thickener, H_2O-binder, stabilizer, and granulation aid), cosmetics (*e.g.*, as a thickener and suspension aid in toothpastes and shampoos), latex paints, and ceramics. CMC absorbs many times its own mass of H_2O to form a stable colloidal mass. Because of the -CO_2H groups, CMC is also a polyelectrolyte (pK_a 4–5, depending on the DS).

In the reactions of alkali cellulose with an alkylene oxide, the clear commercial hydroxyalkylcelluloses are Cell-$(OCH_2CH_2)_nOH$ and hydroxypropylcellulose Cell-$(OCH_2CH(CH_3))_nOH$ [9, 10, 17, 21, 22, 163]. They are obtained in the reaction of alkali cellulose with gaseous alkene oxides or their corresponding liquid chlorohydrins. Hence, Cell-$(OCH_2CH_2)_nOH$ is typically produced from alkali cellulose and ethylene oxide:

$$\text{Cell–O}^{\ominus\oplus}\text{Na} + n\,H_2C\!\!-\!\!CH_2 \quad \longrightarrow \quad \text{Cell–}(OCH_2CH_2)_nOH + NaOH \tag{6.8}$$

Hydroxyalkylation with epoxides is a base-catalyzed substitution reaction that does not require a stoichiometric amount but, in principle, only a catalytic amount of $HO^\ominus$ ions for the cleavage of the epoxy ring [9]. However, hydroxyalkylation is not limited only to the -OH groups originally present in alkali cellulose and the reaction can proceed further at the newly formed primary -OH groups, leading to hydroxyalkyl chains of varying length. Hence, in the case of hydroxylethylation, pendant oxyethylene chains are formed.

Because the ethylene oxide can react with both the original -OH groups of cellulose and/or the -OH of the substituents already reacted with cellulose (*i.e.*, forming an oxyethylene chain), the term "molar substitution" (MS) is used to describe the degree of reaction and the product itself [9]; hence, it denotes the average number of alkylene oxide molecules added per AGU. DS is lower than MS in the products and the ratio MS/DS can be used as a measure of the relative length of the side chains. Additionally, the alkali-catalyzed hydroxyethyl ether formation is accompanied by a reaction of $HO^\ominus$ ions with ethylene oxide to $HOCH_2CH_2OH$ and further to polyglycols ($H(OCH_2CH_2)_nOH$). Usually, only about one-half of the ethylene oxide reacts with alkali cellulose, while the other half is consumed in different side reactions.

Cell-$(OCH_2CH_2)_nOH$ and Cell-$(OCH_2CH(CH_3))_nOH$ are white, odorless, and nontoxic powders, whose solubilities depend highly on the substitution [9]. Products with a MS of 0.05–0.5 are only soluble in alkalis, while those with a MS of 1.5 or more are H_2O soluble. Cell-$(OCH_2CH(CH_3))_nOH$ is more hydrophobic than Cell-$(OCH_2CH_2)_nOH$ and it can be extruded without a softener at 160 °C, while Cell-$(OCH_2CH_2)_nOH$ is not thermoplastic and decomposes in aqueous solutions above 100 °C.

The commercial Cell-$(OCH_2CH_2)_nOH$, with a MS of 1.8–3.5, is used as a thickener, protective colloid, binder, stabilizer, and suspending agent in a variety of industrial applications, including latex paints, construction substances, ceramics, paper chemicals, cosmetics, pharmaceuticals, textiles, and as an aid in polymerization [9]. It can be also used in coatings on food-contact surfaces of metal, paper, or board. Cell-$(OCH_2CH(CH_3))_nOH$ is used for similar purposes, although its use is more limited. Cell-$(OCH_2CH_2)_nOH$ and, particularly, Cell-$(OCH_2CH(CH_3))_nOH$ have also a tendency to form liquid-crystalline aqueous systems that have been comprehensively studied in recent years. The production of Cell-$(OCH_2CH(CH_2CH_3))_nOH$ is of minor importance.

Further modification of hydroxyalkylcelluloses with another functional group can enhance or introduce novel product properties [162, 163]. Carboxymethylhydroxyethylcellulose (CMHEC) is one commercial example of such mixed cellulose ethers. This product is an anionic modification of Cell-$(OCH_2CH_2)_nOH$; it can be manufactured by a reaction of alkali cellulose, either simultaneously or sequentially, with ethylene oxide and $ClCH_2CO_2Na$. Different grades are available and are primarily used in oil recovery applications.

Examples of other commercial mixed cellulose ethers include methylhydroxyethylcellulose (MHEC) or hydroxyethylmethylcellulose (HEMC), ethylhydroxyethylcellulose (EHEC) or hydroxyethylethylcellulose (HEEC), hydrophobically modified hydroxyethylcellulose (HMHEC), cationic hydroxyethylcellulose (cationic HEC), methylhydroxypropylcellulose (MHPC) or hydroxypropylmethylcellulose (HPMC), and hydroxybutylmethylcellulose (HBMC) [9]. Additionally, simple cellulose ethers, with both -CH_3 and -CH_2CH_3 groups, have been manufactured.

The alkylene oxide derivatives of Cell-O-CH_3 are (in parentheses are shown the typical values of DS and MS) [9]: MHEC (DS_{methyl} 1.3–2.2 and $MS_{hydroxyethyl}$ 0.06–0.5), MHPC (DS_{methyl} 1.1–2.0 and $MS_{hydroxypropyl}$ 0.1–1.0) as well as HBMC (DS_{methyl} ≥1.9 and $MS_{hydroxybutyl}$ ≥0.04). These products are used like hydroxyalkylcelluloses for construction materials (*e.g.*, cements and mortars), agricultural and food products (*e.g.*, mayonnaise and dressings), cosmetics and pharmaceuticals (*e.g.*, tablets and formulations), and latex paints. The most common DS range of the alkylene oxide derivatives of Cell-O-CH_2CH_3 is 2.2–2.7 and they are soluble in many organic solvents. For example, EHEC is used in inks and lacquers. Additionally, the alkylene oxide derivatives of Cell-O-$CH_2CH_2CH_3$ and Cell-O-$CH_2C_6H_5$ are produced to some extent.

In the reactions of alkali cellulose with an α,β-unsaturated compound, which contains a strongly electron-attracting group (reaction time is some hours at 30–50 °C in the presence of NaOH), Cell-O-$CH_2CH_2C{\equiv}N$ is the most common product [9, 21, 22]. The cellulose anion (Cell-$O^\ominus$) attacks the carbon atom with a small positive charge in acrylonitrile ($CH_2{=}CH{-}C{\equiv}N$) to form a resonance-stabilized intermediate anion that then adds a proton from H_2O, leading to the formation of Cell-O-$CH_2CH_2C{\equiv}N$ under the simultaneous liberation of an $HO^\ominus$ ion. All reaction steps are reversible and because of the regeneration of $HO^\ominus$ ions, no alkali is consumed. The process is accompanied with the consumption of $CH_2{=}CH{-}C{\equiv}N$ in many side reactions, such as the formation of di(cyanoethyl)ether or 3,3′-oxydipropionitrile ($O(CH_2CH_2C{\equiv}N)_2$).

The solubility of Cell-O-CH$_2$CH$_2$C≡N depends on the DS [9]. For example, for solubility in alkali, a DS of 0.25–0.5 and an uniform distribution of the substituents are necessary, while products with a DS of about 2.5 are soluble in polar organic solvents. Due to an unusually high dielectric constant and low dissipation factor, they can be used as insulating materials. Cyanoethylated paper also has good thermal and dimensional stabilities.

The reactivity of acrylamide (CH$_2$=CH-CONH$_2$) is lower than that of CH$_2$=CH-C≡N, but it can be added to alkali cellulose in a similar manner, resulting in the formation of Cell-O-CH$_2$CH$_2$CONH$_2$ or the carbamoylethyl ether of cellulose [9]. Cell-O-CH$_2$CH$_2$C≡N and Cell-O-CH$_2$CH$_2$CONH$_2$ both decompose in an aqueous medium of higher alkalinity (NaOH) at elevated temperatures to give, by saponification, the sodium salt of Cell-O-CH$_2$CH$_2$CO$_2$H (Cell-O-CH$_2$CH$_2$CO$_2$Na) as the stable end product. Here, the amide group (-CONH$_2$) is more easily saponified to a -CO$_2$H group than the nitrile group (-C≡N).

Besides the -C≡N and -CONH$_2$ groups, several other substituents are also able to activate the >C=C< bond by the addition reaction with alkali cellulose [9]. Typical examples of other reagents include methacrylonitrile (H$_2$C=C(CH$_3$)C≡N), α-methyleneglutaronitrile (HO$_2$CC(=CH$_2$)(CH$_2$)$_3$C≡N), α-chloroacrylonitrile (H$_2$C=CClC≡N), *trans*-crotonitrile (H$_3$CCH=CHC≡N), and allyl cyanide (H$_2$C=CHCH$_2$C≡N). There are also other common functionalized cellulose ethyl ethers, such as aminoethylcellulose Cell-O-CH$_2$CH$_2$NH$_2$ and sulfoethylcellulose Cell-O-CH$_2$CH$_2$SO$_3$H.

6.5 Specific cellulose products

6.5.1 Cross-linking and graft polymerization

In addition to straightforward manufacture of cellulose derivatives, other potential ways to tailor the properties of cellulose and its common derivatives are through cross-linking and graft copolymerization with certain monomers to obtain new materials for a great variety of applications [9, 17, 21, 22, 165, 166]. One of the most important routes to modify the macromolecular skeleton of cellulose is the formation of covalent cross-links between the cellulose chains in heterogeneous systems. This commercial modification ("fabric finishing") improves the performance of cellulose textiles, for example, crease and wrinkle resistance, wash-and-wear performance, and durable-press properties. However, the rigid three-dimensional network obtained may also be rather brittle. Hence, one purpose of covalent crosslinking is to avoid undesirable changes of cellulose goods in the wet state. Because ester cross-links have a low stability against various alkalis, reagents capable of forming cross-links through ether bonds with the -OH groups of cellulose are preferred. The conventional method of cellulose cross-linking dates back to the beginning of the twentieth century, when the action of formaldehyde (HCHO) on cellulose was studied and its positive effect on the strength of the fibers was noted by Xavier Eschalier in 1906 [167]. The process is a two-step equilibrium reaction *via* a cellulose hemiacetal as a methylolcellulose intermediate (Cell-O-CH$_2$OH):

$$\text{Cell-OH} + \text{HCHO} \leftrightarrow \text{Cell-O-CH}_2\text{OH} \tag{6.9}$$

$$\text{Cell-O-CH}_2\text{OH} + \text{HO-Cell} \leftrightarrow \text{Cell-O-CH}_2\text{-O-Cell} + \text{H}_2\text{O} \tag{6.10}$$

The reaction is usually performed as a wet process in the presence of acids and the cross-linking takes place within a few minutes during the subsequent curing at 100–130 °C [22]. The process has been typically applied to improve the dimensional stability of rayon fibers. However, it has been gradually replaced by other cross-linking agents, primarily methylolated or alkoxy-methylated derivatives of different *N*-containing compounds, such as ureas, cyclic ureas, carbamates, and triazines or acid amides. For example, after impregnation of the cellulose substrate with an aqueous solution containing H$_2$N-CO-NH$_2$/HCHO pre-condensates, cross-linking takes place during a subsequent short heating at 130–160 °C in the presence of acid.

Examples of other cross-linking routes and/or cross-linking agents are as follows [22]:

– formation of ether bonds (R-O-R) with bifunctional etherifying agents (*e.g.*, alkyl halides, epoxides, or divinyl sulfone)
– formation of disulfide bridges (-S-S-) from mercapto groups (-SH) attached to cellulose
– formation of urethane bridges (R-NHCO$_2$-R) in a reaction of cellulose -OH groups with isocyanates (R-N=C=O)
– formation of ester groups (R-O-CO-R) in a reaction of cellulose -OH groups with polycarboxylic acids (*e.g.*, five-membered anhydrides)
– reaction of anionic cellulose derivatives with divalent metal cations
– recombination of cellulose macroradicals formed chemically or by irradiation

The principal aim of graft copolymerization is to find routes to combine the advantages of cellulose with those of synthetic polymers [168–173]. The method was originally presented in 1943 by Ushakov, who synthetized vinyl and allyl ethers of cellulose and then copolymerized these cellulose derivatives with the esters of maleic acid (HO$_2$CCH=CHCO$_2$H, *cis*-form) [174]. Nowadays, a significant number of scientific reports on grafting polymer side chains onto cellulose and other polymers are available. However, most cellulose grafting methods involve polymerization of vinyl monomers of different types of CH$_2$=CH-X (X is an inorganic moiety or an organic substituent). Grafting is conventionally performed by allowing liquid or gaseous monomers to react with solid cellulosic materials; the course of this heterogeneous reaction is strongly dependent on the raw materials. The most known procedure is the so-called "grafting-from approach" in which the growth of polymer chains occurs from initiating sites on the cellulose backbone.

Graft copolymerization of various monomers onto cellulose or its derivatives can be classified into three categories [9, 21, 166, 172, 173]: *i*) free radical polymerization, *ii*) ionic and ring opening polymerization, and *iii*) condensation or addition polymerization. Among all these grafting methods, the first group methods have received the greatest interest and most products available are based on this kind of production. Main reasons for this paramount position include a simple process technology, together with its ability to provide almost an unlimited number of copolymers, as well as the possibility of using a wide range of monomers (*e.g.*, acrylic acid, (meth)acrylic acid esters, (meth)acryla-mides, acrylonitrile, 2-(dimethylamino)ethyl methacrylate, hydroxylacrylates, vinylpyri-dines, *N*-vinyl-2-pyrrolidone, vinyl acetate, butadiene, and styrene), tolerance of different reaction conditions (typically rather "mild"), and the presence of H_2O or other impurities. Ionic and ring opening polymerization of cellulose comprises, besides different cationic and anionic graft polymerization methods, the possibility to use cyclic monomers (*e.g.*, ε-caprolactone) in the presence of catalyst for this purpose. However, these methods, like condensation or addition polymerization methods, are today not commonly applied.

Radical polymerization is a chain reaction process, consisting of three main steps: *i*) initiation, *ii*) propagation, and *iii*) termination [9, 21, 22, 166]. Free radicals on the cel-lulose backbone can be formed by chemical initiators, *i.e.*, by abstraction of a hydrogen atom by means of free radical initiators such as azobis(isobutyronitrile) (AIBN), hydro-gen peroxide (H_2O_2), and benzoyl peroxide (Ph-CO-O-OH), by irradiation with ultraviolet (UV) light, or by irradiation with gamma rays from the source of a synthetic radioactive isotope of cobalt (^{60}Co) or highly accelerated electron beams. However, the radiation degrades cellulose in a disproportionation reaction *via* the cleavage of glycosidic bonds, leading to a loss of mechanical strength of the cellulose fibers.

The term "chain transfer" refers to the pathway in which the abstraction of a hy-drogen atom from cellulose occurs by means of a growing chain radical rather than creating directly, for example, by the catalysts [9, 21, 22, 166]. Chain transfer reactions have been extensively used in cellulose grafting and certain compounds, such as those bearing thiol groups (-SH), to facilitate chain transfer, leading to higher grafting yields. In this case, -SH groups can be introduced into the cellulose by its reaction with the cyclic ethylene sulfide or thiirane (C_2H_4S), followed by the formation of a thiol radical with a simultaneous termination of the chain (R) and initiation of graft polymer formation:

$$\text{Cell-O-CH}_2\text{CH}_2\text{-SH} + R\bullet \rightarrow \text{Cell-O-CH}_2\text{CH}_2\text{-S}\bullet + R \tag{6.11}$$

$$\text{Cell-O-CH}_2\text{CH}_2\text{-S}\bullet + n(\text{CH}_2 = \text{CHX}) \rightarrow \text{Cell-O-CH}_2\text{CH}_2\text{-S-(CH}_2\text{CHX})_n\bullet \tag{6.12}$$

Other methods to facilitate the yield of grafting involve the initiation *via* the use of potassium persulfate ($K_2S_2O_8$) or Fenton's reagent (*i.e.*, $Fe^{2\oplus}/H_2O_2$ system). In the latter case, hydroxy radicals (HO•) are produced according to the following reaction:

$$\mathrm{Fe}^{2\oplus} + \mathrm{H_2O_2} \rightarrow \mathrm{Fe}^{3\oplus} + \mathrm{HO}^{\ominus} + \mathrm{HO} \bullet \tag{6.13}$$

The hydroxy radicals can then react with cellulose, initiating graft copolymerization, or react with monomer, leading to homopolymerization. However, since the monomer concentration in the system, compared to the reactive sites of cellulose, is often relatively low, the reacting species have a greater change to initiate graft copolymerization than homopolymerization of the monomers. Almost a similar system is based on the use of ceric ions (Ce(IV) ions), which produces radicals by direct oxidation of the cellulose chains, thus initiating graft polymerization (the analogous system is Mn(II)/Mn(III)):

$$\mathrm{Cell\text{-}H} + \mathrm{Ce}^{4\oplus} \rightarrow \mathrm{Cell}\bullet + \mathrm{Ce}^{3\oplus} + \mathrm{H}^{\oplus} \tag{6.14}$$

$$\mathrm{Cell}\bullet + \mathrm{M} \rightarrow \text{graft polymer} \tag{6.15}$$

In the free radical copolymerization of cellulose, the radicals formed can then add to monomers to form a covalent bond between the monomer and cellulose. Since a free radical site is also formed on the newly formed branch, some monomers may add subsequently to this site of the branch. The propagation of the branch continues until termination takes places either by a combination of two growing cellulose chains or by a disproportion mechanism, *i.e.*, a hydrogen atom is abstracted by another growing polymer chain. Additionally, termination may occur by chain transfer to monomer, initiator, dead polymer, additives, or impurities.

The term "living polymerization" is used from the early 1980s to describe a chain growth process, where chain breaking reactions such as chain transfer or irreversible termination have been minimized, and it is possible to control the composition and molar-mass distribution of the polymers (*e.g.*, to obtain a narrow range of polydispersity) and tailor the macromolecules with complex architectures [9]. Recent advances include free-radical polymerizations such as "nitroxide-mediated polymerization" (NMP), "atom transfer radical polymerization" (ATRP), and "reversible addition-fragmentation chain transfer (RAFT) polymerization". Of these techniques, the last method can be applied to a wide range of radically polymerizable monomers using reaction conditions that are similar to those of free radical polymerization.

6.5.2 Ion exchangers and chromatographic applications

Cellulose is a high-molar-mass biopolymer and is suitable for many applications that require a rigid matrix with a large surface area. For example, it is possible to introduce different chemical groups into the backbone of cellulose; typical examples comprise the preparation of cation and anion exchangers. Although the cellulose-based ion exchangers are chemically less stable than most of the synthetic ion exchange resins (*i.e.*, typically copolymers of styrene and divinylbenzene) and their capacity is relatively low, they are useful, especially for biochemical applications, such as protein separations [21].

Ion exchange is the most common chromatographic method for separating inorganic and organic ions [9, 21, 175–177]. Ionogenic groups are classified as being weak (*e.g.*, $-CO_2^{\ominus}$, $-OPO_3^{2\ominus}$, aromatic-$O^{\ominus}$ or aromatic-$S^{\ominus}$) or strong (*e.g.*, $-SO_3^{\ominus}$) acids and weak (*e.g.*, $-N^{\oplus}H_3$, $-N^{\oplus}RH_2$ or $-N^{\oplus}R_2H$) or strong (*e.g.*, $-N^{\oplus}R_3$) bases. Strong acid and base groups are highly dissociated and the exchangers containing these groups resemble insoluble strong electrolytes that possess a permanent positive or negative charge. Quaternary ammonium ($-CH_2N^{\oplus}(CH_3)_3$) resins contain strong anion-exchange groups that are effective throughout the pH range 2–12, like strong cation-exchange groups. Similarly, common weak acid and base groups resemble insoluble weak electrolytes and their ability to participate in ion exchange as cation and anion exchangers, respectively, depends on the dissociation of these groups (*i.e.*, on their ionization constants) and the pH of the environment (normally effective in pH ranges of 6–10 and 2–9, respectively). The exchanger is typically packed into a column through which the sample solution flows. In this arrangement, the ion-exchange reaction is intrinsically reversible and goes to completion in the desired manner. Although ion-exchange columns are easy to use, the theory behind their use is rather complicated.

A great number of cellulose-based ion exchangers have been prepared for commercial use (Table 6.8) [9, 21, 177]. Their chemical structure is a hydrophilic cellulose network that contains acidic or basic groups. The most common type of $-CO_2H$-group-containing weak ion exchanger is pre-swollen, micro-granular CMC (the counter ion is usually $Na^{\oplus}$ and the effective pH range is above 4) with a low DS, resulting in a cation exchange capacity of 0.4–0.7 meq/g. CMC with a higher capacity cannot be used in chromatography unless they are cross-linked to prevent extensive swelling. By partially oxidizing the primary -OH groups of cellulose into $-CO_2H$ groups, for example, by nitrogen oxide, another $-CO_2H$-group-containing cellulose cation exchanger can be obtained. Phosphatecellulose represents an intermediate divalent cation exchanger, based on $-OPO_3^{2\ominus}$ groups. It can be prepared by reacting chloromethylphosphonic acid ($ClCH_2PO_3H_2$) with cellulose in the presence of NaOH. Sulfoethylcellulose, like sulfomethylcellulose and sulfopropylcellulose, is a strong cation exchanger with an approximate capacity of 0.5 meg/g. It can be prepared by reacting α-chloroethanesulfonic acid ($ClCH_2CH_2SO_3H$) with cellulose in the presence of NaOH. The generally known polyethyleneiminecellulose (PEI-cellulose) is not chemically modified cellulose, but a complex of cellulose with polyethyleneimine ($(CH_2CH_2NH-)_n$).

Diethylaminoethylcellulose is the most common weakly basic anion exchanger; the counterion is usually $Cl^{\ominus}$ and the effective pH range is below 9 [21, 166, 178]. It can be made from cotton linters, previously cross-linked with HCHO or 1,3-dichloropropanol ($ClCH_2C(OH)CH_2Cl$), using 2-chlorotriethylamine ($ClCH_2CH_2N(CH_2CH_3)_2$) as a reagent. Pre-swollen, microgranular aminoethylcellulose is also a weakly basic anion exchanger that can be obtained from a reaction of cellulose with α-aminoethylsulfonic acid, ($H_2N-CH(CH_3)SO_3H$), in the presence of NaOH. Its capacity is about 0.2 meq/g but it is possible to increase it to about 0.7 meq/g by cross-linking. Strongly basic quaternary cellulose anion exchangers have been prepared by reacting diethylaminoethylcellu-

Table 6.8: Examples of cellulose-based ion exchangers [9].

Cellulose derivative	Functional group	Type
Cation exchangers		
Oxidized	$-CO_2^{\ominus}$	Weak
Carboxymethyl CM	$-OCH_2CO_2^{\ominus}$	Weak
Phosphate P	$-OPO_3^{2\ominus}$	Intermediate
Sulfomethyl SM	$-OCH_2SO_3^{\ominus}$	Strong
Sulfoethyl SE	$-OCH_2CH_2SO_3^{\ominus}$	Strong
Sulfopropyl SP	$-OCH_2CH_2CH_2SO_3^{\ominus}$	Strong
Anion exchangers		
Aminoethyl AE	$-OCH_2CH_2N^{\oplus}H_3$	Weak
Diethylaminoethyl DEAE	$-OCH_2CH_2N^{\oplus}(CH_2CH_3)_2$	Weak
Triethylaminoethyl[a] TEAE	$-OCH_2CH_2N^{\oplus}(CH_2CH_3)_3$	Strong
Queternary aminoethyl QAE	$-OCH_2CH_2N^{\oplus}(CH_2CH_3)_2CH(OH)CH_3$	Strong

[a]Only part of the functional groups are of this type.

lose with alkyl halides under anhydrous conditions to result, for example, in triethylaminoethylcellulose. The commercially available triethylaminoethylcellulose typically resembles the weakly basic diethylaminoethylcellulose, showing only slight conversion to a quarternary anion exchanger. A product with an ion exchange capacity of 0.3–0.4 meq/g has also been made in a reaction between cellulose, NaOH, triethanolamine ($N(CH_2CH_2OH)_3$), and epichlorohydrin or 2-(chloromethyl)oxirane (C_3H_5ClO), and is normally referred to as ECTEOLA-cellulose.

In many applications, the chromatographic separation (resolution) of racemic forms optically active components (enantiomers). The separation is of importance and can be accomplished by means of chiral stationary phases [9, 177]. As regards the separation principles and technical solutions, liquid chromatography (LC) has many different configurations. Paper chromatography is the simplest and cheapest technique in which the stationary phase consists of a special filter paper. In this case, papers are impregnated with ion exchange reagents, or finely distributed ion exchange powder is embedded into the paper, or the paper itself is prepared from a cellulose ion exchanger. In thin-layer chromatography (TLC), the commercial plates exist mostly in the form of pre-coated layers supported on glass, plastic sheets, or aluminum foil, although for special purposes, pre-coated layers, including cellulose and cellulose derivatives (*e.g.*, polyethyleneiminecellulose and diethylaminoethylcellulose), are also used. However, it should be noted that other carbohydrate supports such as cross-linked dextran gels (the products are known as trade name "Sephadex") are very hydrophilic, and like cellulose, easily derivatized with ionogenic functional groups. They have been used for analysis and purification of biological molecules, such as various proteins.

In high-performance liquid chromatography (HPLC) and, in principle, also in TLC, biopolymer-based stationary phases, such as certain cellulose derivatives, have proved to be effective in resolving a wide class of racemic forms [9, 177]. In these cases, the stationary phases exhibit interactions with a particular enantiomer through hydrogen bonding, charge transfer, and inclusion interactions. The term "ion exclusion" describes the mechanism by which ion exchangers are used for the fractionation of neutral and ionic species. This chromatography technique is a mode of HPLC and thus, the same equipment can be used, with the proper eluent, column, and detection.

In medicine, qualitative tests provide results that are either positive or negative for the substance to be detected [9]. They can be performed in a simple way and the results are possible to be read within a few minutes from color change, instead of exact numerical values from more sensitive and expensive quantitative tests. For example, in the detection of human chorionic gonadotropin (hCG) from a sample of urine, a positive test result indicates that the patient is most likely pregnant. In this case, an antibody, specific to the hCG molecule or its polypeptide chain, is chemically bound to a Cell-O-NO$_2$ filter paper and reacts with the urine sample hCG. Another illustrative example is an enzymatic glucose test based on the activity of an enzyme, glucose oxidase. In this case, a firm plastic strip, with a stiff absorbent cellulose area, is impregnated with a buffered mixture of glucose oxidase, together with other reagents. The positive result can be detected as a visible color change. There are also, for several purposes, a great number of similar indicative spot tests that utilize filter paper as a carrier matrix.

Among other specific applications, many modified cellulose-based filters (*e.g.*, cellulose nitrates and organic esters) are used as a collection technique (total particulate/aerosol sampling) for many analytical purposes [9, 177]. In a typical case, metal-containing particles in air-dust or H$_2$O samples are examined by drawing air or H$_2$O through a filter and subsequently analyzing the filter on which the particles are trapped. A similar approach is column chromatography in which a column is packed, for example, with cellulose adsorbent and the sample solution passes through this affinity column. Electrophoresis is defined as the movement of charged particles when placed in an electrical field of varying electrical potential. In this method, due to the current between the cathode and anode, the charged molecule ions move through the pores of the support medium toward an opposite charge. In an ideal case, the support medium does not interact with the charged molecule ions but acts as a filter to retard movement of the molecules with different size and shape. Since commercially available cellulose acetate sheets have a homogeneous micropore structure, paper electrophoresis, based on this cellulose derivative, is significant in clinical diagnostics, for example, in the analysis of hemoglobin or in the separation of other blood proteins, enzymes, and mucopolysaccharides.

6.5.3 Micro-sized and nano-sized celluloses

Nanotechnology is defined as the technology that allows the manipulation of materials to create products, where at least one of the constituent phases has one dimension <100 nm (*i.e.*, between 1 nm and 100 nm) [178–182]. The physical, chemical, or biological properties of materials within the nanoscale dimension area are fundamentally different from those of the bulk material; at <1 nm or below, quantum physics rules and at >100 nm or above, classical physics and chemistry dictate properties of matter [9, 183]. Hence, it is expected that this kind of "hybrid matter" within the nanoscale dimension area can behave rather exceptionally, thus providing the opportunity to develop materials with a variety of advantageous properties (*e.g.*, improved strength, stiffness, abrasion, thermal, compatibility, anti-microbial, optical, and magnetic properties). However, when creating new, highly uniform nanoscale-sized materials and related useful structures, the products must also have novel and unique properties and functions that are repeatable and controllable. In general, the result of years of research in conventional fields, such as materials and colloids science, has gradually led to many successes that are currently attributed to nanotechnology. Due to this versatile development, nanotechnology, as an engineering discipline, has rapidly become and represents a multidisciplinary and transformative field [184]. In spite of the bright utilization prospects of nanotechnology, one of the most essential research topics is still the preparation of various nanocellulose (NC) products from different biomass resources.

Cellulose itself, as a multifunctional raw material, has basically a nanofibrillar structure and self-assembles into well-defined architectures at multiple scales, ranging from the nanoscale to the macroscale [9, 182]. Hence, it is evident that cellulose, based on an abundant and renewable feedstock, has also great potential to be utilized, after modification, in nanotechnological applications. However, the interesting nanotechnological aspects on cellulose and cellulose-containing materials as precursors for useful applications had not been fully realized until the last two decades. In many cases, the research on NC is practically still in its early phase, and to fully exploit cellulose-based nanotechnology, significant research and development investments in the science and engineering are needed. The main challenges include, besides the development and characterization of new products and their applications, the development of production processes that are economically transferable to an industrial scale. In general, the use of cellulose-based nanoproducts and nanotechnology can offer many promising possibilities, for example, in the forest products industry [185–190].

Various terminologies have been used for different NC products that may lead to ambiguities and misunderstanding. NC can be generally classified, based on the appearance and preparation methods, into three major structures, namely *i)* cellulose nanofibrils (CNFs) (or nanofibrillated cellulose (NFC) or cellulose nanofiber (CNF)) and *ii)* cellulose nanocrystals (CNCs) (or crystalline nanocellulose (CNC) or nanocrystalline cellulose (NCC)) together with *iii)* bacterial nanocellulose (BNC) that can be obtained from the biosynthesis by bacterial species [182, 191–196]. As general examples illustrating the wide application range, NC has been used in different composites [197], biomedical applications [198], and environmental remediation as an adsorbent [199]. Additionally, another kind of approach is composite nanoparticles, consisting of lignin and polysaccharides (*e.g.*, Cell-O-(CO)CH$_3$) [200] or cellulose-supported metal nanoparticles, applied in catalysis for organic synthesis [201].

CNFs (traditionally also known as the established "first phase" product, named microfibrillated cellulose (MFC)) are nanoproducts that can be produced by liberating and degrading sub-structural fibrils and cellulose microfibrils from wood pulp fibers under mechanical action with high shear forces and heat (*e.g.*, by the subsequent pretreatment, disintegration, and homogenization) [192, 202–212], Besides bleached kraft pulp and bleached sulfite pulp as the most important industrial sources of cellulose fibers [208, 213–218], CNFs can be also manufactured from a variety of different non-woody raw materials [208, 219–222]. However, the raw materials seem to have only a little influence on the properties of CNFs, even though they play a significant role on the processing energy consumption. In case of the pretreatment [208, 223] of raw materials, there are many process alternatives possible: for example, enzymatic pretreatments with cellobiohydrolases and endoglucanases [224, 225], TEMPO(2,2,6,6-tetramethyl-1-piperidinyloxy)-mediated oxidation pretreatments [226, 227] or various alkaline pretreatments [228–231]. There are also different alternative mechanical processes through which cellulose feedstock materials are required to go through after pretreatment processing: high-pressure homogenization (HPH) [202, 203], microfluidization [232, 233], grinding [206, 234], and other systems, such as cryocrushing [235, 236], high-intensity ultrasonication [237, 238], bead milling [216, 239], and twin-screw extrusion[240]. The microfibrills are visible only with the electron microscope; they are 5–50 nm in width and may have a length of even several μm (typically 100–2000 nm) [9, 195, 202]. The toxicological data from humans and animals provide no evidence that the nanoparticles in case of inhalation or *via* the gastro-intestinal tract, skin or eyes could cause toxic effects in humans (*i.e.*, are metabolically inert) [241–243].

MFC has been discovered in the 1970s and is typically a white, odorless, tasteless, and easily free-flowing powder, with certain advantageous physical properties [202, 203]. Due to its good flowability, compactibility, and compressibility, it is a versatile product in many industrial applications, including rheology modifiers, food ingredients (*e.g.*, as food and beverage texturing agents and dietetic substances), pharmaceuticals (*e.g.*, as a tableting aid and a drug carrier), and chromatography [210–212, 241, 244, 245]. CNFs have reported to gain considerable interest, due to their mechanical robustness, large surface area, and biodegradability [246]. They can be formed into various structures, such as solids, films, foams, and gels (*e.g.*, hydrogels and aerogels), and combined with polymers or other materials to form composites [247–252]. However, the presence of -OH groups on the surface of CNFs creates strong hydrogen bonding between two nanofibrils and to the gel-like structure, once produced, and thus, makes them difficult and costly to dry [208, 253, 254]. On the other hand, the highly hydrophilicity of these materials limits their uses in several applications, such as in paper coating (*e.g.*, increase of dewatering effect) or composites (*e.g.*, tendency to form agglomerates in petrochemical polymers). Hence, the chemical modification of CNFs surfaces is of interest to limit these phenomena and opens up new applications [254–258].

If the nanoproduct is originated from high-quality wood pulp and the crystalline regions of cellulose are partially isolated from the amorphous regions more effectively after mild acid hydrolysis, for example, with HCl (*i.e.*, the structure is only moderately degraded), the term "microcrystalline cellulose" (MCC) is often used [9, 192, 259]. The term is also gradually established for a commercially available material, especially in the pharmaceutical industry and for materials generally produced by chemical hydrolysis, often in combination with mechanical milling of some kind. However, the effective mineral acid (*e.g.*, with concentrated H_2SO_4) hydrolysis of MCC or cellulose fibers at elevated temperatures, followed by vigorous agitation of the slurry and spray drying, leads to highly crystalline rod-like small particles through selective and intensive degradation of the more accessible materials, CNCs [182,195, 260–264]. In the preparation, various pretreatments [265–270] facilitate the hydrolysis, and it has been performed also with HCO_2H [271, 272]. CNCs are 3–20 nm in width and typically have a length of 100–300 nm [195, 260].

The preparation of CNCs disrupts and removes the amorphous regions of cellulose, surrounding and embedding cellulose microfibrils, while leaving the crystalline segments intact [260]. It generally involves four steps [195]: *i*) mechanical size reduction, *ii*) purification by alkali and bleaching treatments, *iii*) controlled chemical treatment, predominantly by acid hydrolysis, and *iv*) mechanical or ultrasound treatment. The so-called "cellulose nanowhiskers" (CNWs) (previously called also as "cellulose nanocrystals" (CNXLs) [273]) with a narrow size distribution (typically 5–10 nm in width and about 200 nm in length [274]) represent the smallest subunits of cellulose and can be prepared by acid hydrolysis using HCl or H_2SO_4, followed by ultrasonic treatment and differential centrifugation [275–281]. CNWs have attracted significant attention with many applications during the last two decades, although the commercial products are not yet common [282–286]. In practice, CNWs are CNCs and are more or less identical products. The idea of the acid hydrolysis process to prepare CNCs originated about 70 years ago [287–290] and the process was later optimized in more detail in the 2000s [291, 292].

Extensive studies have shown that CNCs (and CNWs) have the potential to be used for many applications, including pigments, inks, and cosmetics, reinforced composites, emulsions, manufacturing of improved construction products, coatings and new fillers for papermaking, and biocomposites for bone replacement [128, 192, 293–299]. Also in this case, the surfaces of reinforcement nanocomposites can be modified for different purposes and the properties of polymers can be readily improved by changes in the composition of the reinforcing and matrix phases [205, 276]. Additionally, CNCs are promising nanomaterials for fabricating fluorescent composites [300]. Especially, the very large specific area of CNCs leads to increased interaction with the matrix polymer on molecular level, which results in materials with new properties [277]. Since CNCs have, compared to chemical pulps, a very high tensile strength and Young's modulus, it can be concluded that these materials should generally be very good reinforcing fillers for various composite materials [9].

Due to the rod-like shape of colloidal CNCs (and CNWs) particles, their suspensions display liquid crystalline behavior above a critical concentration [301–303]. At low CNCs concentrations, the suspensions are isotropic, with a random arrangement of rod-shaped particles, while at high concentrations, the suspensions are anisotropic, with the particles packed in a chiral nematic (cholesteric) arrangement. The same phenomenon can be observed when high concentrations of cellulose (>10%) are dissolved in BMIMCl or in NMMO [74]. In these cases, liquid crystalline solutions of cellulose that are optically anisotropic between crossed polarizing filters, and display birefringence, are formed. Oriented suspensions of CNCs were observed in 1959 [304, 305], although spontaneously anisotropic molecular solutions of cellulose derivatives were not discovered until the mid-seventies; it was noted [306] that concentrated aqueous solutions of hydroxypropylcellulose display iridescent colors that change with concentration and viewing angle. Later, a wide range of cellulose derivatives, especially certain cellulose esters, has been found to form both lyotropic and thermotropic cholesteric liquid crystals in different solvents. It can be claimed that the discovery of liquid crystalline derivatives has provided considerable stimulus to the chemical industry. There are many possible applications for liquid crystalline polymers, ranging from their high ability to form high modulus and tensile strength fibers to their versatile use in modern liquid-crystalline display devices and screens.

Bacterial cellulose is typically produced from bacteria (the most efficient producer is *Gluconacetobacter xylinus*), as a separate molecule, and does not require additional processing to remove contaminants, such as lignin and hemicelluloses [9, 182, 183, 206, 307–312]. The synthesis of an extracellular gelatinous mat by *G. xylinus* was reported as early as in 1886, although bacterial cellulose did not attract major attention until the second half of the twentieth century. The cellulose secreted from bacteria offers certain exceptional properties (*e.g.*, high crystallinity up to 85%, high DP, and high H_2O-holding capacity) and produces a very fine and pure fiber network structure, together with higher mechanical strength. The production yields an ultrafine network of ribbons, which contain the clusters of microfibrils; their approximate dimensions are 3–4 nm in width and 70–130 nm in length.

In large-scale bioproduction, cultivated bacteria produce bacterial cellulose conventionally in a culture medium containing carbon and nitrogen sources in an agitated reactor [9]. Bacterial cellulose is mostly produced at the interface of liquid and air since *G. xylinus* demands a high amount of oxygen. Depending on the economic aspects, including, for example, availability of cheap sugar substrates, higher volumetric yields, and higher scale of production, the bacterial cellulose may be a potential alternative to the plant cellulose. Bacterial cellulose has been applied as a source material for nanofibers and utilized, especially for the production of various biocomposites [31, 34, 313–317], and biomedical applications [32, 33, 35, 194, 318–320], including directly as tissues and bone growth. However, it is important to highlight that even prior to processing, many bacterial celluloses possess widths in the nanometer range.

References

[1] Young, R.A., and Rowell, R.M. (Eds.). 1986. *Cellulose – Structure, Modification and Hydrolysis*. John Wiley & Sons, New York, NY, USA. 379 p.

[2] Fengel, D., and Wegener, G. 1989. *Wood – Chemistry, Ultrastructure, Reactions*. Walter de Gruyter, Berlin, Germany. Pp. 66–105.

[3] French, A.D., Bertoniere, N.R., Battista, O.A., Cuculo, J.A., and Gray, D.G. 1993. Cellulose. In: *Kirk-Othmer – Encyclopedia of Chemical Technology, Volume 5*. 4th edition. John Wiley & Sons, New York, NY, USA. Pp. 476–496.

[4] Sjöström, E. 1993. *Wood Chemistry – Fundamentals and Applications*. 2nd edition. Academic Press, San Diego, CA, USA. Pp. 54–62.

[5] Klemm, D., Philipp, B., Heinze, T., Heinze, U., and Wagenknecht, W. 1998. *Comprehensive Cellulose Chemistry – Volume 1, Fundamentals and Analytical Methods*. Wiley-VCH, Weinheim, Germany. 260 p.

[6] Alén, R. 2000. Structure and chemical composition of wood. In: Stenius, P. (Ed.). *Forest Products Chemistry*. Fapet, Helsinki, Finland. Pp. 11–57.

[7] Horii, F. 2001. Structure of cellulose: recent developments in its characterization. In: Hon, D.N.-S., and Shiraishi, N. (Eds.). *Wood and Cellulosic Chemistry*. 2nd edition. Marcel Dekker, New York, NY, USA. Pp. 83–107.

[8] Wiedenhoeft, A.C., and Miller, R.B. 2005. Structure and Function of Wood. In: Rowell. R.M. (Ed.). 2005. *Handbook of Wood Chemistry and Wood Composites*. Taylor & Francis, Boca Raton, FL, USA. Pp. 9–33.

[9] Alén, R. 2011. Cellulose derivatives. In: Alén, R. (Ed.). *Biorefining of Forest Resources*. Paper Engineers' Association, Helsinki, Finland. Pp. 305–354.

[10] Alén, R. 2018. *Carbohydrate Chemistry – Fundamentals and Applications*. World Scientific, Singapore. Pp. 281–301.

[11] Hon, D.N.-S. 2001. Functional natural polymers: A new dimensional creativity in lignocellulosic chemistry. In: Hon, D.N.-S. (Ed.). *Chemical Modification of Lignocellulosic Materials*. Marcel Dekker, New York, NY, USA. Pp. 1–10.

[12] Biermann, C.J. 1996. *Handbook of Pulping and Papermaking*. 2nd edition. Academic Press, San Diego, CA, USA. 348 p.

[13] Statista. 2022. (https://www.statista.com/statistics/1178289/production-of-chemical-pulp-worldwide/).

[14] Sixta, H. 2006. Pulp properties and applications. In: Sixta, H. (Ed.). *Handbook of Pulp, Vol. 2*. Wiley-VCH, Weinheim, Germany. Pp. 1009–1062.

[15] Kumar, H., and Christopher, L. 2017. Recent trends and developments in dissolving pulp production and application. *Cellulose* 24(6):2347–2365.

[16] Reveley, A. 1985. A review of cellulose derivatives and their industrial applications. In: Kennedy, J.F. (Ed.). *Cellulose and Its Derivatives: Chemistry, Biochemistry and Applications*. Ellis Horwood Limited, Chichester, England. Pp. 211–225.

[17] Fengel, D., and Wegener, G. 1989. *Wood – Chemistry, Ultrastructure, Reactions*. Walter de Gruyter, Berlin, Germany. Pp. 482–525.

[18] Fengl, R. 1993. Cellulose esters, inorganic esters. In: *Kirk-Othmer – Encyclopedia of Chemical Technology, Volume 5*. 4th edition. John Wiley & Sons, New York, NY, USA. Pp. 529–540.

[19] Gedon, S., and Fengl, R. 1993. Cellulose esters, organic esters. In: *Kirk-Othmer – Encyclopedia of Chemical Technology, Volume 5*. 4th edition. John Wiley & Sons, New York, NY, USA. Pp. 497–529.

[20]   Majewicz, T.G., and Podlas, T.J. 1993. Cellulose ethers. In: *Kirk-Othmer – Encyclopedia of Chemical Technology, Volume 5*. 4th edition. John Wiley & Sons, New York, NY, USA. Pp. 541–563.

[21]   Sjöström, E. 1993. *Wood Chemistry – Fundamentals and Applications*. 2nd edition. Academic Press, San Diego, CA, USA. Pp. 204–224.

[22]   Klemm, D., Philipp, B., Heinze, T., Heinze, U., and Wagenknecht, W. 1998. *Comprehensive Cellulose Chemistry, Volume 2 – Functionalization of Cellulose*. Wiley-VCH, Weinheim, Germany. 389 p.

[23]   Heinze, T.J., and Glasser, W.G. (Eds.). 1998. *Cellulose derivatives: modification, characterization, and nanostructures*. ACS Symposium Series 688, American Chemical Society, Washington, DC, USA. 361 p.

[24]   Isogai, A. 2001. Chemical modification of cellulose. In: Hon, D.N.-S., and Shiraishi, N. (Eds.). *Wood and Cellulosic Chemistry*. 2nd edition. Marcel Dekker, New York, NY, USA. Pp. 599–625.

[25]   Woodings, C. 2001. *Regenerated Cellulose Fibres, Woodhead Textiles Series No. 18*. Woodhead Publishing Limited, Cambridge, England. 352 p.

[26]   Kamide, K. 2005. *Cellulose and Cellulose Derivatives*. Elsevier, London, England. 652 p.

[27]   Fang, L., and Catchmark, J.M. 2014. Characterization of water-soluble exopolysaccharides from *Gluconacetobacter xylinus* and their impacts on bacterial cellulose crystallization and ribbon assemply. *Cellulose* 21:3965–3978.

[28]   Huang, Y., Zhu, C., Yang, J., Nie, Y., Chen, C., and Sun, D. 2014. Recent advances in bacterial cellulose. *Cellulose* 21:1–30.

[29]   Velásquez-Riaño, M., and Bojacá, V. 2017. Production of bacterial cellulose from alternative low-cost substrates. *Cellulose* 24:2677–2698.

[30]   Andriani, D., Apriyana, A.Y., and Karina, M. 2020 The optimization of bacterial cellulose production and its applications: a review. *Cellulose* 27:6747–6766.

[31]   Lai, C., Zhang, S., Chen, X., and Sheng, L. 2014. Nanocomposite films based on TEMPO-mediated oxidized bacterial cellulose and chitosan. *Cellulose* 21:2757–2772.

[32]   Ullah, H., Santos, H.A., and Khan, T. 2016. Applications of bacterial cellulose in food, cosmetics and drug delivery. *Cellulose* 23:2291–2314.

[33]   Jun, S.-H., Lee, S.-H., Kim, S., Park, S.-G., Lee, C.-K., and Kang, N.-K. 2017. Physical properties of TEMPO-oxidized bacterial cellulose nanofibers on the skin surface. *Cellulose* 24:5267–5274.

[34]   Lv, P., Zhou, H., Zhao, M., Li, D., Lu, K., Wang, D., Huang, J., Cai, Y., Lucia, L.A., and Wei, Q. 2018. Highly flexible, transparent, and conductive silver nanowire-attached bacterial cellulose conductors. *Cellulose* 25:3189–3196.

[35]   Cañas-Gutiérrez, A., Osorio, M., Molina-Ramírez, C., Arboleda-Toro, D., and Castro-Herazo, C. 2020. Bacterial cellulose: a biomaterial with high potential in dental and oral applications. *Cellulose* 27:9737–9754.

[36]   Wolf, O., Crank, M., Patel, M., Marscheider-Weidemann, F., Schleich, J., Hüsing, B., and Angerer, G. 2005. Techno-economic feasibility of large-scale production of bio-based polymers in Europe. *Technical Report EUR 22103 EN*. European Commission, Joint Research Centre (DG JRC) & Institute for Prospective Technological Studies (ipts), European Communities. 256 p.

[37]   Nevell, T.P., and Zeronian, S.H. (Eds.). 1985. *Cellulose Chemistry and its Applications*. Ellis Horwood Limited, Chichester, England. 552 p.

[38]   Liebert, T. 2008. Cellulose solvents: remarkable history and bright future. *Abstracts of papers*. In: *235th ACS National Meeting*, April 6–10, 2008, New Orleans, USA.

[39]   El Seoud, O.A., Fidale, L.C., Ruiz, N., D´Almeida, M.L.O., and Frollini, E. 2008. Cellulose swelling by protic solvents: which properties of the biopolymer and the solvent matter? *Cellulose* 15:371–392.

[40]   Kupczak, A., Bratasz, Ł., Kryściak-Czerwenka, J., and Kozłowski. R. 2018. Moisture sorption and diffusion in historical cellulose-based materials. *Cellulose* 25:2873–2884.

[41]   French, A.D. 2014. Idealized powder diffraction patterns for cellulose polymorphs. *Cellulose* 21:885–896.

[42] Kolpak, F.J., Weih, M., and Blackwell, J. 1978. Mercerization of cellulose: 1. determination of the structure of mercerized cotton. *Polymer* 19(2):123–131.

[43] Halonen, H., Larsson, P.T., and Iversen, T. 2013. Mercerized cellulose biocomposites: a study of influence of mercerization on cellulose supramolecular structure, water retention value and tensile properties. *Cellulose* 20:57–65.

[44] Fischer, S., Voigt, W., and Fischer, K. 1999. The behavior of cellulose in hydrated melts of the composition LiX·nH₂O (X= I⁻, NO₃⁻, CH₃COO⁻, ClO₄⁻). *Cellulose* 6:213–219.

[45] Degroot, W., Carroll, F.I., and Cuculo, J.A. 1986. A C-13-NMR spectral study of cellulose and glucopyranose dissolved in the NH₃/NH₄SCN solvent system. *Journal of Polymer Science, Part A: Polymer Chemistry* 24:673–680.

[46] Hattori, K., Cuculo, J.A., and Hudson, S.M.J. 2002. New solvents for cellulose: hydrazine/thiocyanate salt system. *Journal of Polymer Science Part A: Polymer Chemistry* 40:601–611.

[47] Hattori, M., Koga, T., Shimaya, Y., and Saito, M. 1998. Aqueous calcium thiocyanate solution as a cellulose solvent. structure and interactions with cellulose. *Polymer Journal* 30:43–48.

[48] Dawsey, T.R., and McCormick, C.L. 1990. The lithium chloride/dimethylacetamide solvent for cellulose: a literature review. *Journal of Macromolecular Science. Part C, Reviews in Macromolecular Chemistry and Physics* 30:405–440.

[49] Terbojevich, M., Cosani, A., Conio, G., Ciferri, A., and Bianchi, E. 1985. Mesophase formation and chain rigidity in cellulose and derivatives. 3. Aggregation of cellulose in *N,N*-dimethylacetamide-lithium chloride. *Macromolecules* 18:640–646.

[50] McCormick, C.L., Callais, P.A., and Hutchinson, B.H. Jr. 1985. Solution studies of cellulose in lithium-chloride and *N,N*-dimethylacetamide. *Macromolecules* 18:2394–2401.

[51] McCormick, C.L., and Callais, P.A. 1987. Derivatization of cellulose in lithium chloride and *N,N*-dimethylacetamide solutions. *Polymer* 28:2317–2323.

[52] Ramos, L.A., Assaf, J.M., El Seoud, O.A., and Frollini, E. 2005. Influence of the supramolecular structure and physicochemical properties of cellulose on its dissolution in a lithium chloride/*N,N*-dimethylacetamide solvent system. *Biomacromolecules* 6:2638–2647.

[53] Ciasso, G.T., Liebert, T.F., Frollini, E., and Heinze, T.J. 2003. Application of the solvent dimethyl sulfoxide/tetrabutylammonium fluoride trihydrate as reaction medium for the homogeneous acylation of sisal cellulose. *Cellulose* 10:125–132.

[54] Liebert, T.F., and Heinze, T.J. 2001. Exploitation of reactivity and selectivity in cellulose functionalization using unconventional media for the design of products showing new superstructures. *Biomacromolecules* 2:1124–1132.

[55] Ramos, L.A., Frollini, E., and Heinze, T. 2005. Carboxymethylation of cellulose in the new solvent dimethyl sulfoxide/tetrabutylammonium fluoride. *Carbohydrate Polymers* 60:259–267.

[56] Frey, M.F., Li, L., Xiao, M., and Gould, T. 2006. Dissolution of cellulose in ethylene diamine/salt solvent systems. *Cellulose* 13:147–155.

[57] Tamai, N., Tatsumi, D., and Matsumoto, T. 2004. Rheological properties and molecular structure of tunicate cellulose in LiCl/1,3-dimethyl-2-imidazolidinone. *Biomacromolecules* 5:422–432.

[58] Edgar, K.J., Arnold, K.M., Blount, W.W., Lawniczak, J.E., and Lowman, D.W. 1995. Synthesis and properties of cellulose acetoacetates. *Macromolecules* 28:4122–4128.

[59] Yan, L., and Gao, Z. 2008. Dissolving of cellulose in PEG/NaOH aqueous solutions. *Cellulose* 15:789–796.

[60] Heinze, T., and Liebert, T. 2001. Unconventional methods in cellulose functionalization. *Progress in Polymer Science* 26:1689–1762.

[61] Borbély, E. 2008. Lyocell, the new generation of regenerated cellulose. *Acta Polytechnica Hungaria* 5(3):11–18.

[62] Wendler, F., Persin, Z., Stana-Kleinschek, K., Reischl, M., Ribitsch, V., Bohn, A., Fink, H.-P., and Meister, F. 2011. Morphology of polysaccharide blend fibers shaped from NaOH, N-methylmorpholine-N-oxide and 1-ethyl-3-methylimidazolium acetate. *Cellulose* 18:1165–1178.

[63] Ilyin, S.O., Makarova, V.V., Anokhina, T.S., Ignatenko, V.Y., Brantseva, T.V., Volkov, A.V., and Antonov, S.V. 2018. Diffusion and phase separation at the morphology formation of cellulose membranes by regeneration from N-methylmorpholine N-oxide solutions. *Cellulose* 25:2515–2530.

[64] Qi, H., Yang, Q., Zhang, L., Liebert, T., and Heinze, T. 2011. The dissolution of cellulose in NaOH-based aqueous system by two-step process. *Cellulose* 18:237–245.

[65] Isobe, N., Noguchi, K., Nishiyama, Y., Kimura, S., Wada, M., and Kuga, S. 2013. Role of urea in alkaline dissolution of cellulose. *Cellulose* 20:97–103.

[66] Liu, Z., Zhang, C., Liu, R., Zhang, W., Kang, H., Che, N., Li, P., and Huang, Y. 2015. Effects of additives on dissolution of cellobiose in aqueous solvents. *Cellulose* 22:1641–1652.

[67] Swatloski, R.P., Spear, S.K., Holbrey, J.D., and Rogers, R.D. 2002. Dissolution of cellulose with ionic liquids. *Journal of American Chemical Society* 124:4974–4975.

[68] Heinze, T. 2004. Chemical functionalization of cellulose. In: Dumitriu, S. (Ed.). *Polysaccharide: Structure Diversity and Functional Versatility*. 2nd edition. Marcel Dekker, New York, NY, USA. Pp. 551–590.

[69] Turner, M.B., Spear, S.K., Holbrey, J.D., and Rogers, R.D. 2004. Production of bioactive cellulose films reconstituted from ionic liquids. *Macromolecules* 5:1379–1384.

[70] Wu, J., Zhang, J., Zhang, H., He, J., Ren, Q., and Guo, M. 2004. Homogeneous acetylation of cellulose in a new ionic liquid. *Macromolecules* 5:266–268.

[71] Heinze, T., Schwikal, K., and Barthel, S. 2005. Ionic liquids as reaction medium in cellulose functionalization. *Macromolecular Bioscience* 5:520–525.

[72] Zhang, H., Wu, J., Zhang, J., and He, J. 2005. 1-Allyl-3-methylimidazolium chloride room temperature ionic liquid: a new and powerful nonderivatizing solvent for cellulose. *Macromolecules* 38:8272–8277.

[73] Barthel, S., and Heinze, T. 2006. Acylation and carbanilation of cellulose in ionic liquids. *Green Chemistry* 8:301–306.

[74] Zhu, S., Wu, Y., Chen, Q., Yu, Z., Wang, C., Jin, S., Ding, Y., and Wu, G. 2006. Dissolution of cellulose with ionic liquids and its application: a mini-review. *Green Chemistry* 8:325–327.

[75] Kosan, B., Michels, C., and Meister, F. 2008. Dissolution and forming of cellulose with ionic liquids. *Cellulose* 15:59–66.

[76] Mazza, M., Catana, D.-A., Vaca-Garcia, C., and Cecutti, C. 2009. Influence of water on the dissolution of cellulose in selected ionic liquids. *Cellulose* 16:207–215.

[77] Kosan, B., Schwikal, K., and Meister, F. 2010. Solution states of cellulose in selected direct dissolution agents. *Cellulose* 17:495–506.

[78] Zakrzewska, M.E., Bogel-Łukasik, E., and Bogel-Łukasik, R. 2010. Solubility of carbohydrates in ionic liquids. *Energy & Fuels* 24:737–745.

[79] Mostofian, B., Smith, J.C., and Cheng, X. 2014. Simulation of a cellulose fiber in ionic liquid suggests a synergistic approach to dissolution. *Cellulose* 21:983–997.

[80] Idström, A., Gentile, L., Gubitosi, M., Olsson, C., Stenqvist, B., Lund, M., Bergquist, K.-E., Olsson, U., Köhnke, T., and Bialik, E. 2017. On the dissolution of cellulose in tetrabutylammonium acetate/dimethyl sulfoxide: a frustrated solvent. *Cellulose* 24:3645–3657.

[81] Welton, T. 2018. Ionic liquids: a brief history. *Biophysical Reviews* 10(3):691–706.

[82] Casal, M.F., van den Bruinhorst, A., Zubeir, L.F., Peters, C.J., and Kroon, M.C. 2014. Ionic liquids vs. deep eutectic solvents. In *Proceedings of the 28th ACS National Meeting & Exposition*, August 10–14, 2014, San Francisco, CA, USA.

[83] Gabriel, S., and Weiner, J. 1888. Über einige Abkömmlinge des Propylamins. *Berichte* 21(2):2669–2679.

[84] Walden, P. 1914. Molecular weight and electrical conductivity of several fused salts. *Bulletin de l'Académie Impériale des Sciences de Saint-Petersbourg* 1800:405–422.

[85] Parviainen, A., King, A.W.T., Mutikainen, I., Hummel, M., Selg, C., Hauru, L.K.J., Sixta, H., and Kilpeläinen, I. 2013. Predicting cellulose solvating capabilities of acid-base conjugate ionic liquids. *ChemSusChem* 6:2161–2169.

[86] Stepan, A.M., King, A.W.T., Kakko, T., Toriz, G., Kilpeläinen, I., and Gatenholm, P. 2013. Fast and highly efficient acetylation of xylans in ionic liquid systems. *Cellulose* 20:2813–2824.

[87] King, A.W.T., Xie, H., Fiskari, J., and Kilpeläinen, I. 2014. Reduction of biomass recalcitrance via ionic liquid pre-treatments. In: Ragauskas, A.J. (Ed.). *Materials for Biofuels*. World Scientific, Singapore. Pp. 95–125.

[88] Jogunola, O., Eta, V., Hedenström, M., Sundman, O., Salmi, T., and Mikkola, J.-P. 2016. Ionic liquid mediated technology for synthesis of cellulose acetates using different co-solvents. *Carbohydrate Polymers* 135:341–348.

[89] Fu, D., Mazza, G., and Tamaki, Y. 2010. Lignin extraction from straw by ionic liquids and enzymatic hydrolysis of the cellulosic residues. *Journal of Agricultural and Food Chemistry* 58:2915–2922.

[90] Elgharbawy, A.A., Alam, M.Z., and Moniruzzaman, M. 2016. Ionic liquid pretreatment as emerging approaches for enhanced enzymatic hydrolysis of lignocellulosic biomass. *Biochemical Engineering Journal* 109:252–267.

[91] Saha, K., Dasgupta, J., Chakraborty, S., Antunes, F.A.F., Sikder, J., Curcio, S., Dos Santos, J.C., Arafat, H.A., and da Silva, S.S. 2017. Optimization of lignin recovery from sugarcane bagasse using ionic liquid aided pretreatment. *Cellulose* 24:3191–3207.

[92] Jiang, G., Yuan, Y., Wang, B., Yin, X., Mukuze, K.S., Huang, W., Zhang, Y., and Wang, H. 2012. Analysis of regenerated cellulose fibers with ionic liquids as a solvent as spinning speed in increased. *Cellulose* 19:1075–1083.

[93] Wanasekara, N.D., Michud, A., Zhu, C., Rahatekar, S., Sixta, H., and Eichhorn, S.J. 2016. Deformation mechanisms in ionic liquid spun cellulose fibers. *Polymer* 99:222–230.

[94] Azimi, B., Maleki, H., Gigante, V., Bagherzadeh, R., Mezzetta, A., Milazzo, M., Guazzelli, L., Cinelli, P., Lazzeri, A., and Danti, S. 2022. Cellulose-based fiber spinning processes using ionic liquids. *Cellulose* 29:3079–3129.

[95] Michud, A., Tanttu, M., Asaadi, S., Ma, Y., Netti, E., Kääriäinen, P., Persson, A., Berntsson, A., Hummel, M., and Sixta, H. 2016. Ioncell-F: ionic liquid-based cellulosic textile fibers as an alternative to viscose and lyocell. *Textile Research Journal* 86:543–552.

[96] Zhang, Q., De Oliveira Vigier, K., Royer, S., and Jérôme, F. 2012. Deep eutectic solvents: syntheses, properties and applications. *Chemical Society Reviews* 41(21):7108–7146.

[97] Tang, B., and Row, K.H. 2013. Recent developments in deep eutectic solvents in chemical sciences. *Monatshefte Für Chemie* 144:1427–1454.

[98] Smith, E., Abbott, A., and Ryder, K. 2014. Deep eutectic solvents (DESs) and their applications, *Chemical Reviews* 114(21):11060–11082.

[99] Loow, Y.-L., New, E.K., Yang, G.H., Ang, L.Y., Foo, L.Y.W., and Wu, T.Y. 2017. Potential use of deep eutectic solvents to facilitate lignocellulosic biomass utilization and conversion. *Cellulose* 24:3591–3618.

[100] Lynam, J.G., Kumar, N., and Wong, M.J. 2017. Deep eutectic solvents' ability to solubilize lignin, cellulose, and hemicellulose; thermal stability; and density. *Bioresource Technology* 238:684–689.

[101] Xu, P., Zheng, G.W., Zong, M.H., Li, N., and Lou, W.Y. 2017. Recent progress on deep eutectic solvents in biocatalysis. *Bioresources and Bioprocessing* 4(1):1–18.

[102] Lehto, J. 2021. Use of deep eutectic solvents. In: Alén, R. (Ed.). *Pulping and Biorefining – ForestBioFacts, Digital Learning Environment*. Paperi ja Puu Ltd., Helsinki, Finland.

[103] Smith, E., Abbott, A., and Ryder, K. 2014. Deep eutectic solvents (DESs) and their applications. *Chemical Reviews* 114(21):11060–11082.

[104] Tomé, L., Baião, V., Silva, W., and Brett, C. 2018. Deep eutectic solvents for the production and application of new materials. *Applied Materials Today* 10(3):30–50.

[105] Abbott, A.P., Capper, G., Davies, D.L., Rasheed, R.K., and Tambyrajah, V. 2003. Novel solvent properties of choline chloride/urea mixture. *Chemical Communications* 2003(1):70–71.

[106] Dai, Y., van Spronsen, J., Witkamp, G.J., Verpoorte, R., and Choi, Y.H. 2013. Natural deep eutectic solvents as new potential media for green technology. *Analytical Chimica Acta* 766(3):61–68.

[107] Dai, Y., Witkamp, G.J., Verpoorte, R., and Choi, Y.H. 2013. Natural deep eutectic solvents as a new extraction media for phenolic metabolites in *Carthamus tinctorius* L. *Analytical Chemistry* 85(13):6272–6278.

[108] Paiva, A., Craveiro, R., Aroso, I., Martins, M., Reis, R.L., and Duarte, A.R.C. 2014. Natural deep eutectic solvents – solvents for the 21st century. *ACS Sustainable Chemistry & Engineering* 2(5):1063–1071.

[109] Chen, Z., Bai, X., A.L., and Wan, C. 2018. High-solid lignocellulose processing enabled by natural deep eutectic solvent for lignin extraction and industrially relevant production of renewable chemicals. *ACS Sustainable Chemistry and Engineering* 6(9):12205–12216.

[110] de Los Angeles Fernandez, M., Boiteux, J., Espino, M., Gomez, F.J.V., and Silva, M.F. 2018. Natural deep eutectic solvents-mediated extractions: the way forward for sustainable analytical developments. *Analytica Chimica Acta* 1038(12):1–10.

[111] Gómez, A.V., Biswas, A., Tadini, C.C., Furtado, R.F., Alvese, C.R., and Cheng, H.N. 2019. Use of natural deep eutectic solvents for polymerization and polymer reactions. *Journal of the Brazilian Chemical Society* 30(4):717–726.

[112] Liu, Y., Friesen, J.B., McAlpine, J.B., Lankin, D.C., Chen, S.N., and Pauli, G.F. 2018. Natural deep eutectic solvents: properties, applications, and perspectives. *Journal of Natural Products* 81(3):679–690.

[113] Gómez, A.V., Biswas, A., Tadini, C.C., Furtado, R.F., Alvese, C.R., and Cheng, H.N. 2019. Use of natural deep eutectic solvents for polymerization and polymer reactions. *Journal of the Brazilian Chemical Society* 30(4):717–726.

[114] Espino, M., de Los Angeles Fernandez, M., Gomez, F.J.V., and Silva, M.F. 2016. Natural designer solvents for greening analytical chemistry. *Trends in Analytical Chemistry* 76(2):126–136.

[115] Cui, Q., Liu, J.Z., Wang, L.T., Kang, Y.F., Meng, Y., Jiao, J., and Fu, Y.J. 2018. Sustainable deep eutectic solvents preparation and their efficiency in extraction and enrichment of main bioactive flavonoids from sea buckthorn leaves. *Journal of Cleaner Production* 184(5):826–835.

[116] Florindo, C., Branco, L.C., and Marrucho, I.M. 2017. Development of hydrophobic deep eutectic solvents for extraction of pesticides from aqueous environments. *Fluid Phase Equilibria* 448(9):135–142.

[117] Dietz, C.H.J.T., Erve, A., Kroon, M.C., van Sint Annaland, M., Gallucci, F., and Held, C. 2019. Thermodynamic properties of hydrophobic deep eutectic solvents and solubility of water and HMF in them: measurements and PC-SAFT modeling. *Fluid Phase Equilibria* 489(6):75–82.

[118] Dwamena, A. 2019. Recent advances in hydrophobic deep eutectic solvents for extraction. *Separations* 6(1):9.

[119] van Osch, D.J.G.P., Dietz, C.H.J.T., van Spronsen, J., Kroon, M.C., Gallucci, F., van Sint Annaland, M., and Tuinier, R. 2019. A search for natural hydrophobic deep eutectic solvents based on natural components. *ACS Sustainable Chemistry and Engineering* 7(3):2933–2942.

[120] Krizek, T., Bursova, M., Horsley, R., Kuchar, M., Tuma, P., Cabala, R., and Hlozek, T. 2018. Menthol-based hydrophobic deep eutectic solvents: towards greener and efficient extraction of phytocannabinoids. *Journal of Cleaner Production* 93(8):391–396.

[121] Makos, P., Przyjazny, A., and Boczkaja, G. 2018. Hydrophobic deep eutectic solvents as "green" extraction media for polycyclic aromatic hydrocarbons in aqueous samples. *Journal of Chromatography A* 1570(10):28–37.

[122] Craveiro, R., Aroso, I., Flammia, V., Carvalho, T., Viciosa, M.T., Dionísioa, M., Barreiros, S., Reis, R.L., Duarte, A.R.C., and Paiva, A. 2016. Properties and thermal behavior of natural deep eutectic solvents. *Journal of Molecular Liquids* 215(3):534–540.

[123] Zhu, J., Yu, K., Zhu, Y., Zhu, R., Ye, F., Song, N., and Xu, Y. 2017. Physicochemical properties of deep eutectic solvents formed by choline chloride and phenolic compounds at t = (293.15 to 333.15) K: the influence of electronic effect of substitution group. *Journal of Molecular Liquids* 232(4):182–187.

[124] Francisco, M., van den Bruinhorst, A., and Kroon, M.C. 2012. New natural and renewable low transition temperature mixtures (LTTMs): screening as solvents for lignocellulosic biomass processing. *Green Chemistry* 14(8):2153–2157.

[125] Mbous, Y.P., Hayyana, M., Hayyan, A., Wong, W.F., Hashima, M.A., and Looi, C.Y. 2017. Applications of deep eutectic solvents in biotechnology and bioengineering – promises and challenges. *Biotechnology Advances* 35(2):105–134.

[126] Tang, X., Zuo, M., Li, Z., Liu, H., Xiong, C., Zeng, X., Sun, Y., Hu, L., Liu, S., Lei, T., and Lin, L. 2017. Green processing of lignocellulosic biomass and its derivatives in deep eutectic solvents. *ChemSusChem* 10(13):2696–2706.

[127] Yu, H., Xue, Z., Lan, X., Liu, Q., Shi, R., and Mu, T. 2020. Highly efficient dissolution of xylan in ionic liquid-based deep eutectic solvents. *Cellulose* 27:6175–6188.

[128] Laitinen, O., Ojala, K., Sirviö, J.A., and Liimatainen, H. 2017. Sustainable stabilization of oil in water emulsions by cellulose nanocrystals synthesized from deep eutectic solvents. *Cellulose* 24:1679–1689.

[129] Chen, Y., Shen, K., He, Z., Wu, T., Huang, C., Liang, L., and Fang, G. 2021. Deep eutectic solvent recycling to prepare high purity dissolving pulp. *Cellulose* 28:11503–11517.

[130] Lin, G., Tang, Q., Huang, H., Yu, J., Li, Z., and Ding, B. 2022. Process optimization and comprehensive utilization of recyclable deep eutectic solvent for the production of ramie cellulose fibers. *Cellulose* 29:3689–3701.

[131] Sun, L., Han, J., Tang, C., Wu, J., Fang, S., Li, Y., Mao, Y., Wang, L., and Wang, Y. 2020. Choline chloride-based deep eutectic solvent system as a pretreatment for microcrystalline cellulose. *Cellulose* 29:8133–8150.

[132] Ünlü, A.E., Arıkaya, A., and Takaç, S. 2019. Use of deep eutectic solvents as catalyst: a mini-review. *Green Processing and Synthesis* 8(1):355–372.

[133] Pätzold, M., Siebenhaller, S., Kara, S., Liese, A., Syldatk, C., and Holtmann, D. 2019. Deep eutectic solvents as efficient solvents in biocatalysis. *Trends in Biotechnology* 37(9):943–959.

[134] Williamson, S.T., Shahbaz, K., Mjalli, F.S., AlNashef, I.M., and Farid, M.M. 2017. Application of deep eutectic solvents as catalysts for the esterification of oleic acid with glycerol. *Renewable Energy* 114(12):480–488.

[135] Jeong, K., Lee, M., Nam, M., Zhao, J., Jin, Y., Lee, D.-K., Kwon, S., Jeong, J., and Lee, J. 2015. Tailoring and recycling of deep eutectic solvents as sustainable and efficient extraction media. *Journal of Chromatography A* 1424(21):10–17.

[136] de Faria, E., Do Carmo, R., Cláudio, A., Freire, C., Freire, M., and Silvestre, D. 2017. Deep eutectic solvents as efficient media for the extraction and recovery of cynaropicrin from *cynara cardunculus* L. leaves. *International Journal of Molecular Science* 18(11):2276–2285.

[137] Huang, Y., Feng, F., Jiang, J., Qiao, Y., Wu, T., Voglmeir, J., and Chen, Z.G. 2017. Green and efficient extraction of rutin from tartary buckwheat hull by using natural deep eutectic solvents. *Food Chemistry* 221(4):1400–1405.

[138] Lewin, M., and Goldstein, I.S. (Eds.) 1991. *Wood Structure and Composition*. Marcel Dekker, New York, NY, USA. 488 p.

[139] Lai, Y.-Z. 1996. Reactivity and accessibility of cellulose, hemicelluloses, and lignins. In: Hon, D.N.-S. (Ed.). *Chemical Modification of Lignocellulosic Materials*. Marcel Dekker, New York, NY, USA. Pp. 35–95.

[140] Hon, D.N.-S. 1996. Chemical modification of cellulose. In: Hon, D.N.-S. (Ed.). *Chemical Modification of Lignocellulosic Materials*. Marcel Dekker, New York, NY, USA. Pp. 97–127.

[141] Edgar, K.J. 2004. Cellulose esters, organic. In: Mark, H.F. (Ed.). *Encyclopedia of Polymer Science and Technology, Volume 9, Part 3*. John Wiley & Sons, New York, NY, USA. Pp. 129–158.

[142] Alén, R. 2018. *Carbohydrate Chemistry – Fundamentals and Applications*. World Scientific, Singapore. Pp. 301–324.

[143] Ishizu, A. 1990. Chemical modification of cellulose. In: Hon, N.-S.D., and Shiraishi, N. *Wood and Cellulosic Chemistry*. Marcel Dekker, Marcel Dekker, New York, NY, USA. Pp. 757–800.

[144] Treece, L.C., and Johnson, G.I. 1993. Cellulose acetate. *Chemistry & Industry* 49:224–241.

[145] Malm, C.J., Mench, J.W., Kendall, D.L., and Hiatt, G.D. 1951. Properties of cellulose esters. *Journal of Industrial and Engineering Chemistry* 43:688–691.

[146] Ott, E., Spurlin, H.M., and Graffin, M.W. (Eds.). 1955. *Cellulose and Cellulose Derivatives*, Part III. Interscience Publisher, New York, NY, USA.

[147] Treiber, E.E. 1985. Formation of fibers from cellulose solutions. In: Nevell, T.P., and Zeronian, S.H. (Eds.). *Cellulose Chemistry and Its Applications*. Ellis Horwood, Chichester, England. Pp. 455–479.

[148] Rose, M., and Palkovits, R. 2011. Cellulose-based sustainable polymers: state of the art and future trends. *Macromolecular Rapid Communications* 32:1299–1311.

[149] Segal, L., and Eggerton, F.V. 1961. Some aspects of the reaction between urea and cellulose. *Textile Research Journal* 31(5):460–471.

[150] Cai, J., and Zhang, L. 2005. Rapid dissolution of cellulose in LiOH/urea and NaOH/urea aqueous solutions. *Macromolecular Bioscience* 5(6):539–548.

[151] Harlin, A. 2019. *Cellulose Carbamate: Production and Applications*. VTT Technical Research Centre of Finland, Espoo, Finland. 32 p.

[152] Klemm, D., Philipp, B., Heinze, T., Heinze, U., and Wagenknecht, W. 1998. *Comprehensive Cellulose Chemistry, Volume 2: Functionalization of Cellulose*. Wiley-VCH, Weinheim, Germany. Pp. 161–164.

[153] Kotek, R. 2007. Regenerated cellulose fibers. In: Lewin, M. (Ed.). *Handbook of Fiber Chemistry*. CRC Press, Boca Raton, FL, USA. Pp. 668–771.

[154] Paunonen, S., Kamppuri, T., Katajainen, L., Hohenthal, C., Heikkilä, P., and Harlin, A. 2019. Environmental impact of cellulose carbamate fibers from chemically recycled cotton. *Journal of Cleaner Production* 222:871–881.

[155] Willberg-Keyriläinen, P., Hiltunen, J., and Ropponen, J. 2018. Production of cellulose carbamate using urea-based deep eutectic solvents. *Cellulose* 25:195–204.

[156] Diamantoglou, M., Platz, J., and Vienken, J. 1999. Cellulose carbamates and derivatives as hemocompatible membrane materials for hemodialysis. *Artificial Organs* 23(1):15–22.

[157] Zhang, L., Ruan, D., and Zhou, J. 2001. Structure and properties of regenerated cellulose films prepared from cotton linters in NaOH/urea aqueous solution. *Industrial and Engineering Chemistry Research* 40(25):5923–5928.

[158] Yang, Q., Fukuzumi, H., Saito, T., Isogai, A., and Zhang, L. 2011. Transparent cellulose films with high gas barrier properties fabricated from aqueous alkali/urea solutions. *Biomacromolecules* 12(7):2766–2771.

[159] Fu, F., Yang, Q., Zhou, J., Hu, H., Jia, B., and Zhang, L. 2014. Structure and properties of regenerated cellulose filaments prepared from cellulose carbamate-NaOH/ZnO aqueous solution. *ACS Sustainable Chemistry & Engineering* 2(11):2604–2612.

[160] Weißl, M., Hobisch, M.A., Johansson, L.S., Hettrich, K., Kontturi, E., Volkert, B., and Sprik, S. 2019. Cellulose carbamate derived cellulose thin films: preparation, characterization and blending with cellulose xanthate. *Cellulose* 26:7399–7410.

[161] Yang, Y.-T., Huang, J.-L., Wang, X., Grunlan, J., Song, L., and Hu, Y. 2022. Flame retardant and hydrophobic cotton using a unique phosphorous-nitrogen-silicon-containing coating. *Cellulose* 29:8473–8488.

[162] Sau, A.C., and Majewicz, T.G. 1992. Cellulose ethers, self-cross-linking mixed ether silyl derivatives. *ACS Symposium Series* 476:265–272.

[163] Majewicz, T.G., and Podlas, T.J. 1993. Cellulose ethers. In: *Kirk-Othmer – Encyclopedia of Chemical Technology, Volume 5*. 4th edition. John Wiley & Sons, New York, NY, USA. Pp. 541–563

[164] Stigsson, V. 2006. *Some Aspects on the Carboxymethyl Cellulose Process*. Doctoral Thesis. Karlstad University, Faculty of Technology and Science, Karlstad, Sweden. 61 p.

[165] Nishio, Y. 2006. Material functionalization of cellulose and related polysaccharides via diverse microcompositions. *Advanced Polymer Science* 205(Polysaccharides II):97–151.

[166] Roy, D., Semsarilar, M., Guthrie, J.F., and Perrier, S. 2009. Cellulose modification by polymer grafting: a review: *Chemical Society Reviews* 38:2046–2064.

[167] Eschalier, X. 1906. Process of strengthening cellulose threads, filaments. British Patent 0625647.

[168] Krässig, H. 1971. Graft copolymerization onto cellulose fibers: a new process for graft modification. *Svensk Papperstidning* 74:417–421.

[169] Mansour, O.Y., Nagaty, A., Beshay, A.D., and Nosseir, M.H. 1983. Graft polymerization of monomers onto cellulose and lignocelluloses by chemically induced initiator. *Journal of Polymer Science (In Two Sections)* 21(3):715–724.

[170] Krässig, H.A., and Stannet, V.T. 2006. Graft co-polymerization to cellulose and its derivatives. In: Ferry, J.D., Kern, W., Natta, G., Overberger, C.G., Schulz, G.V., Staverman, A.J., and Stuart, H.A. (Eds.). *Advances in Polymer Science/Fortschritte der Hochpolymeren-Forschung*. Springer-Verlag, Heidelberg, Germany. Pp. 111–156.

[171] Thakur, V.K., Thakur, M.K., and Gupta, R.K. 2013. Graft copolymers from cellulose: synthesis, characterization and evaluation. *Carbohydrate Polymers* 97(1):18–25.

[172] Kang, H., Liu, R., and Huang, Y. 2015. Graft modification of cellulose: methods, properties, and applications. *Polymer* 70:A1–A6.

[173] Thakur, V.K. (Ed.). 2015. *Cellulose-Based Graft Copolymers – Structure and Chemistry*. CRC Press, Boca Raton, FL, USA. 599 p.

[174] Ushakov, S. N. 1943. Copolymerization of unsaturated cellulose derivatives. *Fiz.-Mat. Nauk* 1:35–36. (In Russian).

[175] Helfferich, F. 1995. *Ion Exchange*. Dover Publications, New York, NY, USA. 624 p.

[176] Weiss, J. 1995. *Ion Chromatography*. VCH, Weinheim, Germany. 465 p.

[177] Cazes, J. (Ed.). 2009. *Encyclopedia of Chromatography*. 3rd edition. CRC Press, Boca Raton, FL, USA. 2850 p.

[178] Drexler, K.E. 1981. Molecular engineering: an approach to the development of general capabilities for molecular manipulation. *Proceedings of the National Academies of Science USA* 78:5275–5278.

[179] Chen, H., Weiss, J., and Shahidi. 2006. Nanotechnology in nutraceuticals and functional foods. *Food Technology (Chicago)* 60(3):30–36.

[180] Wegner, T.H., and Jones, P.E. 2006. Advancing cellulose-based nanotechnology. *Cellulose* 13: 115–118.

[181] Kvien, I., and Oksman Niska, K. 2008. Microscopic examination of cellulose whiskers and their nanocomposites. In: *Characterization of Lignocellulosic Materials*. Hu, T.Q. (Ed.). Wiley-Blackwell, Oxford, United Kingdom. Pp. 340–356.

[182] Jonoobi, M., Oladi, R., Davoudpour, Y., Oksman, K., Dufresne, A., Hamzeh, Y., and Davoodi, R. 2015. Different preparation methods and properties of nanostructural cellulose from various natural resources and residues: a review. *Cellulose* 22:935–969.

[183] Kamel, S. 2007. Nanotechnology and its applications in lignocellulosic composites, a mini review. *Express Polymer Letters* 1:546–575.

[184] Lucia, L.A., and Rojas, O.J. 2007. Fiber nanotechnology: a new platform for "green" research and technological innovations. *Cellulose* 14, 539–542.

[185] Wegner, T., and Jones, P. 2005. Nanotechnology for forest products, part 1. *Solutions!* (July):44–46.

[186] Wegner, T., and Jones, P. 2005. Nanotechnology for forest products, part 2. *Solutions!* (August):43–45.

[187] Moon, R.J., Frihart, C.R., and Wegner, T. 2006. Nanotechnology applications in the forest products industry. *Journal of Forest Products* 56(5):4–10.

[188] Kulachenko, A., Denoyelle, T., Galland, S., and Lindström, S.B. 2012. Elastic properties of cellulose nanopaper. *Cellulose* 19:793–807.

[189] González, I., Alcalà, M., Chinga-Carrasco, G., Vilaseca, F., Boufi, S., and Mutjé, P. 2014. From paper to nanopaper: evolution of mechanical and physical properties. *Cellulose* 21:2599–2609.

[190] Parit, M., Aksoy, B., and Jiang, Z. 2018. Towards standardization of laboratory preparation procedure for uniform cellulose nanopapers. *Cellulose* 25:2915–2924.

[191] Brown, R.M., and Montezinos, D. 1976. Cellulose microfibrils – visualization of biosynthetic and orienting complexes in association with plasma-membrane. *Proceedings of the National Academies of Science USA* 73:143–147.

[192] Berglund, L. 2005. Cellulose-based nanocomposites. In: Mohanty, A.K., Misra, M., and Drzal, L.T. (Eds.). *Natural Fibers, Biopolymers, and Biocomposites*. Taylor & Francis, Boca Raton, FL, USA. Pp. 807–832.

[193] Peng, Y., Gardner, D.J., Han, Y., Kiziltas, A., Cai, Z., and Tshabalala, M.A. 2013. Influence of drying method on the material properties of nanocellulose I: thermostability and crystallinity. *Cellulose* 20:2379–2392.

[194] Lu, Y., Tekinalp, H.L., Eberle, C.C., Peter, W., Naskar, A.K., and Ozcan, S. 2014. Nanocellulose in polymer composites and biomedical applications. TAPPI *Journal* 13(6):47–54.

[195] Kargarzadeh, H., Mariano, M., Gopakumar, D., Ahmad, I., Thomas, S., Dufresne, A., Huang, J., and Lin, N. 2018. Advances in cellulose nanomaterials. *Cellulose* 25:2151–2189.

[196] Wang, Y., Wang, X., Xie, Y., and Zhang, K. 2018. Functional nanomaterials through esterification of cellulose: a review of chemistry and application. *Cellulose* 25:3703–3731.

[197] Miao, C., and Hamad, W.Y. 2013. Cellulose reinforced polymer composites and nanocomposites: a critical review. *Cellulose* 20:2221–2262.

[198] Liu, J., Shi, Y., Cheng, L., Sun, J., Yu, S., Lu, X., Biranje, S., Xu, W., Zhang, X., Song, J., Wang, Q., Han, W., and Zhang, Z. 2021. Growth factor functionalized biodegradable nanocellulose scaffolds for potential wound healing application. *Cellulose* 28:5643–5656.

[199] Mahfoudhi, N., and Boufi, S. 2017. Nanocellulose as a novel nanostructured adsorbent for environmental remediation: a review. *Cellulose* 24:1171–1197.

[200] Gericke, M., Bergrath, J., Schulze, M., and Heinze, T. 2022. Composite nanoparticles derived by self-assembling of hydrophobic polysaccharide derivatives and lignin. *Cellulose* 29:3613–3620.

[201] Kamel, S., and Khattab, T.A. 2021. Recent advances in cellulose supported metal nanoparticles as green and sustainable catalysis for organic synthesis. *Cellulose* 28:4545–4574.

[202] Herrick, F.W., Casebier, R.L., Hamilton, J.K., and Sandberg, K.R. 1983. Microfibrillated cellulose: morphology and accessibility. *Journal of Applied Polymer Science: Applied Polymer Symposium* 37:797–813.

[203] Turbak, A.F., Snyder, F.W., and Sandberg, K.R. 1983. Microfibrillated cellulose, a new cellulose product: properties, uses, and commercial potential. *Journal of Applied Polymer Science: Applied Polymer Symposium* 37:815–827.

[204] Pääkkö, M., Vapaavuori, J., Silvennoinen, R., Kosonen, H., Ankerfors, M., Lindström, T., Berglund, L.A., and Ikkala, O. 2008. Long and entangled native cellulose i nanofibers allow flexible aerogels and hierachically porous templates for functionalities. *Soft Matter* 4:2492–2499.

[205] Stenstad, P., Andressen, M., Tanem, B.S., and Stenius, P. 2008. Chemical surface modifications of microfibrillated cellulose. *Cellulose* 15:35–45.

[206] Siró, I., and Plackett, D. 2010. Microfibrillated cellulose and new nanocomposite materials: a review. *Cellulose* 17:459–494.

[207] Tejado, A., Alam, M.N., Antal, M., Yang, H., Ad van de Ven, T.G.M. 2012. Energy requirements for the disintegration of cellulose fibers into cellulose nanofibers. *Cellulose* 19:831–842.

[208] Missoum, K., Belgacem, M.N., and Bras, J. 2013. Nanofibrillated cellulose surface modification: a review. *Materials (Basel)* 6(5):1745–1766.

[209] Hellström, P., Heijnesson-Hultén, A., Paulsson, M., Håkansson, H., and Germgård, U. 2014. The effect of fenton chemistry on the properties of microfibrillated cellulose. *Cellulose* 21:1489–1503.

[210] Vanhatalo, K.M., and Dahl, O.P. 2014. Effect of mild acid hydrolysis parameters on properties of microcrystalline cellulose. *BioResources* 9(3):4729–4740.

[211] Vanhatalo, K.M., Parviainen, K.E., and Dahl, O.P. 2014. Techno-economic analysis of simplified microcrystalline cellulose process. *BioResources* 9(3):4741–4755.

[212] Vanhatalo, K., Lundin, T., Koskimäki, A., Lillandt, M., and Dahl, O. 2016. Microcrystalline cellulose property-structure effects in high-pressure fluidization: microfibril characteristics. *Journal of Material Science* 51:6019–6034.

[213] Spence, K.L., Venditti, R.A., Habibi, Y., Rojas, O.J., and Pawlak, J.J. 2010. The effect of chemical composition on microfibrillar cellulose films from wood pulps: mechanical processing and physical properties. *Bioresource Technology* 101:5961–5968.

[214] Taipale, T., Österberg, M, Nykänen, A., Ruokolainen, J., and Laine, J. 2010. Effect of microfibrillated cellulose and fines on the drainage of kraft pulp suspension and paper strength. *Cellulose* 17:1005–1020.

[215] Bufalino, L., de Sena Neto, A.R., Tonoli, G.H.D., de Souza Fonseca, A., Costa, T.G., Marconcini, J.M., Colodette, J.L., Labory, C.R.G., and Mendes, L.M. 2015. How the chemical nature of Brazilian hardwoods affects nanofibrillation of cellulose fibers and film optical quality. *Cellulose* 22:3657–3672.

[216] Zhang, L., Tsuzuki, T., and Wang, X. 2015. Preparation of cellulose nanofiber from softwood pulp by ball milling. *Cellulose* 22:1729–1741.

[217] Colson, J., Bauer, W., Mayr, M., Fischer, W., and Gindl-Altmutter, W. 2016. Morphology and rheology of cellulose nanofibrils derived from mixtures of pulp fibres and papermaking fines. *Cellulose* 23:2439–2448.

[218] Berglund, L., Anugwom, I., Hedenström, M., Aitomäki, Y., Mikkola, J.-P., and Oksman, K. 2017. Switchable ionic liquids enable efficient nanofibrillation of wood pulp. *Cellulose* 24:3265–3279.

[219] Alemdar, A., and Sain, M. 2008. Isolation and characterization of nanofibers from agricultural residues – wheat straw and soy hulls. *Bioresource Technology* 99:1664–1671.

[220] Bhattacharya, D., Germinario, L.T., and Winter, W.T. 2008. Isolation, preparation and characterization of cellulose microfibers obtained from bagasse. *Carbohydrate Polymers* 73:371–377.

[221] Janoobi, M., Khazaeian, A., Tahir, P.M., Azry, S.S., and Oksman, K. 2011. Characteristics of cellulose nanofibers isolated from rubberwood and empty fruit bunches of oil palm using chemo-mechanical process. *Cellulose* 18:1085–1095.

[222] Pelissari, F.M., Sobral, P.J.A., and Menegalli, F.C. 2014. Isolation and characterization of cellulose nanofibers from banana peels. *Cellulose* 21:417–432.

[223] Copenhaver, K., Li, K., Wang, L., Lamm. M., Zhao, X., Korey, M., Neivandt, D., Dixon, B., Sultana, S., Kelly, P., Gramlich, W.M., Tekinalp, H., Gardner, D.J., MacKey, S., Nawaz, K., and Ozcan, S. 2022. Pretreatment of lignocellulosic feedstocks for cellulose nanofibril production. *Cellulose* 29:4835–4876.

[224] Henriksson, M., Henriksson, G., Berglund, L.A., and Linström, T. 2007. An environmentally friendly method for enzyme-assisted preparation of microfibrillated cellulose (MFC) nanofibers. *European Polymer Journal* 43:3434–3441.

[225] Chen, Y., Fan, D., Han, Y., Li, G., and Wang, S. 2017. Length-controlled cellulose nanofibrils produced using enzyme pretreatment and grinding. *Cellulose* 24:5431–5442.

[226] Isogai, T., Saito, T., and Isogai, A. 2011. Wood cellulose nanofibrils prepared by TEMPO electro-mediated oxidation. *Cellulose* 18:421–431.

[227] Rodionova, G., Eriksen, Ø., and Gregersen, Ø. 2012. TEMPO-oxidized cellulose nanofiber films: effect of surface morphology on water resistance. *Cellulose* 19:1115–1123.

[228] Li, Y., Li, G., Zou, Y., Zhou, Q., and Lian, X. 2014. Preparation and characterization of cellulose nanofibers from partly mercerized cotton by mixed acid hydrolysis. *Cellulose* 21:301–309.

[229] Osong, S.H., Norgren, S., and Engstrand, P. 2016. Processing of wood-based microfibrillated cellulose and applications relating to papermaking: a review. *Cellulose* 23:93–123.

[230] Martins, C.C.N., Dias, M.C., Mendonça, M.C., Durães, A.F.S., Silva, L.E., Félix, J.R., Damásio, R.A.P., and Tonoli, G.H.D. 2021. Optimizing cellulose microfibrillation with NaOH pretreatments for unbleached *Eucalyptus* pulp. *Cellulose* 28:11519–11531.

[231] Wang, C., Luo, L., Zhang, W., Geng, S., Wang, A., Fang, Z., and Wen, Y. 2022. Production of cellulose nanofibrils via an eco-friendly approach. *Cellulose* 29:8623–8636.

[232] Ferrer, A., Filpponen, E., Rodrifgues, A., Laine, J., and Rojas, O.J. 2012. Valorization of residual empty palm fruit bunch fibers (EPFBF) by microfluidization: production of nanofibrillated cellulose and EPFBF nanopaper. *Bioresource Technology EPFBF* 125(1):249–255.

[233] Naderi, A., Lindström, T., Sundström, J., Pettersson, T., Flodberg, G., and Erlandsson, J. 2015. Microfluidized carboxymethyl cellulose modified pulp: a nanofibrillated cellulose system with some attractive properties. *Cellulose* 22:1159–1173.

[234] Wang, T., and Drzal, L.T. 2012. Cellulose-nanofiber-reinforced poly(lactic acid) composites prepared by a water-based approach. *ACS Applied Materials and Interfaces* 4(10):5079–5085.

[235] Frone, A.N., Panaitescu, D.M., and Donescu, D. 2011. Some aspects concerning the isolation of cellulose micro- and nano-fibers. *UPB Science Bulletin, Series B* 73:133–152.

[236] Torstensen, J.Ø., Johnsen, P.-O., Riis, H., Spontak, R.J., Deng, L., Gregersen, Ø.W., and Syverud, K. 2018. Preparation of cellulose nanofibrils for imaging purposes: comparison of liquid cryogens for rapid vitrification. *Cellulose* 25:4269–4274.

[237] Chen, P., Yu, H., Lin, Y., Chen, W., Wang, X., and Ouyang, M. 2013. Concentration effects on the isolation and dynamic rheological behavior of cellulose nanofibers via ultrasonic processing. *Cellulose* 20:149–157.

[238] Wang, S., and Cheng, Q. 2009. A novel process to isolate fibrils from cellulose fibers by high-intensity ultrasonication, part 1. Process optimization. *Journal of Applied Polymer Science* 113(2):1270–1275.

[239] Abe, K. 2016. Nanofibrillation of dried pulp in NaOH solutions using bead milling. *Cellulose* 23:1257–1261.

[240] Ho, T.T.T., Abe, K., Zimmermann, T., and Yano, H. 2015. Nanofibrillation of pulp fibers by twin-screw extrusion. *Cellulose* 22:421–433.

[241] Shlieout, G., Arnold, K., and Müller, G. 2002. Powder and mechanical properties of microcrystalline cellulose with different degrees of polymerization. *AAPS PharmSciTech* 3(2):45–54.

[242] Vartiainen, J., Pöhler, T., Sirola, K., Pylkkänen, L., Alenius, H., Hokkinen, J., Tapper, U., Lahtinen, P., Kapanen, A., Putkisto, K., Hiekkataipale, P., Eronen, P., Ruokolainen, J., and Laukkanen, A. 2011. Health and environmental safety aspects of friction grinding and spray drying of microfibrillated cellulose. *Cellulose* 18:775–786.

[243] Pitkänen, M., Kangas, H., Laitinen, O., Sneck, A., Lahtinen, P., Soledad Persin, M., and Niinimäki, J. 2014. Characteristics and safety of nano-sized cellulose fibrils. *Cellulose* 21:3871–3886.

[244] Tuason, D.C., Krawczyk, G.R., Buliga, G. 2009. Microcrystalline cellulose. In: Imeson, A. (Ed.). *Food Stabilisers, Thickeners and Gelling Agents*. Wiley-Blackwell, Chichester, United Kingdom. Pp. 218–236.

[245] Hindi, S.S.Z. 2017. Microcrystalline cellulose: the inexhaustible treasure for pharmaceutical industry. *Nanoscience and Nanotechnology Research* 4(1):17–24.

[246] Kumar, V., Bollström, R., Yang, A., Chen, Q., Chen, G., Salminen, P., Bousfield, D., and Toivakka, M. 2014. Comparison of nano- and microfibrillated cellulose films. *Cellulose* 21:3443–3456.

[247] Suzuki, K., Okumura, H., Kitagawa, K., Sato, S., Nakagaito, A.N., and Yano, H. 2013. Development of continuous process enabling nanofibrillation of pulp and melt compounding. *Cellulose* 20:201–210.

[248] Varanasi, S., and Batchelor, W.J. 2013. Rapid preparation of cellulose nanofiber sheet. *Cellulose* 20:211–215.

[249] Yildirim, N., Shaler, S.M., Gardner, D.J., Rice, R., and Bousfield, D.W. 2014. Cellulose nanofibril (CNF) reinforced starch insulating foams. *Cellulose* 21:4337–4347.

[250] Osong, S.H., Dahlström, C., Forsberg, S., Andres, B., Engstrand, P., Norgren, S., and Engström, A.-C. 2016. Nanofibrillated cellulose/nanographite composite films. *Cellulose* 23:2487–2500.

[251] Naderi, A. 2017. Nanofibrillated cellulose: Properties reinvestigated. *Cellulose* 24:1933–1945.

[252] Jiang, S., Zhang, M., Li, M., Liu, L., Liu, L., and Yu, J. 2020. Cellulose nanofibril (CNF) based aerogels prepared by a facile process and the investigation of thermal insulation performance. *Cellulose* 27:6217–6233.

[253] Peng, Y., Gardner, D.J., Han, Y., Kiziltas, A., Cai, Z., and Tshabalata, M.A. 2013. Influence of drying method on the material properties of nanocellulose I: thermostability and crystallinity. *Cellulose* 20:2379–2392.

[254] Lamm, M.E., Li, K., Ker, D., Zhao, X., Hinton, H.E., Copenhaver, K., Tekinalp, H., and Ozcan, S. 2022. Exploiting chitosan to improve the interface of nanocellulose reinforced polymer composites. *Cellulose* 29:3859–3870.

[255] Nypelö, T., Pynnönen, H., Österberg, M., Paltakari, J., and Laine, J. 2012. Interactions between inorganic nanoparticles and cellulose nanofibrils. *Cellulose* 19:779–792.

[256] Rodrígues, K., Gatenholm, P., and Renneckar, S. 2012. Electrospinning cellulosic nanofibers for biomedical applications: structure and in vitro biocompatibility. *Cellulose* 19:1583–1598.

[257] Vuoti, S., Talja, R., Johansson, L.-S., Heikkinen, H., and Tammelin, T. 2013. Solvent impact on esterification and film formation ability of nanofibrillated cellulose. *Cellulose* 20:2359–2370.

[258] Hollertz, R., Durán, V.L., Larsson, P.A., and Wågberg, L. 2017. Chemically modified cellulose micro- and nanofibrils as paper-strength additives. *Cellulose* 24:3883–3899.

[259] Leppänen, K., Andersson, S., Torkkeli, M., Knaapila, M., Koteknikova, N., and Serimaa, R. 2009. Structure of cellulose and microcrystalline cellulose from various wood species, cotton and flax studied by X-ray scattering. *Cellulose* 16:999–1015.

[260] Ramires, E.A., and Dufresne, A. 2011. A review of cellulose nanocrystals and nanocomposites. *TAPPI Journal* 10(4):9–16.

[261] Brito, B.S.L., Pereira, F.V., Putaux, J.-L., and Jean, B. 2012. Preparation, morphology and structure of cellulose nanocrystals from bamboo fibers. *Cellulose* 19:1527–1536.

[262] Yue, Y., Zhou, C., French, A.D., Xia, G., Han, G., Wang, Q., and Wu, Q. 2012. Comparative properties of cellulose nano-crystals from native and mercerized fibers. *Cellulose* 19:1173–1187.

[263] Tang, Y., Yang, S., Zhang, N., and Zhang, J. 2014. Preparation and characterization of nanocrystalline cellulose via low-intensity ultrasonic-assisted sulfuric acid hydrolysis. *Cellulose* 21:335–346.

[264] Domingues, A.A., Pereira, F.V., Sierakowski, M.R., Rojas, O.J., and Petri, D.F.S. 2016. Interfacial properties of cellulose nanoparticles obtained from acid and enzymatic hydrolysis of cellulose. *Cellulose* 23:2421–2437.

[265] Yang, H., Alam, M.N., and van de Ven, T.G.M. 2013. Highly charged nanocrystalline cellulose and dicarboxylated cellulose from periodate and chlorite oxidized cellulose fibers. *Cellulose* 20:1865–1875.

[266] Yang, D., Peng, X.-W., Zhong, L.-X., Cao, X.-F., Chen, W., and Sun, R.-C. 2013. Effects of pretreatments on crystalline properties and morphology of cellulose nanocrystals. *Cellulose* 20:2427–2437.

[267] Beltramino, F., Roncero, M.B., Torres, A.L., Vidal, T., and Valls, C. 2016. Optimization of sulfuric acid hydrolysis conditions for preparation of nanocrystalline cellulose from enzymatically pretreated fibers. *Cellulose* 23:1777–1789.

[268] An, X., Wen, Y., Cheng, D., Zhu, X., and Ni, Y. 2016. Preparation of cellulose nano-crystals through a sequential process of cellulase pretreatment and acid hydrolysis. *Cellulose* 23:2409–2420.

[269] Meesupthong, R., Yingkamhaeng, N., Nimchua, T., Pinmanee, P., Mussatto, S.I., Li, B., and Sukyai, P. 2021. Xylanase pretreatment of energy cane enables facile cellulose nanocrystals isolation. *Cellulose* 28:799–812.

[270] Haouache, S., Jimenez-Saelices, C., Cousin, F., Falourd, X., Pontoire, B., Cahier, K., Jérome, F., and Capron, I. 2022. Cellulose nanocrystals from native and mercerized cotton. *Cellulose* 29:1567–1581.

[271] Li, B., Xu, W., Kronlund, D., Määttänen, A., Liu, J., Smått, J.-E., Peltonen, J., Willför, S., Mu, X., and Xu, C. 2015. Cellulose nanocrystals prepared via formic acid hydrolysis followed by TEMPO-mediated oxidation. *Carbohydrate Polymers* 133:605–612.

[272] Du, H., Liu, C., Mu, X., Gong, W., Lv, D., Hong, Y., Si, C., and Li, B. 2016. Preparation and characterization of thermally stable cellulose nanocrystals via a sustainable approach of FeCl$_3$-catalyzed formic acid hydrolysis. *Cellulose* 23:2389–2407.

[273] Choi, Y., and Simonsen, J. 2006. Cellulose nanocrystal-filled carboxymethyl cellulose nanocomposites. *Journal of Nanoscience & Nanotechnology* 6:633–639.

[274] Kvien, I., Tanem, B.S., and Oksman, K. 2005. Characterization of cellulose whiskers and their nanocomposites by atomic force and electron microscopy. *Biomacromolecules* 6:3160–3165.

[275] Azizi Samir, M.A.S., Alloin, F., and Dufresne, F. 2005. Review of recent research into cellulosic whiskers, their properties and their application in nanocomposite fields. *Biomacromolecules* 6:612–626.

[276] Oksman, K., and Mathew, A. 2007. Processing and properties of nanocomposites based on cellulose whiskers. In: *Proceedings of the 9th International Conference on Wood & Biofiber Plastic Composites*, May 21–23, 2007, Madison, WI, USA.

[277] Kvien, I., and Oksman Niska, K. 2008. Microscopic examination of cellulose whiskers and their nanocomposites. In: Hu, T.Q. (Ed.). *Characterization of Lignocellulosic Materials*. Blackwell Publishing, Oxford, United Kingdom. Pp. 340–356.

[278] Bai, W., Holbery, J., and Li, K. 2009. A technique for production of nanocrystalline cellulose with a narrow size distribution. *Cellulose* 16:455–465.

[279] Mao, J., Osorio-Madrazo, A., and Laborie, M.-P. 2013. Preparation of cellulose i nanowhiskers with a mildy acidic aqueous ionic liquid: reaction efficiency and whiskers attributes. *Cellulose* 20:1829–1840.

[280] Nge, T.T., Lee, S.-H., and Endo, T. 2013. Preparation of nanoscale cellulose materials with different morphologies by mechanical treatments and their characterization. *Cellulose* 20:1841–1852.

[281] Nepomuceno, N.C., Santos, A.S.F., Oliveira, J.E., Glenn, G.M., and Medeiros, E.S. 2017. Extraction and characterization of cellulose nanowhiskers from Mandacaru (*Cereus jamacaru* DC.) spines. *Cellulose* 24:119–129.

[282] Mikkonen, K.S., Mathew, A.P., Pirkkalainen, K., Serimaa, R., Xu, C., Willför, S., Oksman, K., and Tenkanen, M. 2010. Glucomannan composite films with cellulose nanowhiskers. *Cellulose* 17:69–81.

[283] Gong, G., Mathew, A.P., and Oksman, K. 2011. Strong aqueous gels of cellulose nanofibers and nanowhiskers isolated from softwood flour. *TAPPI Journal* 10(2):7–14.

[284] Jia, B., Li, Y., Yang, B., Xiao, D., Zhang, S., Rajulu, A.V., Kondo, T., Zhang, L., and Zhou, J. 2013. Effect of microcrystal cellulose and cellulose whisker on biocompatibility of cellulose-based electrospun scaffolds. *Cellulose* 20:1911–1923.

[285] Corrêa, A.C., de Morais Teixeira, E., Carmona, V.B., Teodoro, K.B.R., Ribeiro, C., Mattoso, L.H.C., and Marconcini, J.M. 2014. Obtaining nanocomposites of polyamide 6 and cellulose whiskers via extrusion and injection molding. *Cellulose* 21:311–322.

[286] Wang, W.-J., Wang, W.-W., Shao, Z.-Q. 2014. Surface modification of cellulose nanowhiskers for application in thermosetting epoxy polymers. *Cellulose* 21:2529–2538.

[287] Nickerson, R.F., and Habrle, J.A. 1947. Cellulose intercrystalline structure. *Industrial & Engineering Chemistry* 39:1507–1512.

[288] Rånby, B.G. 1950. Aqueous colloidal solutions of cellulose micelles. *Acta Chemica Scandinavica* 3:649–650.

[289] Rånby, B.G. 1951. Fibrous molecular systems. cellulose and muscle. the colloidal properties of cellulose micelles. *Discussions of the Faraday Society* 11(1):158–164.

[290] Battista, O.A., Coppick, S., Howsman, J.A., Morehead, F.F., and Sisson, W.A. 1956. Level-off degree of polymerization. Relation to polyphase structure of cellulose fibers. *Industrial & Engineering Chemistry* 48(2):333–335.

[291] Bondeson, D., Mathew, A., and Oksman, K. 2006. Optimization of the isolation of nanocrystals from microcrystalline cellulose by acid hydrolysis. *Cellulose* 13(2):171–180.

[292] Hamad, W.Y., and Hu, T.Q. 2010. Structure-process-yield interactions in nanocrystalline cellulose extraction. *Canadian Journal of Chemical Engineering* 88(3):392–402.

[293] Magalhães, W.L.E., Cao, X., Ramires, M.A., and Lucia, L.A. 2011. Novel all-cellulose composite displaying aligned cellulose nanofibers reinforced with cellulose nanocrystals. *TAPPI Journal* 10(4):19–25.

[294] Macdonald, C. 2012. Made-in-Canada, nanotech. *Pulp & Paper Canada* 113(2):12–14.

[295] Incani, V., Danumah, C., and Boluk, Y. 2013. Nanocomposites of nanocrystalline cellulose for enzyme immobilization. *Cellulose* 20:191–200.

[296] Arias, A., Heuzey, M.-C., Huneault, M.A., Ausias, G., and Bendahou, A. 2015. Enhanced dispersion of cellulose nanocrystals in melt-processed polylactide-based nanocomposites. *Cellulose* 22:483–498.

[297] Hai, L.V., Son, H.N., Seo, Y.B. 2015. Physical and bio-composite properties of nanocrystalline cellulose from wood, cotton linters, cattail, and red algae. *Cellulose* 22:1789–1798.

[298] Molnes, S.N., Mamonov, A., Paso, K.G., Strand, S., and Syverud, K. 2018. Investigation of a new application for cellulose nanocrystals: a study of the enhanced oil recovery potential by use of a green additive. *Cellulose* 25:2289–2301.

[299] Tian, X., Hua, F., Lou, C., and Jiang, X. 2018. Cationic cellulose nanocrystals (CCNCs) and chitosan nanocomposite films filled with CCNCs for removal of reactive dyes from aqueous solutions. *Cellulose* 25:3927–3939.

[300] Zhai, S., Chen, H., Zhang, Y., Li, P., and Wu, W. 2022. Nanocellulose: a promising nanomaterial for fabricating fluorescent composites. *Cellulose* 29:7011–7035.

[301] Odijk, T. 1986. Theory of lyotropic polymer liquid crystals. *Macromolecules* 19:2313–2329.

[302] Edgar, C.D., and Gray, D.G. 2002. Influence of dextran on the phase behavior of suspensions of cellulose nanocrystals. *Macromolecules* 35:7400–7406

[303] Zoppe, J.O., Grosset, L., and Seppälä, J. 2013. Liquid crystalline thermosets based on anisotropic phases of cellulose nanocrystals. *Cellulose* 20:2569–2582.

[304] Marchessault, R.H., Morehead, F.F., and Walter, N.M. 1959. Liquid crystal systems from fibrillar polysaccharides. *Nature* 184(suppl. No. 9):632–633.

[305] Gray, D.G. 1983. Liquid crystalline cellulose derivatives. *Journal of Applied Polymer Science. Applied Polymer Symposia* 37:179–192.

[306] Werbowyj, R.S., and Gray, D.G. 1976. Liquid crystalline structure in aqueous hydroxypropyl cellulose solutions. *Molecular Crystals and Liquid Crystals (Letters)* 34:97–103.

[307] Brown, R.M. Jr. 1989. Bacterial cellulose. In: Kennedy, J.F., Phillips, G.O., and Williams, P.A. (Eds.). *Cellulose: Structural and Functional Aspects*. Ellis Horwood, Chichester, England. Pp. 145–151.

[308] Bielecki, S., Krystynowicz, A., Turkiewicz, M., and Kalinowska, H. 2002. Bacterial cellulose. *Biopolymers* 5:37–45.

[309] El-Saied, H., Basta, A.H., and Gobran, R.H. 2004. Research progress in friendly environmental technology for the production of cellulose products (bacterial cellulose and its application). *Polymer-Plastics Technology and Engineering* 43:797–820.

[310] Hirai, A., Inui, O., Horii, F., and Tsuji, M. 2009. Phase separation behavior in aqueous suspensions of bacterial cellulose nanocrystals prepared by sulfuric acid treatment. *Langmuir* 25:497–502.

[311] Soykeabkaew, N., Sian, C., Gea, S., Hishino, T., and Peijs, T. 2009. All-cellulose nanocomposites by surface selective dissolution of bacterial cellulose. *Cellulose* 16:435–444.

[312] Tajima, K., Imai, T., Yui, T., Yao, M., and Saxena, I. 2022. Cellulose-synthesizing machinery in bacteria. *Cellulose* 29:2755–2777.

[313] Nakagaito, A.N., Iwamoto, S., and Yano, H. 2005. Bacterial cellulose: the ultimate nano-scalar cellulose morphology for the production of high-strength composites. *Applied Physics A: Materials Science & Processing* 80:93–97.

[314] Nogi, M., Handa, K., Nakagaito, A.N., and Yano, H. 2005. Optically transparent bionanofiber composites with low sensitivity to refractive index of the polymer matrix. *Applied Physics Letters* 87:1–3.

[315] Yano, H., Sugiyama, J., Nakagaito, A.N., Nogi, M., Matsuura, T., Hikita, M., and Handa, K. 2005. Optically transparent composites reinforced with networks of bacterial nanofibers. *Advanced Materials* 17:153–155.

[316] Ifuku, S., Nogi, M., Abe, K., Handa, K., Nakatsubo, F., and Yano, H. 2007. Surface modification of bacterial cellulose nanofibers for property enhancement of optically transparent composites: dependence on acetyl-group DS. *Biomacromolecules* 8:1973–1978.

[317] Lee, K.-Y., Quero, F., Blaker, J.J., Hill, C.A.S., Eichhorn, S.J., and Bismarck, A. 2011. Surface only modification of bacterial cellulose nanofibers with organic acids. *Cellulose* 18:595–605.

[318] Czaja, W.K., Young, D.J., Kawecki, M., and Brown, R.M. Jr. 2007. The future prospects of microbial cellulose in biomedical applications. *Biomacromolecules* 8:1–12.

[319] Bäckdahl, H., Esquerra, M., Delbro, D., Risberg, B., and Gatenholm, P. 2008. Engineering microporosity in bacterial cellulose scaffolds. *Journal of Tissue Engineering and Regenerative Medicine* 2(6):320–330.

[320] Gao, C., Wan, Y., Lei, X., Qu, J., Yan, T., and Dai, K. 2011. Polylysine coated bacterial cellulose nanofibers as novel templates for bone-like apatite deposition. *Cellulose* 18:1555–1561.

7 Thermochemical conversion

7.1 Introduction

When heating lignocellulosic biomasses even at relatively low temperatures (>100 °C), a variety of degradation reactions starts to take place [1–6]. The thermochemical or thermal conversion of such materials always results in three groups of substances [2, 6]: gases, condensable liquids (bio-oils, tars, or distillates), and solid char or biochar products (in case of wood, charcoal) (Figure 7.1). The relative proportions of these product groups depend essentially on the chosen thermochemical method and its specific reaction conditions. Hence, as indicated in the earlier chapters (see Chapters 1 "Introduction to biorefining" and 3 "Chemical and biochemical conversion"), lignocellulosic biorefineries (e.g., a thermochemical biorefinery) are analogous processes to those conducted in the oil-based industry (i.e., petroleum refineries utilizing fossil resources) and can be defined as the processes of obtaining multiple biofuels as well as a wide range of chemicals and other biomaterials from renewable biomasses [2, 7, 8]. In comparison with the saccharification-based processes (i.e., the selective acid hydrolysis of carbohydrates), the thermochemical conversion processes are generally relatively rapid, avoiding large volumes of water (H_2O) and other external chemicals. However, a major disadvantage is the occurrence of unselective reactions, which produce a great number of products at low individual yields.

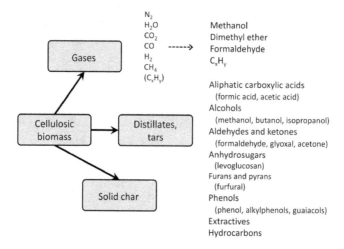

Figure 7.1: Product groups from the thermal conversion of lignocellulosic materials [2]. The relative proportions of these product groups principally depend on the applied method and the reaction conditions. Only some typical examples of individual components in synthesis gas ("syngas") and distillates as well as their further products are given.

https://doi.org/10.1515/9783110608366-007

The thermochemical degradation of lignocellulosic biomass constituents occurs through very complicated reaction pathways [9, 10]. These reactions are strongly affected by the typical operation process parameters, such as temperature, pressure, reaction time (or residence time), and mass transfer [11, 12]. A thermochemical biorefinery applies one or more thermochemical methods to produce energy and other valuable commodities with the potential to increase carbon utilization *via* value-added products [13–15]. In general, it has been shown that the thermochemical conversion of lignocellulosic biomass materials into various energy-containing products (*i.e.*, to process one fuel in order to produce a better one), such as char, bio-oil, and syngas is one of the most potentially beneficial and modern techniques [16]. Hence, the products obtained in different forms from thermochemical conversion are mostly utilized for the production of bioenergy.

The main thermochemical conversion methods comprise, besides direct combustion, pyrolysis (*i.e.*, slow pyrolysis and fast pyrolysis), gasification, hydrothermal carbonization, and liquefaction [1–3, 5, 17–25], whereas torrefaction [26, 27] represents a typical pretreatment process. Table 7.1 shows some characteristics for these methods. Additionally, there is also so-called "flash pyrolysis process", which at 750–1,000 °C, due to the high heating rate (<730 °C/s), low residence time (<0.5 s), and very small particle size (<0.2 mm), gives mostly gaseous products. Hence, this method differs strongly from that of the conventional slow pyrolysis carried out at the low heating rate (0.004 °C/s) with massive pieces of wood [29, 30]. The technologies can be readily applied to harvesting residues and other wood (including bark) and agricultural non-wood resources but also to various by-products of chemical pulping (including tall oil soap, black liquor, and kraft lignin).

As an illustrative example the thermochemical treatment of resinous wood has a long and colorful history [6]. One of the most interesting periods was from the seventeenth to the early twentieth century when wood tar was commercially produced, especially in Scandinavia, in tar pits and later in closed retorts. The vapors produced during such destructive distillation of wood can be condensed to give wood tar and an aqueous layer (*i.e.*, pyroligneous liquor or pyroligneous acid). This aqueous raw liquor from hardwood contains mainly acetic acid (CH_3CO_2H), methanol (CH_3OH), and CH_3CO_2H-derived acetone (CH_3COCH_3). As generally accepted, carbonization refers to processes in which solid char is the principal product of interest, while in wood distillation, the main product is liquid tar.

In general, in the use of renewable energy sources, biomass is of growing importance in satisfying environmental concerns over fossil fuel usage [31]. The term "biomass" is used to describe all biologically produced matter covering all the Earth's living material [30]. The chemical energy contained in the wood and other plant biomass is derived from solar energy using the process of photosynthesis where plants and certain other organisms produce oxygen (O_2) and glucose ($C_6H_{12}O_6$) for the needs of the Earth's biosphere from carbon dioxide (CO_2) and H_2O with the help of solar ra-

Table 7.1: Some characteristics of the thermochemical conversion methods of wood [2, 28].

Technology	Operation conditions[a]	Products
Direct combustion	In the presence of an oxidizing agent greater than the stoichiometric supply of oxygen; temperature range 900–1,200 °C; carbon conversion >99%	Heat and flue gas
Slow pyrolysis	In the complete or near complete absence of an oxidizing agent (air or oxygen); temperature range 450–800 °C; residence time: hours up to several days; average yields: 35% G, 30% L, and 35% S	Charcoal and distillates
Fast pyrolysis	In the complete or near complete absence of an oxidizing agent (air or oxygen); temperature range 450–600 °C; residence time: milliseconds to seconds; average yields: 15% G, 70% L, and 15% S	Bio-oil
Gasification	In the presence of a controlled amount of an oxidizing agent; temperature range 700–1,450 °C; carbon conversion 80–95%; average yields: 85% G, 5% L, and 10% S	Synthesis gas ("syngas")
Hydrothermal carbonization	Occurs in an aqueous environment; temperature range 180–260 °C (at 1–5 MPa); residence time: hours and external heat supply; average yields: 2–5% G, 5–20% L (dissolved in the process water), and 50–80% S	Hydrochar and aqueous coproducts
Liquefaction	Normally occurs in the presence of a reducing atmosphere (CO or H_2); i) solvolysis liquefaction in an organic solvent at 120–300 °C and atmospheric pressure and ii) high-pressure hydrothermal liquefaction in a pressurized (from 5 MPa to 25 MPa) water environment at 250–550 °C; average yields: 15% G, 80% L, and 5% S	Crude-like oil
Torrefaction	In the complete or near complete absence of an oxidizing agent at an atmospheric pressure; temperature range 180–300 °C; reaction time: 30 min to a couple of hours; average yields: 20% G, 0% L, and 80% S	Biochar

[a]G refers to gas, L to liquid, and S to solid.

diation energy (photons) [32]. Bioenergy, obtained from biomass *via* various thermochemical methods, is essentially renewable or carbon neutral. CO_2 released from these methods circulates through the biosphere and is reabsorbed in equivalent stores of biomass through photosynthesis.

In contrast to the production of bioenergy, due to the complicated nature of liquid product mixtures, the production of individual chemicals does not seem economically very attractive. There are, in addition to CH_3CO_2H, CH_3OH, and anhydrosugars (*i.e.,* mainly levoglucosan), many other compounds and compound groups that could be obtained by different techniques from raw pyroligneous liquors, including phenols and food flavoring agents [6, 33–37], However, in some of these cases, certain pretreatments of the raw materials are often necessary. In liquid products, an awesome mixture of different products can be chromatographically separated and identified in detail, and they typically contain various phenols, acids, lactones, alcohols, aldehydes, ketones, anhydrosugars, furans, esters, ethers, and hydrocarbons. Oxidative pyrolysis is typically the first step and thus, an inherent part of combustion processes.

Torrefaction of wood-derived biomass is mainly a pretreatment process for drying such biomass to remove its volatile compounds and, on the other hand, to decompose partially its hemicelluloses [2, 6, 26, 27, 38–46]. It is a mild, relatively short-time pyrolysis at 180–300 °C in the complete or near-complete absence of air. This process enhances the energy density of wood by increasing its carbon content and net caloric value. Therefore, it is expected to become more important in the future, especially as a treatment preceding efficient gasification (*i.e.,* in this case, the moisture content of feedstock should be <20–30%) in which it also results in the reduced generation of smoke. The torrefied wood material is more brittle than the feedstock material and it has intermediate characteristics between coal and the original feedstock. In the case of fast pyrolysis, biomass should normally be dried to a moisture content of about 10%, because a considerable amount of reaction H_2O is also formed during this process. The chemistry behind torrefaction resembles to that taking place in the production of so-called "heat-treated wood" [47, 48]. This brownish material is used for construction purposes, since it has a better dimensional stability compared to intact wood and resists biological stress caused by fungi.

Steam explosion is one kind of thermochemical process, which also represents an example of an unusual type of delignification. In this process, the wood feedstock is subjected to a short treatment (<4 min) at temperatures of 200–230 °C and pressures of <2 MPa, followed by a fast pressure release to atmospheric pressure [6]. This process has not gained any significant success, mainly due to a low quality of the fiber product. However, steam explosion can be an effective hydrolytic pretreatment, especially for non-wood feedstocks followed by various enzymatic and chemical treatments, such as in the fermentative production of ethanol (CH_3CH_2OH) after saccharification of soluble hemicelluloses (see Chapter 3 "Chemical and biochemical conversion"). For example, such a pretreatment at high temperatures and short residence times is of interest for lignocellulosic fractionation, since high temperature favors the solubilization of hemicelluloses, while a short residence time avoids their significant degradation and increases the hemicellulose fraction recovered after washing.

It is also evident that the production of solid fuels from different biomass resources has become also more important [5, 49]. Many raw materials exist in a suitable form and can be burned (>900 °C) directly to provide heat [50]. However, to exploit their best fuel properties, certain biomass resources, such as sawdust, bark, shavings, and harvesting residues, often have to be mechanically upgraded by grinding (*i.e.*, crushing, drying, and milling to an appropriate particle size distribution for obtaining "powdered fuels") and/or densified by pressing them into briquettes and pellets.

The densification of lignocellulosic residues in the form of briquettes and pellets contributes to high volumetric concentrations of energy, improves the handling, and reduces the volume and resulting storage and transport costs [28]. It can be claimed that the briquette is one of cleanest and most environmentally friendly solid fuel on the market today. In general, briquettes are produced in a mechanical or hydraulic press by applying load (about 8 MPa) on a die to feedstock particles at 100–140 °C. The conditions cause the "plastification" of lignin (*i.e.*, lignin acts as an agglomerating element), which is strongly influenced by the moisture content of the feedstock (*i.e.*, typically between 10–15%). Additionally, depending on the properties of feedstock, binder and heat treatment can also be applied.

7.2 Pyrolysis

7.2.1 General aspects

As stated in Section 7.1 "Introduction" biomass pyrolysis or devolatilization is a process by which biomass feedstock is thermochemically degraded in the complete or near-complete absence of an oxidizing agent (air or O_2) [2, 6, 20, 29, 30, 51–55]. Pyrolysis is the simplest and oldest method to convert lignocellulosic biomass materials to another and a better fuel material, which can essentially diversify the energy supply in many applications [56–58]. Hence, *via* pyrolysis it is possible to obtain products with a higher energy density and better properties compared to those of its initial feedstock material. The research on pyrolysis has gradually gained increasing importance; it is also a first step in the combustion and gasification processes and not only an independent process. Additionally, due to the chemical and physical variations in heterogeneous lignocellulosic biomass materials, the reaction scheme during pyrolysis is extremely complex resulting in the formation of a huge number of intermediates and final products with varying yields. In general, various kinetic, heat transfer, and mass transfer models for pyrolysis have been proposed from several points of view for the different pyrolysis processes of wood and other lignocellulosic materials [59–67]. Pyrolysis can also be performed in the presence of various catalysts.

The basic phenomena that take place during pyrolysis are as follows [30]: *i*) heat transfer from a heat source (*i.e.*, leads to an increase in temperature inside the feedstock particles), *ii*) initiation of pyrolysis reactions (*i.e.*, leads to the release of volatiles and the formation of char), *iii*) outflow of volatiles (*i.e.*, leads to heat transfer between the hot volatiles and cooler unpyrolyzed feedstock particles), *iv*) condensation of some of the volatiles in the cooler parts of the feedstock (*i.e.*, leads to the formation of tar), and *v*) autocatalytic secondary pyrolysis reactions due to various interactions.

In the slow heat-up of conventional wood slow pyrolysis with a very long residence time typically to 700–800 °C, the most important product is solid charcoal but volatile products and tar are also formed [2, 51, 52, 55, 68]. In fast pyrolysis, which produces higher amounts of distillates, the heating rate (*e.g.*, 300 °C/min to 500–600 °C) is very high and the residence time is only a few seconds or less [2, 5, 18, 20, 31, 59, 69, 70]. Therefore, in this case, chemical reaction kinetics, mass transfer processes, phase transitions, and heat transfer phenomena play important roles and influence the overall outcome. Since the critical issue is to bring the reacting biomass particle to the optimum process temperature, the particle size should be rather small (105–250 μm). In practice, this also minimizes the formation of charcoal. On the other hand, the volatile products of fast pyrolysis require rapid cooling to minimize secondary reactions that can result, for example, in harmful carbon deposition. Additionally, biomass should normally be dried to a moisture content of about 10% prior to fast pyrolysis, because a considerable amount of reaction H_2O (about 25% of feedstock) is formed in this process.

The condensable liquid product from fast pyrolysis of lignocellulosic biomass (*i.e.*, bio-oil, fast pyrolysis bio-oil, distillate, or tar) is a complex mixture of compounds derived from the fragmentation of various feedstock components [6]. Bio-oil is a potential source of revenues for companies that have biomass residues, such as forest residues, forest industry residues, and some agricultural residues at their disposal [2]. Agricultural residues are especially estimated to represent a considerable potential source for pyrolysis, although it is normally difficult to accumulate them in large quantities at a competitive cost in a single location [5]. The yields of product H_2O and gas are generally higher for agricultural biomass than for wood biomass.

A detailed knowledge of pyrolysis behavior of the three structural lignocellulosic material components (*i.e.*, cellulose, hemicelluloses, and lignin) together with non-structural substances, extractives, as well as the corresponding interaction between these constitutes is essential to understand the overall pyrolysis behavior of lignocellulosic feedstock materials [2, 6]. The formation of pyrolysis products depends both on the structure of the initial feedstock substance and on pyrolysis conditions. However, some resemblance between the original raw material and the pyrolysis products is frequently obvious [71]. In general, during pyrolysis under an inert atmosphere, biomass converts into low-degree-polymerized products by a series of complicated reactions, such as depolymerization, ring-opening, and cleavage of chemical bonds [67].

In spite of the complexity of pyrolytic reactions, the pyrolyzates are obtained through the cleavage of a few chemical bonds from the initial compounds [2, 6, 57, 67, 71–76]. Therefore, the way in which molecule fragments are produced depends on the types of chemical bonds involved and the chemical stability of the resulting smaller molecules. In the pyrolysis of carbohydrates-containing materials, such as lignocellulosic biomass, especially various dehydration reactions also happen under the conditions of slow heat rates, low temperatures, and long residence times resulting in a solid residue that contains aromatic polycyclic structures (*i.e.*, principally the combination of benzene rings).

If the volatile pyrolysis products are not chemically stable at temperatures above 600 °C, they can proceed to different secondary reactions, such as recombination or cracking [71, 75, 77, 78]. The recombination mechanisms play a major role regarding pyrolysis outcomes and these reactions take place to form nonvolatile, higher-molar-mass molecules by combining different volatile product compounds mainly leading to secondary char [71, 74, 77–79]. Cracking reactions result in the low-molar-mass molecules by cleavage of chemical bonds either within the volatile compounds or within the polymer [71, 79–81]. Free-radical pathways dominate common mechanisms for the pyrolysis of organic materials, although alternatives such as concerted molecular processes and ionic reactive intermediates may also occur [82–85].

Thermal analysis comprises a group of techniques in which a property of the sample is monitored in a specified atmosphere against time and temperature, while the temperature of the sample is programmed [86]. Many studies of the mechanisms dealing with the pyrolysis reactions of lignocellulosic biomasses are primarily based on kinetics modelling, practically irradiating the thermogravimetric behavior of these feedstocks by thermogravimetric analysis (TGA, *i.e.*, mass change *vs.* temperature) and by recording the first derivative of the TGA curve (*i.e.*, differential thermogravimetry (DTG)) for more clearly detecting small features/boulders as peaks (*i.e.*, rate of mass loss) on the curve, together with the chemical data on the formation of the major products [67, 87]. Therefore, the mass loss of a sample of the TGA curve under a certain heating rate and the peak height of the DTG curve are directly related to the temperature during the process and the reaction rate at the corresponding temperature, respectively. On the other hand, TGA and differential scanning calorimetry (DSC, *i.e.*, heat flux *vs.* temperature or differential thermal analysis (DTA)) primarily give information, besides about the mass loss of a sample over the whole process, also about the endothermic energy (*i.e.*, requires thermal energy from its surroundings) and exothermic energy (*i.e.*, releases thermal energy from the system to its surroundings) temperature ranges upon heating. Additionally, there is evolved gas analysis (EGA or evolved gas detection (EGD), *i.e.*, volatiles *vs.* temperature) by which the released compounds can be selectively identified [87, 88]. In practice, this can be done, for example, by coupling the gas output of a TGA or DTA instrument to a selective detector, such as a mass spectrometer (MS) or a Fourier transform infrared (FTIR) spectrometer (*i.e.*, TGA-MS and TGA-FTIR, respectively). However, if the evolved gas mixtures are so complicated that preliminary separation is required, a gas chromatograph (GC) can be used for the separation, coupled with MS or FTIR (*i.e.*, TGA-GC-MS and TGA-GC-FTIR, respectively). All these methods are widely applied techniques for studying the thermal stability of polymers.

Pyrolysis of lignocellulosic biomass is a complex phenomenon as several parameters determine the outcome of the process [55]; both physical parameters (*i.e.*, pyrolysis conditions) and chemical parameters are included. TGA is a simple technique for determining the pyrolytic behavior and for activation energies of different materials [87]. Hence, it easy to understand that the pyrolytic behavior of the main structural constituents, cellulose, hemicelluloses, and lignin in lignocellulosic biomass, needs to be mastered to gain insight into the mechanisms of pyrolysis [89]. It has also been shown that cellulose, hemicelluloses, and lignin act independently during pyrolysis and only negligible interactions between these major components can be found [14, 55, 90–93]. However, some interactions of the three main lignocellulosic biomass components may occur by influencing slightly the decomposition rates of these components together in the lignocellulosic biomass sample compared to those rates detected separately for the pure constituents [67].

Figure 7.2 illustrates TGA and DTG curves of the pure samples of cellulose, hemicelluloses, and lignin. In general, at temperatures normally under an inert nitrogen (N_2) atmosphere and at a heating rate of 20 °C, between 100 °C and 230 °C, the rate of mass loss is quite slow [94–98]. The active pyrolysis ranges of 300–400 °C and 230–310 °C are observed for cellulose and hemicelluloses, respectively, whereas lignin decomposes slowly in a wide range of temperatures up to 900 °C with a high solid residue remaining at the end of pyrolysis. In the case of hemicelluloses, a relatively high solid residue is also obtained. Thus, the thermal stability of these components against heating increases in the following order: hemicelluloses > cellulose > lignin. However, the thermochemical properties of the components are influenced to some extent by the presence of inorganic materials as well as the heat treatment applied [36].

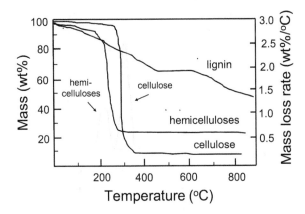

Figure 7.2: Schematic representation of thermogravimetric analysis (TGA)/differential thermogravimetry (DTG) curves of lignocellulosic biomass components under an inert atmosphere [91, 98].

The dissimilarities in the pyrolysis behavior of the major lignocellulosic biomass components can be easily understood based on clear differences in their physicochemical properties. For example, the higher thermal stability of cellulose (*i.e.*, a glucan-type polysaccharide), compared to distinctly lower-molar-mass and amorphous hemicellulo-ses with several types of glycosidic bonds and side groups, is mainly attributed to its unbranched and highly crystalline structure [99]. However, the extent of degradation of cellulose is higher: 80–90% *vs.* 60–70% for the hemicellulose components (*i.e.*, xylan and glucomannan). The clearly lower pyrolysis yield of char for cellulose (*i.e.*, consists only of glucose moieties) compared to that for hemicelluloses suggests more multiple reactions of heterogeneous hemicelluloses and their monosaccharide moieties [95, 100, 101]. Additionally, it has been reported that cellulose undergoes an extensive endother-mic-exothermic sequence immediately above 300 °C [94, 102, 103], whereas, for exam-ple, the hemicellulose component xylan displays clear exothermic behavior [102]. In general, it also seems that glucomannan is slightly more thermally stable than xylan [95]. This is probably due to the fact that pentosans like xylan are somewhat more sus-ceptible to dehydration and hydrolysis reactions than hexosans, like glucomannan.

A great number of kinetic models for the main lignocellulosic biomass compo-nents, especially for cellulose, have been established in large experimental conditions, from slow to fast pyrolysis [105–107]. However, their kinetics and elementary-reaction chemistry of are still debated. Historically, the most common model for cellulose pyrol-ysis (Figure 7.3) is the original multistep Broido-Shafizadeh model (*i.e.*, consists of three coupled first-order reactions and is largely based on mass loss) at low temperatures [111–114], which later has been modified and simplified to some extent by Shafizadeh and Bradbury [114], although other mechanisms have also been considered [75, 115]. In the Broido-Shafizadeh related models, the term "active cellulose" generally refers to an intermediary resulting from a partial depolymerization of cellulose, whereas the term "anhydrocellulose" rather refers to an intermediary formed after varying dehydration reactions. In the case of hemicelluloses, a semi-global reaction model has been intro-duced for xylan [110]. In this model, the pyrolysis produces intermediate condensable volatiles first at low temperatures and then, further volatiles and char are formed at higher temperatures through intermediates. This model is also applicable for gluco-mannan at both high and low temperatures [115]. In the pyrolysis process of lignin, there are several mass loss peaks with a gentle variation tendency due to the wide dis-tribution of bond energy and corresponding activation energies. It has been proposed that for example, in kraft lignin pyrolysis, there are parallel reaction pathways for a short residence time to the production of gas, pyrolysis oil, and char [67, 116, 117].

For a longer residence time, pyrolysis oil and some primary solid products undergo the secondary decomposition into gaseous products and cause an extra reaction path-way. In general, there are great variations reported for the total heat requirements for pyrolysis, the typical value ranges being 230–450 kJ/kg for cellulose, 200–420 kJ/kg for wood species, and 250–500 kJ/kg for non-wood species [75, 118–122]. The pyrolysis of cel-lulose has also been modelled by a first-order reaction with activation energy from 185 kJ/mol to 240 kJ/mol (for xylan 170–185 kJ/mol) [114, 123]. The activation energy for lignin greatly depends on the lignin origin, average values ranging from 50–100 kJ/mol to 250–270 kJ/mol [123–125].

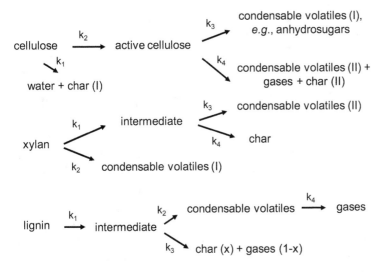

Figure 7.3: Examples of kinetic models for pyrolysis of cellulose, [108, 109] xylan [110], and lignin [67].

7.2.2 Reactions of lignocellulosic biomass components

Cellulose

In general, according to the TGA determinations of cellulose (see Section 2.3.1 "Cellulose"), after a small mass change from the start of temperature rise to about 300 °C, the main degradation of cellulose occurs between 300 °C and 400 °C when most of the evolved products are condensable compounds [126]. For temperatures higher than 400 °C, the residue becomes gradually more aromatic. The following sequences of reactions take place [126, 127]: *i*) formation of the anhydrocellulose and/or active cellulose (150–300 °C), *ii*) depolymerization (300–400 °C), and *iii*) charring process (380–800 °C).

In the first phase (*i*), dehydration reactions, which are highly correlated with char yields, are responsible for most of the mass loss; dehydration reactions can be intermolecular (*i.e.*, between two cellulose chains) or intramolecular (*i.e.*, within a glucopyranose moiety) [126]. Intermolecular dehydration can lead to the formation of additional covalent bonds, whereas intramolecular dehydration results in the formation of double bonds (>C=C<), which can promote, especially with slow heating rates, the formation of different benzene ring structures composing the char [127–129]. Although being relatively rather slow, in this phase various depolymerization reactions are also possible.

In the second fast phase (*ii*), the glycosidic bonds become very reactive, and simultaneously many different types of reactions take place. During this phase more than 80% of condensable volatile compounds (*i.e.*, more than 100 compounds [71]) are formed; for example, anhydrosugars, such as levoglucosan and levoglucosenone as well as furans, such as 5-(hydroxymethyl)furfural, 5-methylfurfural, furfural or 2-furaldehyde, and furfuryl alcohol, together with linear low-molar-mass oxygenated chemicals, such as hydroxyacetone and hydroxyacetaldehyde [71]. Additionally, some gases, such as CO_2, carbon monoxide (CO), and methane (CH_4) are formed. In the third phase (*iii*), various benzene ring structures (start to appear already at 300 °C) are gradually formed in the structure of cellulose char, which is relatively close to that of the lignin char at 400 °C.

As shown in Figure 7.3, in the pyrolysis process of cellulose, parallel and competitive reactions exist in the generation of condensable volatiles and char; the reaction path k_4 overcomes the reaction path k_3 at temperatures above 300 °C. Therefore, three demonstrative reactions, such as dehydration (k_1), depolymerization (k_2 and k_3), and fragmentation (k_4), are the main competitive reactions in different temperature ranges [75, 130–132]. The presence of inorganics, which act as catalysts, strongly increase the production of char and linear low-molar-mass oxygenated products [55, 107, 133, 134]. In general, the longer residence time and the higher reaction temperature result in more secondary cracking reactions [135, 136]. Hence, for example, the residence time should be shortened to maximize the yield of bio-oil.

Figure 7.4 presents a scheme of the main reactions in cellulose pyrolysis. The ring-opening of glucopyranose units takes place to form open-chain structures, followed by dehydration and cyclization to generate 5-(hydroxymethyl)furfural, which further converts through the elimination of the hydroxymethyl group to produce furfural [58, 71, 126, 137–139]. The cleavage of the hydroxymethyl group needs lower energy than the side aldehyde chain. However, 5-(hydroxymethyl)furfural and furfural are more likely yielded through competitive reactions rather than continuous reactions during the pyrolysis of glucose [67].

Levoglucosan is produced by the cleavage of the β-(1→4)-glycosidic linkage in the cellulose chain, followed by an intramolecular rearrangement [58, 71, 126]. Levoglucosanone is formed by the dehydration of levoglucosan in a temperature range of 250–400 °C as well as by dehydration from glucose to produce an intermediate with a carbon-carbon double bond (>C=C<), followed by dehydration *via* a six-membered hydrogen transfer, enol-keto tautomerization, and finally dehydration [67]. The formation of 1,4:3,6-dianhydro-α-D-glucopyranose is due to the double dehydration reactions of the hydroxyl groups in the glucopyranose units [67, 71, 137, 140–144]. A decomposition product of monosaccharides (*e.g.*, D- and D,L-arabinose, D- and D,L-xylose, and D-mannose) as well as alditols (*e.g.*, D- and D,L-arabitol) at above 500 °C is 3-hydroxy-2-penteno-1,5-lactone [145]. Additionally, for clarifying further carbohydrate pyrolysis mechanisms, glycerin (*i.e.*, the formation of acetaldehyde and formaldehyde [146]) and D-glucose (*i.e.*, the formation of C_1 and C_2 carbonyl compounds [147], C_3 and C_4 carbonyl compounds, and a cyclopentenedione isomer [148], and furans [149]) have been studied.

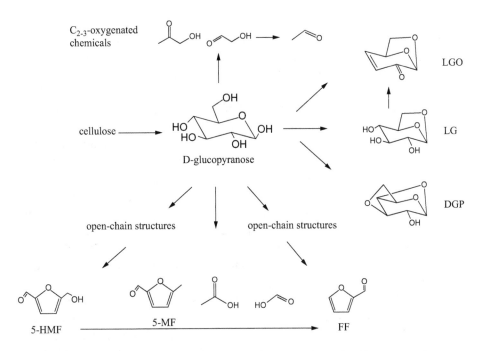

Figure 7.4: Degradation pathway examples of cellulose pyrolysis [55]. LGO = levoglucosenone, LG = levoglucosan, DGP = 1,4:3,6-dianhydro-α-D-glucopyranose, 5-HMF = 5-(hydroxymethyl)furfural, 5-MF = 5-methylfurfural, and FF = furfural. The low-molar-mass compounds are 1-hydroxy-2-propanone or acetol, hydroxyacetaldehyde or glycolaldehyde, and acetaldehyde or ethanal (upper part) and acetic acid and formic acid (lower part).

Hemicelluloses

Despite the variety in the chemical composition of the hemicellulose components (*i.e.*, mainly xylan and glucomannan, see Section 2.3,2 "Hemicelluloses") in lignocellulosic biomass materials [101], the TGA determinations of these constituents show that the main mass loss of hemicelluloses occurs relatively quickly in the temperature range 230–310 °C [95, 126]. In contrast to the linear homopolysaccharide cellulose (*i.e.*, contains only D-glucose moieties), linear hemicelluloses with clearly lower degree of polymerization and lack of crystallinity typically also with some side groups are heteropolysaccharides (*i.e.*, contain different monosaccharide moieties) [150]. Xylan belongs to pentosans (*i.e.*, contains pentose monosaccharide moieties), whereas glucomannan is a hexosan (*i.e.*, contains hexose monosaccharide moieties). Xylan is slightly more unstable than glucomannan and it has been studied more than glucomannan [67, 89, 95, 110, 115, 126, 151–153]. The following sequence of reactions take place [126]: *i*) dehydration and breaking of less stable linkages (xylan 150–240 °C and glucomannan 150–270 °C), *ii*) depolymerization (xylan 240–320 °C and glucomannan 270–350 °C), and *iii*) charring process (xylan 320–800 °C and glucomannan 350–800 °C).

In the first phase (*i*), the release of water, characteristic of dehydration reactions within the polysaccharides, as well as the formation of some typical compounds mainly originated from the substituents of the main chains [95, 102, 126, 139, 154] occurs: CH_3OH from the fragmentation of 4-*O*-methyl-α-D-glucuronic acid groups, formic acid (HCO_2H) from hexuronic acid groups, and CH_3CO_2H from *O*-acetyl groups. Some CO_2 is also released from the carboxylic acid functional groups ($-CO_2H$) of the hemicelluloses.

In the second phase (*ii*), although already to some extent in the previous phase, the glycosidic bonds between monosaccharide moieties become unstable and a rapid depolymerization occurs [126]. The main pyrolysis product of pentoses from xylan is furfural and of hexoses (from glucomannan), 5-(hydroxymethyl)furfural, as well as anhydrosugars (principally including levoglucosan from D-glucose but also levomannosan and levogalactosan from D-mannose and D-galactose, respectively), ketones (*e.g.*, hydroxyacetone, furanones, pyranones, cyclopentanones, and cyclopentenones), some phenols, lactones, and gaseous products (*e.g.*, CO_2, CO, and CH_4) [102, 126]. A significant amount of H_2O, CO_2, and CO is also formed during this phase. However, pyrolysis conditions also largely affect the product distribution [71].

In the third phase (*iii*), as in case of the pyrolysis of cellulose, the structure of residue leads to enhanced amounts of aromatics, when temperature rises above 320 °C and 350 °C for xylan and glucomannan, respectively [126]. It has been shown [102, 110, 115, 152] that the char yield from pentosan pyrolysis is higher than that from hexosan pyrolysis, due to the hydroxymethyl group ($-CH_2OH$) in the hexose structural unit, which can form stabilized low-molar radicals and avoid further repolymerization with each other to form char. Hence, it can be concluded that in contrast to the pyrolysis products from xylan, the pyrolysis products from glucomannan are like those of cellulose [72, 155]. Additionally, at higher temperatures CO, hydrogen (H_2), and CH_4 are liberated [126].

Figure 7.5 summarizes the formation of some major products in the pyrolysis of xylan. The cleavage of the linked lignin fragments in lignin-carbohydrate-complexes occurs and degradation to phenols takes place. After the release of acids (mainly CH_3CO_2H) at temperatures lower than 300 °C depolymerization, pyrolytic ring scission, and side-chain cracking take place [156]. These reactions produce linear ketones, such as hydroxyacetone (*via* carbon chain fracture), furfural, and furanones (by cyclization), and alicyclic ketones, such as cyclopentenones through the combination of double bonds [89, 102, 157–159]. Another major pyrolytic product is 3-hydroxy-2-penteno-1,5-lactone, which is yielded independently from the formation of furfural [160]. The degradation of glycosidic linkages between monosaccharide units of glucomannan results in various anhydrosugars and more stable furan rings (*e.g.*, 5-(hydroxymethyl)furfural, 5-methylfurfural, and furfural) [75, 115].

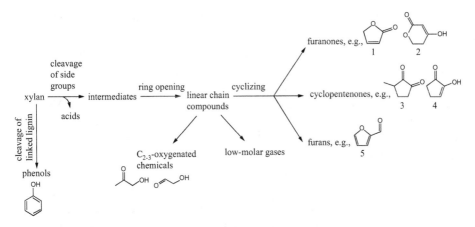

Figure 7.5: Modified decomposition pathways for pyrolysis of xylan [55]. 5H-Furan-2-one (1), 3-hydroxy-2-penteno-1,5-lactone (2), 3-methylcyclopentane-1,2-dione (3), 2-hydroxycyclopent-2-en-1-one (4), and furfural (5). The C_{2-3}-oxygenated compounds are hydroxyacetaldehyde or glycolaldehyde and acetaldehyde or ethanal.

Lignin

Lignin (see Section 2.3.3 "Lignin") is one of the three structural components of plant cell walls and provides rigidity and strength in plant issues [96, 101, 161]. It is a complex, high-molar-mass polymer made by an enzyme-initiated free-radical copolymerization of three phenylpropanoid building units derived from *trans*-coniferyl alcohol, *trans*-sinapyl alcohol, and *trans-p*-coumaryl alcohol (*i.e.*, having different number of methoxyl (-OCH$_3$) groups). The ratio of these precursors depends on the species, *i.e.*, softwoods, hardwoods, and non-woods. Lignin contains less oxygen than lignocellulosic polysaccharides, cellulose, and hemicelluloses and is being effectively studied as a promising source of various pyrolysis bio-oils.

A variety of lignin feedstocks with varying compositions has been used for pyrolysis; for example, kraft lignin [95, 124, 162–166] and lignin samples from other sources (*e.g.*, organosolv lignin, milled wood lignin (MWL), Klason lignin, cellulase lignin, and steam-exploded lignin) [36, 72, 167–174], Due to a large volume production in the pulp and paper industry, a by-product kraft lignin represents a potential feedstock material [2, 5]; kraft lignin dissolved in the cooking liquor (black liquor) is normally used as fuel for the recovery of its energy content, but it is also possible to separate kraft lignin from black liquor, for example, by acid precipitation (see Section 5.3.3 "Fractionation"). Additionally, the pyrolysis of several lignin model compounds has been studied [175–179]. For the most part, model compound pyrolysis studies have concentrated on the thermochemical behavior of specific interunit (*i.e.*, phenylpropane subunits) ether (C-O-C) and carbon-carbon (C-C) linkages found in lignin molecules. Results of these pyrolysis studies have led to information on the reaction kinetics of thermochemical decomposition as well as on the formation of compounds under varying pyrolysis conditions.

The TGA determinations show that lignin decomposes clearly slower over a broader temperature range than cellulose and hemicelluloses; the majority of volatile products are formed over a larger temperature range from 250 °C to 500 °C, with the highest decomposition rate being between 350 °C and 400 °C [95, 126, 168]. The reactions responsible for the main release of volatile compounds are mostly due to the instability of the alkyl side chains and the -OCH$_3$ substituents of aromatic rings and some linkages between phenylpropane units. After these reactions, a charring process, which consists in the rearrangement of the char skeleton in a polycyclic aromatic structure, takes place. In general, under inert atmosphere, aromatic rings are very stable and their concentration within the residue tends to increase throughout the pyrolysis. It should be noted that in contrast to polysaccharide pyrolysis, most of the aromatic rings, which will form the char are already present in the original lignin. The following two main sequences of reactions take place [126]: *i*) conversion of the alkyl side chains and the rupture of some of the linkages between phenylpropane units (150–420 °C) and *ii*) conversion of the short substituents of the aromatic rings and charring process (380–800 °C).

In the first phase (*i*), the reaction of the alkyl side chains of the aromatic rings begins at 180 °C and their hydroxyl groups are implied in various dehydration reactions [67, 126, 181, 182]. Regarding the chemical bonds linking the phenylpropane units, the C-O-C linkages involving the alkyl chains are supposed to be the less stable (at 200–245 °C). The reorganization of the chemical groups formed by the breaking of the ether bonds can result in the formation of oxygenated compounds, such as H$_2$O, CO, and CO$_2$. Hence, as the reaction temperature increases, the devolatilization reaction causes the production of pyrolysis oil and gaseous products *via* cracking the side chains of the phenylpropane units; most of the C-C bonds within and between the alkyl chains become unstable and react (>300 °C). The yields of guaiacols and syringols increase first due to the breakage of C-O-C bonds between structural units and some C-C bonds. Then, as the demethoxylation and demethylation reactions take place, the yields decrease with rising reaction temperature, resulting in the formation of phenols and catechols. These reactions are responsible for the highest decomposition rate of the lignin (350–400 °C) and thus, also for the maximum rate of the formation of phenols.

In the second phase (*ii*), different types of fragmentation reactions lead to the substitution of the -OCH$_3$ by -OH, -CH$_3$, and -H [67, 126]. This kind of fragmentation results in the formation of CH$_3$OH at 400 °C and at 430 °C. CH$_4$ is also formed from -OCH$_3$. Most of the initial bonds between the phenylpropane units are broken for temperatures higher than 450 °C and the more stable bonds, such as 4-*O*-5 and 5–5 are still present. The maximum formation rate of CH$_4$ takes place between 550 °C and 580 °C; it probably releases according to the mechanisms of demethylation. The rupture of the ether bonds 4-*O*-5 is supposed to be a source of the main formation of CO. After

450 °C the char becomes gradually more aromatic, and its formation is closely related to the disappearance of the -OCH$_3$ substituents [73, 182].

The macromolecular structure of lignin depolymerizes to a wide variety of condensable liquids. However, condensable liquids from the conversion of lignin still appear to be less complex than those of coal and contain a higher proportion of phenolic materials [183]. There are three main product groups in lignin pyrolysates [154, 184, 185]: *i*) large-molar-mass oligomers (*i.e.*, pyrolytic lignins), *ii*) monomeric phenolic compounds (*e.g.*, phenols, hydroxylphenols, guaiacols, and syringols), and *iii*) light compounds (*e.g.*, CH$_3$OH and CH$_3$CO$_2$H). Devolatilization and char formation are the two competing reactions, along with some subsequent secondary reactions. At low reaction temperatures, various other products, such as aldehydes and ketones, are formed as a result from the cleavage of glycosidic bonds in lignin-carbohydrate-complexes [71]. Residence time also influences the distribution of pyrolysis products. A too-short residence time results, due to the random breakage of chemical bonds, in the insufficient depolymerization of lignin and forms low yields of gas as well as of heterogeneous liquid products. In contrast, a proper residence time affects devolatilization, leading to increasing yields of pyrolysis oil and gas. However, a too-long residence time favors secondary decomposition of the initial pyrolysis products, causing a decrease in pyrolysis oil yield and an increase in gas yield. Figure 7.6 shows examples of low-molar-mass phenolic compounds as the main pyrolysis products of guaiacyl and syrigyl lignin. Some aromatic hydrocarbons and benzyl alcohols are produced from the removal of the oxygen-containing functional groups as well as from the rearrangement of aromatics with the -OCH$_3$ substituents, respectively [72].

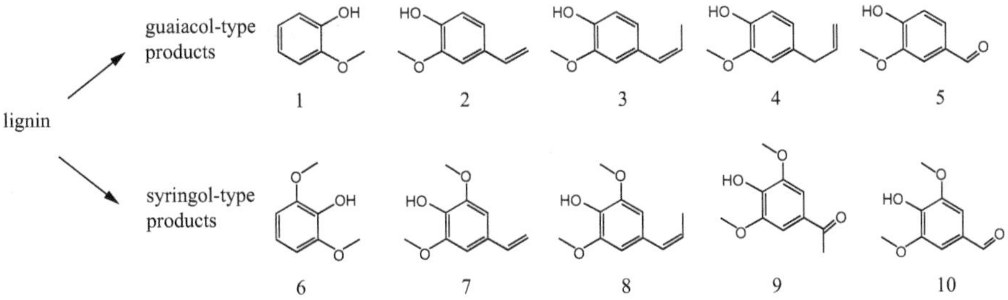

Figure 7.6: Examples of low-molar-mass phenols from lignin pyrolysis [71]. 2-Methoxyphenol (guaiacol) (1), 2-methoxy-4-ethenylphenol (4-vinylguaiacol) (2), 2-methoxy-4-(1-propenyl)phenol (isoeugenol) (3), 2-methoxy-4-(prop-2-en-1-yl)phenol (eugenol) (4), 4-hydroxy-3-methoxybenzaldehyde (vanillin) (5), 2,6-dimethoxyphenol (syringol) (6), 2,6-dimethoxy-4-ethenylphenol (4-vinylsyringol) (7), 2,6-dimethoxy-4-(1-propenyl)phenol (4-propenylsyringol) (8), 4′-hydroxy-3′,5′-dimethoxyacetophenone (acetosyringone) (9), and 4-hydroxy-3,5-dimethoxybenzaldehyde (syringylaldehyde) (10).

Some other parameters, such as the pyrolysis atmosphere and heating rate, also affect the behavior of lignin pyrolysis [71]. For example, during pyrolysis under an air atmosphere, the oxidation reaction is favored, but the release of volatiles occurs under an inert gas atmosphere. Increasing the heating rate can decline the residence time in low temperature ranges, resulting in the high conversion of lignin and lower char yield. Variations in feedstock, extraction processes, and in case of technical lignins, the pulping process also influence the pyrolysis behavior [72, 186, 187].

The complexity of the chemical composition of lignocellulosic biomass materials and various pyrolysis reaction pathways lead to several hundred types of oxygenated compounds with distinct properties [67]. Hence, the complexity of different bio-oils together with high oxygen content, H_2O, and acids contents make it challenging for further utilization as intermediate to produce transportation fuels or biochemicals [188, 189]. However, the addition of a catalyst into the pyrolysis reaction has been investigated as a potential way to optimize the distribution of pyrolysis products. In practice, by using an appropriate catalyst, a specific reaction can be selectively enhanced, resulting, for example, in the increased production of desirable products, such as anhydrosugars and hydrocarbons. Different catalytic configurations, such as inorganic salts, metal oxides, and zeolites, have been used, especially in the pyrolysis of lignin to produce oxygenated aromatics (mainly phenols) or a deoxygenated liquid fraction [36, 173, 190–194].

Extractives

Nonstructural components, organic extractives comprise a diversity of individual compounds with mainly low-molar-masses (see Section 2.4 "Nonstructural constituents"); their amounts strongly depend on the type of lignocellulosic biomass and the location on the plant, the content being usually 3–5% of the wood dry mass [101]. The total amount and the chemical composition of organic extractives are also dependent on the storage time of the wood material. By a broad definition, the extractives are either soluble in neutral organic solvents (*i.e.*, lipophilic and hydrophilic organic compounds) or H_2O (*i.e.*, inorganic compounds) [195]. Although the total amount of inorganics is low, for example, typically <0.5% of the wood dry mass, these compounds together with organic extractives influence the pyrolytic behavior of lignocellulosic biomass and should be also taken into consideration [67, 196]. However, due to a low concentration of extractives in feedstock materials, influences of heating on these compounds have generally received less attention.

The decomposition of organic extractives occurs in the same temperature range as that of cellulose, hemicelluloses, and lignin, and in principle, have a similar pyrogram profile. For example, the main mass loss of the tall oil (*i.e.*, the sodium-free by-product, tall oil soap, from kraft pulping, see Section 4.2.2 "Reactions of the wood chemical constituents") [2] and its main components, fatty acids and resin acids, occurs in the temperature range 250–450 °C with a very low formation of char [95]. However, due to the heterogeneous nature of organic extractives in different wood species and the wide number of chemically different compounds, mechanisms for their thermal decomposition are also complex and there is a multiplicity of possible reaction pathways [67, 85, 195]. It is also likely that at these temperatures, direct evaporation of some low-molar-mass extractives, such as those in a volatile turpentine fraction of softwoods, occurs. Although the pyrolytic decomposition pathways of cellulose, hemicelluloses, and lignin are not altered by the presence of extractives, the presence of extractives gives rise to additional products, influencing to some extent the final product distribution of the whole feedstock sample [67, 85, 101, 197]. On the other hand, it has been reported [67] that the presence of extractives improves the activity of the components and promotes the decomposition of structural constituents. Furthermore, minor amounts of CO_2 are obtained from the extractives during pyrolysis, due to the low mass ratio of oxygen to carbon for this fraction [72]. The extractives-free lignocellulosic biomass produces, besides more CO_2, also more H_2O, CO, and aldehydes, and less acids and alkanes, compared to the raw lignocellulosic biomass [67].

The processes that are traditionally needed for the conversion of vegetable oils (*i.e.*, containing primarily fatty acid esters with glycerol, normally triesters of glycerol or triglycerides) into diesel-type fuels are known either as alcoholysis (*i.e.*, transesterification reaction typically with methanol) or hydrogenation to produce, respectively, fatty acid methyl esters (FAMEs) and hydrocarbons [5, 198–203]. However, commonly practiced transesterification has some drawbacks, mainly including the requirement of a large amount of CH_3OH, which is usually produced from natural gas or other fossil fuel sources. The formation of a low-value by-product, glycerol, is also a problem. Additionally, the heating values of product esters are lower than those of conventional diesel fuels and these types of biodiesel are typically unstable.

Under pyrolysis conditions (200–550 °C) triglycerides are decomposed and free fatty acids are produced [85, 204–209]. Hence, fatty acids are readily released from the glyceridic structures with further degradation to acrolein or homologic derivatives [71, 210]. In general, the rearrangement of the fatty acid substituents produces the saturated or unsaturated carboxylic acids, which further form saturated or unsaturated hydrocarbons *via* rearrangement and decarboxylation. In general, the noncatalytic pyrolysis of triacylglyceride oil mixtures has been an attractive option to produce renewable fuels and other products.

The pyrolysis products of vegetable oils as well as lignocellulosic biomass materials are highly dependent on the chemical composition of the raw material, and the elemental composition of bio-oils typically resembles that of feedstock [5, 71, 72, 85, 211]. It has been shown that the pyrolysis of saponified vegetable oils with calcium hydroxide ($Ca(OH)_2$) (*i.e.*, containing Ca salts of fatty acids) leads to the formation of complex mixtures where either thermochemical decomposition or polymerization can occur [212]. Hence, one possibility is to apply fast pyrolysis to similar raw materials directly or after their alkaline hydrolysis for manufacturing liquid fuels [85, 213–217]. An interesting approach is also the direct pyrolysis of the crude tall oil (CTO) soap [218] for producing biofuel. In this process concept, sodium (Na) could also be recovered simultaneously in the form of sodium carbonate (Na_2CO_3), thus eliminating, for example, the need to add sulfuric acid (H_2SO_4) to liberate chemically bound Na from the -CO$_2$H substituents of extractives. Additionally, the industrial-scale production of biofuels from CTO by hydrogenation has been developed [219]. In this process, CTO is purified, hydrogenated in a chemical reactor, and finally distilled to a clear hydrocarbon-containing liquid product.

The content and distribution of the most abundant alkali metal ions (K$^{\oplus}$ and Na$^{\oplus}$) and alkali earth metal ions (Mg$^{2\oplus}$ and Ca$^{2\oplus}$) vary in different feedstocks. Despite the small amount of inorganic minerals, their notable catalytic effects on the thermochemical decomposition of lignocellulosic biomass materials are well known [67, 196]; they may cause, depending on the ion and compound, a lower decomposition temperature and a thermal decomposition activation energy as well as a higher amount of char. Hence, for example, they affect the product distribution, especially from fast pyrolysis by altering the decomposition pathways of biomass.

The presence of potassium (K) salts generally decreases the temperature of the maximum decomposition and the thermal activation energy [67]. They also catalyze the higher yields of gas, H$_2$O, and char. Additionally, K salts promote the depolymerization and fragmentation of biomass components, which, for example, results in the formation of low-molar-mass compounds, such as 5 H-furan-2-one, furfural, and phenols. However, K salts are unfavorable for the formation of carbonyl compounds and alcohols but promote, mainly due to the good cracking performance, the formation of CO$_2$, CO, and H$_2$O. Alkaline earth metal ions are stronger Lewis acids than alkali metal ions and enhance dehydration reactions for carbohydrate compounds. For example, the presence of MgCl$_2$ and CaCl$_2$ favor the formation of furan ring-containing derivatives, such as furfural, 5-(hydroxymethyl)furfural, and levoglucosenone. It can be noted that ZnCl$_2$ as a typical Lewis acid is also widely adopted in catalytic pyrolysis. It may catalyze the breakdown of C–O and C–C bonds and influences the degradation pathways, including dehydration, depolymerization, and ring-opening.

7.2.3 Production instruments and reactors

Laboratory-scale analytical instruments
The distribution of pyrolysis products depends significantly on pyrolysis conditions, especially on the temperature profile [20, 71, 72, 100, 102, 117, 131]. Hence, the reproducibility of the final pyrolysis temperature *via* a rapid temperature rise and accurate temperature control are the essential parameters in laboratory-scale analytical experiments. In principle, analytical pyrolysis instruments can be divided into three groups according to the method by which heat is introduced to the sample [220, 221]: *i)* resistive heating using platinum filaments (*i.e.*, resistively heated filament pyrolyzers), *ii)* inductive heating (*i.e.*, Curie-point pyrolyzers), and *iii)* isothermal furnace pyrolyzers. The selection of the instrument type depends primarily on personal preference, experimental requirements, running expenses available, and availability.

A resistively heated platinum (Pt) filament pyrolyzer makes normally possible a quick heating of the sample, such as solids, semisolids, and viscous liquids, which are not soluble in a volatile solvent [71]. In this method, an initial pulse of heating at high voltage in a pyroprobe filament unit produces a current through the Pt filament causing it to heat extremely rapidly until the controlled pyrolysis temperature is achieved. The heating rates are selectable, typically between 0.01 °C/ms and 20 °C/ms, temperature ranges are programmable in the range of 1 °C to 1,400 °C, and final hold times can be adjusted from 0.01 s to 99.99 s. The reproducibility for small sample size is generally good.

The pyrolyzate can be readily transferred to analytical instrument (*i.e.*, typically to gas chromatography with flame ionization detection and mass selective detection (GC-FID/MS)) as long as the filament is positioned correctly and the pyroprobe is sealed off from air [71]. However, a heated interface is needed between the pyrolyzer and the GC capillary column; the interface has its own heater to prevent condensation of pyrolyzate compounds and should have minimal volume. For this configuration, a valve is necessary between the pyrolyzer and the column, and the insertion or removal of the pyroprobe filament is possible. In practice, the sample is inserted in a quartz sample tube and centered between quartz wool in nearly the exact spot of the tube for good reproducibility. Additionally, regular maintenance should take place to prevent the contamination of both the sample and the lines [220].

In the Curie-point method, the sample is enclosed in a ferromagnetic metal or deposited on ferromagnetic wire, which is placed in a flow of a GC carrier gas and in a radio frequency field [71]. In practice, the foil of wire is rapidly (<0.5 s) heated to the Curie point of the metal (*i.e.*, to the pyrolysis temperature) with the induction of radio frequency. The Curie-point temperature may vary depending on the specific metal material used. Normally, with this method, less secondary reactions, compared to the resistively heated filament method, are obtained. In contrast, in the isothermal furnace pyrolyzers larger sample sizes, compared to those of other methods, are typically used with poor heat transfer. Additionally, an enhanced number of secondary reactions can take place.

Normally, the laboratory-scale micro-pyrolysis in combination with various detection systems (*e.g.*, Py-GC-FID/MS where the pyrolyzates are directly introduced into the injector of the gas chromatograph) is still one of the most generally used powerful techniques for clarifying in detail the thermochemical behavior of biomass samples under varying conditions as well as the chemical nature of pyrolysis gases and liquids. On the other hand, it is also important to characterize bio-oils from other sources as every bio-oil has different composition depending on the feedstock, applied pyrolysis conditions, and reactor configurations. Therefore, fast and cheap methods are required for pyrolysis liquid characterization. There are a number of analytical possibilities to determine the products from lignocellulosic biomass pyrolysis [80, 222, 223]. In general, these methods give complementary qualitative and quantitative information on a wide variety of products. For example, various characterization schemes have also been developed, mainly based on the solvent extraction of bio-oils [224, 225]. The analysis of the fractions can be then performed separately with suitable analytical methods.

Due to its reasonable price and broad availability, the GC-FID/MS is still one of the most generally used and powerful methods for the analysis of chemically stable pyrolysis products with boiling points below 350 °C [71, 226, 227]. In this case, however, a simple GC analysis normally does not allow identifying and quantifying more than half of the products present, thus giving limited information about the chemical composition. This is mainly due to improper chromatographic resolution, availability of compounds in MS libraries, lack of analytical standards (*i.e.*, those with known response factors for quantification), and inability to analyze nonvolatile products (*e.g.*, sugars and lignin oligomers) as well as the high-molar-mass compounds originating from their condensation in the transfer line between the pyrolyzer and gas chromatograph [220, 221]. Figure 7.7 shows one practical example of a pyrogram profile (Py-GC-MS) for the bio-oil from hardwood (birch) pyrolysis. TIC refers to a total ion chromatogram and is compared to a GC chromatogram. TIC is created by summing up intensities of all mass spectral peaks belonging to the same scan.

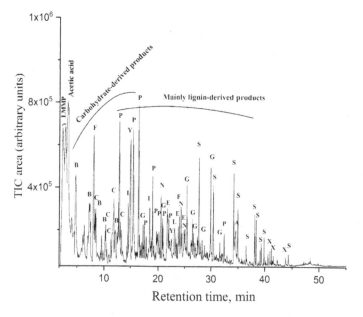

Figure 7.7: The main product groups formed in the pyrolysis experiments (700 °C and 20 s) from birch sawdust [71]. LMMP indicates highly volatile low-molar-mass products. Letters indicate compound groups to which identified products belong: A (anhydrosugar derivatives), B (benzene derivatives), C (cyclopentenone derivatives), E (catechol and benzenediol derivatives), F (furan derivatives), G (guaiacol derivatives), I (indene derivatives), L (lactone derivatives), N (naphthalene derivatives), P (phenol derivatives), S (syringol derivatives), X (fatty acid derivatives), and Y (pyrone derivatives).

On the other hand, simple GC systems have been improved using two-dimensional GC systems (GC × GC or 2D GC methods) coupled with FID or time-of-flight mass spectrometry (TOF-MS) [31, 226–229]. In these methods, each component of a complex mixture is subjected to two independent GC separations (*i.e.*, two columns with different properties), thus increasing the overall chromatographic resolution. However, these methods clearly require higher investment in the instrumentations than simple GC systems.

Further relevant information on pyrolysis liquids can be also obtained by high-performance liquid chromatography (HPLC) combined with MS (*i.e.*, suitable for non-volatile compounds) [70]. For example, HPLC-Orbitrap-MS has been used to identify more than 400 compounds in bio-oils with molar masses mainly between 100 Da and 400 Da [226, 230]. Gel permeation chromatography (GPC), as a type of size-exclusion chromatography (SEC), has been applied to evaluate high-molar-mass compounds with molar masses up to 1,000–2,000 Da together with molar mass distributions [75, 226, 230]. In a conventional mass spectrometer, the sample is bombarded under high vacuum by a beam of electrons to generate positively charged ions [231]. These ions are then forced to travel with a high velocity into the analyzer portion of the mass spectrometer where they will separate according to their mass to charge ratios (*m/e*). This hard ionization, electron ionization (EI) method is, however, not suitable for coupling to HPLC (*i.e.*, HPLS-EI) and is predominantly coupled only with GC (*i.e.*, GC-MS), but there are also available for HPLC soft ionization, MS techniques with different sensitivities and resolutions for specific compounds [37, 227, 232–234]: for example, fast atom bombardment (FAB), chemical ionization (CI), atmospheric-pressure chemical ionization (APCI), electrospray ionization (ESI), laser desorption ionization (LDI), and matrix-assisted laser desorption ionization (MALDI). Additionally, in this case, a tandem mass spectrometry (MS/MS) is a possible application.

FTIR spectroscopy has been widely used to determine the functional chemical groups of pyrolysis products and to obtain information on covalent bonding [20, 235–237]. FTIR spectroscopy has been used for the characterization of gaseous and solid compounds from the pyrolysis of lignocellulosic biomass and its components, often combined with thermogravimetric analysis (TGA-FTIR) [238–240]. In many cases, besides data on functional chemical groups, the elemental analysis of bio-oils is also of great importance and can be obtained by different techniques. Typical elemental and H_2O content of fast pyrolysis bio-oils are as follows [241]: carbon 44–47 wt%, oxygen 46–48 wt%, hydrogen 6–7 wt%, nitrogen 0–0.2 wt%, and H_2O 20–30 wt%.

Various nuclear magnetic resonance (NMR) techniques have been used extensively for analyzing lignocellulosic biomass materials and pyrolysis bio-oils to determine their carbon and proton distributions in different structures [242–247]. The major advantage of two-dimensional NMR (2D NMR) over one-dimensional NMR (1D NMR) is its ability to distinguish between the overlapping signals that exist, especially in larger molecules [248, 249]. 2D NMR gives data plotted in space defined by two frequency axes rather than one. 2D NMR including correlation spectroscopy (COSY, *i.e.*, shows protons that are J-coupled), heteronuclear single-quantum correlation spectroscopy (HSQC, *i.e.*, detects correlations between nuclei of two different types; *e.g.*, $^1H-^{13}C$), and heteronuclear multiple-bond correlation spectroscopy (HMBC, *i.e.*, detects heteronuclear correlations over longer ranges of 2–4 bonds; *e.g.*, $^1H-^{13}C$) have been applied in many applications of pyrolysis oils, for example, to characterize the functional groups, compound groups, and chemical bonds, such as C-H, C-C, and C-O bonds [37, 250, 251]. Low resolution ^{1}H NMR is also a useful method for screening and the semiquantitative analysis of pyrolysis oils [37]. It has been suggested that as a straightforward and cost-effective method, it could be used, for example, in the quality control of bio-oils from small-scale productions.

Larger-scale reactors

A wide range of reactor configurations on bench (1–20 kg/h), pilot (20–200 kg/h), and demonstration (200–2,000 kg/h) scale have been investigated on a variety of feed-stocks and the fast pyrolysis of lignocellulosic biomass has gradually achieved commercial status [4, 31, 54, 58, 70, 241, 252–254]. Most empirical research has mainly focused on technical development and testing of different reactor configurations, although much attention has also been paid to control and improvements of pyrolysis liquid quality and of liquid collection systems. Therefore, the main aims have included process performance and reliability (*e.g.*, effective high heat transfer rates and smooth operation), product characteristics, and scale-up possibilities.

Besides the reactor unit, the fast pyrolysis process consists of feedstock reception, feedstock storage and handling, feedstock drying and grinding, pyrolysis product collection, and storage [31]. Relevant technical issues have also included char and ash separation and product upgrading, when needed. Different reactor configurations of achieving fast pyrolysis are as follows [4, 54, 58, 69, 75, 254]:
- fluidized-bed reactor
- circulating fluidized-bed and transported bed reactors
- ablative pyrolysis reactor
- rotating cone reactor
- vacuum pyrolysis reactor
- other reactor systems

While many reactor configurations have been operated, fluidized-bed and circulating fluidized-bed reactors are the most popular configurations due to their ease of operation and ready scale-up. Some of these configurations are basically also those generally utilized in gasification. The "other reactor systems", including specific reactor configurations, such as entrained flow reactor, fixed bed reactor, microwave pyrolysis reactor, and screw and augur kilns, have also been studied.

Fluidized-bed reactors or "fluid beds", often referred to also as "bubbling fluidized beds" (BFBs) or "bubbling fluid beds" have the advantages of a well-understood technology, are simple in construction and operation, offer good temperature control, and have very efficient heat transfer to feedstock particles (particle sizes of less than 2–3 mm are needed to achieve high feedstock heating rates) arising from the high solids density [4, 54, 254]. Figure 7.8 illustrates a typical configuration of a fluidized-bed reactor and its working principle for lignocellulosic biomass pyrolysis using an electrostatic precipitator for coalescence and collection of aerosols, mainly containing incompletely depolymerized lignin fragments. Pyrolysis liquid yields of 70–75% from wood on a dry-feed basis can usually be obtained [4, 54]. The typical charcoal yield is 10–15% of the wood dry matter (*i.e.*, gas yield also 10–15% of the wood dry matter), corresponding about 25% of the energy of the lignocellulosic feed. Fluidized-bed reactors utilize the inherently good solids mixing; approximately 90% of the heat to the feedstock is transferred by solid-solid heat transfer, while a small contribution from gas-solid convective heat transfer up to 10% is applicable.

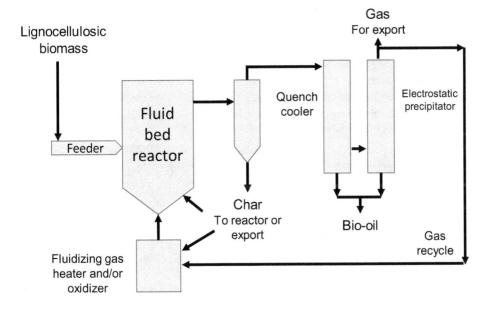

Figure 7.8: Schematic representation of a fluidized-bed reactor for lignocellulosic pyrolysis. Modified from [4, 54].

Circulating fluidized-bed reactors or circulating fluid bed (CFB) reactors and transported bed systems have various features in common with BFBs, except that the residence time of the char is almost the same as that for vapor and gas [4, 253]. Heat supply is typically from the recirculation of heated sand from the secondary char combustor (*i.e.*, either a bubbling or circulating fluid bed). The closely integrated char combustion in the second reactor requires careful control over temperature and heat flux to ensure that the temperature, heat flux, and solids flow match the process and feed requirements. The typical liquid yield is about 75% of the lignocellulosic biomass dry matter [58].

In the ablative pyrolysis system, lignocellulosic biomass particles are introduced tangentially to a heated vortex tube with a jet of inert carrier gas (*i.e.*, vaporized H_2O, steam, or N_2 at >100 m/s) [4, 54, 69, 254]. The particles rapidly slide across the inside surface of the reactor; the wood melts at the heated surface (*i.e.*, contact with it under pressure) and then the molten layer (an oil film) evaporates to pyrolysis vapors similar to those derived from fluid bed systems and also collected in the same way. This process can use larger particles of wood and the rate of reaction is typically limited by the rate of heat supply to the reactor wall. In contrast, in all the other pyrolysis methods, the rate of reaction is mainly limited by the rate of heat transfer through the wood particles, explaining why small particles are required in these cases. Therefore, by keeping the reactor wall temperature in ablative pyrolysis at a high level (at 625 °C), efficient melting of the particles is achieved. However, when necessary, it is also possible to remove partially pyrolyzed feed and char particles tangentially and recycle it to the eductor. Additionally, the rate of reaction is essentially influenced by pressure of the wood onto the heated surface as well as the relative velocity of the wood material. Normally, vapors generated at the surface are transferred away from the reactor by an inert carrier gas, resulting in short vapor residence times. The vortex reactor typically produces, for example, for softwood raw materials bio-oils with a yield of 60–65% from wood on a dry-feed basis together with charcoal (about 10%), gases (about 10%), and H_2O (about 15%).

The rotating cone reactor is a combination of a transport reactor and ablative pyrolysis and operates as a transported bed reactor in which transport is effected by centrifugal forces rather than gas [4, 254, 255]. Centrifugation drives hot sand (550 °C) and biomass up on a rotating heated cone with a rotational speed of 300 rpm, while vapors are collected and processed conventionally. However, because of the absence of carrier gas (as in ablative pyrolysis), the vapor products are not diluted and their flow is minimal; this undiluted and concentrated product flow results in small downstream equipment with related minimal investment costs. Char is burnt in a secondary combustor and the hot sand is recycled to the reactor. One advantage of this type of reactor is that it requires less carrier gas, although this integrated system (*i.e.*, containing a rotating cone pyrolyzer, bubbling bed char combustor, and sand recycling riser) is rather complex compared to other pyrolysis reactors [70]. The typical liquid yield is about 65% of the lignocellulosic biomass dry matter [58].

In the vacuum pyrolysis reactor, the rate of heating is very low compared to the other pyrolysis reactor systems [54]. The effect (*i.e.*, bio-oil quality and yield) is achieved by removing the vapors, as soon as they are formed, by operating under a vacuum (15 kPa) at about 450 °C. The rapid volatilization of the products formed under vacuum minimizes the extent of further decomposition reactions. The typical yield of bio-oil is 55–60% from feedstock on a dry-feed basis with a somewhat higher char yield than typically obtained in other pyrolysis systems. In general, the chemical structure of the pyrolysis products "resembles" more that of the original feedstock material than can be observed in other pyrolysis systems.

7.2.4 Product composition and properties

Main products and product groups

In conventional wood-based approaches (Figure 7.9), a fraction of the non-condensable volatiles ("wood gas" or fuel gas) and the condensable bio-oil can be burned as fuels and solid charcoal is left in the retort [6]. The elemental composition of charcoal or biochar (*i.e.*, contains typically 80% carbon) as well as its yield and physical properties, primarily depend on properties of the feedstock and final carbonization temperatures. Additionally, depending on the end use, the product properties can be greatly influenced by any active post-pyrolysis treatments, such as physical and chemical activation methods (*e.g.*, necessary for increasing porosity in case of active carbon) [256, 257]. The important properties of biochar include the concentrations and carbon- and nitrogen-containing organics and inorganics (*i.e.*, ash content), as well as the total porosity, pore-size distribution, and density. The mineral content of biochar depends on that of the lignocellulosic biomass feedstock and determines both the agricultural lime credit and credits for nutrients, such as phosphorous (P), K, Ca, and magnesium (Mg) in agricultural applications. Typically, freshly made biochar is hydrophobic containing few polar surface, oxygen-containing functional groups, which can be gradually oxidized (*e.g.*, to the -CO$_2$H groups), for example, during aging in soil environments and add cation exchange capacity and transform the biochar from hydrophobic to hydrophilic material [258].

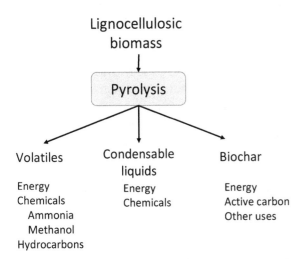

Figure 7.9: Schematic representation of lignocellulosic biomass pyrolysis with possible utilization of the main products.

Since charcoal clearly has higher heat content than dry wood per volume, wood carbonization to yield charcoal for heating and cooking is very common in many developing and less developed countries, often using crude carbonization kilns [6, 37]. A huge amount of forest biomass is being used to produce charcoal; the average estimate of the global charcoal production was 36.7 million tons (MT) in 2019 with the vast majority in Africa (20.2 MT), followed by America (9.7 MT), Asia (6.5 MT), and Europe (0.3 MT) [259]. However, in many cases, especially in Africa, the inefficient thermochemical processes, such as earth pit kilns and earth mounds, are used without any utilization of distillates and gases, thus increasing the release of greenhouse gas (GHG) emissions as well as the risk of environmental contamination.

In general, biochar can be considered a widely used combustible solid with a lower heating value of about 32 MJ/kg [54] and can be burned to generate heat energy in most systems that are burning pulverized coal [256]. Its sulfur content is low and in the industrial combustion applications of biochar, typically no specific technology for removing SO_x from emissions is needed. In contrast, the NO_x emissions are comparable to those coming from coal combustion and require similar abatement technology. In some cases, where bio-char production is based on non-wood raw materials, the high silicon content in the feedstock may cause scaling on the walls of combustion chambers and decrease the usable life of those chambers.

Besides the energy use, biochar has also found many industrial uses as an additive in paints, inks, and medicines as well as an efficient adsorbent for various purposes (*e.g.*, in managing and decontaminating different types of polluted waters) [6]. The use of biochar as soil amendment leads to significant environmental and agronomic benefits; it is a way to sequester large amounts of C and may have other GHG benefits [257]. Additionally, it has been shown that biochar enhances some microbial activity and is known to accelerate nitrification in forest ecosystems probably by adsorbing compounds, such as phenols, inhibiting nitrification in the absence of biochar [256, 260, 261]. Hence, the presence of biochar in soil environments enhances microbial activity and may accelerate the degradation of organic residues and biogenic humic substances, thus boosting productivity through improved plant and soil health without negative effects [262]. In general, some of the C in the biochars produced by low-temperature pyrolysis is bioavailable, whereas C in the high-temperature biochars is either nondegradable by microorganisms or the rate of microbial degradation is exceedingly slow [256, 258].

The most interesting temperature range for making the fast pyrolysis products is between 350 °C and 500 °C [263]. In general, the maximum yields from lignocellulosic biomass can be obtained as follows: *i*) bio-oil; a low-temperature, high-heating-rate, and short-gas-residence time process, *ii*) biochar; a low-temperature and low-heating-rate process, and *iii*) fuel gas; a high-temperature, low-heating-rate, and long-gas-residence time process. For example, at 475–525 °C the typical yields of fast pyrolysis products from wood are as follows [69, 254]: bio-oil 60–70 wt%, biochar 15–20 wt%, and fuel gas 10–15 wt% of the wood dry matter. Variations in pyrolysis product yields are mainly due to differences in the chemical composition and physical properties of the feedstock when operated within a normal fast pyrolysis regime (*i.e.*, fast heat-up of feed, short residence time of solids, and rapid cooling of product vapors). Additionally, the reactor configuration may play a minor role in the bio-oil quality. An example of the chemical composition of fuel gas from pyrolysis at 410 °C is as follows (gas yield 16.0 wt% of the feed dry matter) [264]: CO_2 45.1 vol%, CO 9.4 vol%, hydrocarbons 20.9 vol% (*e.g.*, CH_4 and ethene ($H_2C=CH_2$)), O_2 4.6 vol%, and others 20.0 vol%.

The liquid product, bio-oil is a complex mixture of compounds derived from the fragmentation of cellulose, hemicelluloses, lignin, and extractives. During the biomass pyrolysis, several hundred components have been separated and identified. For simplicity, the dominant GC-amenable pyrolysis products are generally classified into several compound groups and the formation of these monomer-related fragments is determined in different cases. There are a lot of data available on the formation of different compounds under varying pyrolysis conditions, for example, in wood slow pyrolysis [265, 266], in fast pyrolysis of wood and its constituents [71, 72, 267–274], and in fast pyrolysis of various non-wood materials [37, 71, 275–278]. Additionally, pyrolysis has been applied to raw materials that have been pretreated in different ways, such as with a hot-water pretreatment [67, 279–282].

The typical compound groups from wood and non-wood materials include anhydrosugar, cyclopentenone, furan, indene, lactone, and pyrone derivatives originated from carbohydrates and those of guaiacol, phenol, and syringol derivatives from lignin. The characteristic groups of naphthalene and fatty acid derivatives are originated from extractives, whereas, for example, benzene and linear ketone derivatives can be formed from all wood and non-wood constituents. On the other hand, because of the complicated nature of the liquid product mixtures, pyrolysis does not seem very attractive economically when considering the potential production of pure chemicals [250, 283]. In spite of this, for example, it has been shown that high yields of anhydrosugars and phenols are possible by fast pyrolysis techniques, but in these cases, certain pretreatments of the raw materials are necessary. Additionally, the manufacture of other products, such as fertilizers, flavor chemicals, and adhesives are possible from bio-oils [4], and they are also suitable for biological plant protection [284].

Table 7.2 shows the chemical composition of typical bio-oil pyrolysis from pine wood and Table 7.3 gives typical elemental content of wood-derived bio-oil. In general, the bio-oil is highly oxygenated, and it is not very different in elemental composition from the feedstock.

Physical properties of bio-oils

The physical properties that negatively affect pyrolysis oil quality mainly include low heating value, incompatibility with conventional fuels, solids content, high viscosity, and chemical instability [4, 5, 70, 225, 285–288]. Hence, primary bio-oils typically need significant modification before use [4, 188, 189, 254, 285, 289]. For example, the heating

Table 7.2: Example of typical chemical composition of fast pyrolysis bio-oil from pine wood (wt%) [4, 225].

Compounds	Dry product	Wet product
Water	–	23.9
Acids	5.4	4.1
Formic acid	1.4	1.1
Acetic acid	3.3	2.5
Propanoic acid	0.2	0.1
Glycolic acid	0.5	0.4
Alcohols	2.8	2.1
Ethylene glycol	0.3	0.2
Isopropanol	2.5	1.9
Aldehydes and ketones	19.5	14.8
Aliphatic aldehydes	9.3	7.1
Aromatic aldehydes	<0.1	<0.1
Aliphatic ketones	5.2	4.0
Furans	3.3	2.5
Pyrans	1.1	0.8
Sugars	43.5	33.1
1,5-Anhydro-β-D-arabinofuranose	0.3	0.2
1,6-Anhydro-β-D-glucopyranose (levoglucosan)	3.8	2.9
1,4:3,6-Dianhydro-α-D-glucopyranose	0.2	0.1
Catechols	<0.1	<0.1
Lignin-derived phenols	<0.1	<0.1
Guaiacols (methoxyphenols)	3.6	2.7
LMM[a] "lignin-like material"	17.0	12.9
HMM[a] "lignin-like material"	2.5	1.9
Extractives	5.6	4.3

[a]LMM and HMM refer to low-molar-mass and high-molar-mass, respectively.

Table 7.3: Typical elemental content of fast pyrolysis bio-oil [69, 254, 285].

Component	Content, wt%
Carbon	56.4
Hydrogen	6.2
Nitrogen	0.2
Ash	0.1
Oxygen (by difference)	37.1

value can be significantly increased by the removal of oxygen with certain unit processes generally used in the petroleum refining industry, and the other undesired characteristics can be improved by other physical methods as well.

Bio-oils differ significantly from petroleum-based fuels in both chemical composition and physical properties (Table 7.4). They are unstable, acidic, and high-molar-mass products containing a large amount of chemically dissolved water [4]. In contrast, light fuel oil consists mainly of saturated olefinic and aromatic hydrocarbons (C_9-C_{25}) that are not miscible with pyrolysis oils.

Table 7.4: Property examples of pyrolysis oils and light fuel oils [4, 31, 51, 69, 225, 285, 289].

Property	Pyrolysis oils	Light fuel oils
Viscosity (40 °C), cSt	15–40	3.0–7.5
Density (15 °C), kg/dm^3	1.15–1.25	0.89
HHV^a, MJ/kg	22–26	40
LHV^a (as produced), MJ/kg	13–18	
Flash point, °C	40–110	60
Pour point, °C	−9–36	−15
Moisture content, wt%	15–30	0.025
Nitrogen, wt%	0.0–0.4	–
Sulfur, wt%	0.00–0.05	0.2
Solids, wt%	0.01–0.50	–
Ash, wt%	0.01–0.20	0.01
pH value	2–3	Neutral
Stability	Unstable	Stable

^{a}HHV refers to higher heating value and LHV lower heating value.

Although bio-oils can be readily stored, they still can change chemically during storage due to reactions in pyrolysis failing to reach thermodynamic equilibrium [4, 225, 288, 290]. Additionally, H_2O content of bio-oils is typically high, which may lead to the formation of two phases (i.e., lipophilic and hydrophilic phases) during storage. The main chemical changes in pyrolysis liquids during storage are generally an increase in the H_2O-insoluble fraction ("high-molar-mass material") and a decrease in aldehydes and ketones. Furthermore, the H_2O content increases, and the amount of carbohydrates decreases to some extent. In contrast, the content of volatile acids (HCO_2H and CH_3CO_2H) does not change. This "ageing" can be clearly seen as an increase in viscosity (Figure 7.10).

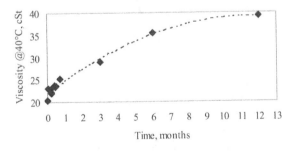

Figure 7.10: Change in viscosity of pyrolysis liquid during storage at room temperature [225].

Crude bio-oils are dark brown and approximate to feedstock lignocellulosic biomasses in elemental composition and they still contain several reactive species, which contribute to their unusual attributes [254]. They are generally very complex mixtures of oxygenated compounds with an appreciable proportion of H_2O from both the original raw material moisture and pyrolysis reactions. Hence, bio-oils have many special characteristics that require consideration for any application. Additionally, the general opinion is that bio-oils offer no significant health, environmental, or safety risks.

Fast pyrolysis bio-oils from lignocellulosic residues are principally an alternative to replace fossil fuels and feedstocks and many examples of power generation from them with or without upgrading have been reported among recent renewable energy processes [4]. The foremost idea behind the use of bio-oils as fuels is that they are much easier to handle and transport than solid biofuels. The most important properties, which may adversely affect the quality of bio-oils are incompatibility with conventional fossil fuels (*i.e.*, due to their high oxygen content), high solids content, high viscosity, and chemical instability [254]. Bio-oils can be upgraded in a great number of ways including *i*) physical, *ii*) catalytic, and *iii*) other chemical processes. In practice, bio-oils may be shipped, stored, and used much like conventional liquids once their specific fuel properties are recognized. Hence, when using mixed with fossil fuels, the properties of pyrolysis oils must be carefully considered in the quality and application of the fuel oil blend obtained.

The physical upgrading methods comprise, for example, hot-vapor filtration (to reduce the ash content), polar solvent addition (*i.e.*, to homogenize and reduce viscosity), and emulsification with diesel oil with the aid of surfactants (to produce stable microemulsions) [254]. The catalytic and chemical upgrading methods mainly include hydrotreating (*i.e.*, rejects oxygen as H_2O by catalytic reaction with H_2), catalytic vapor cracking (*i.e.*, rejects oxygen as CO_2), esterification and related processes (*i.e.*, the improvement of quality properties without substantial deoxygenation), and gasification to syngas (*i.e.*, for the production of H_2 or the synthesis of hydrocarbons and other products).

A great number of companies and other organizations are active in the significant research, demonstration, and commercial processes based on various reactor types with varying throughputs that use both fast and slow pyrolysis processes for production of liquid fuels for different purposes [4, 31, 69, 70, 254, 291–293]. However, bio-oils are still upgraded mostly at a small scale into transport fuel fractions.

7.3 Gasification

7.3.1 General aspects

Lignocellulosic biomass materials can be used as a solid fuel and burned directly to produce energy but they are also suitable, for example, for producing a mixture of gaseous compounds by gasification [2, 3]. The gasification process, performed normally at 800–1,000 °C can be considered as a thermochemical degradation of solid feedstocks and is enhanced by introducing a controlled amount of air, O_2, steam (*i.e.*, vaporized H_2O) or some combination of these [1, 24, 38, 198, 294–304]. Many feedstock materials consist of significant amounts of various volatile constituents, and during this kind of thermochemical conversion the volatile material is directly released by heating. Additionally, a huge amount of other gaseous compounds are formed in a complex combination of chemical reactions between gases, liquids, and solids. The reaction vessel in which gasification occurs is generally known as a gasification reactor or simply as a gasifier.

The gas mixture formed during gasification is called "synthetic gas" or commonly "syngas", although the terms "product gas" and "producer gas" are also used [4]. The latter two terms have a more general meaning for the gas mixture derived from gasification with air to make producer gas, which has higher levels of N_2 and lower concentrations of CO, H_2, CO_2, and CH_4 than syngas [198]. In contrast, the former two terms are especially used in the context when considering this gas mixture as a gaseous feedstock for further chemical conversions or energy. The term "syngas" is used throughout this book. In general, the outlet gas composition from the gasifier depends on the lignocellulosic biomass composition, gasification process, and the gasifying agent.

The chemical composition of syngas may vary over an extreme range depending on the raw material, temperature, and process design but the gaseous product typically contains N_2, steam, CO, CO_2, H_2, and CH_4 in varying proportions [2, 4–6, 198, 305]. Syngas is used either as a fuel or *via* many industrial routes for the production of chemicals, such as H_2, ammonia (NH_3), and CH_3OH, the latter of which can be further processed, for example, to dimethyl ether (DME, H_3COCH_3), methyl *tert*-butyl ether (MTBE), CH_3CO_2H, H_2, formaldehyde (HCHO), olefins, and gasoline. Additionally, a wide range of aliphatic hydrocarbons, *i.e.*, olefins and paraffins, together with oxygenated products, is possible to be manufactured from syngas by catalytic conversion according to the Fischer-Tropsch (F-T) process [306].

Lignocellulosic biomass gasification is similar to coal gasification [198]. It requires lower temperatures than coal gasification, because lignocellulosic biomass is more reactive than coal. Additionally, lignocellulosic biomass materials contain, for example, alkali metals K and Na, as well as alkali earth metals (*e.g.*, Ca) that can cause slagging and fouling problems in conventional gasifiers. Syngas is mainly produced industrially from coal and natural gas.

A clear advantage of gasification of lignocellulosic biomass materials is that all their major constituents, carbohydrates and lignin, are converted into syngas, as compared to the sugar platform routes (*i.e.*, saccharification) where lignin is not converted by hydrolysis or digestion [2, 307]. On the other hand, in these process concepts, lignin is used to generate the energy needed in the production processes. However, there is also an alternative route (*i.e.*, "a hybrid thermochemical and biochemical system") for production of ethanol (CH_3CH_2OH, yields up to 50% have been obtained) including the thermochemical conversion of lignocellulosic biomass materials into syngas followed by biological conversion of gaseous substrates (primarily CO) into the CH_3CH_2OH separated by distillation [308–311]. Microorganisms can also convert CO to other multi-carbon products, such as CH_3CH_2OH, butanol ($CH_3CH_2CH_2CH_2OH$), and butanoic acid ($CH_3CH_2CH_2CO_2H$).

7.3.2 Feedstock materials

It is possible to use for gasification almost any kind of carbonaceous feedstocks (*i.e.*, lignocellulosic-based and fossil-based materials as well as recycled wastes), if they are fed into the gasifier in a desirable size range and moisture content [4, 295, 303, 304]. In biorefinery applications using gasification of lignocellulosic biomass feedstocks, several kinds of fuels can be used at the same time, although fuel properties can set practical restrictions for the use of some fuels. This important flexibility of fuels is a major driving force for the further development of gasification technologies. The factors of fuels affecting gasification include moisture content, particle size, elemental composition, and the amount of volatile matter. In general, the homogeneity of feedstock fuel contributes to the uniform quality of syngas, because fuel quality fluctuations instantly affect the gasifier operation and the chemical composition of syngas.

There are several commercial technologies available for drying lignocellulosic biomass-based and waste-type fuels for combustion or gasification applications [4, 296, 302]. Feedstocks also have to be crushed or pelletized and possibly sorted in some way, prior to feeding them into the gasifier. When gasification is carried out at elevated pressures, the feeding of solid biomass against the higher pressure requires also specific feeding technologies. The same fact holds when removing solid materials, such as ash, from gasification and gas cleaning processes at elevated pressures and temperatures.

The moisture content of a feedstock is inversely proportional to its heating value and to maintain the gasification capacity and efficiency [4, 198, 294, 295], it has to be reduced in most cases before feeding the feedstock into the gasifier. Hence, all the energy used inside the gasifier for the vaporization of the H_2O reduces the maximum throughput of the gasifier. In practice, the higher the used amount of heat in evaporation, the lower the energy left for gasification reactions and thus, the high moisture feedstock content has a clear negative influence on the gasification process efficiency. Suitable feedstock moisture is usually below the range of 20–30%.

Particle size is one key factor determining the properties of a solid feedstock during gasification [4, 294, 295]; gasifiers are usually more capable of processing relatively small particles than larger ones. Hence, it is important that the feedstock material can be crushed with reasonable energy consumption, if its particle size needs to be reduced. Additionally, due to the increasing ratio between particle surface area and particle volume, gas-to-particle heat transfer is strongly correlated toward particles of smaller size. The gasification of small particles is controlled by chemical kinetics but there is a size limit above which heat transfer will become the determining feature. However, in the gasification of lignocellulosic biomass materials, large fuel particles may explode into smaller particles due to fast expansion of gases during pyrolysis.

With respect to the chemical composition of syngas, the properties and elemental composition of a fuel are of importance. The common properties of some typical feedstock materials are shown in Table 7.5, which clearly indicates essential differences. Additionally, there are also some variations in the elemental composition of feedstocks.

Table 7.5: Some characteristics of carbonaceous feedstocks used in gasification [4, 312].

Property	Wood	Bark	Peat	Coal
Moisture (wt%)	45–60	40–70	40–55	10
Ash (wt%, dry)	0.4–0.5	2–3	1–16	14
Volatiles (wt%, dry)	84–88	70–80	70–86	30
LHV[a] (MJ/kg, dry)	18–20	18–23	17–37	26–28
Elemental analysis (wt%, dry)				
Carbon	48–50	50–57	50–57	76–87
Hydrogen	6.0–6.5	5.9–6.8	5.0–6.5	3.5–5.0
Nitrogen	0.5–2.3	0.4–0.8	1.0–2.7	0.8–3.0
Oxygen	38–42	34–41	30–40	3–11
Sulfur	<0.5	<0.5	<1	0.5–5.0
Chlorine	<0.01	<0.03	0.03	<0.1

[a]LHV refers to lower heating value.

High alkali content in feedstocks (*i.e.*, feedstock ash) significantly reduces ash melting temperatures [4] particularly. This can lead to serious fouling and sintering problems, if the temperature level in the gasifier is too high for a specific raw material. Hence, due to the melting properties of flue ashes in high temperature treatments and at different pressure levels, fuel ashes may also play an important role in the processing of feedstocks into gases. Partial melting or sintering of fuel ash indirectly influences the final gas composition, for example, by setting the maximum operating temperature for the fluidized-bed gasifier. In case of entrained bed gasifiers, ash sintering determines the minimum process temperature, because ash has to be removed as liquid.

In different thermochemical applications, the amount of volatile matter in a feedstock comprises a significant factor [4]. For example, the low volatile content in some types of coal (Table 7.5) results in a high char yield after pyrolysis. In the gasifier, high operating temperatures are usually required to ensure the satisfactory reactivity of char. Typically, for lignocellulosic biomass materials with high volatile content, high reactivity, and low char yield, relatively low temperature levels can be used in the fluidized-bed gasifier. Volatile components in wood-based feedstocks comprise about

80% of the total mass of the feedstock, corresponding to about 50% of its combustion energy. However, a high concentration of volatiles generally also contributes to a higher formation of tar (*i.e.*, a mixture of various higher-molar-mass hydrocarbons) during gasification. These tar-based hydrocarbons are in gaseous form at gasification temperatures, but they can condense on colder downstream surfaces in the synthetic gas processing line, thus causing operating problems in the form of blockages. In general, the formation of tar can be reduced by choosing the proper gasification conditions and the gasifier. Furthermore, one approach to decrease the tar content is to add solid catalysts inside the gasifier or to mix alkali metal catalysts with the lignocellulosic biomass feedstock by dry mixing or wet impregnation [198].

7.3.3 Gasification process

In the gasification of lignocellulosic biomass feedstocks, the O_2 fed to the gasifier (*i.e.*, in O_2-blown gasification) corresponds to 20–50% of the stoichiometric amount of full oxidation [4, 294–296]. Partial combustion of feedstocks with O_2 provides the heat required for gasification reactions. Steam can also be used in the fluidized-bed air gasification of these feedstocks to improve the quality of the product gas. Additionally, steam is used as a fluidizing medium, since it also dilutes the gases and keeps the temperature levels inside the gasifier uniform.

The reactivity of feedstocks is one major factor determining the required gasification medium [4, 294–296]. Highly reactive biomass with a high content of volatiles can be efficiently processed by air gasification, whereas coal with a lower content of volatiles and a lower reactivity often requires the use of O_2-blown gasification. The final use of syngas also sets its own requirements for the gasification medium. Excess N_2 is an unwanted compound in chemical syntheses, but is useful when injected into the gas turbine in which it increases the total mass flow of gas through the turbine.

The syngas produced by O_2-blown gasification has a relatively high chemical energy, because there is no diluting N_2 (from air) in the gas [4, 294, 296]. The lower heating values (LHVs), characteristic of syngas from this kind of gasification, are between 7 MJ/m^3_n and 15 MJ/m^3_n. Typical applications for this medium heating-value syngas are gas turbines, production of synthetic or substitute natural gas (SNG), production of H_2, fuel cells, and syntheses of chemicals and liquid fuels [297, 305]. The O_2 generation with existing technologies consumes plenty of electricity and investment costs are considerable. Hence, the O_2-blown gasification process is generally considered suitable for large-scale syngas production or processes demanding high gasification or syngas combustion temperatures.

The thermal conversion of a solid feedstock in the fluidized-bed gasifier proceeds along the following stages, which practically take place partly sequentially and partly in parallel [4, 198, 313]: fast particle heating-up, possible swelling, drying, pyrolysis, volatile combustion, char combustion, char gasification, char fragmentation, and char elutriation. These main phenomena are illustrated in Figure 7.11. When the feedstock particle is exposed to high temperature, it heats up to drying temperature and the H_2O evaporates as steam. After this stage, the dry fuel particle continues to heat up and, due to the increasing temperature, particle pyrolysis or volatilization occurs with the simultaneous escape of volatile compounds with low molar masses (*i.e.*, primary pyrolysis vapors when heated above 500 °C). After the pyrolysis stage, the residual fuel char or charcoal is gradually consumed in various gasification and combustion reactions. However, since these stages take place partly in parallel, a large particle can still be drying from inside, while it is already reacting from outside. In principle, no chemical interactions are taking place among the organic compounds during primary pyrolysis reactions. During secondary reactions, the primary vapors and liquids form besides CO, CO_2, H_2, and H_2O, gaseous and condensed olefins, phenols, and other aromatics in the temperature regime from 700 °C to 850 °C. The final composition of the syngas is a combination of volatile products from all the stages and completes in tertiary reactions at 850–1,000 °C.

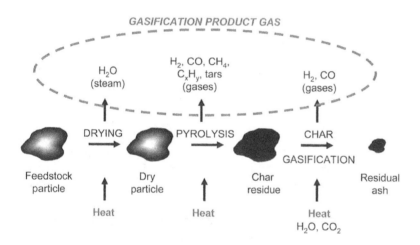

Figure 7.11: The main phenomena during gasification of solid fuel [4].

Most of the H_2O evaporation or drying in the gasifiers occurs due to the high reactor temperatures near the raw material feeding point [4, 294–296]. In general, heat transfer is the most limiting factor in the drying stage and mass transfer is relatively fast, due to the large pressure gradient caused by the expanding steam vapors. A low moisture content in feedstock results in a high gasifier capacity, as less heat is consumed in evaporation. Gasification reactions, as well as those taking place in pyrolysis are mainly endothermic, thus requiring external heat to maintain the process. The exothermic reactions between feedstock and oxidizing media in the gasifier produce most of the heat necessary for these pyrolytic and gasification reactions.

7.3.4 Principal reaction mechanisms

In gasification, depending on the gasification medium, a great variety of different pyrolytic and gasification reactions take place. In principle, the main reactions are basically the same for all carbonaceous fuel types. The exothermic reactions between feedstock C (as well as CO and H_2) and oxidizing media (O_2) produce most of the heat necessary for gasification reactions [4, 294, 295, 313]:

$$C + {}^1\!/_2 O_2(g) \rightarrow CO(g) \qquad \Delta H_{298K} = -110.5 \text{ kJ/mol} \qquad (7.1)$$

$$C + O_2(g) \rightarrow CO_2(g) \qquad \Delta H_{298K} = -393.5 \text{ kJ/mol} \qquad (7.2)$$

$$CO(g) {}^1\!/_2 + O_2(g) \rightarrow CO_2(g) \qquad \Delta H_{298K} = -283.0 \text{ kJ/mol} \qquad (7.3)$$

$$H_2(g) {}^1\!/_2 + O_2(g) \rightarrow H_2O(g) \qquad \Delta H_{298K} = -241.8 \text{ kJ/mol} \qquad (7.4)$$

The most important gasification reactions are heterogeneous and homogeneous reactions between the solid C of the char residue, CO, and CH_4 and H_2O, CO_2, and H_2:

$$C + H_2O(g) \leftrightarrow CO(g) + H_2(g) \qquad \Delta H_{298K} = 131.3 \text{ kJ/mol} \qquad (7.5)$$

$$C + CO_2(g) \leftrightarrow 2CO(g) \qquad \Delta H_{298K} = 172.4 \text{ kJ/mol} \qquad (7.6)$$

$$C + 2H_2(g) \leftrightarrow CH_4(g) \qquad \Delta H_{298K} = -74.6 \text{ kJ/mol} \qquad (7.7)$$

$$CO(g) + H_2O(g) \leftrightarrow CO_2(g) + H_2(g) \qquad \Delta H_{298K} = -41.1 \text{ kJ/mol} \qquad (7.8)$$

$$CH_4(g) + H_2O(g) \leftrightarrow CO(g) + 3H_2(g) \qquad \Delta H_{298K} = 205.9 \text{ kJ/mol} \qquad (7.9)$$

Due to the reversible reactions 7.5→7.9 the outcome of the gasification process is dependent on the chemical equilibrium in the process conditions. When the heating rate of the particles is fast enough, pyrolysis is not yet completely finished when the gasification reactions start to emerge. In fluidized beds, this is possible due to effective gas-to-particle heat transfer. Typical chemical compositions of syngases originated from different feedstocks and processes are shown in Table 7.6.

7.3.5 Production technologies and products from syngas

Reactor types
The thermochemical conversion of lignocellulosic biomass materials to syngas generally includes the following important steps [198]: feedstock storage and transport, size reduction, drying, feeding, gasifigation, syngas conditioning, and ash disposal or recycling. Gas conditioning is a general term for removing unwanted impurities (*e.g.*, tar) from the syngas and usually involves a multistep and integrated approach; a combination of the following main strategies are typically used, *i.e.*, hot-gas conditioning (a chemical destruction method, may include the use of catalysts, see below), hot-gas fil-

Table 7.6: Typical examples of chemical compositions of syngases from different gasification processes (volume %) [4, 314, 315].

Composition	Wood O_2 and steam-blown (dry gas)	Peat O_2 and steam-blown (dry gas)	Wood air-blown (wet gas)
Carbon monoxide (CO)	20–22	15–16	12–14
Carbon dioxide (CO$_2$)	25–35	20–28	12–13
Hydrogen (H$_2$)	23–25	16–28	9–11
Water (H$_2$O)	–	–	12–18
Methane (CH$_4$)	7–9	4–6	3–5
Nitrogen (N$_2$)	1–10	1–10	40–49
LHV[a] (MJ/m^3_n)	8–11	7–10	4–6

[a]LHV refers to lower heating value.

tration, wet scrubbing (a physical method, if syngas is used at atmospheric conditions), and dry/wet-dry scrubbing.

At the moment, "biomass to liquid" (BTL) processes *via* gasification are established technologies and, for example, in recent decades, many pilot-scale and demonstration-scale reactor configurations for varying feedstocks have been initiated [4, 296, 301–304]. The main types of gasifiers include *i*) fluidized-bed reactors or BFBs, *ii*) entrained bed and fixed bed reactors, and *iii*) dual-fluidized bed reactors (see also Section 7.2.3 "Production instruments and reactors").

In the most common fluidized-bed reactors, the gasification of solid feedstocks takes place in a bed of solid particles that is fluidized by an upward flowing gasification medium [4, 296, 301–304]. The behavior of solid particles in the bed resembles boiling liquid. Fluidized-bed reactors can be divided into two categories by their typical fluidization velocities: *i*) velocities between 0.7 m/s and 1.5 m/s in BFBs and *ii*) velocities between 4 m/s and 10 m/s in CFBs.

In entrained bed reactors, the medium can be O_2 or air combined with steam [4, 296, 301–304]. Feedstock and oxidizing gases are injected into the gasifier at high velocity, leading to efficient turbulent mixing. These kinds of gasifiers are suitable especially, for the gasification of fossil fuels. However, the limiting particle size of the feedstock material is about 100 μm and larger feedstock particles with weak grindability cannot be used efficiently. Hence, entrained bed reactors are not very useful for the gasification of lignocellulosic biomass materials. The small particle size needed partly compensates for the short residence time characteristic of entrained bed reactors. They operate at high temperatures (>1,300 °C) and molten ash is removed from the reactor as a slag. Due to the high investment costs of O_2 production facilities, entrained bed reactors are typically cost-effective only in large-scale installations. The biggest challenge is in feeding and removing solid feedstock particles under high pressure and temperature conditions, requiring high-durability materials, which are suitable for such conditions.

Fixed bed gasifiers (updraft and downdraft) are used in small-scale applications for the combined production of power and heat (*i.e.*, in the combined heat and power (CHP) production) [4, 296, 301–304]. These types of gasifiers utilize different zones with temperature fluctuations, because of nonuniform flow and channeling of solids and gases. Hence, the reactors can suffer from ash sintering and agglomeration problems. On the other hand, zones of too low temperatures inside the gasifier are also possible and this can lead to the harmful formation of tars. The channeling effects can escalate in large-scale reactors and, in practice, this risk makes fixed bed reactors unsuitable for large-scale gas production from lignocellulosic biomass materials.

Dual-fluidized bed gasification reactors are generally demonstrated only in small-scale applications for the production of syngas [4, 296, 301–304]. The main principle in these applications is that the heat required for lignocellulosic biomass pyrolysis and endothermic gasification reactions is provided indirectly or externally. This means that feedstock char is partially burnt in a separate combustion chamber. The heat of combustion is then transferred to the gasifier. The foremost advantage of the indirect gasification methods is that char combustion can be carried out separately by using air, instead of O_2. Char combustion also provides high C conversion. Since the inert N_2 (from air) in the flue gases is not mixed with the gasification gases, the syngas produced as a result of gasification has a relatively high concentration of the valuable components (*i.e.*, CO and H_2). The use of air instead of O_2 also eliminates the need for an O_2 production plant, which involves a high investment cost.

Product profile description

Gasification is one of the potential routes to produce electricity, liquid fuels, and chemicals from lignocellulosic biomass materials. It roughly proceeds as a two-stage process (Figure 7.11), which includes the formation of the primary tar and char fractions (*i.e.*, primary pyrolysis) followed by their secondary reactions to form the gaseous final products [316, 317]. However, syngas cleaning problems exist as one of the major challenging technical issues for developing the commercial applications of gasification [305, 318, 319].

Pyrolysis behavior of lignocellulosic biomass materials and their major chemical constituents (*i.e.*, cellulose and hemicellulose polysaccharides as well as lignin) has been investigated under varying conditions mainly focusing on the elucidation of the interaction of these components during tar and gas evolution [316, 318, 320–322]. In general, for example, tar compositions have been found to be quite different from wood polysaccharides and lignin and the evolutions of H_2O-soluble tar and gaseous products (*i.e.*, CO, H_2, CH_4, and $H_2C=CH_2$) are clearly suppressed by the interaction between cellulose and lignin. Furthermore, it has been found that the primary cellulose-derived tar fraction is readily converted to gaseous products, while reducing the amounts of the tar components. On the other hand, the gasification reactivities of the lignin-derived tar fraction have shown to be clearly lower than that of the cellulose-derived tar. Additionally, it has been indicated that the gasification of the spruce wood-derived char is catalyzed by alkali metals (*e.g.*, K) and alkali earth metals (*e.g.*, Ca) present in the lignocellulosic biomass materials [323].

Syngas from the typical pressurized O_2-blown gasification process of lignocellulosic materials can be used as a raw material to manufacture industrially significant products for further utilization. This integration results, for example, in hydrocarbons (*i.e.*, liquid fuels, diesel), SNG, H_2, and various alcohols (*e.g.*, CH_3OH and CH_3CH_2OH). CH_3OH can be further processed, for example, to DME or HCHO [2, 4, 294, 295, 297, 306].

Gasification of coal is an established technology but there is much less experience of gasification of lignocellulosic biomass materials and gas purification [4]. For example, the composition and thermochemical properties of the various types of lignocellulosic biomass materials vary significantly. However, in principle, syngas reactions are not much dependent on the raw material, though there are still relevant differences. Typically, lignocellulosics-based gasification plants are principally one or two orders of magnitude smaller than plants using coal. Since gasification and gas cleaning account for a decisive part of the investment cost, it is challenging to find feasible and cost-effective solutions for small-scale applications [296].

Many impurities are present in the syngas stream and efficient gas purification is necessary before the syngas can be used for typical conversions [4, 296, 306, 319, 324]. The requirements for syngas purity are especially stringent when used for syntheses. Practically, all of these conversion reactions applied are catalyzed by metal catalysts (*e.g.*, iron (Fe), cobalt (Co), ruthenium (Ru), nickel (Ni), rhodium (Rh), and platinum (Pt) supported on CeO_2/SiO_2 and dolomite), which are easily poisoned by various impurities [198]. In this respect, all the typical contaminants such as mechanical particles, alkali metals, acidic and basic gases, and tars are problematic and only very low concentrations of these impurities can be tolerated. However, it is a matter of controversy what the maximum acceptable limits for various contaminants are; one example of the limit data are shown in Table 7.7. The specifications of syngas purity are a question of optimization of the total gasification process and essentially depends on the application. For example, it has been reported [306] that Fe catalysts are much more tolerant to sulfur than Co catalysts and especially, Ru catalysts.

Fischer-Tropsch process

One of the most important conversion processes is the F-T synthesis [6, 198, 306, 325, 326]. The basic idea behind this catalytic process is to convert the essential components of syngas, CO and H_2, into straight-chain hydrocarbons (liquid hydrocarbons). The synthesis is typically carried out at temperatures of 150–300 °C and pressures of one to several tens of atmospheres.

The F-T process is based on an old technology, originally developed at the Kaiser-Wilhelm-Institut für Kohlenforschung (nowadays known as the Max-Planck-Institut

Table 7.7: Requirements for syngas purity [4, 296].

Characteristics	Required purity[a]
Sulfur	<0.15 mg/m^3_n
Nitrogen compounds	<0.015 mg/m^3_n
Halogens	<10 ppb$_V$
Alkalis	<10 ppb$_V$
Heavy metals	"Low"
Tars	<1 ppm$_V$ (below dew point–15 °C)
Particulates	0.1 mg/m^3_n
Share of inert gas	<10 vol%
Ratio H_2/CO	1.4–2.0 (Fe catalyst) and
	≈2 (Co catalyst)

[a]ppm refers to "parts per million" (*e.g.*, 104 ppm = 1%) and ppb to "parts per billion" (1 ppm = 1,000 ppb).

für Kohlenforschung or the Max Plank Institute for Coal Research) in Mühlheim, Germany in the early 1920s by Franz Fischer and Hans Tropsch [6]. They originally used Fe, Co, and Ru catalysts to produce liquid hydrocarbons rich in oxygenated compounds from coal-derived syngas (the Synthol process).

The preceding phase of this breakthrough, the synthesis of CH_4 from the mixture of CO and H_2 (*i.e.*, hydrogenation of CO) over a Ni catalyst, was discovered in 1902 by the French chemists, Paul Sabatier and Jean Baptiste Senderens [6]. All these reactions were also an important step in the use of metal-catalyzed reactions in the large-scale synthesis of organic compounds. One example is catalytic acetylene (HC≡CH) reactions under high pressure for producing a wide range of chemicals; this research area (the "Reppe chemistry") was started by Walter Julius Reppe in 1928.

In Germany, the preferred domestic fuel was originally brown coal (*i.e.*, lignite, subbituminous coal, and some bituminous coal), which was readily available in a country still suffering from the recession after World War I [4]. Later, typical fuels include coal (coal to liquids (CTL)), natural gas (gas to liquids (GTL)), or recently, increasingly BTL.

The first commercial plant in Germany was established in 1936 [4]. During World War II, the Third Reich, being petroleum-poor but coal-rich, manufactured replacement liquid fuels with the F-T process from brown coal [6]. For example, about 19.7 million liters of synthetic low-sulfur fuel (about 95,000 barrels) were produced in 1944 daily in 25 manufacturing plants. Additionally, by this technique, for military purposes, a synthetic lubrication oil and a synthetic rubber copolymer of acrylonitrile (H_2C=CH-C≡N) and 1,3-butadiene (H_2C=CH-CH=CH_2), nitrile rubber or "Buna N", developed in 1934 by Erich Konrad, Eduard Tschunkur, and Helmut Kleiner, and styrene-butadiene rubber ("Buna S") developed in 1929 by Eduard Tschunkur and Walter Bock, were produced.

The F-T conversion process was commercially not very favorable after the war [6, 198]. Coal-based liquid fuel production, the Sasol process, was started in South Africa in 1952 due to a trade embargo, including the import of oil against South Africa, as consequence of the apartheid (racial segregation) policy enforced through a legislation in 1948. In addition to synthetic fuels, a great number of chemical products such as paraffin waxes and other hydrocarbon mixtures, lubricants, and corrugated cardboard coating emulsions are produced. Examples of similar manufacturing units can be found in Qatar, Australia, Malaysia, and USA.

A great variety of catalysts has been tested and used for the F-T process [198, 326]; the most common ones are transition metals such as Co, Fe, and Rh, but Ni can be also used, although it favors the formation of CH_4 (*i.e.*, methanation). In conversion reactions of this type, catalysts are generally one of the key factors – over 80% of the present chemical processes contain a catalytic stage [6]. Catalysts offer generally an obvious way to reduce energy requirements of chemical reactions. A catalyst is commonly defined as an aid material that changes the rate of a chemical reaction without being consumed in the process. Catalytic reactions have a lower rate-limiting free energy of activation than the corresponding non-catalyzed reactions; this results in a lower overall energy required and hence, a higher reaction rate, at a given temperature. Besides this, the use of catalysts typically increases reaction selectivity. The catalyst can be heterogeneous or homogeneous (where catalyst occupies the same phase as the substrate) and various biocatalysts are normally considered as a separate group.

Syngas entering an F-T reactor has to be desulfurized since sulfur-containing impurities deactivate the catalysts needed for F-T reactions [4]. Syngas obtained, for example, from coal-based gasification typically have a H_2/CO ratio of about 0.7, whereas it would be beneficial to produce syngas with a H_2/CO ratio of about 2. This fact has to be taken in consideration in the gasification process, and in the syngas clean up and conditioning chain. However, there are several reactions by which the H_2/CO ratio of syngas can be adjusted. Most important is the catalytic H_2O-gas shift (WGS) reaction, which is commonly used in the chemical industry to produce H_2 as an expense of CO:

$$H_2O + CO \rightarrow H_2 + CO_2 \qquad (7.10)$$

For F-T plants that use CH_4 as the feedstock gas, another important reaction is steam reforming, converting CH_4 into H_2 and CO:

$$H_2O + CH_4 \rightarrow 3H_2 + CO \qquad (7.11)$$

The F-T synthesis is normally completed with Co-based and Fe-based catalysts [4]. The reaction leading to the formation of paraffins is the desired and most dominant reaction in the F-T process, especially over Co catalysts. However, depending on the catalyst and reaction conditions, the contribution of other reactions can also vary remarkably. For example, the WGS reaction (7.10) is catalyzed only by Fe. The share of olefins and oxygen-containing products is also typically higher on Fe catalysts. Co or Ru catalysts used at low temperatures, high pressures, and low space time velocities, are generally considered the best option to produce heavier paraffins from syngas. Co catalysts are more active for the F-T process when the feedstock is natural gas.

In spite of the long history of the F-T synthesis, the detailed reaction mechanism is still somewhat ambiguous. Based on the simple methanation (*i.e.*, CO and H_2 react to form CH_4 and H_2O), the conversion of a mixture of H_2 and CO into aliphatic hydrocarbons is a multistep reaction, with a great number of intermediate compounds [4, 198, 327, 328]. The basic idea is the polymerization of the methylene (-CH_2-) groups, which are formed by several reactions; in principle, one -CH_2- group is produced by $2H_2 + CO \rightarrow (-CH_2-) + H_2O$. Hence, it can be generally concluded that the original catalytic F-T process involves a series of chemical reactions that lead to a variety of hydrocarbons of the basic formula C_nH_{2n+2}, the overall net reaction being:

$$(2n+1)H_2 + nCO \rightarrow C_nH_{2n+2} + nH_2O \tag{7.12}$$

The main reactions of the F-T synthesis are characterized by high exothermicity, which must be taken into account in the process design [4]. Additionally, it is obvious that efficient heat transfer from the process and utilization of the released heat are essential from the viewpoint of total process economics and cost-effectiveness. Inadequate heat transfer from the process leads to higher operating temperatures, which promotes C deposition and catalysts deactivation. Typical reactors include multi tubular fixed-bed, entrained flow, slurry and fluid-bed, and circulating catalyst reactors [326].

Modeling the product distribution in the F-T process has proven very difficult and the direct F-T process does not allow the selective production of materials with a narrow C number range [4, 329]. Hence, further processing of the product is also always required. For example, only petrol (C_5–C_{12} hydrocarbons, max. 51 wt%) or diesel fraction (C_{13}–C_{18} hydrocarbons, max. 21 wt%), with modest selectivity, can be obtained directly. In contrast, except for CH_4, heavy waxes are possible to be produced with high selectivity.

This finding of high selectivity was first utilized in the Shell middle-distillate synthesis (SMDS) process, which converts in two stages, natural gas into hydrocarbon products, low-sulfur diesel fuels, and food-grade wax [4, 329]. In the first stage of the SMDS process, syngas is converted into long-chain hydrocarbon wax (*via* the heavy paraffin synthesis (HPS)). In the second stage, the heavy paraffins are selectively converted into the desired middle distillates, kerosene, and gas oil (*via* the heavy paraffin conversion (HPC)). The HPC stage is a mild hydrocracking process using a dual-functional catalyst. By varying the process conditions, it is possible to tailor the product distribution according to demand. Recent GTL, CTL, and BTL applications have been also almost exclusively based on a corresponding two-stage technology. An obvious drawback of these design concepts is the complexity of the process, resulting in higher investment costs as well as a need for H_2 in the cracking step.

Other products

The production of H_2 is a well known process in the chemical industry [330]. It is usually used for the production of transport fuel and the production of other products, for example, NH_3, diesel, and petrol in crude oil refineries [198, 297, 331]. It is also an essential reactant for a number of lignocellulosic biomass conversion strategies. H_2 can be produced in biomass gasification by maximizing its yield in syngas using a WGS reactor. Then, the other syngas components (CO, CO_2, and H_2O) can be separated and the leftover CO is converted into CH_4. The technical challenges in achieving a stable H_2 economy include improving process efficiencies, lowering the cost of production, and harnessing renewable sources for H_2 production.

SNG consists mainly of CH_4, which can be produced out of cleaned syngas via the very exothermic methanation reaction ($\Delta H_{298K} = -247$ kJ/mol) [4, 297]:

$$CO + 3H_2 \rightarrow CH_4 + H_2O \tag{7.13}$$

SNG can be used in many applications [297, 332], including its utilization as transport fuel, in the same way as biogas produced from anaerobic digestion. However, when compared to liquid biofuels, the energy density of SNG is quite low, unless liquefied.

In addition to F-T diesel, H_2, and CH_4, the production of certain oxo chemicals, mainly alcohols, from syngas has been under strong development. CH_3OH has been produced in many alternative ways [333], but is today almost exclusively produced from syngas, using a Co-Zn-based catalyst at pressures of 50–100 bar and temperatures of 220–300 °C [198, 297, 334]. The yield from biomass to CH_3OH can be 55%. The production is based on the following reaction:

$$CO(g) + 2H_2(g) \rightarrow CH_3OH(g) \tag{7.14}$$

DME can be produced from CH_3OH using Al-based catalyst at a pressure of 26 bar and a temperature of 310 °C via the following reaction [4, 334]:

$$2CH_3OH(g) \rightarrow H_3COCH_3(g) + H_2O(l) \tag{7.15}$$

In the reaction, DME can be separated in gaseous form from the mixture of DME, CH_3CH_2OH, and H_2O. It is a good substitute for liquefied petroleum gas (LPG) having a high cetane number and a high oxygen content to make it an attractive diesel engine fuel for replacing fossil diesel [4]. In terms of selectivity and efficiency, the advantage of producing SNG, CH_3OH, and DME, compared with the F-T process (i.e., the production of diesel), is that all these main products, being small molecules, can be made from syngas without a significant generation of side-products.

Petrol (i.e., "n(-CH_2-)") can be produced from CH_3OH using zeolite catalysts in the CH_4 to gasoline (MTG) process, and in an improved version of it, the Topsoe improved gasoline synthesis (TIGAS) [315, 334]. The production is based via a group of reactions, including CH_3OH synthesis (7.14), DME synthesis (7.15), and WGS reaction (7.10). In the TIGAS process, separate synthesis or storage of CH_3OH is not required because it reacts immediately further to form DME [4]. The raw petrol product has to be separated from light hydrocarbon gases and H_2O. Another advantage of the TIGAS process is that the required H_2/CO ratio for the net reaction is about 1 [315]. Therefore, the air-blown product gas resulting from biomass gasification (with inert N_2 from air) can be used to produce CO and H_2 for the synthesis. This eliminates the need for an O_2 separation plant in commercial applications.

For the CH_3CH_2OH synthesis, the commercial CH_3OH catalysts based on Cu/Co/ZnO can be modified to catalyze the synthesis reaction with a relatively high selectivity (<60%) toward CH_3CH_2OH [4]. Even higher selectivity can be obtained with Rh catalysts at about 275 °C and a pressure of 100 bar. Additionally, mixed alcohols (MAs) synthesis, also known as "higher alcohol synthesis" (HAS), is similar to the F-T and CH_3OH syntheses with the same type of catalyst [335]. The process produces a mixture of alcohols, such as CH_3OH, CH_3CH_2OH, propanols, butanols, and some heavier alcohols. CO can also be converted using a Mo-based catalyst [336].

7.3.6 Gasification of kraft black liquor

The organic solids in the spent liquors of alkaline kraft pulping processes (*i.e.*, black liquors, see Section 5.3 "Black liquor") are primarily composed of degraded lignin fractions and polysaccharide degradation products (mainly aliphatic carboxylic acids), together with minor fractions of residual extractives and hemicellulose residues [2, 5, 218]. In the kraft mill, after the separation of most of the extractives in the form of tall soap, the liquor is conventionally concentrated to 65–75% solids content by evaporation and then combusted in the recovery furnace for the recovery of the cooking chemicals and the generation of energy [337, 338]. There have been several efforts to optimize the combustion of black liquor, together with many improvements of the energy efficiency of pulp and paper mills [339]. However, because of various factors, mainly including both the low thermal efficiency of the energy recovery and the high capital costs of the recovery equipment, increasing effort has also been directed at the possibility of enhancing black liquor conversion to high-value energy.

Since lignocellulosic biomass gasification (*i.e.*, gasification-based biorefineries) was gradually developed to a known technology, there were also activities to significantly improve the efficiency of the energy (*i.e.*, the evident production of electric power) and chemical recovery cycle in kraft pulping by the potential integrated forest biorefinery concept comprising black liquor gasification (BLG) [5, 339, 340]. Hence, the basic assumption was simply that the use of BLG would have the potential to achieve higher overall energy efficiency than the conventional recovery furnace and the benefits of this kind of integration would be evident. Before these relatively recent BLG applications, there have also been similar early approaches to utilize kraft black liquors thermochemically in a different way than their typical combustion in the recovery furnace [341–343].

In principle, two general BLG approaches have been as follows [344]: *i)* atmospheric, air-blown booster gasifier systems, providing incremental capacity for overloaded recovery boilers and *ii)* pressurized O_2-blown gasifier systems, targeted to serve as an alternative to conventional recovery boilers. In BLG, the black liquor is gasified in a separate reactor [5, 345]. Due to certain advantageous properties, the gasification of black liquor is easier and more rapid than many other lignocellulosic biomass feedstocks [346].

With respect to the reaction temperature, two different process modifications have been studied [5, 346, 347]: *i)* "low-temperature gasification" at 600–850 °C where the inorganic compounds (mainly sodium carbonate (Na_2CO_3)) are below their melting points and *ii)* "high-temperature gasification", generally at 900–1,000 °C, which produces a molten mixture of inorganics (smelt). In both cases, the syngas is cleaned to remove particulates and tar and to absorb inorganic species (*i.e.*, hydrogen sulfide (H_2S) and sulfur dioxide (SO_2)) for preventing damage to the gas turbine and for reducing pollutant emissions. The clean syngas is then burned in gas turbines coupled with generators to produce electricity. The hot exhaust gas is passed through a heat exchanger to produce high-pressure steam for a steam turbine or another process. The smelt continuously leaves the bottom of the gasifier, is cooled to recover heat, and is then dissolved into H_2O and introduced to the recovery of cooking chemicals. A lot of small-scale research has been made under varying conditions on BLG concerning, for example, the formation of product gases, tar, and char [348–351] as well as the char reactivity [352] and the fate of char nitrogen in catalytic gasification [353].

In principle, the recovery furnace could be replaced by the BLG process, although, due to some unsolved problems, this technology has so far not been successfully realized on a full scale. In principle, the recovery furnace could be replaced by the BLG process, although, due to some essential problems, this technology has so far not been widely realized on a full scale. Various BLG technologies including, for example, Manufacturing and Technology Conversion International (MTCI Fluidized Bed Gasification), Direct Alkali Regeneration System (DARS), SCA-Billerud Process, BLG with Direct Causticization, and Chemrec Gasification processes, have been studied [4, 345–347, 354, 355].

The recovery furnace and the BLG processes have a rather similar overall efficiency, although the BLG process allows a syngas-based production of CH_3OH (and from it. *e.g.*, DME) and a wide range of aliphatic hydrocarbons, together with oxygenated products [356]; this technical way is less competitive in comparison with the large-scale coal gasification [193]. In the BLG technologies, there are still challenges dealing mainly with the feeding of black liquor against a high pressure and temperature, high requirements for material quality for reactor walls and connections (*i.e.*, the influence of corrosion), high demand for syngas cooling, and the purification of syngas from ash [4, 345]. Furthermore, the recovery of cooking chemicals is not straightforward and only a portion of the sulfur is normally converted to sodium sulfide (Na_2S).

7.4 Hydrothermal carbonization and liquefaction

Hydrothermal carbonization

In the case of solid feedstock materials, the feedstock preparation is an essential stage for any thermochemical conversion method. The important properties in the pretreatment of lignocellulosic biomass are moisture content, size distribution and particle shape, bulk and particle densities, as well as compressive and compact ratio [28, 357]. All these parameters have a clear and different influence on the thermochemical behavior of feedstocks in various conversion methods [358–362].

A fundamental pretreatment of lignocellulosic biomass feedstock for typical thermochemical conversion technologies, such as combustion, pyrolysis, and gasification, described earlier, is drying, since the reduction of moisture content of the feedstock leads to higher efficiencies [28]. Additionally, at high feedstock to air ratios and high moisture contents of the feedstocks in gasification, the WGS reaction (7.10) results in the formation of H_2 and increases the concentration of CO_2. In pyrolysis, the initial moisture content of the feedstock contributes to the H_2O content in bio-oil (also the reaction H_2O is formed), and above a moisture content of about 10%, the bio-oil produced may separate into two phases [293]. Furthermore, the failing to dry the feedstock will generally consume energy to vaporize the H_2O in the feedstock during conversion, and dried lignocellulosic biomass materials are usually preferred in thermochemical processes [28].

Technologies such as hydrothermal carbonization (HTC) or hydrothermal treatment (HTT) upgrade the lignocellulosic in an aqueous environment (i.e., a high moisture content of the feedstock is desired), which is then applied to transform lignocellulosic biomass to hydrochar or biocoal [28, 363, 364]. The process occurs at relatively low temperatures, usually in the range 180–260 °C in a closed reactor under autogenous pressures of 1–5 MPa with long residence times of hours. Hydrochar typically holds 55% to 90% of the initial mass of lignocellulosic biomass, corresponding to 80% to 95% of its energy content. Because the HTC process requires external heat to be supplied, heat integration with a CHP plant may offer an attractive alternative for a biorefinery utilizing lignocellulosic biomass residues. Hydrochar can be used, besides for energy production, also as a valuable product such as soil conditioner, activated carbon, or adsorbent agent.

The principal chemical reactions occurring in the HTC are hydrolysis, dehydration, and decarboxylation [364]. The hydrolysis (including also deacetylation) degrades the polysaccharides of lignocellulosic biomass materials to lower-molar-mass fragments, which further deplete from O_2 with the gradual aromatization by the latter two reactions, with an increase in the heating value of the feedstock [365, 366]. Hence, as a result of HTC, the carbonaceous structure of hydrochar becomes chemically more stable and compact compared to the feedstock lignocellulosic biomass [28]. The HTC liquor that contains organic by-products (e.g., HCO_2H, CH_3CO_2H, acrylic acid ($H_2C=CHCO_2H$), lactic acid ($CH_3CH(OH)CO_2H$), and levulinic acid as well as furans and phenols) can be used for various industrial applications [28, 367–373]. Additionally, the HTC process removes alkali and alkali earth metals from the feedstock, unlike torrefaction.

It is evident that differences in the chemical composition of lignocellulosic biomass materials also lead to significant variations in the decomposition behavior of these feedstocks during the HTC treatment [28]. The HTC process has been applied, for example, to various non-wood materials and agricultural residues [28, 369, 374–377], wood materials [363, 364, 366, 369, 378–380], and wood-derived materials [381–383] as well as food waste and packaging materials [384], municipal solid waste [385], and microalgae [386].

Liquefaction

The common process chain of converting coal into liquid hydrocarbons (*i.e.*, liquid fuels and petrochemicals) is generally known as CTL [387, 388]. The best-known CTL process is the F-T synthesis (see Section 7.3.5 "Production technologies and products from syngas"), representing the indirect coal liquefaction (ICL) technology. In this technology, coal is first converted into a gas mixture, syngas, which is then used, for example, for the production of liquid hydrocarbons. Hence, since the term "liquefaction" is also generally defined as a change from a gas to a liquid by cooling, "indirect liquefaction" involves successive production of a volatile product, for example, by gasification and the subsequent conversion of this intermediate to a liquid fuel. In contrast, the direct coal liquefaction (DCL) technology (*e.g.*, the Bergius process) converts coal directly into liquid hydrocarbon distillate products without having to rely on intermediate steps, by breaking down the mostly aromatic structure of coal with a suitable additive at high temperatures and pressures.

Analogous to the DCL process, lignocellulosic biomass materials can be converted by direct liquefaction, the hydrothermal liquefaction (HTL) process (*i.e.*, the direct BTL process), into various liquefied products through versatile thermochemical reactions [292, 389–394]. In principle, this conversion can be accomplished at high temperatures under high pressures in the presence of a reducing atmosphere. In this case, the process is often called "catalytic liquefaction", whereas a non-catalytic process is simply known as "direct liquefaction". During the HTL process, the wet lignocellulosic feedstock is broken into low-molar-mass molecules, which are unsteady and reactive and can polymerize into oil-like substances with a broad range of molar-mass distribution [391].

The liquefaction-based thermochemical conversion (often, generally referred also to as "thermolysis") is accomplished at high temperatures and varying pressures, typically in the presence of H_2O or an alcohol, such as CH_3CH_2OH and CH_3OH. Direct liquefaction generally refers to *i*) solvolysis liquefaction (SL) in an organic solvent at moderate temperatures (120–300 °C) and atmospheric pressure [389, 393, 395–398] or *ii*) to high-pressure (from 5 MPa to 25 MPa) (*i.e.*, "typical" HTL) in a hot, pressurized water environment at 250–550 °C [389, 393, 394], both for varying reaction times to produce crude-like oil ("biocrude"), which normally needs subsequent upgrading (*e.g.*, a hydrotreating process [399–401] with high pressure H_2 in the presence of a suitable catalyst) to be useful in the current fuel markets. This bio-oil has typically high energy density with a LHV of 33.8–36.9 MJ/kg and contains 5–20 wt% oxygen and renewable chemicals [292, 390]. The HT-like processes under varying conditions have been applied, for example, to lignin [402–406], hardwoods [398, 407], and microalgae [408–411].

Based on these facts above, thermochemical conversion can be generally subdivided into pyrolysis, gasification, and direct liquefaction [391]. Hence, in addition to these methods, indirect liquefaction offers the straightforward production of high-viscosity liquids from solid biomass. However, with pyrolysis, a catalyst is usually unnecessary, whereas direct liquefaction is normally performed in the presence of a suitable catalyst. Additionally, because of the need for more complicated and expensive reactors in liquefaction, there is, at present, less general interest in liquefaction than these other methods.

As indicated in Section 7.3.6 "Gasification of kraft black liquor", due to several factors, including the low thermal efficiency in energy recovery, the limited capacity for electricity generation, and the high capital cost of the recovery furnace, the possibility of enhancing black liquor conversion to high-value forms of energy has attracted increasing interest. Hence, besides gasification, another thermochemical possibility to convert the organic matter in black liquor into crude bio-oil (Table 7.8) is a technique similar to that used in the liquefaction of wood [412–414]. This method is based on liquid-phase thermal treatment under a reducing atmosphere (CO or H_2) at high temperatures (300–350 °C) and pressures (about 20 MPa) with or without the addition of additives (*e.g.*, sodium hydroxide (NaOH)). During such a treatment, an oil-like organic product (a total yield of 40–60% of black liquor organics) separates out from an aqueous phase containing the inorganic material and the residual organics (Figure 7.12). In principle, the desired conversion can be accomplished in either the presence or absence of a reducing atmosphere, although in the latter case, the addition of excess alkali is necessary for producing an oil of acceptable quality (Table 7.9).

Table 7.8: Composition of the dry matter of the typical black liquor feedstock (% of the total dry solid) [414].

Component	Content
Organic constituents	76.8
Lignin	43.2
Aliphatic carboxylic acids	28.2
Others	5.4
Inorganic constituents	23.2
Inorganic compounds	11.9
Sodium[a]	10.7
Sulfur[a]	0.6

[a]Bound chemically to organic material.

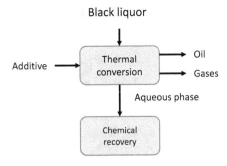

Black liquor

Additive → Thermal conversion → Oil
Thermal conversion → Gases
↓ Aqueous phase

Chemical recovery

Figure 7.12: Schematic representation of the thermochemical treatment process (liquefaction) for black liquor [414].

Table 7.9: Conversion example[a] of black liquor organics into the main product groups (% of the initial total organics) [412, 413].

Product	With alkali addition	Without alkal addition
Bio-oil	40	42
Volatile compounds	36	38
Aqueous organics	24	20

[a]Reactions for about 45 min at 350 °C, alkali (NaOH) addition 27.8%, and reducing atmosphere H_2.

It has been shown [412, 413, 415] that the organic-phase oil product originates mainly from the lignin material of black liquor. About 60% of the heating value of the black liquor is recovered as heating value of the oil product. This organic phase product can be used after refining [406] as fuel for various purposes. It has been also shown [416] that during the treatments, the prominent aliphatic hydroxy carboxylic acids, containing five or six carbon atoms, in the original black liquor significantly degrade into lower-molar-mass acids and volatile compounds. These acids can be recovered and used as organic chemicals and the inorganic components are also possible to recycle in the kraft mill (Figure 7.12). So far, this kind of thermochemical conversion of black liquor has been carried out only in small-scale reactors.

Depending on the source of kraft black liquor (*i.e.*, from softwood or hardwood), the $-OCH_3$ content of kraft lignin varies in the range of 14–20% of the lignin [417]. On heating under pressure, normally with added sodium sulfide (Na_2S) and NaOH at 250–350 °C (kraft cooking temperature is 160–180 °C), the formation of methyl mercaptan or methanethiol (CH_3SH (MM)) takes place *via* the cleavage of the methyl aryl ether bonds (*i.e.*, demethylation with $HS^\ominus$ ions) [218, 417–420]. MM can further react in its ionized form (CH_3S^-) with another $-OCH_3$ group, leading to dimethyl sulfide (CH_3SCH_3 (DMS)). The process also results in the formation of catechol, catechol derivatives, and other low-molar-mass aromatics as well as low-molar-mass aliphatic carboxylic acids, such as CH_3CO_2H, propanoic ($CH_3CH_2CO_2H$), glycolic ($HOCH_2CO_2H$), $CH_3CH(OH)CO_2H$, and oxalic (HO_2CCO_2H) acids. Some of these products can be recovered by extraction with organic solvents (*e.g.*, with diethyl ether ($CH_3CH_2OCH_2CH_3$)). The DMS can be obtained in commercial quantities and it can be oxidized to dimethyl sulfoxide (DMSO (CH_3SOCH_3) used as a solvent) [421].

Many variations of this kind of approach have been reported to produce sulfur-containing chemicals as well as low-molar-mass aromatics, aliphatic acids, and "demethylated lignin" from kraft black liquors [417, 419–423]. One straightforward example has been to add NaOH in amounts of about 18% of the dry matter to softwood kraft black liquor with a dry matter content about 55%, and then after spray drying to a dry matter content of about 95% and heating continuously for 7 min, strong disproportionation and degradation of organic matter took place [419]. In practice, during this treatment, the kraft lignin is completely demethylated, degraded, and hydrogenated, forming, for example, 4-methylcatechol, 4-ethylcatechol, and other catechol derivatives and phenols.

7.5 Production of bioenergy

Because of the increasing demand for energy and growing environmental pressures, the sustainable production of bioenergy from a variety of renewable lignocellulosic biomass resources (*i.e.*, the renewable sources of carbon) has rapidly become increasingly important during the past few decades [2, 6, 424–426]. On the other hand, world energy demand has dramatically increased in the last decades, mainly concerning the use of conventional fossil and renewable resources, but also the use of relatively recent new sources of renewable energy (*e.g.*, the solar photovoltaic (PV) panels that generate electricity and hot H_2O as well as the combined wind and wave energy technologies) have increased remarkably [28, 301, 427, 428].

The majority of the world's energy is supplied by fossil-based fuels, and petroleum-derived distillation fractions are used in a wide range of industrial applications [209]. In 2019, the use of renewable energy sources accounted for 13.8% of the world total primary energy supply (TPES, *i.e.*, 606.5 EJ; 1 EJ = 1,018 Joules (J) = 24 million tons of oil equivalent (Mtoe)), thus corresponding to about 2,000 Mtoe, and it is predicted to reach about 2,700 Mtoe in 2040 [429]. The renewable energy sources include hydroelectricity, biofuels, renewable municipal waste, and solar PV and thermal as well as geothermal and tidal. Nowadays, due to its widespread use in developing countries for heating and cooking applications, solid biofuel/charcoal is by far the largest renewable energy source (58.1%), followed by hydroelectricity (18.2%), wind (6.2%), liquid biofuels (5.1%), geothermal (5.0%), and others with shares lower than 3% [429]. For example, the average estimate of the global charcoal production in 2019 was 36.7 million tons (Mt): in Africa, 20.2 Mt; America, 9.7 Mt; Asia, 6.5 Mt; and Europe, 0.3 Mt [430].

In response to the present trend toward replacing fossil fuels with renewable energy sources and to tackle acute environmental challenges (*e.g.*, to reduce fossil-fuels based GHG emissions, such as CO_2, H_2O vapor, and CH_4 and related to the global warming of the Earth), many countries have set their target to achieve net zero emissions by 2050, although the obvious further trend is also to generally increase energy efficiency [6]. To meet these overall targets, versatile actions are needed in several different areas; for example, it is of importance to utilize effectively existing infrastructure for developing novel bioenergy productions.

The use of biofuels also often offers clear advantages, if both the production and consumption of energy are normally located in the same area [2, 6]. Hence, the most effective conversion route depends on regional possibilities and the availability of lignocellulosic resources. Additionally, the H_2O footprint per unit of bioenergy, besides the society's C footprint, has to be kept at a satisfactory level and the production should not be allowed to compete with that of food, if based on the same feedstock.

As shown in the previous chapters, lignocellulosic biomass materials can be converted, besides also into a variety of chemicals and other bioproducts, also into a great variety of energy products, such as heat, electricity, and transport fuels, through various thermochemical conversion processes [2, 3, 6, 7, 68, 209, 300, 301, 426, 431–441]. Traditionally, the thermochemical methods are roughly divided, according to the process conditions applied, into combustion, pyrolysis, gasification, and liquefaction. The choice of conversion process is dependent on several factors, including economics, the type of biomass feedstock, the desired form of bioenergy, and environmental aspects. Hence, besides the technical aspects, the economic feasibility of these approaches must also be considered. In this conclusion chapter, the aim is to briefly show only various possibilities of utilizing lignocellulosic biomass materials in energy-producing technologies.

In general, the term "biofuels" covers a wide range of fuels, including solid biomasses, liquid fuels, and different biogases. For example, due to the non-homogeneity and high H_2O content of lignocellulosic biomass feedstocks, they often need to be pretreated in some way prior to thermochemical conversion, although many of them already exist in a suitable form and are burned (>900 °C) directly to provide heat [5]. However, to exploit their best fuel properties, certain lignocellulosic resources, such as sawdust, shavings, bark, and harvesting residues, often have to be mechanically upgraded, besides, by drying, by grinding (*i.e.*, crushing and milling to an appropriate particle size distribution for obtaining "powdered fuels") and/or densified by pressing them into briquettes and pellets. Especially, the particle size distribution of raw materials is one of the most important parameters when evaluating their quality for different purposes and their thermochemical behavior.

It is also evident that the production of solid fuels from forest biomass has become more important [301]. Today, 50–55% of the total worldwide utilization of wood is related to its direct combustion for producing energy [2]. Wood-derived material is used in the forest industry for bioenergy production, either in the form of waste wood (*i.e.*, harvesting and thinning residues, sawdust, and bark) or black liquor. Additionally, power and heat plants as well as private households use woody biomass for energy purposes.

The structure of bark is somewhat different from that of wood, and differences between species are much greater in bark than in wood [2, 442, 443]. Bark residues are normally produced as waste by the traditional forest industry in large quantities during debarking and is combusted in a bark-burning furnace, often together with biosludge, for producing energy. Mainly because of this straightforward utilization, bark has found only limited high-volume chemical and other uses. Bark is obviously also being utilized as a raw material, for example, as a part of forest residues in many burning and gasification applications. Additionally, bark could be a potential raw material for vacuum pyrolysis, for example, for producing various phenols [444, 445] and activated carbon [446]. Furthermore, it has been used as such for many other purposes [443], including biochemicals, oil absorption, metals-binding materials, earth filling materials and agricultural soil conditioners, components in various special products, and odor removal in kraft mills (*i.e.*, as a biofilter grafting with suitable microbes) as well as bedding straw in animal husbandry.

Figure 7.13 summarizes the general conversion routes for producing energy and liquid fuels from lignocellulosic biomass materials. In each case, the most effective conversion route depends on the regional possibilities as well as the availability of feedstock resources.

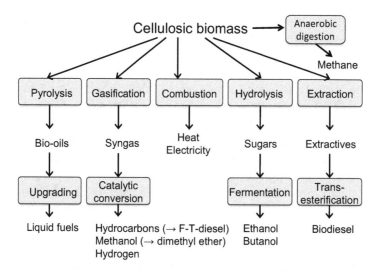

Figure 7.13: Principal conversion routes for lignocellulosic biomass materials to produce various energy sources. F-T refers to Fischer-Tropsch.

CH_3CH_2OH has traditionally been used as a fuel for vehicles in its pure form, but it is normally used as a petrol additive to increase the octane number and to curb vehicle emissions [2]. Biodiesel can also be used as such as a fuel for vehicles, although its main use is as a diesel additive. Biodiesel normally contains, compared to fossil diesel, less carbon and correspondingly, higher amounts of hydrogen and oxygen ("oxygenated fuel"). It is the most common biofuel in Europe, whereas CH_3CH_2OH is widely used, especially in Brazil and the United States.

Along with the established manufacturing of CH_3CH_2OH [447–451], production of $CH_3CH_2CH_2CH_2OH$ [2, 452, 453] may offer a promising full-scale way of converting cellulosic biomass into useful transportation fuel (see Section 3.3 "Use of carbohydrates in hydrolyzates"). $CH_3CH_2CH_2CH_2OH$ can be produced with anaerobic bacteria *Clostridium acetobutylicum* or *C. beijerinckii* from sugars in the ABE $CH_3COCH_3/CH_3CH_2CH_2CH_2OH/CH_3CH_2OH$) fermentation, developed by Azriel Weizmann, when producing CH_3COCH_3 that was important for British industry during World War I [454]. $CH_3CH_2CH_2CH_2OH$ is an excellent fuel extender, being, for example, superior to CH_3CH_2OH; it contains only 22% oxygen (in CH_3CH_2OH 35%), is able to tolerate H_2O contamination, and is easy to blend with petrol in higher concentrations without harm to engines. Following the recent progress in fermentation and separation techniques, this application is becoming more important [455]. Additionally, $CH_3CH_2CH_2CH_2OH$ can also be produced from syngas *via* fermentation, utilizing the strain of *Butyribacterium methylotrophicum*, which converts CO directly to $CH_3CH_2CH_2CH_2OH$, CH_3CH_2OH, butanoic acid ($CH_3CH_2CH_2CO_2H$), and CH_3CO_2H [456].

In addition to pyrolysis and gasification, there are processes where liquid fuels are directly obtained by liquefaction (*i.e.*, the BTL process) or biogases [2, 6, 457]. In naturally occurring ecosystems, organic matter is gradually decomposed by certain microorganisms under conditions ranging from O_2-limiting to O_2-free or to a CH_4-rich gas. Industrial biogas generation represents an interesting way of producing fuel from different wastes [458, 459]. Raw materials conventionally used have been agricultural, animal, or urban wastes. The anaerobic digestation conversion of cellulose-containing feedstocks proceeds roughly as a three-stage process, leading to the formation of CH_4, mixed with a substantial portion of CO_2. In the first stage of this process (*i.e.*, hydrolysis), the feedstock material is enzymatically decomposed to relatively low-molar-mass compounds, such as sugars and long-chain fatty acids (LCFA), which are then further broken down in the second stage (*i.e.*, acidogenesis) to CH_3CO_2H, H_2, and CO_2. In the last stage, CH_4 is formed by the action of methanogens that can also be found, for example, in wetlands (*i.e.*, the formation of march gas) and in the digestive tracts of animals, such as ruminants and humans. For example, biogas from sewage digesters normally contains 55% to 65% CH_4, 35% to 45% CO_2, and < 1% N_2.

Biogas obtained from energy or waste crops can be used as vehicle fuel in natural gas light-duty or heavy-duty vehicles because the same engine and vehicle configurations are used with both fuels [459]. When biogas is used as vehicle fuel, typical reductions in air emissions compared to those from diesel fuel are typically from 60% to 85% in NO_X, from 60% to 80% in particulates, and from 10% to 70% in CO. However, when biogas is used in vehicles, H_2S and halogenated compounds, normally found in biogas, can cause corrosion in engines. Primary methods of biogas refining include H_2O scrubbing and pressure swing adsorption, although more sophisticated techniques have also been developed. The use of this kind of biofuel offers a good local opportunity since both the production and consumption of energy can occur in the same area, unlike when using fossil fuels.

Various transesterification strategies include considerations to utilize, in addition to relatively pure vegetable oil source materials [198, 202, 203, 211], tall oil components from kraft pulping as a source of raw material for producing biodiesel or additives for petrodiesel [2, 213, 217, 460]. Additionally, one possibility is to apply fast pyrolysis to vegetable oil materials directly or after their alkaline hydrolysis for manufacturing liquid fuels (mainly biodiesel) [85, 209, 216]. In using tall oil soap, Na could be recovered simultaneously in the form of Na_2CO_3, thus eliminating, for example, the need to add sulfur-containing H_2SO_4 to liberate chemically bound Na from the $-CO_2Na$ groups of extractives.

Energy production should not be allowed to compete with that of food [6]; *i.e.*, the raw materials are clearly suitable for both purposes. Examples of such cases are the so-called "first-generation biofuels", such as CH_3CH_2OH obtained from maize and biodiesel obtained from vegetable oils after transesterification, *i.e.*, the production of fatty acid methyl esters (FAMEs). In contrast, the production of syngas-based biofuels ("second-generation biofuels"), including F-T diesel, CH_3OH, and DME as well as H_2, is under strong development to solve various technical problems in the corresponding production technologies.

One potential new source of energy is certain fast-growing algae (*i.e.*, up to 30 times more energy per area than land crops can be obtained), and the production of biodiesel ("algae fuel") from this feedstock ("farming algae") is currently attracting increasing interest [461, 462]. In general, photosynthetic algae, including microalgae (single celled algae) as well as cyanobacteria ("blue-green algae"), together with seaweeds, are effective to produce various lipids and long-chain hydrocarbons by utilizing sunlight energy [463].

References

[1]   Brown, R.C. 2006. Biomass refineries based on hybrid thermochemical-biological processing – an overview. In: Kamm, B., Gruber, P.R., and Kamm, M. (Eds.). *Biorefineries – Industrial Processes and Products, Volume 1*. Wiley-VCH, Weinheim, Germany. Pp. 227–252.

[2]   Alén, R. 2011. Principles of biorefining. In: Alén, R. (Ed.). *Biorefining of Forest Resources*. Paper Engineers' Association, Helsinki, Finland. Pp. 55–114.

[3]   Brown, R.C. (Ed.). 2011. *Thermochemical Processing of Biomass*. John Wiley & Sons, Chichester, United Kingdom. 348 p.

[4]   Konttinen, J., Reinikainen, M., Oasmaa, A., and Solantausta, Y. 2011. Thermochemical conversion of forest biomass. In: Alén, R. (Ed.). *Biorefining of Forest Resources*. Paper Engineers' Association, Helsinki, Finland. Pp. 262–304.

[5]   Alén, R. 2015. Pulp mills and wood-based biorefineries. In: Pandey, A., Höfer, R., Taherzadeh, M., Nampoothiri, K.M., and Larroche, C. (Eds.). *Industrial Biorefineries & White Biotechnology*. Elsevier, Amsterdam, The Netherlands. Pp. 91–126.

[6]   Alén, R. 2018. Production of biofuels. In: Alén, R. *Carbohydrate Chemistry – Fundamentals and Applications*. World Scientific, Singapore. Pp. 462–472.

[7]   Clark, J., and Deswarte, F. 2008. The biorefinery concept – an integrated approach. In: Clark, J.H., and Deswarte, F.E.I. (Eds.). *Introduction to Chemicals from Biomass*. John Wiley & Sons, Chichester, United Kingdom. Pp. 1–20.

[8]   Capolupo, L., and Faraco, V. 2016. Green methods of lignocellulose pretreatment for biorefinery development. *Applied Microbiology and Biotechnology* 100(22):9451–9467.

[9]   Elliott, D.C., Biller, P., and Ross, A.B. 2015. Hydrothermal liquefaction of biomass: Developments from batch to continuous process. *Bioresource Technology* 178:147–156.

[10]   Amini, E., Safdari, M.-S., DeYoung, J.T., Weise, D.R., and Fletcher, T.H. 2019. Characterization of pyrolysis products from slow pyrolysis of live and dead vegetation native to the southern united states. *Fuel* 235:1475–1491.

[11]   Al Arni, S. 2017. Comparison of slow and fast pyrolysis for converting biomass into fuel. *Journal of Renewable Energy* 124:1–5.

[12]   Dhyani, V., and Bhaskar, T. 2018. A comprehensive review on the pyrolysis of lignocellulosic biomass. *Journal of Renewable Energy* 129:695–716.

[13]   Piccolo, C., and Bezzo, F. 2009. A techno-economic comparison between two technologies for bioethanol production from lignocellulose. *Biomass and Bioenergy* 33:478–491.

[14]   Mohr, A., and Raman, S. 2013. Lessons from first generation biofuels and implications for the sustainability appraisal of second generation biofuels. *Energy Policy* 63:144–122.

[15]   Isikgor, F.H., and Becer, C.R. 2015. Lignocellulosic biomass: A sustainable platform for the production of bio-based chemicals and polymers. *Polymer Chemistry* 6:4497–4559.

[16]   Kan, T., Strezov, V., and Evans, T.J. 2016. Lignocellulosic biomass pyrolysis: A review of product properties an effects of pyrolysis parameters. *Renewable and Sustainable, Energy Reviews* 57:1126–1140.

[17]   Williams, P.T., and Besler, S. 1996. The influence of temperature and heating rate on the slow pyrolysis of biomass. *Renewable Energy* 7(3):233–250.

[18]   Bridgwater, A.V. 1999. Principles and practice of biomass fast pyrolysis process for liquids. *Journal of Analytical and Applied Pyrolysis* 51(1–2):3–12.

[19]   Tillman, D.A. 2000. Biomass cofiring: The technology, the experience, the combustion consequences. *Biomass and Bioenergy* 19(6):365–384.

[20]   Bridgwater, A.V. (Ed.). 2002. *Fast Pyrolysis of Biomass: A Handbook, Volume 2*. CPL Press, Newbury, United Kingdom. 426 p.

[21]   Demirbaş, A. 2007. Combustion systems for biomass fuel. *Energy Sources Part A* 29(4):303–312.

[22]   Basu, P. 2010. *Biomass Gasification and Pyrolysis: Practical Design and Theory*. Elsevier, Oxford, United Kingdom. 376 p.

[23]   Funke, A., and Ziegler, E. 2010. Hydrothermal carbonization of biomass: A summary and discussion of chemical mechanisms for process engineering. *Biofuels, Bioproducts and Biorefining* 4(2):160–177.

[24]   Libra, J.A., Ro, K.S., Kammann, C., Funke, A., Berge, N.D., Neubauer, Y., Titirici, M.-M., Füchner, C., Bens, O., Kern, J., and Emmerich, K.-H. 2010. Hydrothermal carbonization of biomass residuals: A comparative review of the chemistry, processes and applications of wet and dry pyrolysis. *Biofuels* 2(1):71–106.

[25]   Reza, M.T., Uddin, M.H., Lynam, J.G., and Coronella, C.J. 2014. Hydrothermal carbonization: Reaction chemistry and water balance. *Biomass Conversion and Biorefinery* 4(4):311–321.

[26]   Van der Stelt, M., Gerhauser, H., Kiel, J., and Ptasinski, K. 2010. Biomass upgrading by torrefaction for the production of biofuels: A review. *Biomass and Bioenergy* 35(9)3748–3762.

[27]   Chen, W.H., Peng, J., and Bi, X.T. 2015. A state-of-the-art review of biomass torrefaction, densification and applications. *Renewable and Sustainable Energy Reviews* 44:847–866.

[28]   Mendoza, C.L.M. 2021. *Assessment of Agro-Forest and Industrial Residues Potential as an Alternative Energy Source*. Doctoral Thesis. Laboratory of Sustainable Energy Systems, Lappeenranta-Lahti University of Technology LUT, Lappeenranta, Finland. 91 p.

[29]   Demirbaş, A., and Arin, G. 2002. An overview of biomass pyrolysis. *Energy Sources* 24:471–482.

[30]   Babu, B.V. 2008. Biomass pyrolysis: A state-of-the-art review. *Biofuels, Bioproducts and Biorefining* 2:393–414.

[31]   Bridgwater, A.V., and Peacocke, G.V.C. 2000. Fast pyrolysis processes for biomass. *Renewable and Sustainable Energy Reviews* 4:1–73.

[32]   Alén, R. 2018. Carbohydrate biosynthesis. In: Alén, R. *Carbohydrate Chemistry – Fundamentals and Applications*. World Scientific, Singapore. Pp. 131–146.

[33]   Piskorz, J., Radlein, D., Scott, D.S., and Czernik, S. 1989. Pretreatment of wood and cellulose for production of sugars by fast pyrolysis. *Journal of Analytical and Applied Pyrolysis* 16:127–142.

[34]   Shafizadeh, F., Furneaux, R.H., Cochran, T.G., Scholl, J.P., and Sakai, Y. 1979. Production of levoglucosan and glucose from pyrolysis of cellulosic materials. *Journal of Applied Polymer Science* 23:3525–3539.

[35]   Radlein, D. 2002. Study of levoglucosan production – a review. In: Bridgwater, A.V. (Ed.). *Fast Pyrolysis of Biomass: A Handbook, Volume 2*. CPL Press, Newbury, United Kingdom. Pp. 205–241.

[36]   de Wild, P. 2011. *Biomass Pyrolysis for Chemicals*. Doctoral Thesis. University of Groningen, The Netherlands. 163 p.

[37]   Salami, A. 2021. *Biorefining of Lignocellulosic Biomass and Chemical Characterization of Slow Pyrolysis Distillates*. Doctoral Thesis. University of Eastern Finland, Kuopio, Finland. 143 p.

[38]   Amidon, T. 2006. Forest biorefinery: A new business model. *Pulp and Paper Canada* 107(3):19.

[39]   Acharys, B., Sule, I., and Dutta, A. 2012. A review on advances of torrefaction technologies for biomass processing. *Biomass Conversion and Biorefinery* 2:349–369.

[40]   Chen, W.-H., and Kuo, P.-C. 2011. Isothermal torrefaction kinetics of hemicellulose, cellulose, lignin and xylan using thermogravimetric analysis. *Energy* 36:6451–6460.

[41]   Shoulaifar, T.K., DeMartini, N., Ivaska, A., Fardim, P., and Hupa, M. 2012. Measuring the concentration of carboxylic acid groups in torrefied spruce wood. *Bioresource Technology* 123:338–343.

[42]   Shoulaifar, T.K., DeMartini, N., Zevenhoven, M., Verhoeff, F., Kiel, J., and Hupa, M. 2013. Ash-forming matter in torrefied birch wood: Changes in chemical association. *Energy & Fuels* 27:5684–5690.

[43]   Kymäläinen, M., Havimo, M., Keriö, S., Kemell, M., and Solio, J. 2014. Biological degradation of torrefied wood and charcoal. *Biomass and Bioenergy* 71:170–177.

[44]   Fagernäs, L., Kuoppala, E., and Arpiainen, V. 2015. Composition, utilization and economic assessment of torrefaction condensates. *Energy & Fuels* 29:3134–3142.

[45]   Shoulaifar, T.K., DeMartini, N., Karlström, O., and Hupa, M. 2016. Impact of organically bonded potassium on torrefaction: Part 1. experimental. *Fuel* 165:544–552.

[46]   Shoulaifar, T.K., DeMartini, N., Karlström, O., Hemming, J., and Hupa, M. 2016. Impact of organically bonded potassium on torrefaction: Part 1. Modeling. *Fuel* 168:107–115.

[47]   Kotilainen, R. 2000. *Chemical Changes in Wood during Heating at 150–260 °C*. Doctoral Thesis. Laboratory of Applied Chemistry, University of Jyväskylä, Jyväskylä, Finland. p. 57.

[48]   Alén, R., Kotilainen, R., and Zaman, A. 2002. Thermochemical behaviour of norway spruce (*picea abies*) at 180–225 °C. *Wood Science and Technology* 36:163–171.

[49]   Arshadi, M., and Sellstedt, A. 2008. Production of energy from biomass. In: Clark, J.H., and Deswarte, F.E.I. (Eds.). *Introduction to Chemicals from Biomass*. Wiley & Sons, New York, NY, USA. Pp. 143–178.

[50]   Jenkins, B.M., Baxter, L.L., and Koppejan, J. 2011. Biomass combustion. In: Brown, R.C. (Ed.). *Thermochemical Processing of Biomass*. John Wiley & Sons, Chichester, United Kingdom. Pp. 13–46.

[51]   Soltes, E.J., and Elder, T.J. 1981. pyrolysis. In: Goldstein, I.S. (Ed.). *Organic Chemicals from Biomass*. CRC Press, Boca Raton, FL, USA. Pp. 63–99.

[52]   Shafizadeh, F. 1982. Introduction to pyrolysis of biomass. *Journal of Analytical and Applied Pyrolysis* 3:283–305.

[53]   Venderbosch, R.H., and Prins, W. 2011. Fast pyrolysis. In: Brown, R.C. (Ed.). *Thermochemical Processing of Biomass*. John Wiley & Sons, Chichester, United Kingdom. Pp. 124–156.

[54]   Vamvuka, D. 2011. Bio-oil, solid and gaseous biofuels – an overview. *International Journal of Energy Research* 35:835–862.

[55]   Wang, S., and Luo, Z. 2017. *Pyrolysis of Biomass*. Walter de Gruyter, Berlin, Germany. 268 p.

[56]   Meier, D., and Faix, O. 1999. State of the art of applied fast pyrolysis of lignocellulosic materials – a review. *Bioresource Technology* 68:71–77.

[57]   Demirbaş, A. 2009. Pyrolysis of biomass for fuels and chemicals. *Energy Sources, Part A* 31(12): 1028–1037.

[58]   Shen, D., Xiao, R., Gu, S., and Zhang, H. 2013. The overview of thermal decomposition of cellulose in lignocellulosic biomass. In: Kadla, J., and Van De Ven, T.G.M. (Eds.). *Cellulose – Biomass Conversion*. Intech Open Book Series. Pp. 193–226.

[59]   Di Blasi, C. 1996. Kinetic and heat transfer control in the slow and flash pyrolysis of solids. *Industrial and Engineering Chemistry Research* 35:37–46.

[60]   Fisher, T., Hajaligol, M., Waymack, B., and Kellog, D. 2002. Pyrolysis behavior and kinetics of biomass derived materials. *Journal of Analytical and Applied Pyrolysis* 62:331–349.

[61]   Babu, B.V., and Chaurasia, A.S. 2003. Modeling for pyrolysis of solid particle: Kinetics and heat transfer effects. *Energy Convers Management* 44:2251–2275.

[62]   Babu, B.V., and Chaurasia, A.S. 2004. Pyrolysis of biomass: Improved models for simultaneous kinetics & transport of heat, mass, and momentum. *Energy Conversion and Management* 45:1297–1327.

[63]   Babu, B.V., and Chaurasia, A.S. 2004. Heat transfer and kinetics in the pyrolysis of shrinking biomass particle. *Chemical Engineering Science* 59:1999–2012.

[64]   Tanoue, K.-I., Hinauchi, T., Oo, T., Nishimura, T., Taniguchi, M., and Sasauchi, K.-I. 2007. Modeling of heterogeneous chemical reactions caused in pyrolysis of biomass particles. *Advanced Powder Technology* 18(6):825–840.

[65]   Di Blasi, C. 2008. Modeling chemical and physical processes of wood and biomass pyrolysis. *Progress in Energy and Combustion Science* 34:47–90.

[66]   White, J.E., Catallo, W.J., and Legendre, B.L. 2011. Biomass pyrolysis kinetics: A comparative critical review with relevant agricultural residue case studies. *Journal of Analytical and Applied Pyrolysis* 91:1–33.

[67]   Wang, S., Dai, G., Yang, H., and Luo, Z. 2017. Lignocellulosic biomass pyrolysis mechanism: A state-of-the art review. *Progress in Energy and Combustion Science* 62:33–86.

[68]   Alén, R. 1990. Conversion of cellulose-containing materials into useful products. In: Kennedy, J.F., Phillips, G.O., and Williams, P.A. (Eds.). *Cellulose Sources and Exploitation – Industrial Utilization, Biotechnology, and Physico-Chemical Properties*. Ellis Horwood, Chichester, England. Pp. 453–464.

[69]   Bridgwater, A.V., Meier, D., and Radlein, D. 1999. An overview of fast pyrolysis of biomass. *Organic Geochemistry* 30:1479–1493.

[70]   Bridgwater, A.V., Czernik, S., and Piskorz, J. 2002. The status of biomass fast pyrolysis. Bridgwater, A.V. (Ed.). 2002. *Fast Pyrolysis of Biomass: A Handbook, Volume 2*. CPL Press, Newbury, United Kingdom. Pp. 1–22.

[71]   Ghalibaf, M. 2019. *Analytical Pyrolysis of Wood and Non-Wood Materials from Integrated Biorefinery Concepts*. Doctoral Thesis. University of Jyväskylä, Laboratory of Applied Chemistry, Jyväskylä, Finland. 106 p.

[72]   Alén, R., Kuoppala, E., and Oesch, P. 1996. Formation of the main degradation compound groups from wood and its components during pyrolysis. *Journal of Analytical and Applied Pyrolysis* 36:137–148.

[73]   Balat, M. 2008. Mechanisms of thermochemical biomass conversion processes. Part 1: Reactions of pyrolysis. *Energy Sources, Part A* 30(7):620–635.

[74]   Moldoveanu, S.C. 2010. The chemistry of the pyrolytic process. In: Moldoveanu, S.C. (Ed.). *Techniques and Instrumentation in Analytical Chemistry*. Elsevier, Amsterdam, The Netherlands. Pp. 7–48.

[75]   Van de Velden, M., Baeyens, J., Brems, A., Janssens, B., and Devil, R. 2010. Fundamentals, kinetics and endothermicity of the biomass pyrolysis reaction. *Renewable Energy* 35(1)232–242.

[76]   Collard, F.X., Blin, J., Bensakhria, A., and Valette, J. 2012. Influence of impregnated metal on the pyrolysis conversion of biomass constituents. *Journal of Analytical and Applied Pyrolysis* 95:213–226.

[77]   Morf, P., Hasler, P., and Nussbaumer, T. 2002. Mechanisms and kinetics of homogeneous secondary reactions of tar from continuous pyrolysis of wood chips. *Fuel* 81(7):843–853.

[78]   Wei, L., Xu, S., Zhang, L., Liu, C., Zhu, H., and Liu, S. 2006. Characteristics of fast pyrolysis of biomass in a free fall reactor. *Fuel Processing Technology* 87(10):863–871.

[79]   Neves, D., Thunman, H., Matos, A., Tarelho, L., and Gomez-Barea, A. 2011. Characterization and prediction of biomass pyrolysis products. *Progress in Energy and Combustion Science* 37(5)611–630.

[80]   Evans, R.J., and Milne, T.A. 1987. Molecular characterization of pyrolysis of biomass. 1. Fundamentals. *Energy & Fuels* 1(2):123–138.

[81]   Blanco López, M.C., Blanco, C.G., Martínez-Alonso, A., and Tascón, J.M.D. 2002. Composition of gases released during olive stone pyrolysis. *Journal of Analytical and Applied Pyrolysis* 65(2):313–322.

[82]   Carey, F.A. 2000. Free-radical reactions. In: Carey, F.A., and Sundberg, R.J. (Eds.). *Advanced Organic Chemistry, Part A: Structure and Mechanisms*. 4th edition. Kluwer Academic/Plenum Publishers, New York, NY, USA. Pp. 663–734.

[83]   Poutsama, M.L. 2000. Fundamental reactions of free radicals relevant to pyrolysis reactions. *Journal of Analytical and Applied Pyrolysis* 54(1–2):5–35.

[84]   Savage, P.E. 2000. Mechanisms and kinetics models for hydrocarbon pyrolysis. *Journal of Analytical and Applied Pyrolysis* 54(1–2):109–126.

[85]   Lappi, H. 2012. *Production of Hydrocarbon-Rich Biofuels from Extractives-Derived Materials*. Doctoral Thesis. University of Jyväskylä, Laboratory of Applied Chemistry, Jyväskylä, Finland. 111 p.

[86]   Nguyen, T., Zavarin, E., and Barrall II, E.M. 1981. Thermal analysis of lignocellulosic materials. Part I. unmodified materials. *Journal of Macromolecular Science – Reviews in Macromolecular Chemistry* C20(1):1–65.

[87]   Brown, M.E. (Ed.). 2001. *Introduction to Thermal Analysis*. Kluwer Academic Publishers, New York, NY, USA. 264 p.

[88]   Fedelich, N. 2019. *Evolved Gas Analysis – Hyphenated Techniques*. Carl Hanser Verlag, Munich, Germany. 214 p.

[89]   Shen, D.K., Gu, S., and Bridgwater, A.V. 2010. The thermal performance of the polysaccharides extracted from hardwood: Cellulose and hemicellulose. *Carbohydrate Polymers* 82:39–45.

[90]   Svenson, J., Pettersson, J.B.C., and Davidsson, K.O. 2004. Fast pyrolysis of the main components of birch wood. *Combustion Science and Technology* 176(5,6):977–990.

[91]   Yang, H., Yan, R., Chen, H., Zheng, C., Lee, D.H., and Liang, D.T. 2006. In-depth investigation of biomass pyrolysis based on three major components: Hemicellulose, cellulose and lignin. *Energy & Fuels* 20(1):388–393.

[92]   Biagini, E., Barontini, F., and Tognotti, L. 2006. Devolatilization of biomass fuels and biomass components studied by TG/FTIR technique. *Industrial & Engineering Chemistry Research* 45(13):4486–4493.

[93] Liu, Q., Zhong, Z., Wang, S., and Luo, Z. 2011. Interactions of biomass components during pyrolysis: A TG-FTIR study. *Journal of Analytical and Applied Pyrolysis* 90:213–218.

[94] Elder, T. 1991. Pyrolysis of wood. In: Hon, D.N.-S., and Shiraishi, N. (Eds.). *Wood and Cellulose Chemistry*. Marcel Dekker, New York, NY, USA. Pp. 665–692.

[95] Alén, R., Rytkönen, S., and McKeough, P. 1995. Thermogravimetric behavior of black liquors and their organic constituents. *Journal of Analytical and Applied Pyrolysis* 31(C):1–13.

[96] Faravelli, T., Frassoldati, A., Migliavacca, G., and Ranzi, E. 2010. Detailed kinetic modeling of the thermal degradation. *Biomass and Bioenergy* 34:290–301.

[97] Chen, W.H., and Kuo, P.C. 2011. Isothermal torrefaction kinetics of hemicellulose, cellulose, lignin and xylan using thermogravimetric analysis. *Energy* 36(11):6451–6460.

[98] Dhyani, V., and Bhaskar, T. 2017. A comprehensive review on the pyrolysis of lignocellulosic biomass. *Renewable Energy* 245:1–22.

[99] Moriana, R., Zhang, Y., Mischnick, P., Li, J., and Ek, M. 2014. Thermal degradation behavior and kinetic analysis of spruce glucomannan and its methylated derivatives. *Carbohydrate Polymers* 106(1):60–70.

[100] Stefanidis, S.D., Kalogiannis, K.G., Iliopoulou, E.F., Michailof, C.M., Pilavachi, P.A., and Lappas, A.A. 2014. A study of lignocellulosic biomass pyrolysis via the pyrolysis of cellulose, hemicellulose and lignin. *Journal of Analytical and Applied Pyrolysis* 105:143–150.

[101] Alén, R. 2000. Structure and chemical composition of wood. In: Stenius, P. (Ed.). *Forest Products Chemistry*. Fapet, Helsinki, Finland. Pp. 11–57.

[102] Patwardhan, P.D., Brown, R.C., and Shanks, B.H. 2011. Product distribution from the fast pyrolysis of hemicellulose. chemistry and sustainability, *Energy and Materials* 4(5):636–643.

[103] Williams, P.T., and Besler, S. 1993. Thermogravimetric analysis of components of biomass. In: Bridgwater, A.T. (Ed.). *Advances in Thermochemical Biomass Conversion. Vol. 2. Pyrolysis*. Springer, Dordrecht, The Netherlands. Pp. 771–783.

[104] Werner, K., Pommer, L., and Broström, M. 2014. Thermal decomposition of hemicelluloses. *Journal of Analytical and Applied Pyrolysis* 110:130–137.

[105] Antal, M.J. Jr., and Varhegyi, G. 1995. Cellulose pyrolysis kinetics: The current state of knowledge. *Industrial & Engineering Chemistry Research* 34:703–717.

[106] Liao, Y.-f., Wang, S.-r., and Ma, X.-q. 2004. Study of reaction mechanisms in cellulose pyrolysis. *Preprints of Papers – American Chemical Society, Division of Fuel Chemistry* 49(1):407–411.

[107] Lédé, J. 2012. Cellulose pyrolysis kinetics: An historic review on the existence and role of intermediate active cellulose. *Journal of Analytical and Applied Pyrolysis* 94:17–32.

[108] Bradbury, A.G.W., Sakai, Y., and Shafizadeh, F. 1979. A kinetic model for pyrolysis of cellulose. *Journal of Analytical and Applied Pyrolysis* 23(11):3271–3280.

[109] Liu, Q., Wang, S.R., Wang, K.G., Guo, X.J., Luo, Z.Y., and Cen, K.F. 2008. Mechanism of formation and consequents evolution of active cellulose during cellulose pyrolysis. *Acta Physico-Chimica Sinica* 24(11):1957–1963.

[110] Di Blasi, C., and Lanzetta, M. 1997. Intrinsic kinetics of isothermal xylan degradation in inert atmosphere. *Journal of Analytical and Applied Pyrolysis* 40,41:287–303.

[111] Broido, A., and Weinstein, M. 1971. Kinetics of solid-phase cellulose pyrolysis. In: Wiedemann, H.G. (Ed.). *Proceedings of 3rd International Conference on Thermal Analysis*. Birkhauser Verlag, Basel, Switzerland. Pp. 285–296.

[112] Broido, A. 1976. Kinetics of solid phase cellulose pyrolysis. In: Shafizadeh, F., Sarkanen, K., and Tillman, D.A. (Eds.). *Thermal Uses and Properties of Carbohydrates and Lignins*. Academic Press, New York, NY, USA. Pp. 19–36.

[113] Várhegyi, G., Antal, M.J. Jr., Jakab, E., and Szabó, P. 1997. Kinetic modeling of biomass pyrolysis. *Journal of Analytical and Applied Pyrolysis* 42:73–87.

[114] Órfão, J.J.M., Antunes, F.J.A., and Figueiredo, J.L. 1999. Pyrolysis kinetics of lignocellulosic materials – three independent reactions model. *Fuel* 78:349–358.

[115] Branca, C., Di Blasi, C., Mango, C., and Hrablay, I. 2013. Products and kinetics of glucomannan pyrolysis. *Industrial & Engineering Chemistry Research* 52(14):5030–5039.

[116] Cho, J., Chu, S., Dauenhauer, P.J., and Huber, G.W. 2012. Kinetics and reaction chemistry for slow pyrolysis of enzymatic hydrolysis lignin and organosolv extracted lignin derived from maplewood. *Green Chemistry* 14(2):428–439.

[117] Adam, M., Ocone, R., Mohammad, J., Berruti, F., and Briens, C. 2013. Kinetic investigations of kraft lignin pyrolysis. *Industrial & Engineering Chemistry Research* 52(26):8645–8654.

[118] Koufopanos, C.A., Papayannakos, N., Maschio, G., and Lucchesi, A. 1991. Modeling of the pyrolysis of biomass particles – studies on kinetics, thermal and heat-transfer effects. *Canadian Journal of Chemical Engineering* 69:907–915.

[119] Stenseng, M., Jensen, A., and Dam-Johansen, K. 2001. Investigation of biomass pyrolysis by thermogravimetric analysis and differential scanning calorimetry. *Journal of Analytical and Applied Pyrolysis* 58,59:765–780.

[120] Fisher, T., Hajaligol, M., Waymack, B., and Kellogg, D. 2002. Pyrolysis behavior and kinetics of biomass derived materials. *Journal of Analytical and Applied Pyrolysis* 62:331–349.

[121] Smolders, K., and Baeyens, J. 2004. Thermal degradation of PMMA in fluidized beds. *Waste Manag* 24:849–857.

[122] Luo, Z.Y., Wang, Z.R., and Cen, K.F. 2005. A model of wood flash pyrolysis in fluidized bed reactor. *Renewable Energy* 30:377–392.

[123] Cai, J., Wu, W., Liu, R., and Huber, G.W. 2013. A distributed activation energy model for the pyrolysis of lignocellulosic biomass. *Green Chemistry* 15:1331–1340.

[124] Caballero, J.A., Font, R., and Marcilla, A. 1996. Study of the primary pyrolysis of kraft lignin at high heating rates: Yields and kinetics. *Journal of Analytical and Applied Pyrolysis* 36:159–178.

[125] Ferdous, D., Dalai, A.K., Bej, S.K., and Thring, R.W. 2002. Pyrolysis of lignins: Experimental and kinetic studies. *Energy & Fuels* 16(6):1405–1412.

[126] Collard, F.-X., and Blin, J. 2014. A review on pyrolysis of biomass constituents: Mechanism and composition of the products obtained from the conversion of cellulose, hemicelluloses and lignin. *Renewable and Sustainable Energy Reviews* 38:594–608.

[127] Shafizadeh, F., and Sekiguchi, Y. 1983. Development of aromaticity in cellulosic chars. *Carbon* 21(5):511–516.

[128] Boon, J.J., Pastorova, I., Botto, R.E., and Arisz, P.W. 1994. Structural studies on cellulose pyrolysis and cellulose chars by pyms, pygcms, FTIR, NMR and by wet chemical techniques. *Biomass and Bioenergy* 7(1–6):25–32.

[129] Milosavljevic, I., Oja, V., and Suuberg, E.M. 1996. Thermal effects in cellulose pyrolysis: Relationship to char formation processes. *Industrial & Engineering Chemistry Research* 35:653–662.

[130] Arseneau, D.F. 1970. Competitive reactions in the thermal decomposition of cellulose. *Canadian Journal of Chemistry* 49:632–638.

[131] Piskorz, J., Radlein, D., and Scott, D.S. 1986. On the mechanism of the rapid pyrolysis of cellulose. *Journal of Analytical and Applied Pyrolysis* 9(2):121–137.

[132] Boukis, L.P. 1997. *Fast Pyrolysis of Biomass in a Circulating Fluidized Bed Reactor*. Doctoral Thesis. University of Aston, Birmingham, United Kingdom.

[133] Richards, G.N., and Zheng, G. 1991. Influence of metal ions and of salts on products from pyrolysis of wood: Applications to thermochemical processing of newsprint and biomass. *Journal of Analytical and Applied Pyrolysis* 21(1,2):133–146.

[134] Kawamoto, H., Morisaki, H., and Saka, S. 2009. Secondary composition of levoglucosan in pyrolytic production from cellulose biomass. *Journal of Analytical and Applied Pyrolysis* 85(1,2):247–251.

[135] Luo, Z., Wang, S., Liao, Y., and Cen, K. 2004. Mechanism study of cellulose rapid pyrolysis. *Industrial & Engineering Chemistry Research* 43(18):5605–5610.

[136] Drummond, A.R.F., and Drummond, I.W. 1996. Pyrolysis of sugar cane bagasse in a wire-mesh reactor. *Industrial & Engineering Chemistry Research* 35(4):1263–1268.

[137] Wang, S., Guo, X., Liang, T., Zhou, Y., and Luo, Z. 2012. Mechanism research on cellulose pyrolysis by Py-GC/MS and subsequent density functional theory studies. *Bioresource Technology* 104:722–728.

[138] Nowakowski, D.J., and Jones, J.M. 2008. Uncatalysed and potassium-catalysed pyrolysis of the cell-wall constituents of biomass and their model compounds. *Journal of Analytical and Applied Pyrolysis* 83:12–25.

[139] Ponder, G.R., and Richards, G.N. 1994. A review of some recent studies on mechanisms of pyrolysis of polysaccharides. *Biomass and Bioenergy* 7(1–6):1–24.

[140] Scheirs, J., Camino, G., and Tumiatti, W. 2001. Overview of water evolution during the thermal degradation of cellulose. *European Polymer Journal* 37(5):933–942.

[141] Shin, E.J., Nimlos, M.R., and Evans, R.J. 2001. Kinetic analysis of the gas-phase pyrolysis of carbohydrates. *Fuel* 80(12):1697–1709.

[142] Lin, Y.C., Cho, J., Tompsett, G.A., Westmoreland, P.R., and Huber, G.W. 2009. Kinetics and mechanism of cellulose pyrolysis. *The Journal of Physical Chemistry* 113(46):20097–20107.

[143] Shen, D., and Gu, S. 2009. The mechanism for thermal decomposition of cellulose and its main products. *Bioresource Technology* 100(24):6496–6504.

[144] Lu, R., Sheng, G.P., Hu, Y.Y., Zheng, P., Jiang, H., Tang, Y., and Yu, H.Q. 2011. Fractional characterization of bio-oil derived from rice husk. *Biomass Bioenergy* 35(1):671–678.

[145] Räisänen, U., Pitkänen, I., Halttunen, H., and Hurtta, M. 2003. Formation of the main degradation compounds from arabinose, xylose, mannose and arabinitol during pyrolysis. *Journal of Thermal Analysis and Calorimetry* 72(2):481–488.

[146] Paine III, J.B., Pithawalla, Y.B., Naworal, J.D., and Thomas, C.E. Jr. 2007. Carbohydrate pyrolysis mechanisms from isotopic labeling. Part 1: The pyrolysis of glycerin: Discovery of competing fragmentation mechanisms affording acetaldehyde and formaldehyde and the implications for carbohydrate pyrolysis. *Journal of Analytical and Applied Pyrolysis* 80:297–311.

[147] Paine III, J.B., Pithawalla, Y.B., and Naworal, J.D. 2008. Carbohydrate pyrolysis mechanisms from isotopic labeling. Part 2. The pyrolysis of D-glucose: General disconnective analysis and the formation of C_1 and C_2 carbonyl compounds by electrocyclic fragmentation mechanisms. *Journal of Analytical and Applied Pyrolysis* 82:10–41.

[148] Paine III, J.B., Pithawalla, Y.B., and Naworal, J.D. 2008. Carbohydrate pyrolysis mechanisms from isotopic labeling. Part 3. The pyrolysis of D-glucose: Formation of C_3 and C_4 carbonyl compounds and a cyclopentenedione isomer by electrocyclic fragmentation mechanisms. *Journal of Analytical and Applied Pyrolysis* 82:42–69.

[149] Paine III, J.B., Pithawalla, Y.B., and Naworal, J.D. 2008. Carbohydrate pyrolysis mechanisms from isotopic labeling. Part 4. The formation of furans. *Journal of Analytical and Applied Pyrolysis* 83:37–63.

[150] Alén, R. 2018. Polysaccharides and their derivatives. In: Alén, R. *Carbohydrate Chemistry – Fundamentals and Applications*. World Scientific, Singapore. Pp. 280–341.

[151] Shafizadeh, F., McGinnis, C.D., and Philpot, C.W. 1972. Thermal degradation of xylan and related model compounds. *Carbohydrate Research* 25:23–33.

[152] Šimkovic, L., Varhegyi, G., Antal, M.J. Jr., Ebringerová, A., Szekely, T., and Szabo, P. 1988. Thermogravimetric/mass spectrometric characterization of the thermal decomposition of (4-*O*-methyl-D-glucurono)-D-xylan. *Journal of Applied Polymer Science* 36(3):721–728.

[153] Yang, H., Yan, R., Chen, H., Lee, D.H., and Zheng, C. 2007. Characteristics of hemicellulose, cellulose and lignin pyrolysis. *Fuel* 86:1781–1788.

[154] Zhu, X., and Lu, Q. 2010. Production of chemicals from selective fast pyrolysis of biomass. In: Momba, M.N.B., and Bux, F. (Eds.). *Biomass*. Sciyo, Croatia. Pp. 147–164.

[155] Di Blasi, C., Branca, C., and Galgano, A. 2010. Biomass screening for the production of furfural via thermal decomposition. *Industrial & Engineering Chemistry Research* 49(6):2658–2671.

[156] Prins, M.J., Ptasinski, K.J., and Janssen, F.J.J.G. 2006. Torrefaction of wood. Part 2. analysis of products. *Journal of Analytical and Applied Pyrolysis* 77(1):35–40.

[157] Wu, Y., Zhao, Z., Li, H., and He, F. 2009. Low temperature pyrolysis characteristics of major components of biomass. *Journal of Fuel Chemistry and Technology* 37(4):427–432.

[158] Peng, Y., and Wu, S. 2010. The structural and thermal characteristics of wheat straw hemicellulose. *Journal of Applied Polymer Science* 88(2):134–139.

[159] Dong, C., Zhang, Z., Lu, Q., and Yang, Y. 2012. Characteristics and mechanism study of analytical fast pyrolysis of poplar wood. *Energy Conversion and Management* 57:49–59.

[160] Ohnishi, A., Katō, K., and Takagi, E. 1977. Pyrolytic formation of 3-hydroxy-2-penteno-1,5-lactone from xylan, xylo-oligosaccharides, and methyl xylopyranosides. *Carbohydrate Research* 58(2):387–395.

[161] Bocchini, P., Galletti, G.C., Camarero, S., and Martinez, A.T. 1997. Absolute quantitation of lignin pyrolysis products using an internal standard. *J Chromatography A* 773:227–232.

[162] Iatridis, B., and Gavalas, G.R. 1979. Pyrolysis of a precipitated kraft lignin. *Industrial & Engineering Chemistry Product Research and Development* 18(2):127–130.

[163] Petrocelli, F.P., and Klein, M.T. 1984. Model reaction pathways in kraft lignin pyrolysis. *Macromolecules* 17:161–169.

[164] Gardner, D.J., Schultz, T.P., and McGinnis, G.D. 1985. The pyrolytic behavior of selected lignin preparations. *Journal of Wood Chemistry and Technology* 5(1):85–110.

[165] Goheen, D.W. 1971. Low molecular weight chemicals. In: Sarkanen, K.V., and Ludwig, C.H. (Eds.). *Lignins – Occurrence, Formation, Structure and Reactions*. Wiley-Interscience, New York, NY, USA. Pp. 797–831.

[166] Brodin, I., Sjöholm, E., and Gellerstedt, G. 2010. Carbon fibres from kraft lignin. *Journal of Analytical and Applied Pyrolysis* 87:70–77.

[167] Azadi, P., Inderwildi, O.R., Farnood, R., and King, D.A. 2013. Liquid fuels, hydrogen and chemicals from lignin: A critical review. *Renewable and Sustainable Energy Reviews* 21:505–523.

[168] Gardner, D.J., Schultz, T.P., and McGinnis, G.D. 1985. The pyrolytic behavior of selected lignin preparations. *Journal of Wood Chemistry and Technology* 5(1):85–110.

[169] Avni, E., Coughlin, R.W., Solomon, P.R., and King, H.H. 1983. Lignin pyrolysis in heated grid apparatus: Experiment and theory. In: Scott, P.C., Ratcliffe, C.T., and Radding, S.B. (Eds.). *American Chemical Society – Division of Fuel Chemistry* 28(5):307–318.

[170] Liu, Q., Wang, S., Zhen, Y., Luo, Z., and Cen, K. 2008. Mechanism study on wood lignin pyrolysis by using TG-FTIR analysis. *Journal of Analytical and Applied Pyrolysis* 82:170–177.

[171] Jiang, G., Nowakowski, D.J., and Bridgwater, A.V. 2010. Effect of the temperature on the composition of lignin pyrolysis products. *Energy & Fuels* 24:4470–4475.

[172] Beis, S., Mukkamala, S., Hill, N., Joseph, J., Baker, C., Jensen, B., Stemmler, E., Wheeler, C., Frederick, B., and van Heiningen, A. 2010. Fast pyrolysis of lignins. *BioResources* 5(3):1408–1424.

[173] Zhang, M., Resende, F.L.P., and Moutsoglou, A. 2014. Catalytic fast pyrolysis of aspen lignin via Py-GC/MS. *Fuel* 116:358–369.

[174] Wang, S., Lin, H., Ru, B., Sun, W., Wang, Y., and Luo, Z. 2014. Comparison of the pyrolysis behavior of pyrolytic lignin and milled wood lignin by using TG-FTIR analysis. *Journal of Analytical and Applied Pyrolysis* 108:78–85.

[175] Brunow, G., Lundquist, K., and Gellerstedt, G. 1999. Lignin. In: Sjöström, E., and Alén, R. (Eds.). *Analytical Methods in Wood Chemistry, Pulping, and Papermaking*. Springer, Heidelberg, Germany. Pp. 77–124.

[176] Amen-Chen, C., Pakdel, H., and Roy, C. 2001. Production of monomeric phenols by thermochemical conversion of biomass: A review. *Bioresource Technology* 79(3):277–299.

[177] Klein, M.T., and Virk, P.S. 2008. Modeling of lignin thermolysis. *Energy & Fuels* 22:2175–2182.

[178] Kawamoto, H., Ryoritani, M., and Saka, S. 2008. Different pyrolytic cleavage mechanisms of β-ether bond depending on the side-chain structure of lignin dimers. *Journal of Analytical and Applied Pyrolysis* 81:88–94.

[179] Liu, J.-Y., Wu, S.-B., and Lou, R. 2011. Chemical structure and pyrolysis response of β-O-4 lignin model polymer. *BioResources* 6(2):1079–1093.

[180] Brebu, M., and Vasile, C. 2010. Thermal degradation of lignin – a review. *Cellulose Chemistry and Technology* 44(9):353–363.

[181] Vuori, A.I., and Bredenberg, J.B. 1987. Thermal chemistry pathways of substituted anisols. *Industrial & Engineering Chemistry Research* 26(2):359–365.

[182] Hosoya, T., Kawamoto, H., and Saka, S. 2008. Secondary reactions of lignin-derived primary tar components. *Journal of Analytical and Applied Pyrolysis* 83(1):78–87.

[183] Goldstein, I.S. 1975. Perspectives on production of phenols and phenolic acids from lignin and bark. *Applied Polymer Symposium* 28:259–267.

[184] Bai, X., Kim, K.H., Brown, R.C., Dalluge, E., Hutchinson, C., Lee, Y.J., and Dalluge, D. 2014. Formation of phenolic oligomers during fast pyrolysis of lignin. *Fuel* 128:170–179.

[185] Kotake, T., Kawamoto, H., and Saka, S. 2014. Mechanisms for the formation of monomers and oligomers during the pyrolysis of a softwood lignin. *Journal of Analytical and Applied Pyrolysis* 105:309–316.

[186] Zhao, X., and Liu, D. 2010. Chemical and thermal characteristics of lignins isolated from siam weed stem by acetic acid and formic acid delignification. *Industrial Crops and Products* 32(3):284–291.

[187] de Wild, P.J., Huijgen, W.J.J., and Heeres, H.J. 2012. Pyrolysis of wheat straw-derived organosolv lignin. *Journal of Analytical and Applied Pyrolysis* 93:95–103.

[188] Bridgwater, A.V. 2011. Upgrading fast pyrolysis liquids. In: Brown, R.C. (Ed.). *Thermochemical Processing of Biomass*. John Wiley & Sons, Chichester, United Kingdom. Pp. 157–199.

[189] Ben, H. 2014. Upgrade of bio-oil to bio-fuel and bio-chemical. In: Ragauskas, A.J. (Ed.). *Materials for Biofuels*. World Scientific, Singapore. Pp. 229–266 190.

[190] Luo, Z., Wang, S., and Guo, X. 2012. Selective pyrolysis of organosolv lignin over zeolites with product analysis by TG-FTIR. *Journal of Analytical and Applied Pyrolysis* 95:112–117.

[191] Jackson, M.A., Compton, D.L., and Boateng, A.A. 2009. Screening heterogeneous catalysts for the pyrolysis of lignin. *Journal of Analytical and Applied Pyrolysis* 85:226–230.

[192] Ma, Z., Troussard, E., and van Bokhoven, J.A. 2012. Controlling the selectivity to chemicals from lignin via catalytic fast pyrolysis. *Applied Catalysts A: General* 423,424:130–136.

[193] Peng, C., Zhang, G., Yue, J., and Xy, G. 2014. Pyrolysis of lignin for phenols with alkaline additive. *Fuel Processing Technology* 124:212–221.

[194] Nair, V., and Vinu, R. 2016. Production of guaiacols via catalytic fast pyrolysis of alkali lignin using titania, zirconia and ceria. *Journal of Analytical and Applied Pyrolysis* 119:31–39.

[195] Alén, R. 2011. Structure and chemical composition of biomass feedstocks. In: Alén, R. (Ed.). *Biorefining of Forest Resources*. Paper Engineers' Association, Helsinki, Finland. Pp. 17–54.

[196] Mészáros, E., Jakab, E., and Várhegyi, G. 2007. TG/MS, Py-GC/MS and THM-GC/MS study of the composition and thermal behavior of extractive components of *robinia pseudoacacia*. *Journal of Analytical and Applied Pyrolysis* 79:61–70.

[197] Artok, L., and Schobert, H.H. 2000. Reaction of carboxylic acids under coal liquefaction conditions. 1. Under nitrogen atmosphere. *Journal of Analytical and Applied Pyrolysis* 54(1):215–233.

[198] Huber, G.W., Iborra, S., and Corma, A. 2006. Synthesis of transportation fuels from biomass: Chemistry, catalysts, and engineering. *Chemical Reviews* 106:4044–4098.

[199] Doll, K.M., Sharma, B.K., Suaret, P.A., and Erhan, S.Z. 2008. Comparing biofuels obtained from pyrolysis, of soybean oil or soapstock, with traditional soybean biodiesel: Density, kinematic viscosity, and surface tensions. *Energy & Fuels* 22:2061–2066.

[200] Junming, X., Jianchun, J., Yanju, L., and Jie, C. 2009. Liquid hydrocarbon fuels obtained by the pyrolysis of soybean oils. *Bioresource Technology* 100(20):4867–4870.

[201] Pandey, A. (Ed.). 2009. *Handbook of Plant-based Biofuels*. CRC Press, Boca Raton, FL, USA. 297 p.

[202] Christopher, L.P., Kumar, H., and Zambare, V.P. 2014. Enzymatic biodiesel: Challenges and opportunities. *Applied Energy* 119:497–520.

[203] Christopher, L.P., and Kumar, H. 2015. Clean and sustainable biodiesel production. In: Yan, J. (Ed.). *Handbook of Clean Energy Systems*. John Wiley & Sons. Chichester, United Kingdom. Pp. 1–16.

[204] Kitamura, K. 1971. Studies of the pyrolysis of triglycerides. *Bulletin of the Chemical Society of Japan* 44:1606–1609.

[205] Irwin, W.J. 1993. *Analytical Pyrolysis: A Comprehensive Guide*. Marcel Dekker, New York, NY, USA. 600 p.

[206] Pakdel, H., Zhang, H.G., and Roy, C. 1993. Detailed chemical characterization of biomass pyrolysis oils, polar fractions. In: Bridgwater, A.V. (Ed.). *Advances in Thermochemical Biomass Conversion*. 2nd edition. Blackie Academic and Professional, London, United Kingdom. Pp. 1068–1085.

[207] Pakdel, H., Zhang, H.G., and Roy, C. 1994. Production and characterization of carboxylic acids from wood, Part II: High molecular weight fatty and resin acids. *Bioresource Technology* 47(1):45–53.

[208] Idem, R.O., Katikaneni, S.P.R., and Bakhshi, N.N. 1996. Thermal cracking of canola oil: Reaction products in the presence and absence of steam. *Energy & Fuels* 10(6):1150–1162.

[209] Maher, K.D., and Bressler, D.C. 2007. Pyrolysis of triglyceric materials for the production of renewable fuels and chemicals. *Bioresource Technology* 98:2351–2368.

[210] Alencar, J.W., Alves, P.B., and Craveiro, A.A. 1983. Pyrolysis of tropical vegetable oils. *Journal of Agricultural and Food Chemistry* 31(6):1268–1270.

[211] Kumar, H., Lappi, H., and Alén, R. 2018. Current and potential biofuel production from plant oils. *BioEnergy Research* 11(3):592–613.

[212] Fortes, I.C.P., and Baugh, P.J. 1994. Study of calcium soap pyrolysates derived from macauba fruit (*Acronomia sclerocarpa* M.). derivatization and analysis by GC/MS and CI-MS. *Journal of Analytical and Applied Pyrolysis* 29(2):153–167.

[213] Arpiainen, V. 2001. *Production of Light Fuel Oil from Tall Oil Soap Liquids by Fast Pyrolysis Techniques*. Licentiate Thesis. University of Jyväskylä, Laboratory of Applied Chemistry, Jyväskylä, Finland. 51 p. (In Finnish).

[214] Lee, S.Y., Hubbe, M.A., and Saka, S. 2006. Prospects for biodiesel as a byproduct of wood pulping – a review. *BioResoures* 1:150–171.

[215] Demirbaş, A. 2008. Production of biodiesel from tall oil. *Energy Sources, Part A Recovery Utilization and Environmental Effects* 30:1896–1902.

[216] Lappi, H., and Alén, R. 2009. Production of vegetable oil-based biofuels – thermochemical behavior of fatty acid sodium salts during pyrolysis. *Journal of Analytical and Applied Pyrolysis* 86:274–280.

[217] Lappi, H., and Alén, R. 2011. Pyrolysis of crude tall oil-derived products. *BioResources* 6:5121–5138.

[218] Alén, R. 2000. Basic chemistry of wood delignification. In: Stenius, P. (Ed.). *Forest Products Chemistry*. Fapet, Helsinki, Finland. Pp. 58–104.

[219] Mannonen, S. 2012. UPM – producing fuels of the future from wood-based raw materials. In: *Proceedings of The 4th Nordic Wood Biorefinery Conference (NWBC 2012)*, October 23–25, 2012, Helsinki, Finland. Pp. 121–124.

[220] Brettell, T.A. 2004. Forensic science applications of gas chromatography. In: Grob, R.L., and Barry, E.F. (Eds.). *Modern Practice of Gas Chromatography*. 4th edition. John Wiley & Sons, New York, NY, USA. p. 1064.

[221] Kusch, P., Knupp, G., and Morrisson, A. 2005. Analysis of synthetic polymers and copolymers by pyrolysis-gas chromatography/mass spectrometry. In: Bregg, R.K. (Ed.). *Horizons in Polymer Research*. Nova Science Publisher, New York, NY, USA. 199 p.

[222] Evans, R.J., and Milne, T. 1987. Molecular characterization of the pyrolysis of biomass. 2. Applications. *Energy & Fuels* 1(4):311–319.

[223] Bahng, M., Mukarakate, C., Robichaud, D.J., and Nimlos, M.R. 2009. Current technologies for analysis of biomass thermochemical processing: A review. *Analytical Chimica Acta* 651(2):117–138.

[224] Sipilä, K., Kuoppala, E., Fagernäs, L., and Oasmaa, A. 1998. Characterization of biomass-based flash pyrolysis oils. *Biomass and Bioenergy* 14(2):103–113.

[225] Oasmaa, A. 2003. *Fuel Oil Quality Properties of Wood-based Pyrolysis Liquids*. Doctoral Thesis. University of Jyväskylä, Laboratory of Applied Chemistry, Jyväskylä, Finland. 46 p.

[226] Undri, A., Abou-Zhad, M., Briens, C., Berruti, F., Rosi, L., Bartoli, M., Frediani, M., and Frediani, P. 2015. A simple procedure for chromatographic analysis of bio-oils from pyrolysis. *Journal of Analytical and Applied Pyrolysis* 114:208–221.

[227] Staš, M., Kubička, D., Chudoba, J., and Popspíšil, M. 2014. Overview of analytical methods used for chemical characterization of pyrolysis bio-oil. *Energy & Fuels* 28:385–402.

[228] Liu, Z., and Phillips, J.B. 1991. Comprehensive two-dimensional gas chromatography using on-column thermal modulator interface. *Journal of Chromatographic Science* 29(6):227–231.

[229] Frysinger, G.S., and Gaines, R.B. 2002. Forensic analysis of ignitable liquids in fire debris by comprehensive two-dimensional gas chromatography. *Journal of Forensic Sciences* 47(3):471–482.

[230] Castellvi Barnes, M., Lange, J.P., van Rossum, G., and Kersten, S.R.A. 2015. A new approach for bio-oil characterization based on gel permeation chromatography preparative fractionation. *Journal of Analytical and Applied Pyrolysis* 113:444–453.

[231] Williams, D.H., and Fleming, I. 1966. *Spectroscopic Methods in Organic Chemistry*. McGraw-Hill Publishing Company, Berkshire, England. Pp. 130–172.

[232] Chiaberge, S., Leonardis, I., Fiorani, T., Cesti, P., Reale, S., and De Angelis, F. 2014. Bio-oil from waste: A comprehensive analytical study by soft-ionization FTICR mass spectrometry. *Energy & Fuels* 28(3):2019–2026.

[233] Wang, S.Z., Fan, X., Zheng, A.-L., Lu, Y., Wei, X.Y., and Zhao, Y.P. 2014. Evaluation of the oxidation of rice husks with sodium hypochlorite using gas chromatography-mass spectrometry and direct analysis in real time-mass spectrometry. *Analytical Letters* 47(1):77–90.

[234] Sanquineti, M.M., Hourani, N., Witt, M., Sarathy, S.M., Thomsen, L.A., and Kuhnert, N. 2015. Analysis of impact of temperature and saltwater on *nannochloropsis salina* bio-oil production by ultra-high resolution APCI FT-ICR MS. *Algal Research* 9:227–235.

[235] Bordoloi, N., Narzari, R., Chutia, R.S., Bhaskar, T., and Kataki, R. 2015. Pyrolysis of mesua ferrea and pongamia glabra seed cover: Characterization of bio-oil and its sub-fractions. *Bioresource Technology* 178:83–89.

[236] Elkasabi, Y., Mullen, C.A., Jackson, M.A., and Boateng, A.A. 2015. Characterization of fast-pyrolysis bio-oil distillation residues and their potential applications. *Journal of Analytical and Applied Pyrolysis* 114:179–186.

[237] Fan, X., Zhu, J.-L., Zheng, A.-L., Wei, X.-Y., Zhao, Y.-P., Cao, J.-P., Zhao, W., Lu, Y., Chen, L., and You, C.-Y. 2015. Rapid characterization of heteroatomic molecules in a bio-oil from pyrolysis of rice husk using atmospheric solids analysis probe mass spectrometry. *Journal of Analytical and Applied Pyrolysis* 115:16–23.

[238] Bassilakis, R., Carangelo, R.M., and Wójtowicz, M.A. 2001. TG-FTIR analysis of biomass pyrolysis. *Fuel* 80(12):1765–1786.

[239] Wu, S., Shen, D., Hu, J., Xiao, R., and Zhang, H. 2013. TG-FTIR and Py-GC-MS analysis of a model compounds of cellulose – glyceraldehyde. *Journal of Analytical and Applied Pyrolysis* 101:79–85.

[240] Stankovikj, F., and Garcia-Perez, M. 2017. TG-FTIR method for the characterization of bio-oils in chemical families. *Energy & Fuels* 31(2):1689–1701.

[241] Jahirul, M.I., Rasul, M.G., Chowdhury, A.A., and Ashwath, N. 2012. Biofuels production through biomass pyrolysis – a technological review. *Energies* 5(12)4952–5001.

[242] Smets, K., Adriaensens, P., Vandewijngaarden, J., Stals, M., Cornelissen, T., Schreurs, S., Carleer, R., and Yperman, J. 2011. Water content of pyrolysis oil: Comparison between Karl Fischer titration, GC/MS corrected azeotropic distillation and 1H NMR spectroscopy. *Journal of Analytical and Applied Pyrolysis* 90(2):100–105.

[243] Strahan, G.D., Mullen, C.A., and Boateng, A.A. 2011. Characterizing biomass fast pyrolysis oils by 13C NMR and chemometric analysis. *Energy & Fuels* 25(11):5452–5461.

[244] Dalitz, F., Steiwand, A., Raffelt, K., Nirschl, H., and Guthausen, G. 2012. 1H NMR techniques for characterization of water content and viscosity of fast pyrolysis oils. *Energy & Fuels* 26(8):5274–5280.

[245] Ben, H., and Ragauskas, A.J. 2013. Comparison for the compositions of fast and slow pyrolysis oils by NMR characterization. *Bioresource Technology* 147:577–584.

[246] Demiral, İ., and Kul, Ş.Ç. 2014. Pyrolysis of apricot kernel shell in a fixed-bed reactor: Characterization of bio-oil and char. *Journal of Analytical and Applied Pyrolysis* 107:17–24.

[247] Naik, D.V., Kumar, V., Prasad, B., Poddar, M.K., Behera, B., Bal, R., Khatri, O.P., Adhikari, D.K., and Garg, M.O. 2015. Catalytic cracking of jatropha-derived fast pyrolysis oils with VGO and their NMR characterization. *RSC Advances* 5(1):398–409.

[248] Aue, W.P., Bartholdi, E., and Ernst, R.R. 1976. Two-dimensional spectroscopy. application to nuclear magnetic resonance. *The Journal of Chemical Physics* 64(5):2229–2246.

[249] Giraudeau, P. 2014. Quantitative 2D liquid-state NMR. *Magnetic Resonance in Chemistry* 52(6):259–272.

[250] Ben, H., and Ragauskas, A.J. 2011. Heteronuclear single-quantum correlation spectroscopy (HSQC-NMR) fingerprint analysis of pyrolysis oils. *Energy & Fuels* 25(12):5791–5801.

[251] Huang, F., Pan, S., Pu, Y., Ben, H., and Ragauskas, A.J. 2014. ^{19}F NMR spectroscopy for the quantitative analysis of carbonyl groups in bio-oils. *RSC Advances* 4(34):17743–17747.

[252] Fortin, M., Beromi, M.M., Lai, A., Tarves, P.C., Mullen, C.A., Boateng, A.A., and West, N.M. 2015. Structural analysis of pyrolytic lignins isolated from switchgrass fast-pyrolysis oil. *Energy & Fuels* 29(12):8017–8026.

[253] Verma, M., Godbout, S., Brar, S.K., Solomatnikova, O., Lemay, S.P., and Larouche, J.P. 2012. Biofuels production from biomass by thermochemical conversion technologies. *International Journal of Chemical Engineering* 2012:1–18.

[254] Bridgwater, A.V. 2012. Review of fast pyrolysis of biomass and product upgrading. *Biomass and Bioenergy* 38:68–94.

[255] Prins, W., and Wagenaar, B.M. 1997. Review of rotating cone technology for flash pyrolysis of biomass. In: Kaltscmitt, M., and Bridgwater, A.V. (Eds.). *Biomass Gasification and Pyrolysis, State of the Art and Future Prospects*. CPL Press, Newbury, United Kingdom. Pp. 316–326.

[256] Laird, D.A., Brown, R.C., Amonette, J.E., and Lehmann, J. 2009. Review of the pyrolysis platform for coproducing bio-oil and biochar. *Biofuels, Bioproducts & Biorefining* 3:547–562.

[257] Siipola, V., Tamminen, T., Källi, A., Lahti, R., Romar, H., Rasa, K., Keskinen, R., Hyväluoma, J., Hannula, M., and Wikberg, H. 1018. Effects of biomass type, carbonization process, and activation method on the properties of bio-based activated carbons. *BioResources* 13(3):5976–6002.

[258] Cheng, C.H., Lehmann, J., and Engelhard, M.H. 2008. Natural oxidation of black carbon in soils: Changes in molecular form and surface charge along a climosequence. *Geochimica et Cosmochimica Acta* 72:1598–1610.

[259] FAO. 2019. *World Food and Agriculture – Statical Pocketbook 2019*. Food and Agriculture Organization of the United Nations. Rome, Italy.

[260] Warnock, D.D., Lehmann, J., Kuyper, T.W., and Rilling, M.C. 2007. Mycorrhizal responses to biochar in soil – concepts and mechanisms. *Plant & Soil* 300:9–20.

[261] DeLuca, T.H., MacKenzie, M.D., Gundale, M.J., and Holben, W.E. 2006. Wildfire-produced charcoal directly influences nitrogen cycling in ponderosa pine forests. *Soil Science Society of American Journal* 70:448–453.

[262] Wardle, D.A., Nilsson, M.C., and Zackrisson, O. 2008. Fire-derived charcoal causes loss of forest humus. *Science* 320:629.

[263] Balat, M., Balat, M., Kirtay, E., and Balat, H. 2009. Main routes for the thermo-conversion of biomass into fuels and chemicals. Part 1: Pyrolysis systems. *Energy Conversion and Management* 50:3147–3157.

[264] Küçük, M.M., and Demirbaş, A. 1997. Biomass conversion processes. *Energy Conversion and Management* 38:151–165.

[265] Fagernäs, L., Kuoppala, E., Tiilikkala, K., and Oasmaa, A. 2012. Chemical composition of birch wood slow pyrolysis products. *Energy & Fuels* 26:1275–1283.

[266] Fagernäs, L., Kuoppala, E., and Simell, P. 2012. Polycyclic aromatic hydrocarbons in birch wood slow pyrolysis products. *Energy & Fuels* 26:6960–6970.

[267] Radlein, D., Piskorz, J., and Scott, D.S. 1991. Fast pyrolysis of natural polysaccharides as a potential industrial process. *Journal of Analytical and Applied Pyrolysis* 19:41–63.

[268] Kuroda, K.-i., and Yamaguchi, A. 1995. Classification of Japanese softwood species by pyrolysis-gas chromatography. *Journal of Analytical and Applied Pyrolysis* 33:51–59.

[269] Branca, C., Giudicianni, P., and Di Plasi, C. 2003. GC/MS characterization of liquids generated from low-temperature pyrolysis of wood. *Industrial & Engineering Chemistry Research* 42:3190–3202.

[270] Custodis, V., Hemberger, P., Ma, Z., and van Bokhoven, J. 2004. Mechanism of pyrolysis of lignin: Studying model compounds. *Journal of Physical Chemistry B* 118:8524–8531.

[271] Ohra-aho, T., Tenkanen, M., and Tamminen, T. 2005. Direct analysis of lignin and lignin-like components from softwood kraft pulp by py-gc/ms techniques. *Journal of Analytical and Applied Pyrolysis* 74(1,2):123–128.

[272] Butt, D.A.E. 2006. Formation of phenols from the low-temperature fast pyrolysis of radiata pine (*pinus radiata*). Part I. influence of molecular oxygen. *Journal of Analytical and Applied Pyrolysis* 76:38–47.

[273] Ohra-aho, T., and Linnekoski, J. 2015. Catalytic pyrolysis of lignin using analytical pyrolysis-GC-MS. *Journal of Analytical and Applied Pyrolysis* 113:186–192.

[274] Kawamoto, H. 2017. Lignin pyrolysis reactions. *Journal of Wood Science* 63(2):117–132.

[275] Terrón, M.C., Fidalgo, M.L., González, A.E., Almendros, G., and Galletti, G.C. 1993. Pyrolysis-gas chromatography/mass spectrometry of wheat straw fractions obtained by alkaline treatments used in pulping processes. *Journal of Analytical and Applied Pyrolysis* 27:57–71.

[276] Müller-Hagedorn, M., and Bockhorn, H. 2007. Pyrolytic behaviour of different biomasses (angiosperms) (maize plants, straws, and wood) in low temperature pyrolysis. *Journal of Analytical and Applied Pyrolysis* 79:136–146.

[277] Hodgson, E.M., Nowakowski, D.J., Shield, I., Riche, A., Bridgwater, A.V., and Clifton-Brown, J.C. 2011. Variation in *miscanthus* chemical composition and implications for conversion by pyrolysis and thermo-chemical bio-refining for fuels and chemicals. *Bioresource Technology* 102:3411–3418.

[278] Butler, E., Devlin, G., Meier, D., and McDonnell, K. 2013. Characterisation of spruce, salix, miscanthus and wheat straw. *Bioresource Technology* 131:202–209.

[279] Mante, O.D., Amidon, T.E., Stipanovic, A., and Babu, S.P. 2014. Integration of biomass pretreatment with fast pyrolysis: An evaluation of electron beam (EB) irradiation and hot-water extraction (HWE). *Journal of Analytical and Applied Pyrolysis* 110:44–54.

[280] Ghalibaf, M., Lehto, J., and Alén, R. 2017. Fast pyrolysis of hot-water-extracted and delignified silver birch (*betula pendula*) sawdust by Py-GC/MS. *Journal of Analytical and Applied Pyrolysis* 127:17–22.

[281] Ghalibaf, M., Ullah, S., and Alén, R. 2018. Fast pyrolysis of hot-water-extracted and soda-AQ-delignified okra (*Abelmoschus esculentus*) and miscanthus (*miscanthus x giganteus*) stalks by Py-GC/MS *Biomass and Bioenergy* 118:172–179.

[282] Ghalibaf, M., Lehto, J., and Alén, R. 2018. Fast pyrolysis of hot-water-extracted and delignified norway spruce (*picea abies*) sawdust by Py-GC/MS. *Wood Science and Technology* 53(1):87–100.

[283] Xu, F., Luo, J., Jiang, L., and Zhao, Z. 2022. Improved production of levoglucosan and levoglucosenone from acid-impregnated cellulose via fast pyrolysis. *Cellulose* 19:1463–1472.

[284] Tiilikkala, K., and Setälä, H. 2009. Birch tar oil – a new innovation as biological plant protection product. *Proceedings of the 2nd Nordic Wood Biorefinery Conference, NWBC-2009*, September 2–4, 2009, Helsinki, Finland. Pp. 120–122.

[285] Bridgwater, A.V., and Cottam, M.-L. 1992. Opportunities for biomass pyrolysis liquids production and upgrading. *Energy & Fuels* 6(2):113–120.

[286] Fagernäs, L. 1995. Chemical and Physical Characterization of Biomass-based Pyrolysis Oils – Literature Review. *VTT Research Notes 1706*. Technical Research Centre of Finland, Espoo, Finland. 118 p.

[287] Oasmaa, A., Leppämäki, E., Koponen, P., Levander, J., and Tapola, E. 1997. Physical characterisation of biomass-based pyrolysis liquids – application of standard fuel oil analyses. *VTT Publications 306*. Technical Research Centre of Finland, Espoo, Finland. 88 p.

[288] Oasmaa, A., and Peacocke, C. 2001. A guide to physical property characterisation of biomass-derived fast pyrolysis liquids. *VTT Publications 450*. Technical Research Centre of Finland, Espoo, Finland Espoo. 65 p.

[289] Chiaramonti, D., Oasmaa, A., and Solantausta, Y. 2007. Power generation using fast pyrolysis liquids from biomass. *Renewable and Sustainable Energy Reviews* 11:1056–1086.

[290] McCormick, R.L., and Westbrook, S.R. 2010. Storage stability of biodiesel and biodiesel blends. *Energy & Fuels* 24:690–698.

[291] Baldauf, W., and Balfanz, U. 1997. Upgrading of fast pyrolysis liquids at Veba Oel. In: Kaltscmitt, M., and Bridgwater, A.V. (Eds.). *Biomass Gasification and Pyrolysis, State of the Art and Future Prospects*. CPL Press, Newbury, United Kingdom. Pp. 392–398.

[292] Elliott, D.C. 2007. Historical developments in hydroprocessing bio-oils. *Energy & Fuels* 21(3):1792–1815.

[293] Oasmaa, A., Solantausta, Y., Arpiainen, V., Kuoppala, E., and Sipilä, K. 2010. Fast pyrolysis bio-oils from wood and agricultural residues. *Energy & Fuels* 24(2):1380–1388.

[294] Brink, D.L. 1981. Gasification. In: Goldstein, I.S. (Ed.). *Organic Chemicals from Biomass*. CRC Press, Boca Raton, FL, USA. Pp. 45–62.

[295] Higman, C., and van der Burgt, M. 2003. *Gasification*. Elsevier Science, Burlington, MA, USA. 391 p.

[296] Olofsson, I., Nordin, A., and Söderlind, U. 2005. Initial review and evaluation of process technologies and systems suitable for cost-efficient medium-scale gasification for biomass to liquid fuels. *ETPC Report 05-02*. Umeå University and Mid Swedish University, Sweden. 90 p.

[297] Zwart, R.W.R., Boerrigter, H., Deurwaarder, E.P., van der Meijden, C.M., and van Paasen, S.V.B. 2006. Production of synthetic natural gas (SNG) from biomass. *ECN-E-06-018 Report*. Energy Research Centre of the Netherlands. 62 p.

[298] Pervaiz, M., and Sain, M. 2006. Biorefinery: Opportunities and barriers for petrochemical industries. *Pulp and Paper Canada* 107(6):31–33.

[299] Mabee, W.E., and Saddler, W.E. 2006. The potential of bioconversion to produce fuels and chemicals. *Pulp and Paper Canada* 107(6):34–37.

[300] Clements, L.D., and Van Dyne, D.L. 2006. The lignocellulosic biorefinery – a strategy for returning to a sustainable source of fuels and industrial organic chemicals. In: Kamm, B., Gruber, P.R., and Kamm, M. (Eds.). *Biorefineries – Industrial Processes and Products, Volume 1*. Wiley-VCH, Weinheim, Germany. Pp. 115–128.

[301] Arshadi, M., and Sellstedt, A. 2008. Production of energy from biomass. In: Clark, J.H., and Deswarte, F.E.I. (Eds.). *Introduction to Chemicals from Biomass*. John Wiley & Sons, New York, NY, USA. Pp. 143–178.

[302] Anon. 2009. Review of technologies for gasification of biomass and wastes. *Final Report E4tech*. London, United Kingdom. 126 p.

[303] Neathery, J.K. 2010. Biomass gasification. In: Crocker, M. (Ed.). *Thermochemical Conversion of Biomass to Liquid Fuels and Chemicals*. RSC Publishing, Cambridge, United Kingdom. Pp. 67–94.

[304] Bain, R.L., and Broer, K. 2011. Gasification. In: Brown, R.C. (Ed.). *Thermochemical Processing of Biomass*. John Wiley & Sons, Chichester, United Kingdom. Pp. 47–77.

[305] Dayton, D.C., Turk, B., and Gupta, R. 2011. Syngas cleanup, conditioning, and utilization. In: Brown, R.C. (Ed.). *Thermochemical Processing of Biomass*. John Wiley & Sons, Chichester, United Kingdom. Pp. 78–123.

[306] Steynberg, A., and Dry, M. 2004. *Fischer-Tropsch Technology*. Elsevier, Oxford, United Kingdom. 722 p.

[307] Viikari, L., and Alén, R. 2011. In: Alén, R. (Ed.). *Biorefining of Forest Resources*. Paper Engineers' Association, Helsinki, Finland. Pp. 225–261.

[308] Foust, T.D., Aden, A., Dutta, A., and Phillips, S. 2009. An economic and environmental comparison of a biochemical and a thermochemical lignocellulosic ethanol conversion processes. *Cellulose* 16:547–565.

[309] Badger, P.C. 2002. Ethanol from cellulose: A general review. In: Janick, J., and Whipkey, A. (Eds.). *Trends in New Crops and New Uses*. ASHS Press, Alexandria, VA, USA. Pp. 17–21.

[310] Clausen, E.C., and Gaddy, J.L. 1993. Ethanol from biomass by gasification/fermentation. *American Chemical Society* 38:855–861.

[311] Gaddy, J.L. 2000. Biological production of ethanol from waste gases with *Clostridium ljungdahlii*. U.S. Patent 6,136,577.

[312] Alakangas, E., Hurskainen, M., Laatikainen-Luntama, J., and Korhonen, J. 2016. Indigenous fuels in Finland. *VTT Technology 272*. VTT Technical Research Centre of Finland Ltd., Espoo, Finland. 222 p.

[313] de Jong, W. 2005. *Nitrogen Compounds in Pressurized Fluidized Bed Gasification of Biomass and Fossil Fuels*. Doctoral Thesis. Delft University of Technology, Delft, The Netherlands. 283 p.

[314] Kurkela, E., Simell, P., McKeough, P., and Kurkela, M. 2008. Production of synthesis gas and clean fuel gas. *VTT Publications 682*. VTT, Espoo, Finland. 54 p. (In Finnish).

[315] Nielsen, P.E.H. 2009. From biomass to liquid products and power – the topsøe TIGAS process. *Proceedings of 4th BtLtec Conference*. September 24–25, 2009, Graz, Austria.

[316] Hosoya, T., Kawamoto, H., and Saka, S. 2008. Pyrolysis gasification reactivities of primary tar and char fractions from cellulose and lignin as studied with a closed ampoule reactor. *Journal of Analytical and Applied Pyrolysis* 83:71–77.

[317] Di Blasi, C. 2009. Combustion and gasification rates of lignocellulosic chars. *Progress in Energy and Combustion Science* 35:121–140.

[318] Hosoya, T., Kawamoto, H., and Saka, S. 2007. Pyrolysis behaviors of wood and its constituent polymers at gasification temperature. *Journal of Analytical and Applied Pyrolysis* 78:328–336.

[319] Kaisalo, N. 2017. *Tar Reforming in Biomass Gasification Gas Cleaning*. Doctoral Thesis. Department of Chemical and Metallurgical Engineering, Aalto University, Espoo, Finland. 64 p.

[320] Fushimi, C., Katayama, S., and Tsutsumi, A. 2009. Elucidation of interaction among cellulose, lignin and xylan during tar and gas evolution in steam gasification. *Journal of Analytical and Applied Pyrolysis* 86:82–89.

[321] Yoon, H.C., Pozivil, P., and Steinfeld, A. 2012. Thermogravimetric pyrolysis and gasification of lignocellulosic biomass and kinetic summative law for parallel reactions with cellulose, xylan, and lignin. *Energy & Fuels* 26:357–364.

[322] Oluoti, K., Richards, T., Doddapaneni, T.R.K., and Kanagasabapathi, D. 2014. Evaluation of the pyrolysis and gasification kinetics of tropical wood biomass. *BioResources* 9(2):2179–2190.

[323] Perander, M., DeMartini, N., Brink, A., Kramb, J., Karlström, O., Hemming, J., Moilanen, A., Konttinen, J., and Hupa, M. 2015. Catalytic effect of ca and k on CO_2 gasification of spruce wood char. *Fuel* 150:464–472.

[324] Leibold, H., Hornung, A., and Seifer H. 2008. HTHP syngas cleaning concept of two stage biomass gasification for FT synthesis. *Power Technology* 180:265–270.

[325] Jacobs, G., and Davis, B.H. 2010. Conversion of biomass to liquid fuels and chemicals via the Fischer-Tropsch synthesis route. In: Crocker, M. (Ed.). *Thermochemical Conversion of Biomass to Liquid Fuels and Chemicals*. RSC Publishing, Cambridge, United Kingdom. Pp. 95–124.

[326] de Klerk, A. 2013. Fischer-Tropsch process. In: *Kirk-Othmer Encyclopedia of Chemical Technology*. Wiley-VCH, Weinheim, Germany.

[327] Davis, B. 2009. Fischer-Tropsch synthesis: Reaction mechanisms for iron catalysts. *Catalysis Today* 141: 25–33.

[328] Gates, B.C. 1993. Extending the metal cluster-metal surface analogy. *Angewandte Chemie International Edition in English* 32:228–229.

[329] Sie, S.T., Senden, M.M.G., and van Wechem, H.M.H. 1991. Conversion of natural gas to transportation fuels via the shell middle distillate synthesis process (SMDS). *Catalysis Today* 8:371–394.

[330] Neary, R.M. 1974. Industrial gases. In: Kent, J.A. (Ed.). *Riegel's Handbook of Industrial Chemistry*. 7th edition. Van Nostrand Reinhold Company, New York, NY, USA. Pp. 514–536.

[331] Lau, F.S., Bowen, D.A., Dihu, R., Doong, S., Hughes, E.E., Remick, R., Slimane, R., Turn, S.Q., and Zabransky, R. 2002. Techno-economic analysis of hydrogen production by gasification of biomass. *Project Report*. U.S. Department of Energy (DOE), Washington DC, USA.

[332] Tunå, P. 2008. *Substitute Natural Gas from Biomass Gasification*. Master Thesis. Lund University Department of Chemical Engineering, Lund, Sweden.

[333] Haberstroh, W.H., and Collins, D.E. Synthetic organic chemicals. In: Kent, J.A. (Ed.). *Riegel's Handbook of Industrial Chemistry*. 7th edition. Van Nostrand Reinhold Company, New York, NY, USA. Pp. 772–822.

[334] Zhang, W. 2010. Automotive fuels from biogas via gasification. *Fuel Processing Technology* 91(8):866–876.

[335] Anon. 2009. Review of technologies for gasification of biomass and wastes. *Final Report, E4tech*. London, United Kingdom.

[336] Consonni, S., Kafosky, R.E., and Larson, E.D. 2009. A gasification-based biorefinery for the pulp and paper industry. *Chemical Engineering Research and Design* 87:1293–1317.

[337] Vakkilainen, E.K. 2007. *Kraft Recovery Boilers – Principles and Practice*. Helsinki University of Technology, Energy Engineering and Environmental Protection. Espoo, Finland. p. 246.

[338] Tran, H. (Ed.). 2020. *Kraft Recovery Boilers*. 3rd edition. TAPPI Press, Atlanta, GA, USA. p. 375.

[339] Vakkilainen, E.K., Kankkonen, S., and Suutela, J. 2008. Advanced efficiency options: Increasing electricity generating potential from pulp mills. *Pulp and Paper Canada* 109(4):14–18.

[340] Larson, E.D., Consonni, S., Katofsky, R.E., Iisa, K., and Frederick, W.J. Jr. 2008. An assessment of gasification-based biorefining at kraft pulp and paper mills in the united states, Part A. Background and assumptions. *TAPPI Journal* 7(11):8–14.

[341] Prahacs, S., Barclay, H.G., and Bhatia, S.P. 1971. A study of the possibilities of producing synthetic tonnage chemicals from lignocellulosic residues. *Pulp and Paper Magazine of Canada* 72(6):69–83.

[342] Timpe, W.G., and Evers, W.J. 1973. The hydropyrolysis recovery process – a new approach to kraft chemical recovery. *TAPPI* 56(8)100–103.

[343] Coheen, D.W., Orle, J.V., and Wither, R.P. 1976. Indirect pyrolysis of kraft black liquors. In: Shafizadeh, F., Sarkanen, K.V., and Tillman, D.A. (Eds.). *Thermal Uses and Properties of Carbohydrates and Lignins*. Academic Press, New York, NY, USA. P. 320.

[344] Landälv, I. 2007. Black liquor gasification and conversion of the pulp mill into a biorefinery. *Proceedings of the Seminar on Biorefining for the Pulp and Paper Industry*. December 10–11, 2007, Arlanda, Stockholm, Sweden. P. 4.

[345] Hamaguchi, M., Cardoso, M., and Vakkilainen, E. 2012. Alternative technologies for biofuels production in kraft pulp mills – potential and prospects. *Energies* 5:2288–2309.

[346] Bajpai, P. 2013. *Biorefinery in the Pulp and Paper Industry*. Elsevier, Amsterdam, The Netherlands. p. 103.

[347] Naqvi, M., Yan, J., and Dahlquist, E. 2010. Black liquor gasification integrated in pulp and paper mills: A critical review. *Bioresource Technology* 101:8001–8015.

[348] Whitty, K., Backman, R., and Hupa, M. 1998. Influence of char formation conditions on pressurized black liquor gasification rates. *Carbon* 30(11):1683–1692.

[349] Sricharoenchaikul, V., Frederic, W.J. Jr., and Agrawal, P. 2002. Black liquor gasification characteristics. 1. Formation and conversion of carbon-containing product gases. *Industrial & Engineering Chemistry Research* 41:5640–5649.

[350] Sricharoenchaikul, V., Frederic, W.J. Jr., and Agrawal, P. 2002. Black liquor gasification characteristics. 2. Measurement of condensable organic matter (tar) at rapid heating conditions. *Industrial & Engineering Chemistry Research* 41:5650–5658.

[351] Sricharoenchaikul, V., Frederic, Wm.J. Jr., and Agrawal, P. 2003. Carbon distribution in char residue from gasification of kraft black liquor. *Biomass and Bioenergy* 25:209–220.

[352] Saviharju, K., Moilanen, A., and van Heiningen, A.R.P. 1998. New high-pressure gasification rate data for fast pyrolysis of black liquor char. *Journal of Pulp and Paper Science* 24(7):231–236.

[353] Vähä-Savo, N., DeMartini, N., and Hupa, M. 2013. Fate of char nitrogen in catalytic gasification – formation of alkali cyanate. *Energy & Fuels* 27:7108–7114.

[354] Hrbek, J. 2011. Biomass gasification opportunities in forest industry. *Proceedings of IEA Bioenergy, Task 33, Thermal Gasification of Biomass, Workshop.* October 19, 2011, Piteå, Sweden. P. 30.

[355] Barry, D.A., and LeBlanc, R.J. 2009. Entrained flow gasification biorefinery offers step-change in results – a black liquor gasification development plant in Piteå, Sweden, has been operating for more than 10,000 h. *Paper360°* (August):34–35.

[356] Naqvi, M., Yan, J., and Fröling, M. 2010. Bio-refinery system of DME and CH_4 production from black liquor gasification in pulp mills. *Bioresource Technology* 101:937–944.

[357] Pandey, A., Bhaskar, T., Stöcker, M., and Sukumaran; R. 2015. *Recent Advances in Thermochemical Conversion of Biomass.* Elsevier, Amsterdam, The Netherlands. 504 p.

[358] Rhen, C., Öhman, M., Gref, R., and Wästerlund, I. 2007. Effect of raw material composition in woody biomass pellets on combustion characteristics. *Biomass and Bioenergy* 31(1):66–72.

[359] Lu, H., Robert, W., Peirce, G., Ripa, B., and Baxter, L.L. 2008. Comprehensive study of biomass particle combustion. *Energy & Fuels* 22(4):2826–2839.

[360] Hermández, J.J., Aranda-Almansa, G., and Bula, A. 2010. Gasification of biomass wastes in an entrained flow gasifier: Effect of the particle size and the residence time. *Fuel Processing Technology* 91(6):681–692.

[361] Vidal, B.C., Dien, B.S., Ting, K.C., and Singh, V. 2011. Influence of feedstock particle size on lignocellulose conversion – a review. *Applied Biochemistry and Biotechnology* 164(8):1405–1421.

[362] Mlonka-Medrala, A., Magdziarz, A., Dziok, T., Sieradzka, M., and Nowak, W. 2019. Laboratory studies on the influence of biomass particle size on pyrolysis and combustion using TG GC/MS. *Fuel* 252:635–645.

[363] Sermyagina, E., Saari, J., Kaikko, J., and Vakkilainen, E. 2015. Hydrothermal carbonization of coniferous biomass: Effect of process parameters on mass and energy yields. *Journal of Analytical and Applied Pyrolysis* 113:551–556.

[364] Paczkowski, S., Knappe, V., Paczkowska, M., Diaz Robles, L.A., Jaeger, D., and Pelz, S. 2021. Low-temperature hydrothermal treatment (HTT) improves the combustion properties of short-rotation coppice willow wood by reducing emission precursors. *Energies* 14(24):8229–8342.

[365] Kaltschmitt, M., Hartmann, H., and Hofbauer, H. (Eds.). 2016. *Energie aus Biomasse: Grundlagen, Techniken und Verfahren.* 3rd edition. Springer Vierweg, Wiesbaden, Germany. 1867 p.

[366] Knappe, V., Paczkowski, S., Tejada, J., Diaz Robles, L.A., Gonzales, A., and Pelz, S. 2018. Low temperature microwave assisted hydrothermal carbonization (MAHC) reduces combustion emission precursors in short rotation coppice willow wood. *Journal of Analytical and Applied Pyrolysis* 134:162–166.

[367] Sevilla, M., and Fuertes, A.B. 2009. The production of carbon materials by hydrothermal carbonization of cellulose. *Carbon* 47:2281–2289.

[368] Funke, A., and Ziegler, F. 2010. Hydrothermal carbonization of biomass: A summary and discussion of chemical mechanisms for process engineering. *Biofuels, Bioproducts and Biorefining* 4:160–177.

[369] Hoekman, S.K., Brock, A., Robbins, C., Zielinska, B., and Felix, L. 2013. Hydrothermal carbonization (HTC) of selected woody and herbaceous biomass feedstocks. *Biomass Conversion and Biorefinery* 3(2):113–126.

[370] Kruse, A., Funke, A., and Titirici, M.-M. 2013. Hydrothermal conversion of biomass to fuels and energetic materials. *Current Opinion in Chemical Biology* 17:515–521.

[371] Reza, M.T., Uddin, M.H., Lynam, J.G., Hoekman, S.K., and Coronella, C.J. 2014. Hydrothermal carbonization of loblolly pine: Reaction chemistry and water balance. *Biomass Conversion and Biorefinery* 4:311–321.

[372] Eseyin, A.E., and Steele, P.H. 2015. An overview of the applications of furfural and its derivatives. *International Journal of Advanced Chemistry* 3(2):42–47.

[373] Kruse, A., and Dahmen, N. 2018. Hydrothermal biomass conversion: Quo vadis? *Journal of Supercritical Fluids* 134:114–123.

[374] Schneider, D., Escala, M., Supawittayayothin, K., and Tippayawong, N. 2011. Characterization of biochar from hydrothermal carbonization of bamboo. *International Journal of Energy and Environment* 2(4):647–652.

[375] Chen, W.-H., Ye, S.-C., and Sheen, H.-K. 2012. Hydrothermal carbonization of sugarcane bagasse via wet torrefaction in association with microwave heating. *Bioresource Technology* 118:195–203.

[376] Oliveira, I., Blöhse, D., and Ramke, H.G.G. 2013. Hydrothermal carbonization of agricultural residues. *Bioresource Technology* 142:138–146.

[377] Wang, G., Zhang, J., Lee, J.-Y., Mao, X., Ye, L., Xu, W., Ning, X., Zhang, N., Teng, H., and Wang, C. 2020. Hydrothermal carbonization of maize straw for hydrochar production and its injection for blast furnace. *Applied Energy* 266:114818.

[378] Garotte, H., Domingues, H., and Parajo, J.C. 2001. Study on the deacetylation of hemicellulose during the hydrothermal processing of eucalyptus wood. *Holz als Roh- und Werkstoff* 59:53–59.

[379] Hashaikeh, R., Fang, Z., Butler, I.S., Hawari, J., and Kozinski, J.A. 2007. Hydrothermal dissolution of willow in hot compressed water as a model for biomass conversion. *Fuel* 86:1614–1622.

[380] Sevilla, M., Maciá-Agulló, J.A., and Fuertes, A.B. 2011. Hydrothermal carbonization of biomass as a route for the sequestration of CO_2: Chemical and structural properties of the carbonized products. *Biomass and Bioenergy* 35(7):3152–3159.

[381] Titirici, M.M., Thomas, A., Yu, S.-H., Müller, J.-O., and Antonietti, M. 2007. A direct synthesis of mesoporous carbons with bicontinuous pore morphology from crude plant material by hydrothermal carbonization. *Chemistry of Materials* 19:4205–4212.

[382] Sevilla, M., and Fuertes, A.B. 2009. The production of carbon materials by hydrothermal carbonization of cellulose. *Carbon* 47(9):2281–2289.

[383] Parshetti, G.K., Hoekman, K., and Balasubramanian, R. 2013. Chemical, structural and combustion characteristics of carbonaceous products obtained by hydrothermal carbonization of palm empty fruit bunches. *Bioresource Technology* 135:683–689.

[384] Li, L., Diederick, R., Flora, J.R.V., and Berge, N.D. 2013. Hydrothermal carbonization of food waste and associated packaging materials for energy source generation. *Waste Management* 33:2478–2492.

[385] Berge, N.D., Ro, K.S., Mao, J., Flora, J.R.V., Chappell, M.A., and Bae, S. 2011. Hydrothermal carbonization of municipal waste streams. *Environmental Science Technology* 45(13):5696–5703.

[386] Heilmann, S.M., Davis, H.T., Jader, L.R., Lefebvre, P.A., Sadowsky, M.J., Schendel, F.J., von Keitz, M.G., and Valentas, K.J. 2010. Hydrothermal carbonization of microalgae. *Biomass and Bioenergy* 34(6)875–882.

[387] Höök, M., and Kjell, A. 2010. A review on coal to liquid fuels and its coal consumption. *Journal of Energy Research* 34(10):848–864.

[388] Kaneko, T., Derbyshire, F., Makino, E., Gray, D., Tamura, M., and Li, K. 2012. Coal liquefaction. In: *Ullmann's Encyclopedia of Industrial Chemistry*. Wiley-VCH, Weinheim, Germany.

[389] Chornet, E., and Overend, R.P. 1985. Biomass liquefaction: An overview. In: Overend, R.P., Milne, T.A., and Mudge, L.K. (Eds.). *Fundamentals of Thermochemical Biomass Conversion*. Springer, Dordrecht, Germany. Pp. 967–1002.

[390] Goudriaan, F., and Peferoen, D.G.R. 1990. Liquid fuels from biomass via a hydrothermal process. *Chemical Engineering Science* 45(8):2729–2734.

[391] Demirbaş, A. 2000. Mechanisms of liquefaction and pyrolysis reactions of biomass. *Energy Conversion and Management* 41:633–646.

[392] Akhtar, J., and Amin, N.A.S. 2011. A review on process conditions for optimum bio-oil yield in hydrothermal liquefaction of biomass. *Renewable and Sustainable Energy Reviews* 15(3):1615–1624.

[393] Alma, M.H., Salan, T., Altuntas, E., and Karaoğul, E. 2013. Liquefaction processes of biomass for the production of valuable chemicals and biofuels: A review. *Proceedings of Conference on Joint International Convention Forest Products Society and Society of Wood Science and Technology*. Jyväskylä, Finland.

[394] Elliott, D.C., Biller, P., Ross, A.B., Schmidt, A.J., and Jones, S.B. 2015. Hydrothermal liquefaction on biomass: Developments from batch to continuous process. *Bioresource Technology* 178:147–156.

[395] Chen, S., D'cruz, I., Wang, M., Leitch, M., and Xu, C. 2010. Highly efficient liquefaction of woody biomass in hot-compressed alcohol – water co-solvents. *Energy & Fuels* 24(9):4659–4667.

[396] Fan, S.-P., Zakaria, S., Chia, C.-H., Jamaluddin, F., Nabihah, S., Liew, T.-K., and Pua, F.-L. 2011. Comparative studies of products obtained from solvolysis liquefaction of oil palm empty fruit bunch fibres using different solvents. *Bioresource Technology* 102(3):3521–3526.

[397] Peng, X., Ma, X., Lin, Y., Wang, X., Zhang, X., and Yang, C. 2016. Effect of process parameters on solvolysis liquefaction of *chlorella pyrenoidosa* in ethanol – water system and energy evaluation. *Energy and Conversion Management* 117:43–53.

[398] Zhai, Q., Li, F., Wang, F., Xu, J., Jiang, J., and Cai, Z. 2018. Liquefaction of poplar biomass for value-added platform chemicals. *Cellulose* 25:4663–4675.

[399] Anastasakis, K., Biller, P., Madsen, R., Glasius, M., and Johannsen, I. 2018. Continuous hydrothermal liquefaction of biomass in a novel pilot plant with heat recovery and hydraulic oscillation. *Energies* 11:2695–2718.

[400] Castello, D., Haider, M.S., and Rosendahl, L.A. 2019. Catalytic upgrading of hydrothermal liquefaction biocrudes: Different challenges for different feedstocks. *Renewable Energy* 141:420–430.

[401] Haider, M.S., Castello, D., and Rosendahl, L.A. 2021. The art of smooth continuous hydroprocessing of biocrudes obtained from hydrothermal liquefaction: Hydrodemetallization and propensity for coke formation. *Energy & Fuels* 35(13):10611–10622.

[402] McDermott, J.B., Klein, M.T., and Obst, J.R. 1986. Chemical modeling in the deduction of process concepts: A proposed novel process for lignin liquefaction. *Industrial & Engineering Chemistry Process Design and Development* 25:885–889.

[403] Vuori, A., and Bredenberg, J.B-son. 1988. Liquefaction of kraft lignin. 1. Primary reactions under mild thermolysis conditions. *Holzforschung* 42(3):155–161.

[404] Vuori, A., and Niemelä, M. 1988. Liquefaction of kraft lignin. 2. Reactions with a homogeneous lewis acid catalyst under mild reaction conditions. *Holzforschung* 42(5):327–334.

[405] Aglevor, F.A., and Bocock, D.G.B. 1989. The origins of phenol produced in the rapid hydrothermolysis and alkaline hydrolysis of hybrid poplar lignins. *Journal of Wood Chemistry and Technology* 9(2):167–188.

[406] Oasmaa, A., Alén, R., and Meier, D. 1993. Catalytic hydrotreatment of some technical lignins. *Bioresource Technology* 45:189–194.

[407] Borreca, M., Niemelä, K., and Sixta, H. 2013. Effect of hydrothermal treatment intensity on the formation of degradation products from birchwood. *Holzforschung* 67(8):871–879.

[408] Brown, T.M., Duan, P., and Savage, P.E. 2010. Hydrothermal liquefaction and gasification of *nannochloropsis* sp. *Energy & Fuels*. 24:3639–3646.

[409] Barreiro, D.L., Gómez, B.R., Hornung, U., Kruse, A., and Prins, W. 2015. Hydrothermal liquefaction of microalgae in continuous-stirred-tank reactor. *Energy & Fuels* 29(10):6422–6432.

[410] Guo, B., Walter, V., Hornung, U., and Dahmen, N. 2019. Hydrothermal liquefaction of *Chlorella vulgaris* and *Nannochloropsis gaditana* in a continuous stirred tank reactor and hydrotreating of biocrude by nickel catalysts. *Fuel Processing Technology* 191:168–180.

[411] Guo, B., Yang, B., Silve, A., Akaberi, S., Scherer, D., Papachristou, I., Frey, W., Hornung, U., and Dahmen, N. 2019. Hydrothermal liquefaction of residual microalgae biomass after pulsed electric field-assisted valuables extraction. *Algal Research* 43:101650.

[412] Alén, R., McKeough, P., Oasmaa, A., and Johansson, A. 1989. Thermochemical conversion of black liquor in the liquid phase. *Journal of Wood Technology* 9:265–276.

[413] McKeough, P., Alén, R., Oasmaa, A., and Johansson, A. 1990. Thermochemical conversion of black liquor organics into an oil product. I. formation of the major product fractions. *Holzforschung* 44(6):445–448.

[414] Alén, R. 1992. Thermochemical conversion of black liquor organics into oil and chemical feedstocks. In: Kennedy, J.F., Phillips, G.O., and Williams, P.A. (Eds.). *Lignocellulosics – Science, Technology, Development and Use*. Ellis Horwood, Chichester, West Sussex, England. Pp. 803–808.

[415] Alén, R., and Oasmaa, A. 1989. Thermochemical conversion of hydroxy carboxylic acids in the liquid phase. *Holzforschung* 43(3):155–158.

[416] Alén, R., and Oasmaa, A. 1988. Conversion of glucoisosaccharinic acid by heating under pressure. *Acta Chemica Scandinavica* B42(8):563–566.

[417] Goheen, D.W. 1971. Low molar weight chemicals. In: Sarkanen, K.V., and Ludwig, C.H. (Eds.). *Lignins – Occurrence, Formation, Structure and Reactions*. Wiley-Interscience, New York, NY, USA. Pp. 797–831.

[418] Allan, G.G., and Mattila, T. 1971. High energy degradation. In: Sarkanen, K.V., and Ludwig, C.H. (Eds.). *Lignins – Occurrence, Formation, Structure and Reactions*. Wiley-Interscience, New York, NY, USA. Pp. 575–596.

[419] Enkvist, T. 1975. Phenolics and other chemicals from kraft black liquors by disproportionation and cracking reactions. *Applied Polymer Symposium* 28:285–295.

[420] Hearon, W.M., MacGregor, W.S., and Coheen, D.W. 1962. Sulfur chemicals from lignin. *TAPPI* 45(1):28A–36A.

[421] Enkvist, T., and Alfredsson, B. 1953. Low molecular substances of lignin origin in black liquors from sulphate and soda pulping of spruce wood. *TAPPI* 36(5):211–216.

[422] Enkvist, T., Turunen, J., and Ashorn, T. 1962. The demethylation and degradation of lignin or spent liquors by heating with alkaline reagents. *TAPPI* 45(2):128–135.

[423] Enkvist, T., Ashorn, T., and Hästbacka, K. 1962. New aspects on the reaction of lignin in kraft pulping. *Paperi ja Puu* 44(8):395–404.

[424] Bauen, A., Bernes, G., Junginger, M., Londo, M., Ball, R., Bole, T., Chudziak, C., Faaij, A., and Mozaffarian, H. 2009. Bioenergy –a sustainable and reliable energy source – a review of status and prospects. *Main Report*. IEA Bioenergy, Roturia, New Zealand. p. 108.

[425] Vertès, A.A., Qureshi, N., Blaschek, H.P., and Yukawa, H. (Eds.). 2010. *Biomass to Biofuels – Strategies for Global Industries*. John Wiley & Sons, New York, NY, USA. p. 559.

[426] Ioelovich, M. 2015. Recent findings and energetic potential of plant biomass as a renewable source of biofuels – a review. *BioResources* 10(1):1879–1914.

[427] Feldman, D., and Barbose, G. 2014. *Photovoltaic System Pricing Trends – Historical, Recent, and Near-Term Projections – 1014 Edition*. SunShot, U.S. Department of Energy, Washington, DC, USA.

[428] Kesari, J.P., Gupta, A., Shukla, K., and Garg, P. 2019. A review of the combined wind and wave energy technologies. *International Journal of Engineering Trends and Technology* 67(5):131–136.

[429] IEA. 2021. Renewables information: Overview – comprehensive historical review and current market trends in renewable energy. *Statistic Report – August 2021*. International Energy Agency, Paris, France.

[430] FAOSTAT. 2020. *Global Food and Agricultural Statistics*. Food and Agriculture Organization of the United Nations, Statistics Division. Italy, Rome.

[431] Reed, T.B. 1975. Biomass energy for production of fuel and fertilizer. *Applied Polymer Symposium* 28:1–9.

[432] Herrick, F.W., and Hergert, H.L. 1977. Utilization of chemicals from wood: Retrospect and prospect. In: Loewus, F.A., and Runecles, V.C. (Eds.). *The Structure, Biosynthesis, and Degradation of Wood, Recent Advances in Phytochemistry, Volume 11*. Plenium Press, New York, NY, USA. Pp. 443–515.

[433] Seidl, R.J. 1980. Energy from wood: A new dimension in utilization. *TAPPI* 63(1):26–29.

[434] Goldstein, I.S. (Ed.). 1981. *Organic Chemicals from Biomass*. CRC Press, Boca Raton, FL, USA. 310 p.

[435] Sinsky, A.J. 1983. Organic chemicals from biomass: An overview. In: Wise, D.L. (Ed.). *Organic Chemicals from Biomass*. The Benjamin/Cummins Publishing Company, London, England. Pp. 1–67.

[436] Chen, N.Y., Degnan, T.F. Jr., and Koenig, L.R. 1986. Liquid fuel from carbohydrates. *ChemTech* (August):506–511.

[437] Ragauskas, A.J., Williams, C.K., Davison, B.H., Britovsek, G., Cairney, J., Eckert, C.A., Frederick, W.J. Jr., Hallett, J.P., Leak, D.J., Liotta, C.L., Mielenz, J.R., Murphy, R., Templer, R., and Tschaplinski, T. 2006. The path forward for biofuels and biomaterials. *Science* 311:484–489.

[438] Kamm, B., Kamm, M., Gruber, P.R., and Kromus, S. 2006. Biorefinery systems – an overview. In: Kamm, B., Gruber, P.R., and Kamm, M. (Eds.). *Biorefineries – Industrial Processes and Products, Volume 1*. Wiley-VCH, Weinheim, Germany. Pp. 3–40.

[439] Dale, B.E., and Kim, S. 2006. Biomass refining global impact – the biobased economy of the 21st century. In: Kamm, B., Gruber, P.R., and Kamm, M. (Eds.). *Biorefineries – Industrial Processes and Products, Volume 1*. Wiley-VCH, Weinheim, Germany. Pp. 41–66.

[440] Crocker, M. (Ed.). 2010. *Thermochemical Conversion of Biomass to Liquid Fuels and Chemicals*. RSC Publishing, Cambridge, United Kingdom. 532 p.

[441] Ragauskas, A. (Ed.). 2013. *Materials for Biofuels*. World Scientific Publishing Company, Singapore. 380 p.

[442] Goldstein, I.S. 1981. Composition of biomass. In: Goldstein, I.S. (Ed.). *Organic Chemicals from Biomass*. CRC Press, Boca Raton, FL, USA. Pp. 9–18.

[443] Hemingway, R.W. 19821. Bark: Its chemistry and prospects for chemical utilization. In: Goldstein, I.S. (Ed.). *Organic Chemicals from Biomass*. CRC Press, Boca Raton, FL, USA. Pp. 189–248.

[444] Piskorz, J. 2002. Fundamentals, mechanisms and science of pyrolysis. In: Bridgwater, A.V. (Ed.). *Fast Pyrolysis of Biomass: A Handbook, Volume 2*. CPL Press, Newbury, England. Pp. 103–125.

[445] Murwanashyaka, J.N., Pakdel, H., and Roy, C. 2002. Fractional vacuum pyrolysis of biomass and separation of phenolic compounds by steam distillation. In: Bridgwater, A.V. (Ed.). *Fast Pyrolysis of Biomass: A Handbook, Volume 2*. CPL Press, Newbury, England. Pp. 407–418.

[446] Cao, N., Darmstadt, H., and Roy, C. 2001. Activated carbon produced from charcoal obtained by vacuum pyrolysis of softwood bark residues. *Energy & Fuels* 15:1263–1269.

[447] Gírio, F.M., Fonseca, C., Carvalheiro, F., Duarte, L.C., Marques, S., and Bogel-Łukasik, R. 2010. Hemicelluloses for fuel ethanol: A review. *Bioresource Technology* 101:4775–4800.

[448] Joshi, B., Bhatt, M.R., Sharma, D., Joshi, J., Malla, R., and Lakshmaiah, S. 2011. Lignocellulosic ethanol production: Current practices and recent developments. *Biotechnology and Molecular Biology Review* 6(8):172–182.

[449] Gupta, A., and Verma, J.P. 2015. Sustainable bio-ethanol production from agro-residues: A review. *Renewable and Sustainable Energy Reviews* 41:550–567.

[450] Ulaganathan, K., Goud, B.S., Reddy, M.M., Kumar, V.P., Balsingh, J., and Radhakrishna, S. 2015. Proteins for breaking barriers in lignocellulosic bioethanol production. *Current Protein and Peptide Science* 16(1):100–134.

[451] Yusuf, A.A., and Inambao, F.L. 2019. Bioethanol production techniques from lignocellulosic biomass as alternative fuel: A review. *International Journal of Mechanical Engineering and Technology* 10(6)34–71.

[452] Saint Remi, J.C., Rémy, T., Van Hunskerken, V., van de Perre, S., Duerinck, T., Maes, M., De Vos, D., Gobechiya, E., Kirschhock, C.E.A., Baron, G.V., and Denayer, J.F.M. 2011. Biobutanol separation with the metal-organic framework ZIF-8. *ChemSusChem* 4:1074–1077.

[453] Morone, A., and Pandey, R.A. 2014. Lignocellulosic biobutanol production: Gridlocks and potential remedies. *Renewable and Sustainable Energy Reviews* 37:21–35.

[454] Qureshi, N., Saha, B.C., and Cotta, M.A. 2007. Butanol production from wheat straw hydrolysate using *clostridium beijerinckii*. *Bioprocess and Biosystems Engineering* 30:419–427.

[455] Qureshi, N., and Blaschek, H.P. 2010. *Clostridia* and process engineering for energy generation. In: Vertès, A.A., Qureshi, N., Blaschek, H.P., and Yukawa, H. (Eds.). *Biomass to Biofuels – Strategies for Global Industries*. John Wiley & Sons, New York, NY, USA. Pp. 347–358.

[456] Worden, R.M., Grethlein, A.J., Jain, M.K., and Datta, R. 1991. Production of butanol and ethanol from synthesis gas via fermentation. *Fuel* 70:615–619.

[457] Carvalheiro, F., Duarte, L.C., and Gírio, F.M. 2008. Hemicellulose biorefineries: A review on biomass pretreatments. *Journal of Scientific and Industrial Research* 67:849–864.

[458] Lehtomäki, A. 2006. *A Biogas Production from Energy Crops and Crop Residues*. Doctoral Thesis. Department of Biological and Environmental Science, University of Jyväskylä. Jyväskylä, Finland. 91 p.

[459] Rasi, S. 2009. *Composition and Upgrading to Biomethane*. Doctoral Thesis. Department of Biological and Environmental Science, University of Jyväskylä. Jyväskylä, Finland. 76 p.

[460] Lee, S.Y., Hubbe, M.A., and Saka, S. 2006. Prospects for biodiesel as a byproduct of wood pulping – a review. *BioResourses* 1(1):150–171.

[461] Yusuf, C. 2007. Biodiesel from microalgae. *Biotechnology Advances* 25:294–306.

[462] Ruohonen, L., and Tamminen, T. 2009. Microbes and algae for biodiesel production – microfuel. In: Mäkinen, T., and Alakangas, E. (Eds.). *BioRefine Programme 2007–2012, Yearbook 2009*. Tekes, Helsinki, Finland. Pp. 13–28.

[463] Huesemann, M., Roesjadi, G., Benemann, J., and Blaine Metting, F. 2010. Biofuels from microalgae and seaweeds. In: Vertès, A.A., Qureshi, N., Blaschek, H.P., and Yukawa, H. (Eds.). *Biomass to Biofuels – Strategies for Global Industries*. John Wiley & Sons, New York, NY, USA. Pp. 165–184.

Index of selected compounds

"F" refers to "Figure" and "T" to "Table"

https://doi.org/10.1515/9783110608366-008

Index of general subjects

"F" refers to "Figure" and "T" to "Table"

https://doi.org/10.1515/9783110608366-009